INDUSTRIAL WASTEWATER AND SOLID WASTE ENGINEERING

INDUSTRIAL WASTEWATER AND SOLID WASTE ENGINEERING

Edited by

Vincent Cavaseno
and the Staff of Chemical Engineering

CHEMICAL ENGINEERING

McGraw-Hill Publications Co., New York, N.Y.

Library of Congress Cataloging in Publication Data

Main entry under title:

Industrial wastewater and solid waste engineering.

    1.   Factory and trade waste.   I.   Chemical
engineering.
TD897.I435   638.5'4   80-12608
ISBN 0-07-606663-0 (paper)
     0-07-010694-0 (case)

CONTENTS

PREFACE

Of all the areas in which chemical engineers work, none has captured the public's attention more than pollution control, for nothing is more important than the protection of the precious environment that sustains our life. This important and politically sensitive issue has given birth to a broad range of technological developments for the abatement and disposal of air and water pollutants and solid wastes. Nowhere has this been more evident than in the chemical process industries, where the potential for toxic pollution is greater than in perhaps any other industrial segment, and where the total annual expenditure for pollution control has hovered around the $2-billion mark.

This volume, "Industrial Wastewater and Solid Waste Engineering," opens with an introductory section on industrial water pollution control, followed by an instrumentation section. Next are sections on primary, secondary and tertiary wastewater treatment methods. Industrial solid waste control takes the spotlight next, with an introduction, plus sections that cover treatment and disposal of hazardous and non-hazardous wastes. These are especially timely since promulgation of the Resource Conservation and Recovery Act (RCRA) will coincide with the publication of this book.

There is a companion volume, "Industrial Air Pollution Engineering," that presents details on methods and treatments to control air-borne particulates and gaseous pollutants.

The presentation of material in those two volumes naturally reflects the approach taken in *Chemical Engineering*. Thus, they are a source of practical chemical engineering technology, written by a multitude of authors, that is virtually unmatched in the literature. The articles contained in these books will be of value not only to process and operating engineers, who will find much useful information on bringing chemical processes to the highest level of environmental compatibility at the lowest cost, but also to design engineers, who can profit from the valuable experience embodied herein in designing environmentally sound plants from the grassroots up.

One caution: Since changing governmental regulations have played a significant role in the development of environmental technology, the date of original publication is included with each article. The reader is advised to take this date into account when reading an article that cites regulations. Many articles that were published when different standards were in effect have been included because of the value of the technical information they contain.

Part I
INDUSTRIAL WATER POLLUTION CONTROL

Section 1
INTRODUCTION

Chemical-industry costs of water pollution abatement
Refinery wastewater treatment and reuse
The effects of water-pollution control on energy consumption
Industrial wastewater pumps

Chemical-industry costs of water pollution abatement

Pollution abatement is an important factor in the conduct of the chemical industry. The Federal Water Pollution Control Act will cost that industry an estimated $4.1 billion per year by 1983. While there may be delays, such costs will eventually be borne. Chemical engineers can play an important role in deciding what to do about them.

James Cummings-Saxton, International Research & Technology Corp.

□ This article addresses the economic implications of one aspect of the overall environmental-engineering problem—water pollution abatement in compliance with the standards for effluent limitation specified in Public Law 92-500.* The data were developed in a study sponsored by the National Commission on Water Quality (NCWQ).†

The chemical industries are classed within major group 28 of the Standard Industrial Classification (SIC). The principal water-using members are: organic chemicals (SIC 286), inorganic chemicals (SIC 281), and plastics and synthetic materials (SIC 282), which when combined accounted for 86% of the water used in chemicals manufacture in 1973.

Economic effects of treating the plant effluents were appraised for applying the Best Practicable Technology (BPT) and Best Available Technology (BAT) to existing plants; and the Best Available Demonstrated Technology (BADT) to new plants. The EPA has identified specific levels of effluent purity that are judged acceptable under each of these three standards for selected chemicals and production processes.

BPT standards must be achieved by July 1, 1977; and BAT standards by July 1, 1983. BADT standards are effective immediately.

*The Federal Water Pollution Control Act Amendments of 1972.

†The work summarized in this article was performed under the author's direction at International Research and Technology Corp. (I.R.&T.) Others involved in the project include: Albert Loridan, Stavios Xiarchos, Nickolaus Leggett, Beverly Moore, and Caroline Norberg. The complete report of the project can be obtained from the National Technical Information Service (NTIS) in Springfield, Va.: "The Economic Impact of Water Pollution Abatement Costs upon the Chemical Processing Industries," IRT-413-R, Mar. 1976; NTIS PB-252 289/4WP, $12 paper back, $2.25 microfiche. Full references in report.

Originally published November 8, 1976.

Overall effects on the industry

Effluent-purity constraints often impose changes in the production processes that will necessitate an adjustment of the investment requirements and operating and maintenance costs associated with a desired production capacity. The actual cost of pollution abatement is affected by the time-rate of investment—adjusted by inflation and regional location, as they affect taxing and operating considerations. The cost—capital and operating—is passed on to the industry's customers through an adjustment of product price.

Since not all companies incur the same cost increase, and since they may have differing degrees of flexibility in allocating the cost of pollution abatement, this pricing step has an effect on competition within the industry, including a potential expansion of the share of the market occupied by the larger companies, an increase in profit-margin differences among producers, increased difficulties in entering a new market, and regional impacts due to plant closings or curtailments in capacity.

Cost data

In seeking to identify these effects, the study has used cost estimates prepared by Catalytic, Inc. (Philadelphia, Pa.), and Procon, Inc. (Des Plaines, Ill.). Catalytic, Inc., established the abatement cost estimates for inorganic chemicals, organic chemicals, and miscellaneous chemicals. Procon, Inc., provided estimates for plastics and synthetic materials.

Each contractor made in-depth estimates for chemicals selected from among those for which EPA has established effluent guidelines. The chemicals were appraised as to their types of process, plant context, and waste-treatment alternatives. Catalytic evaluated 38 inorganic chemicals and 56 organic chemicals. Procon evaluated 13 plastics and synthetic-material products in a similar manner.

In addition, Catalytic estimated the total abatement costs for several groups of chemicals in order to cover the chemical industry group (SIC 28). Group evaluations were conducted for miscellaneous chemicals and for those inorganic and organic chemicals not considered in depth.

Catalytic introduced an important aspect of the methodology with its concept of Generalized Plant Configurations (GPC), developed in response to two primary problems:

(1) Most chemicals are produced in an integrated facility that concurrently manufactures a number of other products, sometimes in excess of 1,000; and the treatment costs for each product are very much a function of the number, type, and production volume of those other products.

(2) Effluent guidelines have been developed and raw-waste load (RWL) data are available for only a small fraction of the total number of chemical products, e.g., for 84 of the more than 5,000 organic products.

The different plant complexes implied by condition (1) are such that each of the 840 inorganic plants and the 320 organic plants in the U.S. is essentially unique. There are, of course, many other factors, such as plant location, plant age and level of technology, plant man-

---

1973 water usage among chemicals and allied products, billions of gallons   Table I

SIC code	Industry group and industry	Water intake	Gross water usage (including recirculation)	Usage distribution, % Process	Usage distribution, % Cooling and condensing	Usage distribution, % Sanitary and boiler	Treated prior to discharge, % *
28	Chemicals and allied products	4,176	11,099	9.4	86.3	4.3	40.3
281	Inorganic chemicals	808	1,884	14.9	81.6	3.5	24.7
282	Plastic materials, synthetics	568	1,993	13.4	82.1	4.5	32.8
283	Drugs	61	224	(D)	(D)	(D)	26.8
284	Soaps, cleaners, toilet goods	41	91	26.7	56.7	16.6	11.6
2851	Paper and allied products	5	10	(D)	(D)	(D)	17.0
286	Organic chemicals	2,220	5,101	9.1	85.7	5.2	52.6
287	Agricultural chemicals	298	1,482	10.4	79.6	10.0	25.0
289	Miscellaneous chemical products	175	314	19.6	75.8	4.6	14.1

* Estimated on the basis of that portion of intake water whose usage was identified.
(D): Data withheld (in Reference 5) to avoid disclosing operations of individual companies.

Chemicals and allied products water-pollution abatement expenditures for 1973   Table II

Industry Group	Expenditures for new treatment plants $1,000*	Annual cost to operate and maintain existing water treatment plant and equipment, $1,000†	Depreciation (included in annual cost data), $1,000	Equipment leasing (included in annual cost data), $1,000
28	157,123	252,768	36,790	957
281	16,705	51,266	7,479	258
282	44,171	49,349	7,672	237
283	1,595	9,938	1,441	46
284	3,095	2,342	391	(D)
2851	21	881	67	(D)
286	75,085	103,943	16,109	195
287	14,033	26,896	2,941	172
289	2,418	8,171	690	23
All Industries	**511,062**	**865,514**	**122,053**	**5,463**

* Does not include land.

† Includes estimated costs of materials, parts, fuel, power, labor, depreciation (or amortization), equipment leasing costs, and water treatment service costs associated with private contractors. Does not include payments to governmental units for public sewerage.

(D): Data withheld (in Reference 5) to avoid disclosing operations of individual companies.

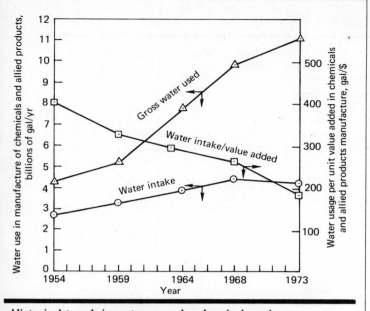

Historical trends in water usage by chemicals and allied products   **Fig. 1**

agement, etc., that further distinguish plants from one another. Few plant waste loads can be completely characterized by the existing raw-waste load data base.

Catalytic introduced the GPCs in order to provide for each product a set of manufacturing contexts that could possibly be related to realistic treatment costs. The first step was to identify all real plants in which each product is produced. The data were then used to identify a selected number of real plants that together produce all the products and represent the alternative production configurations.

By condition (2), however, the number of Phase 1 and Phase 2 products (EPA designations) manufactured in real plants is usually only a small fraction of the total number of products of those plants. In order to artificially increase output in terms of its identifiable RWL, the model assumed additional products that were compatible with the existing real product/processes, as exemplified by their joint occurrence in other real plants. Material balances were performed on each plant to make sure that the scale of the added processes was realistic. These resulting aggregate product/process sets were termed Generalized Plant Configurations.

Historical chemical-industry behavior

The trends in water usage are important from an abatement standpoint. They indicate whether the increasing costs for water have motivated a reduction in the amount of water used per unit production. It is found that while gross water usage increased by 160% over 1954–1973 (Fig. 1), water intake increased by only 56%; and the output of chemical products increased by 244%, so that water intake per unit output consistently declined, with an accelerated reduction occuring during the most recent five-year period. Clearly, the increased cost of water is having an effect on water usage in chemicals manufacture.

In broader perspective, the chemicals and allied products industries rank second among the U.S. industrial water users in terms of water intake. In 1973, they used 4,176 billion gal, accounting for 27.8% of the industrial water usage that year. For comparison, the largest water user, primary metals, employed 4,941 billion gal, while the third largest user, paper and allied products, employed 2,145 billion gal—only 58% as much as chemicals.

Among chemicals and allied products, the organics account for the largest usage (Table I), representing 53.2% of the water intake and 46.0% of the gross usage. The difference between water intake and gross usage reflects the extent of internal recirculation. For the overall chemical industry group (SIC 28), the ratio of gross usage to intake is 2.66, indicating that, on the average, water is recirculated 1.66 times. Table I shows that the fraction of water reused corresponds closely to the fraction employed for cooling and condensing.

Overall, 40.3% of the water is treated prior to discharge, with organics again leading the way with treatment of 52.6% of the discharged water.

Relatively little information is available on expenditures made for pollution abatement prior to 1973, since companies were not accustomed to accounting for them separately. Also, there was no systematic procedure for combining individual company data to arrive at a national total. Frequently, it is difficult to differentiate between pollution-abatement and production-related investments.

The principal source of data is a series of annual surveys initiated by McGraw-Hill in 1967, which indicate that a total of $792 million had been invested in water pollution abatement by the end of 1968. This compares reasonably well with an estimate by the Manufacturing Chemists Assn. that 129 U.S. member companies had invested $757 million through 1967.

More recently, systematic procedures have been established to account for pollution abatement expenditures on a national scale. Survey results by the Bureau of Economic Analysis have been reported in the Survey of Current Business for the years 1972, 1973, 1974 and 1975. In 1973, the Bureau of the Census conducted the first annual Pollution Abatement Expenditures (PAE) Survey. Cost data were included for the first time in the 1972 issue of the Census series, "Water Use in Manufacturing."

Data from the Census report (Table II) show that in 1973 organic chemicals accounted for 47.8% of the total chemical industry expenditures, which is consistent with a higher percentage of water treated prior to discharge (Table I).

Data organization

Once the incremental increases in production costs for representative situations had been estimated by Catalytic, Inc., and Procon, Inc., the aforementioned competitive relationships within the industry made necessary several computational steps to translate the plant-level costs into aggregate industry expenditures. The present and future production capacity for each chemical had to be estimated. The obsolescence of existing plant capacity had to be taken into account. And

Water-pollution-abatement capital investment requirements by chemicals, $1,000*   Table III

Chemical name	Total BPT	Total BADT up to 1977	Total BAT	Total BADT 1978-83	Chemical name	Total BPT	Total BADT up to 1977	Total BAT	Total BADT 1978-83
Acetaldehyde	33,771.0	3,639.0	2,339.0	9,924.6	Manganese sulfate	5,418.0	4,731.1	2,033.0	12,420.0
Acetic acid	3,762.0	597.2	669.0	2,292.9	Methanol	4,762.0	2,401.2	1,417.0	6,787.4
Acetic esters	52.0	28.0	0.0	79.0	Methyl amines	370.0	162.6	1,361.0	471.0
Acetone	14,713.0	3,284.4	11,608.0	9,627.0	Methyl chloride	5,182.0	3,445.2	3,040.0	9,711.0
Acetylene	396.0	883.0	0.0	1,878.8	Methyl ethyl ketone	13,283.0	4,936.5	6,927.0	14,246.8
Acrylic acid	234.0	67.0	25.0	318.2	Methyl methacrylate	3,978.0	1,877.8	1,032.0	5,305.4
Acrylic fibers	7,921.0	4,410.6	6,054.0	13,138.7	Monosodium				
Acrylonitrile	31,832.0	21,818.6	43,230.0	62,446.2	glutamate	2,424.0	1,466.8	466.0	3,991.4
Adiponitrile	2,583.0	1,288.2	0.0	3,628.5	Naphthenic acid	4,624.0	2,595.0	1,231.0	7,211.6
Alkyd resins	1,547.0	446.5	0.0	1,425.2	Nitric acid	1,135.0	1,504.5	928.0	4,028.6
Aluminum chloride	47.0	29.3	0.0	78.6	Nitrogen	1,602.0	994.4	0.0	2,660.5
Aluminum fluoride	3,374.0	1,939.6	15.0	5,151.2	Nylon 6 and 66	8,302.0	8,276.1	15,423.0	24,746.7
Aluminum sulfate	11,880.0	5,664.0	0.0	15,028.6	Oxygen	3,309.0	1,412.4	0.0	3,767.7
Aniline	453.0	214.3	1,169.0	607.8	p-Cresol	1,643.0	842.4	335.0	2,460.2
Barium carbonate	1,841.0	1,489.4	685.0	3,845.5	p-Xylene	83.0	39.1	27.0	111.6
Benzene	15,300.0	7,355.2	10,700.0	20,789.2	Pentaerythritol	3,570.0	1,450.4	0.0	4,036.6
Bisphenol A	2,984.0	761.4	1,418.0	2,149.8	Perchloroethylene	1,512.0	818.1	528.0	2,300.8
Borax	1,068.0	579.4	0.0	1,537.8	Phenol	26,311.0	12,043.1	23,378.0	34,114.5
Boric acid	908.0	678.8	0.0	1,790.5	Phthalic anhydride	2,980.0	1,648.5	365.0	4,647.6
Butadiene	19,506.0	5,215.4	5,609.0	16,208.1	Plasticizers	50,482.0	26,998.4	11,610.0	76,277.8
Calcium hydroxide	5,792.0	3,042.0	0.0	8,072.5	Polyester fibers	24,384.0	11,243.3	1,422.0	33,781.9
Calcium oxide	48,887.0	25,172.4	0.0	66,781.0	Polyester resins	13,411.0	11,208.7	782.0	34,161.0
Caprolactam	7,046.0	3,671.7	1,366.0	10,332.0	Polypropylene	5,414.0	5,096.2	2,418.0	15,248.8
Carbon dioxide	4,519.0	3,182.4	741.0	8,470.9	Polypropylene fibers	23,375.0	15,430.1	2,313.0	45,944.6
Carbon monoxide	7,379.0	10,073.7	4,591.0	27,396.9	Polystyrene	12,634.0	7,690.1	3,439.0	23,128.7
Cellophane	9,169.0	5,821.9	2,538.0	17,494.9	Polyvinyl chloride	17,329.0	32,057.5	6,674.0	106,331.1
Chlorine	10,270.0	6,956.5	7,273.0	18,745.6	Potassium sulfate	1,479.0	762.2	0.0	2,022.4
Chlorobenzene	1,428.0	760.0	262.0	2,142.4	Precipitated calcium				
Chloromethanes	449.0	261.2	119.0	737.2	carbonate	479.0	237.7	311.0	631.6
Chlorotoluene	2,259.0	1,461.2	478.0	4,121.0	Propylene	17,105.0	9,213.4	4,556.0	25,949.7
Copper sulfate	108.0	425.6	364.0	1,136.0	Propylene glycol	988.0	670.2	57.0	1,918.8
Cumene	16,159.0	7,416.2	0.0	20,969.8	Propylene oxide	75,700.0	48,778.7	36,373.0	138,191.7
Cyclohexane	191.0	102.1	237.0	289.7	Rayon	35,201.0	26,822.2	20,979.0	81,797.0
Dimethyl					SAN and ABS resins	691.0	1,186.7	112.0	4,181.8
terephthalate	24,394.0	12,045.2	4,118.0	34,005.0	Silver nitrate	20.0	97.5	35.0	216.3
Dimethylamine	58.0	31.6	7.0	88.0	Sodium bicarbonate	1,318.0	647.7	0.0	1,697.2
Dyes and dye					Sodium carbonate	21,100.0	0.0	20,800.0	0.0
intermediates	65,788.0	29,827.5	21,804.0	84,918.9	Sodium chloride	160,142.0	83,225.4	9,968.0	220,808.1
Ethylbenzene	10,100.0	3,110.0	3,521.0	8,873.8	Sodium dichromate	6,575.0	2,617.8	2,712.0	6,976.0
Ethylene	42,127.0	29,174.4	11,204.0	83,267.0	Sodium				
Ethylene dichloride	5,040.0	1,921.7	9,709.0	6,116.9	hydrosulfide	1,550.0	937.4	111.0	2,446.0
Ethylene glycol	10,763.0	4,585.4	1,721.0	13,194.9	Sodium				
Ethylene oxide	4,653.0	2,300.1	807.0	6,525.5	hydrosulfite	1,642.0	2,100.0	1,764.0	5,585.0
Ferric chloride	875.0	388.2	0.0	1,021.8	Sodium hydroxide	10,678.0	6,192.0	7,562.0	16,482.6
Formaldehyde	4,264.0	1,798.7	461.0	5,104.6	Sodium silicate				
Formic acid	1,834.0	626.5	86.0	1,835.3	(water glass)	10,261.0	9,337.3	3,737.0	24,811.9
Hexamethylene					Sodium				
diamine	2,887.0	1,638.7	1,509.0	4,609.5	silicofluoride	4,849.0	5,608.6	3,197.0	15,009.0
Hexamethylene					Sodium sulfite	1,803.0	1,149.6	381.0	3,062.0
tetramine	376.0	135.5	1,483.0	385.7	Sodium thiosulfate	122.0	248.3	257.0	654.0
High-density					Styrene	27,750.0	18,754.5	6,459.0	53,292.7
polyethylene	1,824.0	2,601.7	1,255.0	7,831.1	Sulfur dioxide	4,410.0	4,325.5	686.0	11,492.4
Hydrochloric acid	2,082.0	1,451.8	93.0	3,894.4	Sulfuric acid	21,461.0	10,644.0	3,720.0	28,345.6
Hydrofluoric acid	29,610.0	19,410.3	1,955.0	51,648.3	Terephthalic acid	5,741.0	3,074.0	916.0	8,690.8
Hydrogen cyanide	7,002.0	16,920.9	0.0	45,121.6	Tetraethyl lead	10,332.0	5,392.4	5,937.0	15,181.7
Hydrogen peroxide	2,999.0	2,859.4	876.0	7,549.9	Titanium dioxide	125,100.0	60,719.0	71,200.0	161,182.1
Isobutylene	16,506.0	8,978.4	11,003.0	25,233.0	Toluene	11,500.0	5,022.5	8,100.0	14,100.0
Isopropanol	28,299.0	9,257.4	7,442.0	27,513.0	Vinyl acetate	398.0	236.6	184.0	668.9
Maleic anhydride	8,423.0	6,258.4	22,093.0	18,018.2	Xylene	9,400.0	3,896.4	6,600.0	10,994.0

*1973 dollars

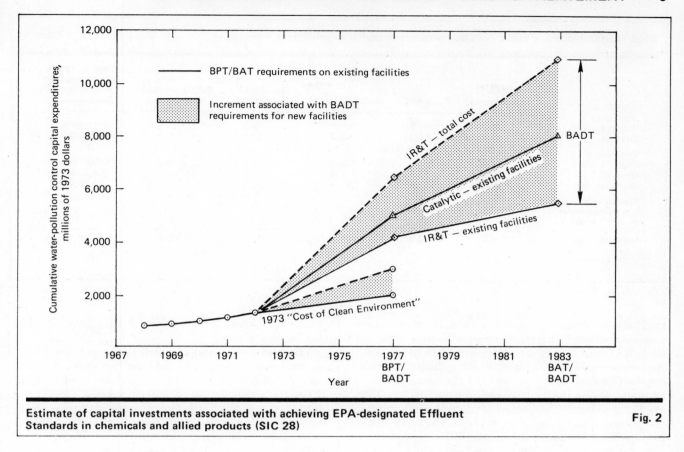

Estimate of capital investments associated with achieving EPA-designated Effluent
Standards in chemicals and allied products (SIC 28)   Fig. 2

the effect of regional differences in taxing and operating costs had to be considered. These required the development of a chemical input-output table, a plant data matrix, and an estimate of obsolescence.

Chemical input-output table (CIOT)

Because the interrelationship among chemicals obscures the pathways by which products emerge for sale outside the industry, it is difficult to translate an estimated demand for consumer goods (e.g., automobiles) into associated demands for chemicals. Some chemicals (e.g., ethylene) function merely as intermediates, using coal, oil, natural gas, natural ore, etc., as feedstocks, and being themselves consumed as feedstocks for subsequent chemicals. Other chemicals (e.g., ethylene glycol) act both as intermediates and as final products. And still others (e.g., plastic resins) serve primarily as the last step in the sequence before going on to other industries.

Consequently, a chemical input/output table (CIOT) was essential for assessing the demand-driven growth in production capacity for each chemical. The CIOT is a matrix within which sales are identified for each chemical, both for the production of other chemicals and for other applications. The other sectors are identified as the 185 sectors of the Inforum Input/Output model developed by Clopper-Almon at the University of Maryland (although chemicals are actually sold to only about 70 of these sectors). Inforum projections were translated by the CIOT into the demands for specific chemicals.

With CIOT (as with all input/output approaches)

the constant relationships of the data over time form an important characteristic. In spite of the many changes taking place in the chemicals industry, the CIOT coefficients remain relatively constant. Factors like new technology only gradually change some of the CIOT coefficients. For example, ethylene has steadily displaced acetylene in the manufacture of VCM, hence the CIOT coefficient for acetylene-to-VCM has diminished with time. Acetylene also illustrates a typical feature that dampens such effects—acetylene is suffering similar losses in all its markets, across the board, so that the proportional distribution of its output has not changed so much as the changes in quantity.

In addition to demand projections, CIOT can estimate the effects of pollution-abatement costs on prices for chemical end-products sold to other industries and to the consumer. These products must carry the costs of their precursor chemicals, unless those costs are to be accommodated within the industry in the form of reduced profits. Thus, there tends to be a passing forward, along the production sequence, of process-related cost variations.

From this perspective, the pathways to each end product look like a genealogical tree. In the absence of CIOT, the multiplicity of these pathways would overwhelm attempts at a systematic evaluation of cost propagation. (It should be noted that abatement costs were developed by the contractors for only 107 chemicals, whereas CIOT includes 374 chemicals. The other 267 chemicals function as passthroughs, exhibiting increased production costs due to increases in raw-materi-

Total chemical industry water-pollution-abatement capital expenditures, millions of $*						Table V
	Prior expenditures	Total BPT	BADT to 1977	Total BAT (incremental)	BADT to 1983 (cumulative)	Total to 1983, excluding prior expenditures (cumulative)
Inorganics						
Phases 1 and 2	270.7	601.6	297.7	146.0	791.6	1,268.5
General	41.1	91.3	45.2	22.2	120.2	192.6
Organics						
Phases 1 and 2	165.1	757.5	326.3	297.1	928.8	1,818.3
General	382.7	1,755.7	756.3	688.6	2,152.7	4,214.3
Miscellaneous	178.0	816.5	351.7	320.2	1,001.1	1,959.8
Plastics and synthetic materials						
Phases 1 and 2	—	161.2	132.3	63.4	409.2	633.8
Industry total	**1,037.6**	**4,183.8**	**1,909.5**	**1,537.5**	**5,403.6**	**10,087.3**

*1973 dollars

be $5,890 million (1973 dollars). Thus, the total water-pollution-abatement requirements to comply with the BPT, BAT and BADT standards are 189% of the 1975 capital expenditures planned by the industry. Deducting the 17% expended prior to 1973, the expenditures to be made during the 11-year period from 1973 to 1983 are 171% of the 1975 planned total. Seen another way, the average annual expenditures from 1973 amount to $917 million/yr, or 15.6% of the 1975 expenditures.

As for actual expenditures, the Bureau of Economic Analysis reports that those made by the chemical industry for water-pollution abatement were (1973 dollars): $213 million in 1973, $231 million in 1974, and a planned $252 million in 1975.

Since the Bureau of Economic Analysis includes in its data only companies whose primary output is chemicals and excludes those that produce chemicals as a secondary product, the Bureau's data is not directly comparable with the IR&T/Catalytic estimates. On the basis of shipments reported, the Bureau's data covers only 85% of the industrial organic chemicals (SIC 2869), 79% of the industrial inorganic chemicals (SIC 2819) and 79% of the plastics materials and resins (SIC 2821).

However, the wide difference between the observed and expected levels of abatement spending is significant. The accumulated capital represents 74% of the planned expenditures for 1975 for all plants, and 42% for existing plants.

On this basis, it is apparent that the industry faces a severe financial load over the next two years, if it is to comply with the BPT/BADT standards, and over the 1978–1983 time frame, if it is to meet the BAT and BADT standards. Furthermore, water-pollution-abatement costs are only a portion of the environment-related costs the companies must bear. There are requirements related to air pollution control, solid-waste disposal, the Occupational Safety and Health Act, the Toxic Substances Act, and the Safe Drinking Water Act. For example, the Bureau of Economic Analysis indicates that, for the chemical industry, air-pollution-abatement expenditures are 70% of those for water-pollution abatement, and solid-waste related expenditures are 17% of those for water.

One possible response to the sharp increase in environment-related expenditures may be a corresponding reduction in new plant expenditures. As noted previously, the abatement requirements for new plants represent 44% of the projected abatement costs in 1976/77; thus a reduction in new plant construction both frees capital and reduces the total abatement requirements to be met. Alternatively, business considerations may dictate that new plant construction proceed, with abate-

Water-pollution-abatement operating and maintenance costs for 107 chemicals, millions of $*						Table VI
	Total BPT	BADT to 1977	Total to 1977	Total BAT (incremental)	BADT to 1983 (cumulative)	Total to 1983 (cumulative)
Inorganics [38]	158.8	94.1	252.9	147.2	250.3	397.5
Organics [56]	120.8	63.2	184.0	275.0	191.8	466.8
Plastics and synthetic materials [13]	25.0	18.8	43.8	21.8	58.3	80.1
Total	**304.6**	**176.1**	**480.7**	**444.0**	**500.4**	**944.4**

*1973 dollars

Water-pollution-abatement capital requirements for 107 chemicals, millions of $*						Table IV
	Total BPT	BADT to 1977	Total to 1977	Total BAT (incremental)	BADT to 1983 (cumulative)	Total to 1983 (cumulative)
Inorganics [38]	523.1	297.7	820.8	146.0	791.6	1,460.7
Organics [56]	658.7	326.3	985.0	297.1	928.8	1,884.6
Plastics and synthetic materials [13]	161.2	132.3	293.5	63.4	409.2	633.8
Total	1,343.0	756.3	2,099.3	506.5	2,129.6	3,979.1

*1973 dollars

als cost, but not contributing additional abatement-related costs.)

Plant data matrix

The second analytical tool, the plant data matrix, includes for each plant of the 107 chemicals studied in depth: company affiliation, process employed, throughput, and regional location (by state and EPA region). The estimated abatement costs for each chemical (usually for several throughputs, plant contexts, and the principal processes) were matched against the plant data matrix to estimate the specific costs incurred at each plant. Knowing the regional locations, the state taxing provisions for industrial abatement equipment were taken into account in arriving at the annual cost. The national totals for capital requirements, operating and maintenance costs, and annual costs could then be determined for each chemical.

Plant obsolescence

Since effluent standards differ between new and existing plants, the age of a plant is important to its pollution abatement expense. While existing plants must attain a level termed BPT by July 1, 1977, and BAT by July 1, 1983, new plants must immediately conform to a level termed BADT, but need not improve their performance beyond this for ten years after completion.

At present, BADT standards usually fall somewhere between BPT and BAT, actually exceeding BAT in a few cases. All standards are subject to periodic review, however, and it is likely that after 1977 the BADT standards will be made at least as stringent as BAT, based on the assumption that an existing plant faces compliance difficulties that can be avoided in a new plant.

Also, the necessary investment in pollution abatement leads to the possibility that marginally profitable plants may be closed more rapidly than normal; or it is possible that a plant which has fallen below the break-even point, but which could easily comply with the pollution standards, might extend its economic life. The net effect will probably be a more rapid depreciation cycle, because standards are presently only slightly more severe for new plants than for existing plants.

Expenditures for pollution abatement

Requirements for the 107 chemicals are given in Table III. Data include total BPT, BAT, BADT to 1977, and total BADT to 1983. Totals for all 107 chemicals are shown in Table IV. For these chemicals, the capital expenditures through 1977 represent 53% of the total expenditures required through 1983 in compliance with established standards.

These 107 selected chemicals represent only a segment of each of the three industries, and were chosen by the contractors primarily on the basis of importance in terms of production volume and pollution.

Catalytic, Inc., developed a method for estimating the abatement costs for the remainder of the industry through an extension of the costs for those chemicals studied in depth. Capital costs for the 107 chemicals have been generalized following the Catalytic ground-rules, and the results are presented in Table V. As suggested by Catalytic, a 15% allowance for sewer separation costs has been added into the BPT capital requirements. Estimates of the fraction BPT standards achieved by facilities (facilities existing in 1972 were used to isolate costs directly associated with the 1972 Amendments. As a percentage of total BPT costs, those estimates are: inorganics, 45%; and organics, 22%.

Fig. 2 compares International Research and Technology Corp. estimates for total water-pollution-abatement investment with those of Catalytic (for inorganics and organics only) and those of the EPA's 1973 report, "Cost of a Clean Environment." The IR&T and Catalytic estimates for BPT costs are both significantly higher than those of the 1973 EPA report. It is understood that the low level of the 1973 EPA estimate was partially due to coverage differences and to the preliminary nature of the estimates. The as-yet-unpublished 1975 estimates by EPA are said to be somewhat higher than those of 1973, but still fall considerably short of the IR&T/Catalytic estimates.

The difference between the IR&T and the Catalytic estimates is strictly due to IR&T's accounting for plant obsolescence and for new production facilities to support demand growth, and a 6.3% difference associated with the inclusion of 13 plastics and synthetic materials. The approach and data base for the two calculations is otherwise the same.

The IR&T/Catalytic estimates are:
■ $6,093 million through 1977, including 31% for new plants and 17% spent before Jan. 1973.
■ $11,125 million total through 1983, with 45% being spent between 1978 and 1983 and 48% going toward meeting BADT requirements for new facilities.

For comparison, the planned 1975 expenditures for all new plants and equipment in the chemical industry were reported by the Bureau of Economic Analysis to

Total chemical industry water-pollution-abatement annualized costs, millions of $*						Table VII
	Prior expenditures	Total BPT	BADT to 1977	Total BAT (cumulative)	BADT to 1983 (cumulative)	Total to 1983, excluding prior expenditures (cumulative)
Inorganics						
Phases 1 and 2	106.1	235.9	138.9	298.5	369.6	562.0
General	16.1	35.8	21.1	45.3	56.1	85.3
Organics						
Phases 1 and 2	43.4	199.4	109.3	459.0	311.4	727.0
General	100.6	462.2	253.4	1,064.0	721.8	1,685.2
Miscellaneous	46.8	215.0	117.8	494.9	335.7	783.8
Plastics and synthetic materials						
Phases 1 and 2	—	68.7	49.2	98.7	152.0	250.7
Industry total	313.0	1,217.0	689.7	2,460.4	1,946.6	4,094.0

*1973 dollars

ment expenditures met by a corresponding increase in aggregate capital expenditures. Water-pollution-abatement expenditures alone would require a 29% increase over 1975 capital expenditures, if the same level of new-plant investment were maintained.

Operating and maintenance costs

O&M costs form one of the two components of annual cost, interest and amortization associated with the capital investment being the other. O&M costs for water-pollution-abatement were determined in much the same way as investment requirements for each of the 107 chemicals. The component costs include: labor, chemicals, energy, maintenance, overhead, monitoring, and sludge handling and disposal. Data for the 107 chemicals studied in depth are summarized in Table VI. By 1983 O&M costs for organics will slightly exceed those for inorganics, but inorganics will bear 53% of the O&M costs associated with meeting the earlier BPT requirements by 1977. The costs for plastics and synthetic materials are considerably smaller, being only 17% of those for organics. Thirteen plastics and synthetic-materials products were studied by Procon, Inc., on the basis that these are the major products faced with substantial capital and O&M water-pollution-abatement costs.

Total annual costs

The annual costs for the 107 chemicals were generalized to estimate the costs for the entire chemical industry (Table VII). These yearly costs associated with Public Law 92-500 are found to be $4,094 million by 1983. Inorganics represent 15.8% of this total, organics 78.1%, and the 13 plastics and synthetic materials only 6.1%. Expenditures made prior to 1973 represent merely 7.4% of the burden. Expenditures to be made in 1977, excluding those associated with pre-1973 facilities, represent 38.9% of the expenditures required in 1983.

Summary

These data demonstrate that pollution abatement is an important factor in the conduct of the chemical

industry. The costs of abatement compose a significant fraction of each year's financial expenditures. It is helpful when considering these expenditures to remember that they directly reflect an increased price for what were until recently essentially free goods—water, air and land.

A similar type of phenomenon is the sharp increase that has occured in the prices of natural gas and oil. Other raw materials can be expected to follow this pattern.

The possibility of delaying or modifying the 1972 Act's provisions is now under consideration. Any change, however, will be designed to minimize disruptive impact on our economic system, not to circumvent the attainment of acceptably clean water. Costs of the magnitude identified in this article will eventually be borne, the primary question being one of timing.

Chemical engineers can make an important contribution, however. New plants account for 54% of the abatement capital expenditures through 1983, and 48% of the annualized abatement costs (O&M, amortization, interest, etc.) in that year. These yet-to-be-built facilities can readily incorporate improved process designs that minimize residual generation. Process improvements can likewise be made in existing facilities. In effect, there is now an economic incentive for the chemical engineer to do some of the things he has always wanted to do.

The author

James Cummings-Saxton is Vice President at International Research & Technology Corp., 1501 Wilson Blvd., Arlington, VA 22209. Assessing the interactive effects—economic and social—of developing technologies is his primary area of expertise. He holds a Ph.D. in chemical engineering from the University of California at Berkeley, a B.E.S. in chemical engineering from Johns Hopkins University, and is working toward an M.A. in economics at Virginia Polytechnic Institute.

Refinery wastewater treatment and reuse

Familiarity with the latest technology and with current regulations is necessary to remain competent in the area of process design. Although petroleum refineries are specifically singled out in this article, the processes described apply to many other industries as well.

Kenneth E. Bush, The Frantz Co.

☐ A tremendous amount of material has been written about wastewater treatment and reuse in recent years. This is partly due to the constantly changing technology in the field. Controversy surrounding the promulgation of regulations by the Environmental Protection Agency (EPA), and subsequent revisions, have added to the confusion. The purpose of this article is to attempt to tie this mass of information together for the engineer who is not involved in this field on a day-to-day basis, but still needs to be knowledgeable.

Rules and regulations

The Water Pollution Control Act of 1972, which is the primary motivating force in current effluent-treatment programs, requires the application of: "best practicable control technology currently available" (BPCTCA) by July 1, 1977; "best available technology economically achievable" (BATEA) by July 1, 1983; "best available demonstrated technology" (BADT) for all new sources; and a national goal of "zero" discharge of pollutants to navigable waters by 1985. Congress left it to EPA to define these technologies and to formulate the effluent-limitation guidelines.

EPA published the effluent limitation guidelines for each degree of technology [1], as well as the current revised guidelines [2], which are expressed in pounds of each particular pollutant per 1,000 bbl of feedstock. The pollutant parameters selected are: BOD_5 (five-day biochemical oxygen demand); TSS (total suspended solids); COD (chemical oxygen demand); oil and grease; phenolic compounds; ammonia as N; sulfides; total chromium; hexavalent chromium; and pH.

To calculate the effluent limitation for a particular refinery, first select the category of technology that is applicable to the refinery. For an existing refinery, BPCTCA is satisfactory until July 1, 1983, after which BATEA is necessary. For a new refinery, it is necessary to apply BADT guidelines.* The pounds of each pollutant are read from a table provided in the guidelines, and multiplied by the capacity of the refinery in 1,000 bbl/d. These numbers are then multiplied by a size factor and process factor to obtain the final allowable effluent composition in lb/d of pollutant discharge.

* Each of these guidelines is found in the May 20, 1975, *Federal Register.*

The size factor varies from 1.02 for capacities of less than 24,900 bbl/d to 1.57 for refineries of 150,000 bbl/d or larger. Tables listing the size factors are also provided in the guidelines. Process factors—listed as a function of the process configuration—are calculated by taking into account the various processes, as well as the relative throughput for each process.†

Waste-treatment processes

One of the functions of a process engineer is to decide which treatment process, or combination of processes, will best perform the necessary job of cleaning up the wastewater effluent involved, because waste-treatment facilities contribute significantly to the total cost of a capital project. Since the cost involves also land and manpower, future needs must also be considered.

This is often difficult to do, because regulations change or new regulations come into effect. For this reason, it is wise many times to institute a development program, so that as wastewater-quality requirements become more stringent, treatment processes can be modified, or new ones added to improve the water quality. As a refinery expands, a similar program may be followed to increase wastewater-handling capacity.

An effluent-treatment system involves one or more stages, depending on the quality of the raw effluent and the required pollutant reduction. The various unit processes are categorized in the following manner [3]:

Primary	Intermediate	Secondary/tertiary
Sulfide/ammonia stripping	Flotation	Activated sludge
API separators	Coagulation-precipitation	Trickling filters
Tilted-plate separators	Equalization	Waste-stabilization ponds
Liquid-liquid extraction		Aerated lagoons
		Filtration
Filtration for oil removal		
pH control		Carbon adsorption
		Chemical oxidation

† A sample calculation is provided on p. 16570 of the May 9, 1974, *Federal Register.*

Originally published April 12, 1976.

The purpose of the primary and intermediate processes is to remove certain pollutants from wastewater before the water goes to the secondary- and tertiary-treatment processes, because high concentrations of suspended solids or oil may be harmful to the processes.

Sulfide/ammonia stripping, oil removal

Significant amounts of H_2S and NH_3 are found in refinery wastewaters due to the breakdown of organic sulfur and nitrogen compounds during the various refining processes. These compounds can be removed by air or steam stripping. Details of design parameters are provided by Ref. 4 and 5.

A 20-tray stripper using two pounds of steam per gallon of feed removes 99% of the H_2S, 75% of the NH_3, and 40% of the phenol present in a sour-water stream. This stripped water is normally utilized as feedwater to the crude-oil desalters.

For refineries that are large enough to justify recovery of the H_2S and NH_3 as products, Ref. 6 provides reported capital and operating costs for the process, where a rate of return of 25% on total investment is said to result. However, there is a 75% return on the incremental investment over a conventional stripper and incinerator. Also reported are commercial units ranging in capacity from 23 to 280 gal/min of sour-water feed, in which a rate of about 150 gal/min is probably necessary to justify product recovery. Examples of several sour-water strippers are provided in Ref. 7.

The traditional method of oil separation in refinery wastewater is via the API gravity separator, which is sized to allow most of the free oil to float to the surface, and the heavier solids to fall to the bottom. These separators are normally an integral part of the operation, due to the amount of recoverable oil in refinery wastewaters.

Several parameters determine the effectiveness of the separator, among which are water temperature; density and size of the oil droplets; and the type of solids in the water. However, only free oil will be separated; emulsions are not broken. Removal efficiencies of 60–99% for oil and 10–50% for suspended solids have been reported [8]. Details of design of API separators are provided in Ref. 9.

Another type of gravity separator finding increased use in the refining industry is the tilted-plate separator. This unit is made up of one or more modules consisting of several corrugated plates tilted at a 45° angle. As the water flows between the plates, the oil droplets collect on the underside and move to the top of the module. Improved efficiencies at less cost and space than those needed by the API separators have been reported.

Wastewaters that contain a high amount of free oil are normally treated in the gravity separator before they mix with other, non-oily streams. This allows the separator to be sized for a smaller throughput volume. Sometimes, the addition of chemicals improves the separation.

Two other processes that have more-limited utilization for oil and solids removal are liquid-liquid extraction and filtration. These processes are normally applicable to the treatment of certain particular types of wastewater.

Control of pH, dissolved-air flotation

Since many waste streams are highly acidic or alkaline—and would therefore be detrimental to biological processes—it is necessary to control the pH of refinery wastewaters before they proceed to biological treatment. Sometimes, phosphoric acid or ammonia addition is beneficial for the dual purpose of controlling pH and supplying the necessary nutrients for downstream biological processes (see further on, where biological processes are discussed).

Dissolved-air flotation is normally considered an intermediate or secondary treatment process for further removal of oil and suspended solids from API-separator effluents, prior to biological treatment. Ref. 10 reports how the American Oil refinery at Whiting, Ind., used a flotation unit—which included a series of bioflotation ponds—to treat effluent from secondary treatment. Apparently, chemical coagulation and air flotation of the effluent from the aerated lagoon produced a final effluent that met tertiary-treatment goals.

EPA reports [8] removal efficiencies of 70–85% for oil, 50–85% for suspended solids, 20–70% for BOD, and 10–60% for COD, for dissolved-air flotation units. However, the API Manual [11] refers to a technical article where better removal efficiencies are reported.

Pilot-plant tests on flocculation-flotation have shown 97% removal of oil, 75% of solids, and 80% of BOD and COD. Chemical flocculating agents used to aid flotation are activated silica, and aluminum and ferric salts. These agents also improved the effectiveness of the flotation unit in removing oil emulsions.

Since the effectiveness of dissolved-air flotation units varies substantially with the characteristics of the particulate matter, it is imperative that laboratory and pilot tests be run to obtain the necessary design criteria. Small pilot units can be rented from vendors. Instructions and illustrations for laboratory tests are provided by the API Manual [11].

In the dissolved-air flotation process, the waste stream is saturated with air under a pressure of several atmospheres. This is done by pumping the stream to 40–50 psig, and feeding compressed air at the pump suction. The stream is then held in a retention tank for several minutes so that the air will dissolve. As the stream exits from the retention tank, it flows through a pressure-reducing valve—which drops the pressure to atmospheric—and then into a flotation tank.

When the pressure is released, the air comes out of solution, forming minute bubbles within the liquid. These bubbles, which are about 30–120 microns in diameter, form on the surfaces of the suspended particles. An aggregate is thus formed, which rapidly rises to the surface and is skimmed from the flotation tank. Design details are provided in the API Manual [11] and in Ref. 12.

Package units are available from various vendors. Ref. 10 cites the use of surplus boxes in the API separator as flocculation chambers, and indicates that coagulant concentration proved to be important. For instance, a concentration of 30–50 ppm of alum was sufficient to promote coagulation without deterioration of effluent quality due to excessive coagulant.

Equalization

Refinery wastewater flowrate may vary significantly from time to time. It is very important for the effectiveness of biological processes that the volume and composition of the feed be fairly constant. Therefore, a holding point with sufficient residence time to even out major fluctuations is needed. This not only attenuates normal process variations but helps to minimize the shock of a major spill.

Secondary and tertiary treatment

The following wastewater-treatment processes are those that improve the quality of wastewater to the point where it can be reused within the refinery, or discharged without any adverse environmental reaction. The extent to which the process is used depends on the quality of the raw effluent, the pollutants contained in the raw effluent, and the method of ultimate disposal or reuse of this water.

Activated-sludge process

Activated-sludge procedures are used extensively for coagulating and removing nonsettleable colloidal solids, as well as to stabilize organic matter. EPA reports [8] removal efficiencies of 80–99% for BOD, 50–95% for COD, 60–85% for suspended solids, 80–99% for oil, 95–99+% for phenol, 33–99% for ammonia, and 97–100% for sulfides. Thus, a high-quality effluent is obtainable from a properly designed and operated activated-sludge system.

This particular process is an aerobic biological procedure containing, within a reaction tank, a high concentration of microorganisms, which is maintained by recycling activated sludge. Oxygen is supplied to the wastewater in the reaction tank either by mechanical aerators or a diffused-air system. Inasmuch as the microorganisms remove the organic materials by biochemical synthesis and oxidation reactions, the converted organic matter must be removed by sedimentation, prior to final discharge.

The main components of the process are the aeration, or reaction, vessel and the sedimentation tank. Sludge removed from the sedimentation tank is recycled to the aeration vessel to maintain the required concentration of microorganisms. Since a portion of the sludge must be discarded, it is first dewatered and then used as landfill.

Nutrients—primarily nitrogen and phosphorus—are required to maintain a healthy growth of microorganisms within the system. These elements are needed in minimum amounts of 5 lb nitrogen and 1 lb phosphorus for every 100 lb of BOD to be removed [13]. Generally, refinery wastewater contains enough ammonia to supply adequate nitrogen, but it may be deficient in phosphorus. Some of the sources of these elements are [14]:

■ Water condensate from catalytic cracking units and normally containing ammonia and other nitrogen compounds.

■ Spent phosphoric acid catalyst from polymerization units; the phosphates can be leached from the catalyst with water.

■ Boiler or cooling-tower blowdown, which contains phosphates.

Now and then, it may also be necessary to adjust the pH of the waste stream and add nutrients. Nitrogen and phosphorus in acid or alkaline compounds can achieve both of these purposes.

An inadequate supply of nutrients results in poor stabilization of the wastes and will stimulate fungi growth. This results in poor sludge-settling characteristics, and longer periods of aeration.

The oxygen required by the activated-sludge system varies between 0.6 and 1.5 lb O_2/lb of BOD removal. For design purposes, an oxygen demand of about 1.0 lb O_2/lb of BOD is often used; however, this should be determined experimentally before designing a particular system.

Trickling filters

The trickling filter is an aerobic biological device that is used extensively in the refining industry. It may be used as a secondary treating system by itself, whenever a high-quality effluent is not required. This filter may also be used upstream of an activated-sludge unit, to reduce the loading or to attenuate the organic loading on the unit.

A trickling-filter system consists of the filter bed with a wastewater distributor and a sedimentation tank. The filter is usually a bed of broken rock or coarse aggregate; recently, plastic sheets have come into use as filter media.

Rocks ranging in size from 2–4 in are best, so as to get a maximum amount of surface while maintaining good ventilation through the bed.

Beds are normally from 3–4 ft deep, and can be either circular or rectangular. If plastic media are used, they can be up to 40 ft deep. Sprinklers are used to evenly distribute the wastewater being fed to the filter over the bed's surface.

A sedimentation tank is needed for clarification of the effluent, due to the biologic growth that sloughs from the filter bed.

EPA reports removal efficiencies of 60–85% for BOD, 30–70% for COD, 60–85% for suspended solids, and 50–80% for oil. The efficiency of a particular filter depends significantly on its loadings.

There are two general groupings of trickling filters, based on their organic loadings. Low-rate filters have organic loadings less than 15 lb of BOD/1,000 ft³ of filter volume per day, and hydraulic loadings of 2–4 million (MM) gal/(acre)(d). High-rate filters have organic loadings greater than 15 lb of BOD/1,000 ft³ of filter volume per day, with a hydraulic loading of 10–40 MM gal/(acre)(d).

The API Design Manual [15] reports a relationship between organic loading and BOD removal efficiency. Only at very low loadings can a high-quality effluent be obtained, and it is at such loadings that the cost of the trickling filter is higher than other comparative processes. Consequently, this filter is feasible only as a "roughing" device, rather than as a complete treating system. Design equations that have been proposed by several investigators for trickling filters are provided in Ref. 18.

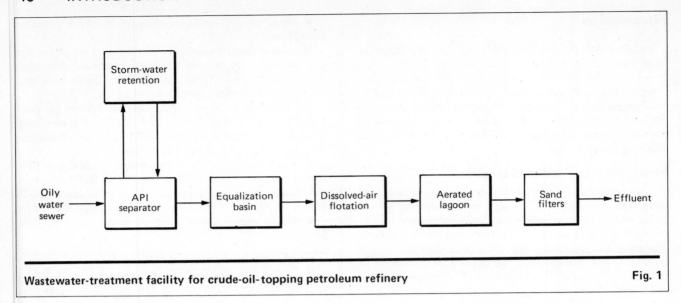

Wastewater-treatment facility for crude-oil-topping petroleum refinery   **Fig. 1**

Waste-stabilization ponds

A stabilization pond is simply a large shallow pond in which bacteria stabilize the wastewater fed to the pond. Stabilization ponds are classified as aerobic, aerobic-anaerobic, and anaerobic, depending on the type of biological activity. Since refinery wastes are not normally suited to anaerobic treatment, we will discuss only aerobic, or oxidation, ponds.

Oxidation ponds are normally about 4 ft deep. Shallow ponds tend to have excessive weed growth, and deep ponds do not get adequate oxygen transfer to maintain aerobic conditions. Oxygen—supplied by natural surface aeration and by the photosynthetic action of the algae present in the pond—is used by bacteria for the aerobic degradation of organic matter. In turn, algae utilize the nutrients given off by bacteria.

Organic loadings of oxidation ponds reported in the API Design Manual [16] range from 20–100 lb BOD/(acre)(d). Recommended loadings for ponds handling raw wastes are about 40 lb BOD/(acre)(d). For ponds downstream of other treatment processes, loadings as low as 10 lb/(acre)(d) should be used. Retention times range from 1 to 90 days. However, the API Manual points out that for a pond to be effective, it should have a retention time of at least 7 days. Approximately 30 days is recommended for an organic loading of 50 lb BOD/(acre)(d).

EPA reports [8] the following removal efficiencies for oxidation ponds: 40–95% for BOD, 30–65% for COD, 20–70% for suspended solids, and 50–90% for oil. The differences in removal efficiencies are due to the wide variations in loadings and retention times reported above. Consequently, it is difficult to develop any general design parameters for oxidation ponds. Any parameter must be functionally related to temperature, solar intensity, wind, organic loading, and the nature of the waste. Methods for relating these factors are provided in Ref. 17.

The effectiveness of oxidation ponds is also influenced by other factors such as temperature and turbidity of the water, as well as by emulsions present. Since the microbial activity in the pond is decreased at low temperatures, if ice forms over the water surface, this further aggravates the problem by decreasing both the oxygen and light transmission into the water. If this condition is prolonged, the bacteria will convert to anaerobic metabolism, which may result in an odor problem whenever the ice melts. Any turbidity or emulsion in the water also reduces the light transmission, thereby inhibiting the photosynthetic action of the bacteria.

Because a large land area is required for an oxidation pond, such ponds are impractical, unless land is plentiful and cheap. Their primary advantage is lack of maintenance or operator attention, because there is no mechanical equipment involved.

Aerated lagoons

Aerated lagoons operate generally by the same basic principles as do oxidation ponds. Mechanical-aeration equipment, associated with aerated lagoons, results in a higher concentration of bacteria than is present in oxidation ponds. Consequently, land requirements are less for aerated lagoons than for oxidation ponds having equal loadings. Retention time is usually 3–10 d. Removal efficiencies are 75–95% for BOD, 60–85% for COD, 40–65% for suspended solids, 70–90% for oil, 90–99% for phenol, and 95–100% for sulfides [8].

Aeration of the lagoon keeps most of the solids in suspension. One method of decreasing the suspended solids in the effluent is to have a sedimentation section in the lagoon, where the water is calm enough and the residence time is adequate to allow sedimentation of the solids. Otherwise, it may be necessary to include a settling tank.

Since, like the oxidation pond, an aerated lagoon is sensitive to temperature, in colder climates the performance is hampered during the winter months. Some sources [19] recommend that solids be returned to the lagoon during the winter to improve performance, so as to operate the lagoon essentially the same as the acti-

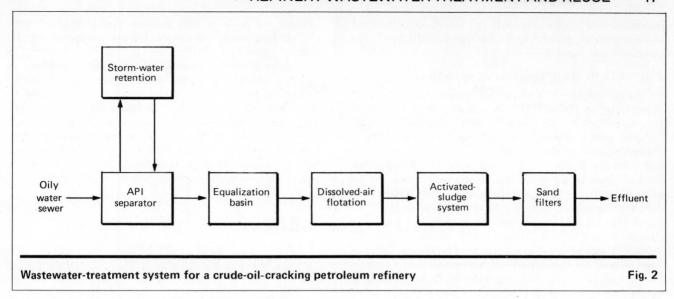

Wastewater-treatment system for a crude-oil-cracking petroleum refinery   **Fig. 2**

vated-sludge process. In a development program, it is very convenient for the aerated lagoon to be the first step, so that simply by adding clarifiers, sludge-return pumps, and additional aeration equipment, the lagoon becomes an activated-sludge system.

Filtration

Filtration is used only when it is necessary to obtain a very high-quality effluent or recycle stream. The effluent from activated-sludge systems contains organic matter in suspended or colloidal form that can be filtered; typically, the effluent ranges in content between 5 and 50 mg/l, which can be even further reduced to around 3 to 20 mg/l in a granular-media filter.

The results of filtering an activated-sludge effluent through a dual-media filter containing anthracite and sand are reported in Ref. 20. The solids in the effluent from a pilot-activated sludge unit are said to have a bimodal particle-size distribution. The smaller particles were in the 1–5-micron range, while the larger ones were in the 50–180-micron range. This makes the dual- or tri-media filters superior to the single-medium sand filter, where the larger particles tend to plug the filter, running times are shortened, and smaller particles are allowed to pass. With a dual-media filter, much more efficient solids removal and longer filter runs are experienced.

Typical dual-media filters are: anthracite and sand; activated-carbon and sand; resin beds and sand; and resin beds and anthracite. Filtration rates are in the 2–10-gpm/ft² range, with 5–6 gpm/ft² being typical. It is recommended that a pilot study be done for a particular system, because filter performance is very dependent on the particular characteristics of the specific liquid feed.

Carbon adsorption

There has been a tremendous amount of development in recent years in the use of activated carbon for the removal of dissolved organic material from wastewater. Carbon adsorption is normally used for final polishing of the effluent from a biological-treatment process, but it is feasible only if a very high-quality effluent is needed. Typically, the effluent from an activated-sludge system might be treated in an adsorption unit to further reduce the BOD to 3–10 mg/l, the oil content to less than 1 mg/l, and phenol to almost zero.

For most commercial-scale adsorption units, it is economical to regenerate the spent carbon for reuse. It is reported [21] that whenever more than 500–1,000 lb/d of carbon are used, a thermal regeneration system (1,300–1,800°F) is justified.

A carbon-adsorption unit consists of the adsorbers in which the wastewater stream contacts the activated-carbon bed; a transport system for moving the carbon from the adsorbers to the regenerator and back; and a regeneration system. The adsorbers may be arranged as fixed beds in parallel, or a moving carbon bed in which carbon is periodically removed from the bottom of the bed, and fresh carbon is added at the top. The API Design Manual [22] refers to the following design parameters for adsorbers: flowrate, 5–10 gpm/ft²; bed depth, 10 ft minimum; and contact time, 15–38 min.

Of the various methods of carbon regeneration that have been used, thermal regeneration is the most widely applicable, because it may have multiple-hearth furnaces, rotary kilns, or fluidized-bed furnaces. Other types include chemical, solvent and biological regeneration [23].

Operating a carbon-adsorption unit is reported [21] to be about 30–35¢/1,000 gal. The variation in cost is due to ranges in carbon dosage, carbon makeup rate, and the price of carbon. Reported ranges of carbon dosages are 800–2,000 lb/MM gal, with carbon makeup rates ranging from 5–10%, and the delivered price of carbon being between 26¢ and 32¢/lb. Typically, direct operating costs are about 12¢/1,000 gal.

Chemical oxidation

The primary application of chemical oxidation is for the reduction of phenols and cyanides in waste streams, by oxidants such as chlorine or ozone. These processes, however, are applicable only to small, concentrated streams, where conventional biological oxidation processes are not feasible.

Factors affecting treatment systems

The complexity of a treatment system varies with the complexity of the refinery. A simple case would be a treatment facility for a small refinery having a crude-oil unit with a desalter, a small reformer, and a desulfurization unit, as shown in Fig. 1.

Inasmuch as the storm-water retention pond stabilizes the flow to the separator, it also serves as a source of makeup water for the cooling tower and boiler. The API separator and the dissolved-air flotation unit remove free oil and suspended solids. Effluent from the dissolved-air flotation is treated in an aerated lagoon to reduce dissolved organic materials. The final effluent is then filtered for suspended-solids removal.

A refinery having topping, reforming and cracking operations is the next order of complexity. Fig. 2 shows a typical wastewater-treating facility for a cracking refinery. Note that the only change from Fig. 1 is that the aerated lagoon has been replaced by the activated-sludge system. An aerated lagoon can be modified to operate as an activated-sludge system by adding a sedimentation tank and sludge-recycle pumps. This might be a consideration for a refinery having an aerated lagoon, and needing to increase the efficiency of organic-material removal.

Another step in improving effluent quality might be the addition of carbon-adsorption facilities. This is usually justified only in cases where an effluent of the highest quality is required, or if total recycle and reuse of wastewater is being practiced.

Any design of wastewater-treatment facilities has to consider the national goal of zero discharge of pollutants to navigable waters by 1985. Theoretically, effluent streams can be treated for reuse within a refinery, but eventually something must leave the system, either as a liquid brine or as dry salt. For a zero-discharge system, a demineralization process must be used. Current technology available for this purpose is the use of either multiple-effect evaporation or solar-evaporation ponds.

Because solar-evaporation ponds require a large amount of land, they are not feasible in many cases. And since multiple-effect evaporation has a high energy demand, it is not particularly attractive. Three processes currently in the developmental stage show considerable promise for future water treatment: reverse osmosis, electrodialysis, and ion exchange.

A zero-discharge refinery has been proposed [24] in which the final effluent would be recycled back as cooling-tower makeup. The cooling-tower blowdown would be treated for heavy-metal removal before going to an evaporator, and finally to an incinerator. The final discharge from the incinerator would be dry salt and ash.

Other changes within a refinery are included in this proposal. One is the addition of a preliminary, crude-oil desalting step, where the crude oil is initially contacted with neutralized spent caustic. The spent caustic dissolves a portion of the salt from the crude oil. The effluent water from this desalter then combines with the cooling-tower blowdown for concentration of the salt.

Although each wastewater-treatment problem is unique, similar cases can be compared and ideas obtained as to what treatment process can be used to solve a particular problem. In many instances, it may be necessary to set up a pilot plant to determine what process can best do the job. Some equipment suppliers have portable pilot-plant facilities that can be rented. Such facilities are quite beneficial for determining the optimum system for a special pollution problem.

References

1. *Federal Register*, Vol. 39, No. 91, May 9, 1974.
2. *Federal Register*, Vol. 40, No. 90, May 20, 1975.
3. Carnes, B.A., Ford, D.L., and Brady, S., "Treatment of Refinery Waste-waters for Reuse," National Conference on Complete WateReuse, Washington, D.C., Apr. 1973.
4. "Wastewater Engineering: Collection, Treatment, Disposal," pp. 638-644, Metcalf and Eddy, Inc., New York.
5. Weisberg, E., and Stockton, D.L., "WateReuse in a Petroleum Refinery," National Conference on Complete WateReuse, Apr. 1973.
6. Annesen, R., and Gould, G., Sour-Water Processing Turns Problem Into Payout, *Chem. Eng.*, Mar. 22, 1971, pp. 67-69.
7. "API Manual on Disposal of Refinery Wastes, Volume of Liquid Wastes," Chap. 15, American Petroleum Institute (API), Washington, D.C.
8. "Development Document for Proposed Effluent Limitations Guidelines and New Source Performance Standards for Petroleum Refining," Dec. 1973, p. 110, Environmental Protection Agency, Washington, D.C.
9. "API Manual on Disposal of Refinery Wastes, Volume of Liquid Wastes," Chap. 5 and 6, API, Washington, D.C.
10. Franzen, A., Skogan, V., and Grutsch, J., *Chem. Eng. Progr.*, Vol. 68, No. 8, Aug. 1972, pp. 65-72.
11. "API Manual on Disposal of Refinery Wastes, Volume of Liquid Wastes," Chap. 9, API, Washington, D.C.
12. "Wastewater Engineering: Collection, Treatment, Disposal," pp. 299-301, Metcalf and Eddy, Inc., New York.
13. Dahlstrom, D., Lash, L., and Boyd, J., Biological and Chemical Treatment of Industrial Wastes, *Chem. Eng. Progr.*, Vol. 66, No. 11, Nov. 1970, pp. 41-48.
14. "API Manual on Disposal of Refinery Wastes, Volume of Liquid Wastes," Chap. 13, pp. 13-14, API, Washington, D.C.
15. Ibid., Chap. 13, pp. 13-41.
16. Ibid., Chap. 13, pp. 13-54.
17. "Wastewater Engineering: Collection, Treatment, Disposal," pp. 556-564, Metcalf and Eddy, Inc., New York.
18. Ibid., pp. 533-542.
19. Ibid., p. 543.
20. Ibid., pp. 644-650.
21. Hutchins, R. A., *Chem. Eng. Progr.*, Vol. 69, No. 11, Nov. 1973, p. 48.
22. "API Manual on Disposal of Refinery Wastes, Volume of Liquid Wastes," Chap. 10, pp. 10-39, API, Washington, D.C.
23. Loven, A. W., *Chem. Eng. Progr.*, Vol. 69, No. 11, Nov. 1973, pp. 56-62.
24. Porter, J.W., Blake, J.H., and Milligan, R.T., "Zero Discharge of Wastewater from Petroleum Refineries," National Conference on Complete WateReuse, Washington, D.C., Apr. 1973.

The author

Kenneth E. Bush is Manager, Process Engineering, for The Frantz Co., Suite 400, Bank of Woodlake Bldg., 2600 South Gessner, Houston, TX 77063. His responsibilities include process design of petroleum refineries and chemical plants, as well as wastewater-treatment and energy-efficiency studies. A registered professional engineer in Texas, he is also chairman of the State Affairs Subcommittee of the Critical Issues Committee of the AIChE South Texas Section. He holds B.S. and M.S. degrees in chemical engineering, the former from Texas A&M University, the latter from the University of Houston.

The effects of water-pollution control on energy consumption

The costs of pollution control take up a larger percentage of total operating costs in the high-energy, basic industries than for industry as a whole.

J. I. Stevens *and* **C. L. Kusik,** *Arthur D. Little, Inc.*

☐ When the amendments to the Water Pollution Control Act were passed in 1972, the timetable was set for (1) installing by 1977 the best practicable technology available, (2) installing by 1983 the best available technology economically achievable, and (3) reducing to zero by 1985 the discharge of pollutants to receiving waters. Since then, of course, the world has experienced an energy crisis, in which hitherto unreckoned supplies of fuel have been found limited. Thus an important concern of the Water Pollution Control Act and its goals has come to be the cost of the Act in terms of energy.

One way to get some perspective on the energy associated with pollution control comes from examining our national energy consumption patterns (Table I), where generated electrical energy is expressed in terms of an approximate fossil-fuel equivalence of 10,500 Btu/kWh, thus allowing for a generating efficiency of about 32.5%. This table indicates four major sectors: Total Industrial, Household/Commercial, Transportation, and Electrical Generation. Of these, the first, second and fourth will bear the energy costs of water pollution control.

Municipal sewage treatment

Although the Household/Commercial sector is smaller than Total Industrial, Transportation, or Electrical Generation so far as purchased fuel is concerned,

it is also the largest consumer of electricity (58%); and when generated electricity is allocated to the other three sectors, Household/Commercial consumes 35.0% of the purchased fuel equivalent.

The major pollution-control problem met in this sector is the treatment of wastewaters in municipally operated sewage-treatment plants. Although not all of the treating plants needed to meet the 1977 goals of the Act are yet under construction, it is nevertheless possible to get an idea of the energy consequences for the country as a whole.

In recent articles [2,3], Hagan and Roberts presented estimates of energy requirements for a number of different municipal-sewage-treatment systems. Using their estimates for activated-sludge treatment and assuming that the entire U.S. population were served with this system, we project that 0.075 Quad (1.0 Quad = 10^{15} Btu) would be required annually for municipal sewage treatment. Since this energy consumption is principally in the form of electricity, and since the Household/Commercial sector uses 9.7 Quads equivalent in electricity (Table I), the net effect is an increase in electrical energy use of $100(0.075/09.7) = 0.8\%$.

The potential for reducing energy consumption in municipal wastewater treatment is minimal. Rather, energy consumption for such treatment will increase significantly in the next ten years, as more and more

	Purchased fuels		Consumption of electricity, 10^{12} Btu/yr		Total fuel equivalent	
Sector	10^{12} Btu	%	@ 3,412 Btu/kWh	@ 10,500 Btu/kWh	10^{12} Btu/yr	%
Total industrial	20,294	29.4	2,329	7,166	27,460	40.1
Manufacturing	14,329	20.8	1,756	5,403	19,732	28.8
Nonmanufacturing	5,965	8.6	573	1,763	7,728	11.3
Household/commercial	14,281	20.7	3,160	9,725	24,006	35.0
Transportation	16,971	24.6	18	55	17,026	24.9
Electrical generation	17,443	25.3	—	—	—	—
Total	68,989	100.0	5,507	16,946	68,492	100.0

Distribution of energy-consumption by economic sector* — Table I

*In 1971.

Originally published August 15, 1977.

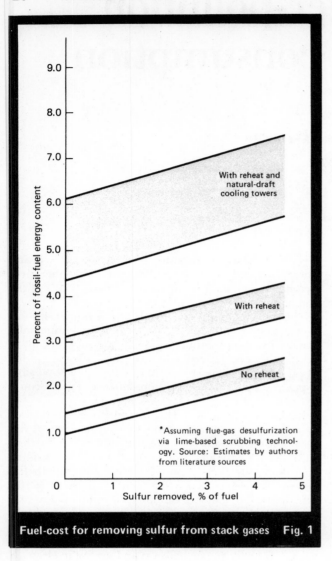

Fuel-cost for removing sulfur from stack gases   Fig. 1

facilities go into operation, and pressure increases from regulations being developed under the Solid Waste Disposal Act.

In many areas, requirements for receiving-water quality will make more energy-intensive treatment necessary and/or land restrictions will prevent the relatively low energy-consuming landfill method of sludge disposal. If alum treatment for phosphorus removal, or nitrification/denitrification for nitrogen removal were required, for example, energy needs for wastewater treatment might increase by a factor of 2.5.

We estimate that in the foreseeable future total energy demand for municipal sewage treatment will probably be in the range of 0.1 to 0.3% of total annual energy use.

Electric power generation

Increased electric power for pollution control has a direct corollary in the increased burden on energy sources for flue-gas desulfurization by lime-based scrubbing within the electric power industry [4,13]. The size of this energy requirement is shown in Fig. 1, as a percentage of the energy in the fossil fuel that would be required for varying levels of sulfur removal. It will not be unusual for those generating plants based on coal or oil to spend 4 to 5% of their total energy input for pollution control.

With electric power generation consuming about 25% of the purchased fuels in the U.S. (Table I), and with generating plants turning increasingly to fossil fuels, especially coal [15], pollution control for these plants may well take 0.7 to 1.2% of total demand for U.S. energy.

Industrial processing

The Industrial sector, including nonmanufacturing industries, accounts for about 40% of total U.S. energy demand (Table I). Predictions concerning energy use

Five energy-intensive manufacturing industries	Purchased fuels only		Purchased fuels and electricity valued on thermal basis[a]		Purchased fuels and electricity valued on fossil-fuel basis[b]	
	10^{12} Btu	%	10^{12} Btu	%	10^{12} Btu	%
Primary-metals industry	3,613	25.2	4,030	25.1	4,896	24.8
Petroleum & coal products	2,443	17.0	2,783	17.3	3,489	17.7
Chemicals & allied products	2,877	20.1	2,956	18.4	3,120	15.8
Paper & allied products	1,196	8.3	1,315	8.2	1,562	7.9
Stone, clay & glass products	1,291	9.0	1,367	8.5	1,525	7.7
Total of five	11,420	79.6	12,451	77.5	14,592	73.9
Other manufacturing	2,909	20.3	3,634	22.6	5,140	26.0
Total manufacturing	14,329	99.9[c]	16,085	100.1[c]	19,732	99.9[c]

Distribution of energy consumption within the manufacturing sector   Table II

[a] Purchased electricity valued at its thermal equivalence of 3,412 Btu/kWh.
[b] Purchased electricity valued at an approximate fossil-fuel equivalence of 10,500 Btu/kWh.
[c] Failure to sum to 100% due to rounding error.
Source: FEA Project Independence Blueprint, Vol. 3, 1974. Based largely on purchased fuels except for primary-metals industries and petroleum and coal products, where captive consumption of energy from byproducts is included.

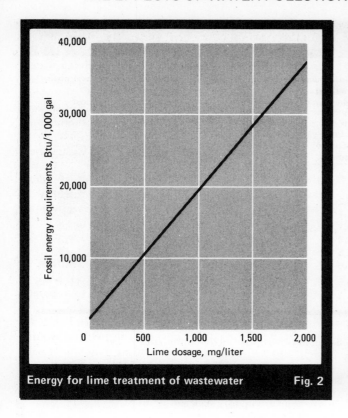

Energy for lime treatment of wastewater   Fig. 2

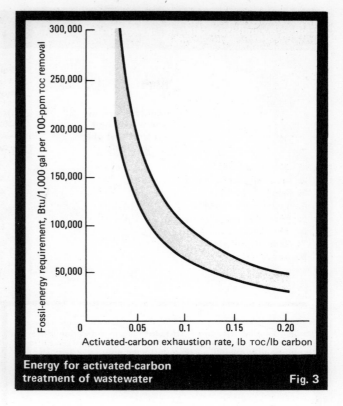

Energy for activated-carbon treatment of wastewater   Fig. 3

for pollution control in this sector are difficult to make because of the variation among plants.

Almost 80% of the energy used in manufacturing is concentrated in five industries—primary metals; petroleum and coal products; chemicals and allied products; paper and allied products; and stone, clay and glass products—as shown in Table II. Not only are these industries the leading energy users, but they also have some of the greatest pollution-control problems.

At the present time, we perceive that air pollution control requires the greatest amount of energy [9]. This energy demand, of course, varies widely, depending on the technology—from relatively low amounts for electrostatic precipitators, to the relatively high amounts for large pressure-differential wet scrubbers such as the venturi type.

Similar wide variations exist in the energy needed to meet water-effluent limitations, such as low energy for simple sedimentation of suspended solids, versus high energy for effective removal of soluble organic compounds.

Effluent limitations for "priority pollutants" could significantly affect the future course of industrial wastewater treatment. These limitations must be established by the U.S. Environmental Protection Agency (EPA) as a result of the settlement agreement of June 7, 1976, reached in the U.S. District Court for the District of Columbia between the Natural Resources Defense Council, Inc., et al., and EPA.

When facing stringent pollution-control regulations, industrial plants have three options: close down, install pollution-control equipment and make process modifications, or replace polluting processes with less-polluting processes.

Many obsolescent or marginal producers in various industries have closed down (e.g. iron foundries, partially because of air pollution regulations). By far the greatest number, however, have added the necessary pollution-control equipment, resulting in a generally dramatic increase in the proportion of new investment money devoted to pollution control systems. It is not now unusual to find a significant number of large corporations committing 15% or more of the annual construction funds for such systems. Although the problem of capital spending for pollution control equipment has received most attention in the technical press, operating costs, especially where large amounts of energy are needed, will certainly receive increasing attention in the future.

In general, the 1977 goals for industrial wastewater treatment are based on biological systems. Because treatment of industrial wastewaters differs from that of municipal sewage, requiring longer retention time, increased mixing and aeration, etc., the energy requirements for industrial wastewater are generally greater. Prather and Young [1] report values of approximately 4.6 kWh/1,000 gal, equivalent to 48,000 Btu/1,000 gal, for a system treating petroleum-refinery wastewater, compared to about 9,800 Btu/1,000 gal for municipal sewage [2,3]. Energy consumption for other industrial wastewaters can vary widely, but we estimate the average will lie in the range of 2.5 to 6.0 kWh/1,000 gal equivalent.

High-energy waste treatments

Industry representatives have been pointing out the high energy requirements for pollution control, while environmentalists have been equally active in minimiz-

Energy for pollution control* in iron & steel industry   Table III

	Production-energy cost as a percent of total cost of shipments	Pollution energy cost, percent of total production-energy cost		Pollution electric-power cost, percent of total electric-power cost	
		Stage I (including in-place control systems)	Stage II (including Stage I systems)	Stage I (including in-place control systems)	Stage II (including Stage I systems)
1. Multi-integrated	12.9	3.8	99.9	14.0	34.4
2. Individual integrated	9.3	4.2	12.7	12.1	34.4
3. Semi-integrated	6.1	4.2	14.6	4.2	14.6
4. Medium alloy steel	4.3	6.4	18.0	7.7	21.6
5. High alloy steel	1.8	4.1	17.0	4.1	17.0
6. Mini-mills	7.4	5.3	20.9	5.3	20.9
7. Carbon steel reroller	1.5	2.5	5.0	2.5	5.1
8. Alloy steel reroller	0.3	2.6	3.6	4.4	6.1
Weighted average	8.4	4.1	10.9	11.1	28.9

*Not included are control costs for carbon monoxide, nitrogen oxides, and potential toxic or hazardous materials; additional water-quality-related costs, monitoring and analysis costs, and unusual site-specific costs; and OSHA costs. Storm-runoff costs have been estimated separately and are excluded except where noted.

Source: Steel and The Environment: A Cost Impact Analysis— Report to the American Iron and Steel Institute by Arthur D. Little, Inc.

Selected pollution-control-compliance stages

	Stage I	Stage II
Solid waste disposal (essentially as practiced in 1972)	X	X
Air (to meet 1975 primary standards)	X	X
Air—fugitive emissions		X
Water—1977 EPA guidelines	X	X
Water—1983 EPA guidelines		X

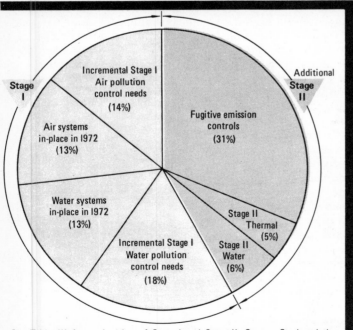

See Table III for explanation of Stage I and Stage II. Source: Steel and the Environment: A Cost Impact Analysis — Prepared for The American Iron & Steel Institute by Arthur D. Little, Inc.

Distribution of pollution control investment for the iron & steel industry by type of pollution in Stage I and Stage II   Fig. 4

Summary of 1971 energy purchased in selected industry sectors   Table IV

Industry sector	Quads, 10^{15} Btu/yr
Blast furnaces and steel mills	3.49[1]
Petroleum refining	2.96[2]
Paper and allied products	1.59
Olefins	0.984[3]
Ammonia	0.63[4]
Aluminum	0.59
Textiles	0.54
Cement	0.52
Glass	0.31
Alkalis and chlorine	0.24
Phosphorus and phosphoric acid production	0.12[5]
Primary copper	0.081
Fertilizers (excluding ammonia)	0.078

[1] Estimate for 1967 reported by FEA Project Independence Blueprint, p. 6-2, U.S. Govt. Printing Office, Nov. 1974.
[2] Includes captive consumption of energy from process by-products (FEA Project Independence Blueprint).
[3] Olefins only, includes energy of feedstocks: Arthur D. Little, Inc., estimates.
[4] Ammonia feedstock energy included: Arthur D. Little, Inc., estimates.
[5] Arthur D. Little, Inc., estimates.
Source: 1972 Census of Manufactures, FEA Project Independence Blueprint, USGPO, Nov. 1974, and Arthur D. Little, Inc., estimates.

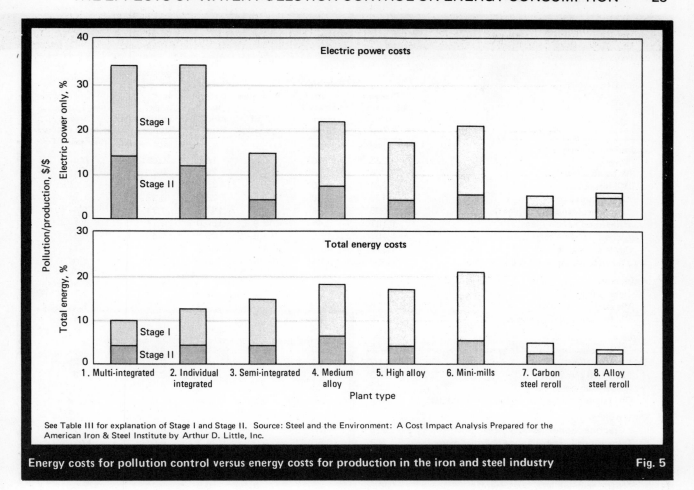

See Table III for explanation of Stage I and Stage II. Source: Steel and the Environment: A Cost Impact Analysis Prepared for the American Iron & Steel Institute by Arthur D. Little, Inc.

Energy costs for pollution control versus energy costs for production in the iron and steel industry   **Fig. 5**

Energy use for water pollution control in the pulp and paper industry   **Table V**

Subcategory*	Process energy usage (million Btu/ton)	Process energy for water pollution control, %		
		Pretreatment	BPT†	BAT†
Papergrade sulfite	40.1	1.1	4.8	6.2
Dissolving sulfite	43.8	0.7	3.5	5.9
De-ink	18.3	5.7	6.1	
Dissolving kraft	39.3	0.7	2.7	6.0
Market kraft	37.5	0.7	2.3	5.5
BCT kraft	41.9	1.3	2.8	5.6
Fine kraft	42.9	1.2	2.3	4.9
Groundwood (Chem.—mech.)	37.8	0.5	2.4	4.3
Groundwood (CMN)	36.6	0.6	1.8	3.9
Groundwood fine	37.6	0.6	1.6	3.7
Soda	41.9	1.1	2.7	5.5
Nonintegrated fine	17.9	1.1	2.3	3.0
Nonintegrated tissue	12.7	2.7	6.4	6.8
Nonintegrated tissue (FWP)	13.9	2.5	5.7	6.5
Average		1.1	3.3	5.2

*See ref. 10, Arthur D. Little, Inc., estimates, assuming: (1) 85% thermal efficiency for steam generation, and (2) 10,500 Btu equivalent per kWh.
†BPT = best practical technology.
BAT = best available technology.

Process and pollution costs, and energy usages, for selected processes						Table VI
Process	**Process operating costs, $/ton**	**Pollution control operating costs, $/ton**	**Pollution costs, % of operating**	**Process energy usage, million Btu/ton**	**Pollution control energy usage, million Btu/ton**	**Pollution-control energy usage, % of operating energy**
Alumina and aluminum						
Bayer	235	1.40	0.6	14.5	0.06	0.4
Nitric acid leaching	226	19.00	8.4	26.8	0.70	2.6
HCl leaching	321	5.00	1.6	39.2	0.02	0.05
Toth alumina process	179	0.80	0.5	28.6	0.29	1.0
*Hall-Heroult (existing plant)	698	44	6.3	187.8	1.71	0.9
Hall-Heroult (new plant)	1,137	44	3.9	150.0	1.71	1.1
Alcoa chloride electrolysis	1,107	16.5	1.5	135.1	0.40	0.3
Bayer & Hall-Heroult	1,372	46.7	3.4	164.6	1.80	1.1
Toth alumina & Alcoa chloride	1,032	37.4	3.6	100.0	1.00	1.0
Ammonia						
*Via natural gas	97	1.1	1.1	37.0	0.7	1.1
Via coal gasification	136	9.8	7.2	35.5	0.6	1.7
Via heavy-oil gasification	146	4.6	3.1	35.0	0.5	1.4
Cement						
*Natural-gas or oil-fired long kiln	47.3	2.0	4.2	5.6	0.07	1.3
Flash calciner	43.7	1.4	3.2	4.2	0.05	1.2
Fluid-bed cement	44.3	2.1	4.7	5.0	0.10	2.0
Suspension preheater	43.7	1.4	2.4	4.2	0.05	1.2
Conversion to coal from oil and natural gas	45.6	2.3	5.0	5.6	0.07	1.3
Chloro-alkali						
*Graphite-anode diaphragm cell	115	0.6	0.5	40.6	0.01	0.02
Dimensionally stable anode polymer membrane	103	0.50	0.5	34.7	0.01	0.03
Modern mercury cell	134	3.60	2.7	39.5	0.01	0.03
Fertilizers						
*Nitric acid—no NO_x abatement	56.5	—	0	0.9	—	0
Nitric acid with catalytic reduction	56.5	6.2	11.0	0.9	1.1	122.0
Nitric acid with molecular sieve	56.5	4.0	7.1	0.9	0.4	44.4
Nitric acid with Grande Paroisse	56.5	2.3	4.1	0.9	0.1	11.1
Nitric acid with CDL/Vitok	56.5	1.6	2.8	0.9	0.3	33.3
Nitric acid with Masar	56.5	1.9	3.4	0.9	0.1	11.1
Glass						
*Natural-gas fired—cold charge	72.7	7.7	10.6	7.0	0.6	8.6
Coal gasification & glassmaking	91.7	23.1	25.2	8.6	0.9	10.5
Direct coal firing	67.2	10.7	15.9	7.0	0.7	10.0
Electric melting	76.0	5.5	7.2	8.2	0.03	0.4
Coal, hot gas generation & glassmaking	85.2	9.9	11.6	8.6	0.8	9.3
Batch preheat & glassmaking	72.9	5.8	8.0	5.7	0.4	7.0

ing total energy needs. However, the high-energy-consuming wastewater-treatment processes have yet to be installed, either for industrial effluents or for municipal wastes. Among the most likely candidates for industrial wastewater treatment are adsorption by activated carbon or increased use of chemicals, such as lime. Selective application of such liquid-system processes as electrodialysis, ultrafiltration, reverse osmosis, and evaporation can also be expected.

Anticipated energy use for lime treatment and activated-carbon treatment are shown in Fig. 2 and 3, respectively [6,7,8]. The dramatic increase in energy for activated-carbon treatment is illustrated as a function of the exhaustion rate for the total organic carbon (TOC)

in the wastewater. Hager [5] has reported the results of 220 laboratory tests of activated carbon on a wide variety of wastewaters. From his data, we have calculated the average exhaustion rate to be 0.17 lb TOC/lb carbon, with a standard deviation of 0.13 lb TOC/lb carbon. Because the exhaustion rate found in actual operation is usually significantly less than such laboratory-determined values, it is apparent that a large number of industrial wastes will use significant amounts of activated carbon.

Our estimates of energy consumption in Fig. 3 were based on:

■ 5,000 to 8,000 Btu/lb of carbon regenerated for auxiliary fuels.

Process and pollution costs, and energy usages, for selected processes (cont'd) Table VI

Process	Process operating costs, $/ton	Pollution control operating costs, $/ton	Pollution costs, % of operating	Process energy usage, million Btu/ton	Pollution control energy usage, million Btu/ton	Pollution-control energy usage, % of operating energy
Iron and steel						
*Combustion of off-gases	—	1.1	—	—	0.15	—
CO collection from BOP	—	0.7	—	—	(0.36)	—
*Coke oven, blast furnace, BOP	134	9.3	6.9	11.2	0.7	6.3
Direct reduction—electric furnace	140	1.1	0.8	17.7	0.8	4.5
*Desulfurization in blast furnace	101	4.6	4.6	9.7	0.3	3.1
External desulfurization of blast furnace, hot metal	101	4.4	4.4	9.4	0.3	3.2
Olefins						
*Ethane-propane coil cracking	194	1.0	0.5	84.2	0.03	0.04
Naphtha coil cracking	256	2.2	0.9	124	0.09	0.07
Gas-oil coil cracking	250	3.4	1.4	167	0.16	0.10
Paper & allied products						
*Conventional kraft	252	38	15.0			
Alkaline oxygen pulping	222	34	15.3			
Rapson effluent-free kraft process	254	5.0	2.0			
*Refiner mechanical pulp (RPM)	104	3.0	3.0			
De-inking of waste paper as substitute for mechanical pulping	148	11.0	7.4			
Thermo-mechanical pulping	110	2.0	2.0			
Phosphorus						
*Electrothermal phosphoric acid	326	12.0	3.7	88.0	0.06	0.07
Chemical cleanup of wet-process acid	286	8.3	2.9	2.8	0.14	5.0
Solvent extraction of wet-process acid	241	11.9	4.9	4.4	0.8	18.2
*Conventional wet-process acid	254	6.9	2.7	2.6	0.14	5.4
Strong acid production of wet-process acid	248	6.9	2.8	2.2	0.14	6.4
Primary copper						
*Reverb/converter green feed	300	49.5	16.2	23.9	3.7	15.5
Outokumpu flash smelting	293	48.3	16.5	8.7	6.3	72.4
Noranda process	290	48.3	16.7	6.2	6.3	101.6
Mitsubishi process	299	48.3	16.2	5.7	6.3	110.5
Oxygen in flash smelting	216	48.3	22.4	6.9	6.3	91.3
*Reverb/converter calcine feed } *Electro refining }	542	58.2	10.7	24.4	4.7	19.3
Arbiter process	602	33.1	5.5	45.0	NIL	—

* Denotes baseline process for comparison with process immediately following.

■ 1 lb of steam/lb of carbon regenerated (1,150 Btu/lb).

■ Losses of 5 to 10% per regeneration cycle.

■ Energy content of activated carbon of 19,000 Btu/lb.

A highly concentrated waste, containing over 1,000 mg/liter of TOC, with a low exhaustion rate, becomes very energy-demanding.

While the energy demand involved in activated-carbon adsorption may exceed by as much as an order of magnitude that incurred in traditional wastewater treatment, less-conventional treatment processes may be even more demanding. Direct energy use is rarely less than 150,000 Btu/1,000 gal for reverse osmosis; 1 to 12 million Btu/1,000 gal for evaporative systems, such as vapor recompression and multiple-effect evaporation; and more than 20 million Btu/1,000 gal, if thermal incineration for destruction of hazardous components is required. Faced with such energy demands, it is apparent that industrial-wastewater-treatment technologies and process controls will have to be examined carefully to determine the most cost-effective methods.

Industrywide studies of iron and steel and of pulp and paper—two of the top five users of energy—may give some perspective on pollution control costs. Table III shows that pollution control can add up to 11% of the total energy costs in the iron and steel industry. Since energy costs at the time of the study (1972 was the

Critical areas of industrial pollution control   Table VII

Alumina: Acid leaching of clays would create a significantly greater water and solid-waste problem because of the larger amounts of nitrates and chlorides than in the present Bayer process.

Aluminum: The major problem will continue to be air pollution, though this will be somewhat lessened.

Ammonia: Switching to heavy oil or coal will increase overall environmental problems because of the sulfur, metals and ash content of the alternative fuels. Air pollution and solid-waste disposal will be the problem areas.

Cement: Air pollution control will remain the most important problem.

Chlor-alkali: Steps taken to cut energy consumption may be expected also to reduce pollution control problems, which are already low in this industry.

Copper: The new pyrometallurgical processes that utilize the heat of oxidation of sulfur and iron to supply a part of the process energy tend to produce a steady stream of concentrated sulfur dioxide. Over 90% of the SO_2 may then be recovered at about the same cost as the 50 to 70% recovery in conventional smelters. This higher capture rate will make the energy used for air pollution control a greater portion of the overall energy use.

The hydrometallurgical processes are not expected to create solid-waste or wastewater problems significantly different from those presently faced in other industrial sectors.

Nitric acid: More-stringent control of nitric acid emissions will significantly increase air-pollution control costs and energy use. Installation of control equipment will force a careful evaluation of the energy cycle.

Glass: Substitution of a lower fuel form will have little effect on the industry's air and water pollution control problems. Moving major emission points away from the glass plant to the coal-gasification or electric-power plants will have the net effect of more air pollution because of the dirtier fuels.

Iron and steel: Air pollution problems will probably decrease with recovery of carbon monoxide from the basic-oxygen pro-

cess. However, sulfur oxide emissions may increase because of external desulfurization of hot metal, since the blast furnace is an effective sulfur remover.

Air pollution control is generally less of a problem with the electric-furnace process for direct reduction than with the blast furnace, but energy consumption is nearly 40% higher. On the other hand, the need for high-grade metallurgical coal is eliminated. The industry will proceed cautiously until the process technology is proved successful.

Olefins: The high energy content of the raw materials and products [11,12] means that energy for pollution control is a small fraction of total energy involved. For ethylene, the major volume product, the use of heavier feedstocks, naphtha and straight-run gas-oil, will mean a significant increase in valuable byproducts such as butadiene. These byproducts must find ready markets, however, in order for the substitution to be economically viable. High-sulfur feedstock could cause an indirect environmental impact by increasing the sulfur content of the byproduct fuel oil, thereby necessitating feedstock desulfurization or flue-gas desulfurization at the point of use.

Petroleum refining: Attempts to improve yields of light-end products through greater use of asphalt would significantly increase the potential for air pollution problems from sulfur oxide emissions.

Because of the high cost of controlling such emissions from a large number of small sources, a possible alternative is to sell the asphalt to electric generating plants equipped with flue-gas-desulfurization systems. This approach, however, may conflict with government efforts to reduce the amount of petroleum used by electric utilities.

Phosphoric acid (detergent grade): The solvent extraction method of hydrochloric acid leaching produces a large amount of calcium chloride brine that presents a serious water pollution problem. Neutralization/precipitation cleanup would be the most environmentally acceptable process, although it produces nearly twice as much sludge as solvent extraction.

Pulp and paper: Adoption of the Rapson process and alkaline-oxygen pulping will achieve a significant reduction in air and water pollution problems.

base-line year) were 8.4% of the total of products shipped, it follows that pollution-related energy costs could account for about 1% of the total.

The wide variation in energy costs for pollution control in different segments of the iron and steel industry (Table III and Fig. 5) indicates clearly that the impact of these costs will not be distributed uniformly. The large increase in electric power costs between Stage I and Stage II of Fig. 4 are attributable principally to those for the control of fugitive airborne emissions and for thermal pollution control. Since the iron and steel industry's annual energy use is about 3.5 Quads, pollution control in that industry alone might approximate 0.5% of total U.S. energy consumption.

Table V shows that wastewater pollution control in the pulp and paper industry, the third largest industrial

user of energy in the U.S., could be equivalent to approximately 0.1% of total U.S. energy consumption, or about the same as that for municipal sewage treatment. Similar data for the second-largest U.S. industrial user of energy, petroleum refining, were not available, but energy use for pollution control in refining is probably less than 0.1% of the total. Based on a similar rationale, annual energy use for all industrial pollution control is probably about 1% of the total U.S. energy budget.

We recently examined the effects of replacing energy-intensive processes with more-efficient ones, either in terms of total energy use or of substituting a lower for a higher energy form for the industrial sectors shown in Table IV [15]. (With proven gas reserves in the U.S. limited, much of our petroleum imported, and tremendous coal reserves, gas has the highest form-value, followed by oil and coal.)

Process replacement

We examined in detail over 80 processes, of which 55 are new or recently installed, and estimated energy and pollution control requirements for both conventional and new technologies. In all cases, the plant size chosen for estimating energy consumption costs for operating processes and pollution control was typical for that sector of the industry. Table VI is a compilation of the estimated energy consumption and costs for those industry sectors where the most direct comparisons can be made. In establishing the energy estimates, the study concentrated on process changes, and did not consider the potential for energy conservation that might result from:

- Improved waste-heat utilization.
- Improved maintenance and operation.
- Power generation.
- Steam generation by alternative fuels.
- Fuel substitution in fired process-heaters.
- Substitution of scrap for virgin materials.

Because industries that use large amounts of energy tend to be mature, are capital intensive, have relatively long-lived plants and equipment, and have little potential for substituting raw materials, any new processes will have little significant impact on the type of pollutants, although the size of the discharges can change. Some of the areas of concern for various industries are noted in Table VII. Despite these concerns, however, recent installations of new process technology indicate increasing improvements, such as:

Basic product	New technology	Company/location
Ammonia	Coal gasification	TVA/Tenn.
Blister copper	Flash smelting	Phelps Dodge/N.M.
Copper	Electric furnace	Copper Hill, Tenn.
Iron and steel	Direct reduction	Azcon/St. Joe Minerals, Tenn. Stelco/Griffith Mine, Ont.
Kraft pulp	Alkaline-oxygen process	Weyerhaeuser/Everett, Wash.
Kraft pulp	Rapson process	Great Lakes Paper/ Thunder Bay, Ont.
Mechanical pulp	Thermochemical pulping	Continental Forest Industries/Augusta, Ga.

The survey of energy-intensive industries [15] indicates that managers are closely reviewing plant operations and processes even where new technology has not been installed.

In conclusion, only a few new processes will differ significantly from present ones in the amount of energy employed for pollution control. Industry will use an increasing proportion of its total energy budget for pollution control, especially as greater regulation of wastewater and solid waste disposal comes into effect. The energy requirements for controlling pollution to presently established standards usually do not represent a large percentage of the total energy use. Finally, the costs of pollution control represent a larger percentage of total operating costs for the high-energy, basic industries than for industry at large. The problem is heightened by constraints on the availability of capital, and reluctance to incur costs for a non-productive system.

References

1. Prather, B. V., and Young, E. P., Energy for Wastewater Treatment, *Hydrocarbon Proc.*, May, 1976, pp. 88–91.
2. Hagan, R. M., and Roberts, E. B., Energy requirements in Water Supply, Use and Conservation, Part 1, *Water & Sewage Works*, Nov., 1976, pp. 64–67.
3. Ibid., Energy Requirements for Wastewater Treatment, Part 2, Dec., 1976, pp. 52–57.
4. Hirst, E., The Energy Cost of Pollution Control, *Environment*, Vol. 15, No. 8, Oct., 1973, pp. 37–42.
5. Hager, D. G., A Survey of Industrial Wastewater Treatment by Granular Activated Carbon, Fourth Joint Chem. Eng. Conference, AIChE and Canadian Soc. for Chem. Eng., Oct., 1973.
6. Rizzo, J. L., and Shephard, A. R., Treating Industrial Wastewaters with Activated Carbons, *Chem. Eng.*, Jan. 3, 1977, pp. 95–100.
7. Matsumoto, Z., and Humasaki, K., Regenerate Granular Carbon, *Hydrocarbon Proc.*, May, 1976, pp. 157–160.
8. Hutchins, R. A., Activated Carbon Regeneration: Thermal Regeneration Costs, *Chem. Eng. Prog.*, Vol. 71, No. 5, May, 1975, pp. 80–86.
9. Kusik, C. L., others, Energy Use and Air Pollution Control in New Process Technology, AIChE Annual Meeting, Aug.–Sept., 1976.
10. Ibid., Development Document for Advance Notice of Proposed or Promulgated Rate Making for Effluent Limitations Guidelines and New Source Performance Standards for the Bleached Kraft, Groundwood, Sulfur, Soda, De-ink, and Non-integrated Paper Mills Segment of the Pulp and Paperboard Mills—Point Source Category, U.S. EPA 440/1-75-047, Aug., 1975.
11. Ponder, C., (ed.), 1969 Petrochemical Handbook Issue, *Hydrocarbon Proc.*, 11, 1969.
12. Saxton, J. C., others, Federal Findings on Energy for Industrial Chemicals, *Chem. Eng.*, Sept. 2, 1974, p. 71.
13. LaMantia, C. R., others, Operating Experience—CEA/ADL Dual Alkali Prototype System at Gulf Power/Southern Services, Inc., EPA Flue Gas Desulfurization Symposium, New Orleans, La., Mar. 8–11, 1976.
14. Does Pollution Control Waste Too Much Energy, *Bus. Wk.*, Mar. 29, 1976, p. 72.
15. Environmental Considerations of Selected Energy Conserving Manufacturing Process Options, EPA–600/7-76-034, *a* through *o*, Dec., 1976.

The authors

James I. Stevens is a senior staff member of Arthur D. Little, Inc., Acorn Park, Cambridge, MA 02140, where he specializes in technical, economic and regulatory aspects of pollution control. Formerly technical director of Infilco, he holds B.Ch.E. and M.Ch.E. degrees from the University of Louisville. A member of the Water Pollution Control Federation, Air Pollution Control Assn., AIChE and ACS, he is a registered professional engineer in Arizona and Massachusetts.

Charles L. Kusik is a senior staff member of Arthur D. Little, Inc., where he specializes in technical and economic aspects of process technology, and has directed many studies on steel-making, pulp and paper, industrial chemicals, copper and other nonferrous metals. A registered professional engineer in Massachusetts, he holds B.S. and M.S. degrees from M.I.T. and a D.Sc. degree in chem. eng. from New York University, and is a member of the Metallurgical Society of the A.I.M.E., the AIChE and the ACS.

Industrial wastewater pumps

This is an overview of the available types of industrial wastewater-treatment pumps, with emphasis on process considerations of effluent transfer, chemical addition, slurry and sludge transport, and sampling systems.

Jacoby A. Scher, *Fluor Engineers and Constructors, Inc.*

☐ Within the last decade, industrial wastewater treatment has taken a much more important role in overall plant facilities. Increasingly stringent effluent requirements promulgated by all agencies have resulted in a surge of more sophisticated waste-treatment methods. Now, even contaminated storm-water must be treated, and more emphasis must be placed upon intricate gathering and treatment networks.

The problem of gathering, transporting and treating waste liquids in existing plants is much more complex than for a grassroots facility. Frequently, the wastewater-treatment system is relegated to a distant corner of available plot space, where ease of hydraulic transport is not necessarily a prime consideration. Liquids must be moved, chemicals supplied, slurries processed and other liquid or semiliquid services provided.

More often than not, pumps of many types and functions are used to ensure that the process operates. When examining a particular pump application, it is useful to check with others who have used pumps in similar situations. Also, pump manufacturers can be of great help in matching the right pump for the service involved.

Processing units generally dispose of wastewater by gravity into underground sewers or into open channels. Surface runoff of contaminated stormwater flows by gravity to a centralized collection sump. Cooling water and boiler blowdowns, aqueous water-treatment and/or slurried-waste streams—as well as other utility-type wastes—must be segregated or combined, and moved to the treatment area.

Various types of chemical solutions, whether acids or bases for neutralization, coagulants for clarification, conditioning chemicals for sludge dewatering, etc., must be provided at the various unit waste-treatment operations. Slurries such as lime, biological mixed liquor, or similar suspensions must be transported. Sludges of varying viscosities and composition must be mechanically dewatered or moved to thickeners, pits, etc., for concentration or drying. Finally, continuous samples for process and effluent quality control must be obtained. All of these manipulations usually require pumps of various types and construction, that must be manufactured for a wide variety of services.

Types of pumps

Of the many different types of pumps, the ones listed below are the most generally used for industrial wastewater-treatment systems. They are all listed for general information. Some are so common that they will not be discussed here, but others will be described in detail because of the uniqueness of their design and application. The various kinds are:

Centrifugal—This is a very general type of pump. In this article, the pump referred to is the American Voluntary Standard (AVS) type of overhung pump. It is so commonly used that it will not be dealt with here.

Sump—This is essentially a vertical centrifugal pump, which can be either the single-stage, volute type, or the single or multistage turbine type. The latter is an adaptation of the deep-well kind that is used for pumping water. Sump pumps, as shown in Fig. 1 (F/1), are designed to be installed in a pit or sump where an established water level exists.

Submersible—This is a centrifugal pump especially constructed so that the entire pump is submerged, including the motor (F/2). They are frequently designed to pass large particles, and can run dry for reasonable periods without damage.

Piston (positive-displacement)—This pump utilizes a cam-driven piston to directly move the fluid. It is available in a wide range of capacities and is capable of

Originally published October 6, 1975.

attaining discharge pressures as high as 10,000 psig.

Diaphragm—This is an especially constructed positive-displacement pump. The diaphragm acts as a piston when it is motivated into reciprocal motion by either mechanical linkage, compressed air, or a fluid from an external pulsating force. The diaphragm acts as a seal between the motivating force and the liquid being pumped. F/3 shows a cutaway view of the pump, which indicates how the liquid is draw in and expelled by the diaphragm.

Rotary—Common pumps in this category are the gear and the progressive-cavity or screw pump. Defined by the Hydraulic Institute, a rotary pump consists of a fixed casing containing gears, cams, screws, plungers or similar elements actuated by rotation of the drive shaft. The rotary pump of primary interest is the progressive-cavity or screw pump. It consists of a rotor turning within a stator whereby cavities are formed moving toward the discharge that transports the slurry. These pumps (F/4) can be accompanied by units* that grind certain materials to convert them to slurries.

*Called MAZ-O-RATOR grinders, made by Robbins & Myers, Inc., Springfield, Ohio.

Peristaltic—This is a special type of rotary pump that utilizes a piece of tubing wrapped around a cam moving eccentrically. As the cam rotates, the liquid is mechanically squeezed through the tubing (F/5).

All rotating equipment require some kind of seal between the shaft and the housing. Two types are common: the packed stuffing box and the mechanical seal. The packed gland has multilayers of packing placed inside a compression ring, which is gradually tightened as the packing wears away.

The lubrication for this seal is the fluid being pumped, which must be allowed to constantly drip from the stuffing box. Since the compression ring must never be tightened to the point where dripping ceases, this continuous drip prevents stuffing damage. If the pumped liquid contains suspended material, the packing can be quickly eroded and a severe leak may develop. A packed stuffing seal is fine for clear, rather innocuous fluids. It should not be used for corrosive, hazardous or abrasive solutions.

A mechanical seal has one stationary face and one rotating member held together by springs, a backup plate and other appurtenances. The close coupling provides a true mechanical seal that prevents leakage of fluid from the housing. If the solution being pumped is particularly hazardous or toxic, a double mechanical seal may be chosen, which has a nontoxic liquid between the seals to buffer any leakage. Also, to prevent scoring or seal crystallization, a mechanical seal with a flushed face may be selected. Water is circulated between the two seal faces to keep them clean and to extend their service life. Mechanical seals are very efficient in their operation if applied properly. However, they are more expensive than stuffed packing; constant replacement can add to maintenance costs.

A pump without a seal is also available. Called "canned rotor pump," it permits liquid to circulate through the motor bearings and then return to the general flow. A thin plate prevents fluid from contacting the motor windings; no shaft seal is required as

Sum pumps are installed where water levels exist (F)1

USS Oilwell

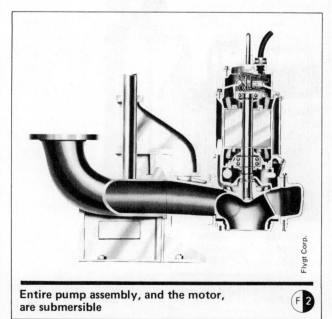

Entire pump assembly, and the motor, are submersible (F)2

Flygt Corp.

there is no leakage whatever. This pump is more compact than the conventional ones. For pumps smaller than 30 hp, no foundations are needed because shaft-alignment problems are not encountered (F/6).

Since most wastewater-pumping applications—such as that of stormwater, cooling tower, boiler blowdowns, clarifier effluent, etc.—are within a pH range of 6–9, the liquid generally does not contain hazardous or corrosive chemicals in high concentration. In such instances, pumps can be made of carbon steel, cast iron or other mild materials of construction.

In services where chemicals are added, or when skimmed oil, ballast-water treatment, or other corrosive fluids are transported, other materials of construction are needed that are more resistant to various chemicals. The accompanying table indicates typical materials of construction for some liquids frequently encountered in industrial wastewater treatment.

Effluent transfer

Collection systems for wastewater treatment are unlike process transport systems. Instead of pressurized enclosed-pump networks, wastewater effluents are usually drained into gravity sewers that flow to a common low-point collection area. Wastewater flows are also much more hydraulically irregular (e.g., stormwater runoff) than process flows. Equalization of flows prior to treatment is necessary because the various waste-treatment unit processes are rate dependent and sensitive to shock mass or hydraulic loadings.

Collection sumps and equalization ponds are utilized to smooth out fluctuations in gravity flows. Sump design and pump-capacity selection must be closely investigated so that the entire anticipated flow range can be adequately handled. Pump-capacity selection is im-

portant because a pump that continually cycles on/off/on may burn out a motor; and a pump that runs very infrequently may not function at all when it is needed. Often, if a collection sump experiences a flow range over several orders of magnitude, various different pumps with different capacities (set to actuate by sequential level control) may be the optimum configuration.

A typical installation recently encountered had a sump that collected a continuous effluent stream of 250 gpm, with additional stormwater flows up to 12,500 gpm. Four sump pumps were provided: one of 250 gpm to pump the continuous stream; two of 3,000 gpm each, to discharge low-intensity storm runoffs; and one of 6,500 gpm that would operate only under high-intensity storm flows.

Vertical sump pumps, whether of the centrifugal volute type or turbine type, are ideally suited to pump liquids from an open or closed collecting sump. Capacities for the volute type range from a few to several thousand gpm, with typical heads of 150–200 ft. Vertical turbine pumps with multistages have typical capacities of 100,000+ gpm and discharge pressures up to 1,000 psig or even higher.

These pumps can be operated by a float-type level controller, which prevents the pump from running if the level drops too low in the sump, but activates the pump when the liquid level rises. Vertical sump pumps require a minimum submergence above the intake bell to prevent vortex formation. Also, an intake screen is usually provided to prevent trash from being sucked into the pump. Sumps should be designed with a well, or boot, below the main sump floor into which the pump will fit. When this is done, the sump can be completely emptied, except for the capacity of the boot.

A submersible-type pump can perform several services in effluent transfer. It can also operate over a wide range of liquid levels, and can be obtained in capacities up to several thousand gpm and heads up to 100+ psig. This unit is somewhat capacity-limited, because the entire pump and motor are housed in a casing and placed on the bottom.

The advantages of the submersible type are that it can pump completely dry for a period of time, and can also pump large suspended solids if equipped with an open type of impeller. Maintenance is somewhat difficult because the entire pump assembly must be lifted from the bottom. It pumps very well from ponds that have widely fluctuating liquid levels; it can even pump a pit or pond dry.

Chemical feed

Many of the unit processes associated with waste-treatment systems require addition of chemicals for optimization. Typical additions include:

Acid and/or base for pH adjustment—These chemicals are usually sulfuric or hydrochloric acid, sodium hydroxide, sodium carbonate or lime solution.

Nutrients for biological treatment—To satisfy biological phosphorus and nitrogen needs, phosphoric acid, forms of sodium phosphate, gaseous or aqueous ammonia, or ammonium-salt solutions are added.

Coagulants for suspended solids control—Typical of these

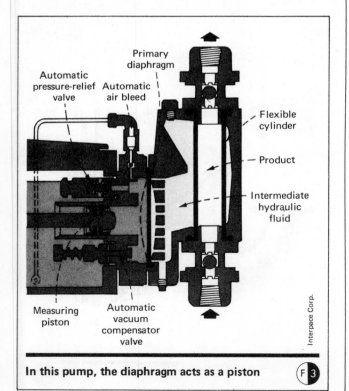

Automatic pressure-relief valve

Automatic air bleed

Primary diaphragm

Flexible cylinder

Product

Intermediate hydraulic fluid

Measuring piston

Automatic vacuum compensator valve

Interpace Corp.

In this pump, the diaphragm acts as a piston   F 3

Chemical resistance data for pumps*

Material	Epoxy	Polypropylene	Polyvinyl chloride Type 1	Nylon	Stainless steel 316	Stainless steel 304	Titanium	Buna N	Viton
Acetaldehyde	A	A	D	A	A	A	A	D	A
Acetic acid, glacial[1]	—	A	D	A	A	B	B	D	D
Acetic acid	A	A	A	D	A	B	B	C	C
Acetone	A	A	D	A	A	A	A	D	D
Alcohols									
Butyl	A	B	A	A	A	A	—	A	A
Ethyl	A	A	A	A	A	A	A	A	A
Hexyl	A	—	—	A	A	A	—	A	A
Isopropyl	A	A	—	A	A	A	—	C	A
Methyl	A	A	A	A	A	A	B	A	A
Aluminum chloride, 20% solution	A	A	A	A	C	D	B	A	A
Aluminum fluoride	A	A	A	A	C	—	A	A	—
Aluminum hydroxide	A	—	A	A	A	A	—	A	A
Aluminum sulfate	A	A	A	A	C	C	A	A	A
Ammonia, anhydrous	A	A	A	A	A	B	B	B	D
Ammonia, liquid	A	A	A	—	A	A	—	B	A
Ammonium nitrate	A	A	—	—	A	A	—	C	—
Ammonium phosphate									
Dibasic	A	—	A	A	A	A	—	A	A
Monobasic	A	—	A	A	A	A	—	A	A
Tribasic	A	—	A	A	A	A	—	A	A
Ammonium sulfate	A	A	A	A	B	B	A	A	A
Aromatic hydrocarbons	A	—	D	—	A	—	—	D	A
Benzene[2]	A	D	D	A	A	A	A	D	A
Bromine[2]	D	D	D	D	D	D	A	D	A
Calcium carbonate	—	—	A	—	A	A	A	A	A
Calcium chloride	A	A	A	A	A	A	A	A	A
Calcium hydroxide	A	A	A	A	A	A	A	B	A
Calcium hypochlorite	A	A	A	D	C	D	A	B	—
Calcium sulfate	A	—	A	A	A	A	—	B	A
Chlorine, anhydrous, liquid	C	D	D	D	D	D	B	C	A
Chlorox (bleach)	A	D	A	D	A	A	—	C	A
Citric acid	A	A	A	A	A	A	A	B	A
Copper chloride	A	A	A	A	D	D	A	A	A
Copper sulfate, 5% solution	A	A	A	D	A	A	A	A	A
Cresols[2]	A	D	D	—	A	A	—	C	A
Cresylic acid	A	—	A	D	A	A	—	C	A
Ethylene glycol[4]	A	D	A	A	A	A	—	A	A
Ferric chloride	A	A	A	D	D	D	A	A	A
Ferric nitrate	A	A	A	—	A	A	—	A	A
Ferric sulfate	A	A	A	A	C	C	A	B	A
Ferrous chloride	A	A	A	D	D	D	A	B	A
Ferrous sulfate	A	A	A	D	A	A	A	B	A
Formaldehyde	A	A	A	—	A	A	B	C	A
Fuel oils	A	A	A	A	A	C	—	A	A
Gasoline[1,4]	A	D	D	A	A	A	—	A	A
Hydraulic oils									
Petroleum[1]	A	D	—	A	A	A	—	A	A
Synthetic[1]	A	D	—	A	A	A	—	C	A
Hydrochloric acid									
20% solution[4]	A	C	A	D	D	D	C	B	A
37% solution[4]	A	C	A	D	D	D	C	A	A
Hydrofluoric acid									
20% solution[1]	A	A	D	D	D	D	D	C	A
50% solution[1,2]	A	D	D	D	D	D	D	C	A
75% solution[1,2]	A	D	D	D	D	D	D	D	B
Hydrogen peroxide	C	A	A	A	C	C	B	A	A
Hydrogen sulfide, aqueous solution	A	A	A	A	C	B	A	B	A
Kerosene[2]	A	D	A	A	A	A	—	A	A
Lime	A	—	A	—	A	A	A	—	—
Lubricants	A	A	A	A	A	A	—	A	A
Magnesium hydroxide	A	A	A	A	A	A	A	A	A

(table continued)

Material	Epoxy	Polypropylene	Polyvinyl chloride Type 1	Nylon	Stainless steel 316	Stainless steel 304	Titanium	Buna N	Viton
Magnesium sulfate	A	A	A	A	B	B	A	A	A
Nitric acid									
5-10% solution	A	A	A	A	A	A	A	D	A
20% solution	B	A	A	A	A	A	A	D	A
50% solution[2]	D	D	A	A	B	B	A	D	A
concentrated	D	D	D	A	D	D	A	D	D
Phenol (carbolic acid)	C	A	A	D	A	A	B	D	A
Phosphoric acid									
10-40% solution	A	A	A	D	A	B	B	C	A
40-100% solution	A	A	A	C	B	C	C	C	A
Potassium cyanide									
solutions	A	—	A	A	A	A	—	A	A
Potassium hydroxide	A	A	A	A	B	B	B	B	A
Potassium permanganate	A	A	A	D	B	B	A	A	A
Sea water	A	A	A	A	A	A	A	A	A
Sodium aluminate	A	—	—	—	A	—	B	A	A
Sodium bisulfite	A	A	A	D	A	A	A	A	A
Sodium carbonate	A	A	A	A	B	B	A	A	A
Sodium chloride	A	A	A	A	A	A	A	A	A
Sodium hydroxide									
20% solution	A	A	A	A	B	B	A	A	A
50% solution	A	A	A	A	D	D	A	D	A
80% solution	A	A	A	A	D	D	A	D	B
Sodium hypochlorite, up to 20% solution[3]	C	D	A	A	C	C	A	C	A
Sodium metaphosphate[2]	A	D	—	A	A	—	—	A	—
Sodium metasilicate	A	—	—	—	A	—	—	—	A
Sodium polyphosphate, mono, di, tribasic	A	—	—	—	A	A	—	A	A
Sodium silicate	A	A	A	A	A	A	A	A	A
Sodium sulfate	A	A	A	A	A	A	A	A	A
Sulfate liquors	A	A	—	—	C	C	—	—	—
Sulfur dioxide[2]	A	D	D	D	A	A	A	C	A
Sulfur trioxide, dry	A	—	—	—	C	—	—	C	A
Sulfuric acid									
up to 10% solution	A	A	A	D	C	D	A	C	A
10-75% solution[2]	C	D	A	D	D	D	D	D	A
Sulfurous acid	A	A	A	D	A	C	A	C	A
Water, acidic from mines	A	A	A	A	A	A	—	A	A
White liquor (pulp mill)	A	A	A	A	A	A	—	—	—
White water (paper mill)	A	A	—	A	A	A	—	—	—
Zinc sulfate	A	A	A	A	B	B	A	A	A

*The ratings for these materials are based upon chemical resistance only. Added consideration must be given to pump selection when the chemical is abrasive or viscous. Pressure should also be considered; maximum internal pressure or pumping head is 12-15 lb. The chemical-effect rating is as follows:
A = no effect, excellent; B = minor effect, good; C = moderate effect, fair; D = severe effect, not recommended.
[1] Polyvinyl chloride, satisfactory to 72°F.
[2] Polypropylene, satisfactory to 72°F.
[3] Polypropylene, satisfactory to 120°F.
[4] Buna N, satisfactory for "O" rings.

are sodium aluminate, aluminum chloride or sulfate, and ferric or ferrous chlorides or sulfates. These chemicals form flocs under controlled water-chemistry conditions and assist gravity sedimentation, dissolved-air flotation, and sludge thickening. They are also used in raw-water treatment and sludge conditioning.

Coagulant aids—To supplement the primary coagulants in floc formation, various aids may be added, such as natural or artificial polymers, or sodium metasilicate. Aids are also used in raw-water treatment and sludge conditioning, either in conjunction with primary coagulants or by themselves, under proper conditions.

Miscellaneous—Aqueous chlorine solution, or sodium or calcium hypochlorite for bacterial disinfection;

methanol for denitrification; and potassium permanganate for chemical oxidation are a few of the other types of controlled chemical additions used in waste treatment.

Chemical addition—used also in process and utilities control—is a highly developed science, for which the diaphragm and piston-type pumps are commonly used.

Since most chemical-addition quantities are in the gallons per hour range, the diaphragm pump is ideal for this service. A plunger or cam, pushing against an internal fluid (usually a petroleum-based oil), generally actuates the diaphragm.

Check valves on the suction and discharge ends ensure consistent addition. Discharge-capacity control is

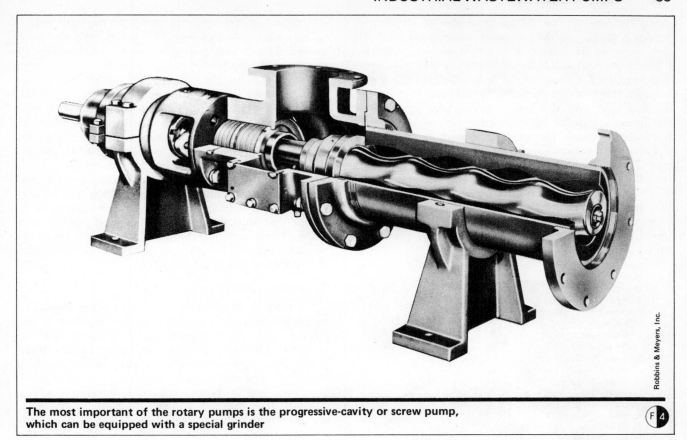

Robbins & Meyers, Inc.

The most important of the rotary pumps is the progressive-cavity or screw pump, which can be equipped with a special grinder   F 4

established by adjusting the column of hydraulic fluid that bypasses the diaphragm cavity. The stroke adjustment—usually from zero to 100%—can be performed when the pump is operating by manual, electronic or pneumatic control. Capacities of these pumps range from zero to several thousand gallons per hour, with discharge pressures available up to 1,000 + psig.

Since the flow is pulsed because of the motion of the plunger, chemical addition is not uniform. One way to circumvent this phenomenon is to use two or three heads on the same pump to smooth out the flow. Multiple heads also raise the addition volume.

Diaphragm pumps are able to move highly viscous polymer solutions and are available in a wide range of metallic or plastic construction. Pumps of this type are quite reliable if suspended solids are not present in the liquid. Any maintenance required is fairly easily performed. This pump is truly the workhorse of industrial-waste treatment, with respect to chemical addition.

If larger capacities or higher discharge pressures are needed, a piston pump may be used. This pump has essentially the same pulsed-delivery characteristics as the diaphragm pump, except that the plunger interacts directly with the fluid. Multiple heads and pulsation dampers can be had to smooth out the flow. This pump is common where large volumes of chemicals (e.g., acid or caustic for neutralization) must be delivered.

Other pumps occasionally used for chemical addition are the so-called "canned" pump and the peristaltic pump. The former is a sealed, leak-proof centrifugal pump with delivery capacity in the gallons per minute range, while the latter is a very low-capacity pump that

has flexible tubing wrapped around a rotating cam. Peristaltic pumps are primarily for laboratory or pilot-scale facilities, while the canned pump delivers larger volumes of hazardous or toxic chemicals.

Slurry and sludge transport

There are many sources of slurries and sludges in wastewater treatment. Stringent effluent requirements dictate that suspended solids be removed to a level of 10–20 mg/l, and sometimes even lower. Biological treatment produces a mixed liquor that must be pumped for solids recycling. Excess biomass must be digested, thickened and dewatered. Typical solids concentrations of these streams range from 0.5% to 15%.

Chemical-physical treatment generates slurry streams of varying chemical constituency, solids content, particle size and abrasiveness. Sludges from the petroleum and organic chemical industry are usually processed in the waste-treatment area for oil removal.

The entire metals industry produces sludges that vary widely in composition. Even the fossil-fueled power industry, with its air-pollution-control scrubbers, must handle and dispose of large quantities of slurries and sludge. The list is endless, with each sludge or slurry being a little bit different from the others.

Misapplication of pumps for slurry or sludge transport has been common because the chemistry, particle size, abrasion characteristics, and other parameters have not been investigated in detail. Frequently, the liquids are non-Newtonian.

Rotary pumps, particularly the progressive-cavity or screw pump, have found wide application in slurry or sludge transport. The pumping action of all rotary pumps is similar. As the pumping elements are rotated, they open on the inlet side to create a void, which is filled by the fluid as it is forced by atmospheric pressure. The continued turning of the rotors encloses the fluid between the rotating elements themselves, or between the elements and the pump casing. At this point, the rotors are under inlet pressure until the enclosed portion opens into the outlet chamber.

Particles in suspension do not affect the pump operation. The general rule is that if the fluid is able to move into the pumping elements, it can also be forced out. These pumps are self-priming up to 28 ft of suction lift, they provide a uniform rate of flow, and they are generally nonfouling. Pumping capacities are available up to several hundred gpm, with discharge pressures up to 300–400 psig.

Barnant Corp.

**In a peristaltic pump, the liquid is
mechanically squeezed through the tubing**   F 5

Crane Co.

This "canned" rotor pump operates without a seal   F 6

Although it is common to pump slurries with 5–15% solids content, sludges with 70% solids have also been moved. The operating speed of the pump is influenced mostly by the properties of the suspension. The greater the solids content, the greater the required horsepower and the lower the operating speed. For any material, however, the throughput per motor revolution is constant. The abrasive properties of a suspension affect the pump speed; capacities are lower with highly abrasive materials. Maintenance can be expected to be greater than for the other pumps already discussed.

Other pumps for slurry movement include the rotary-gear, piston and diaphragm pumps. The rotary-gear pump is similar to the progressive-cavity pump, except that intermeshing gears are used instead of a screw. The rotary-gear kind can pump a wide range of materials such as oils, tar, polymers, or any liquid that does not contain hard solids.

Sampling systems

An often overlooked factor in wastewater treatment is the continuous or semicontinuous sampling of the various streams for operational control or effluent-quality reporting. Some samplers continuously bleed liquid from a pumped stream, while others obtain a sample at a preselected time interval. If at all possible, the easiest way to obtain a representative sample is from a small slipstream off a pump discharge. Frequently, samples must be obtained from open-channel or sewer flows, or from pits, ponds or lagoons. In these instances, pumping to the sampling device is ordinarily required.

The type of pump selected generally depends upon the sampling-apparatus configuration, as well as on the characteristics of the stream to be sampled. Flowrate to the sampler, suspended solids, entrained gases, chemical ions in solution, and other considerations may affect the pump selection. Typical installations use centrifugal pumps, various rotary pumps (including the peristaltic kind), and diaphragm pumps. Each case and its pump selection must be investigated individually.

References

1. "Controlled Volume Pumps and Systems," Milton Roy Co. brochures.
2. Edwards, James A., Pumps for Pollution Control, *Pollution Eng.*, Nov. 1974, p. 26.
3. Aieks, Tyler G., "Pump Selection and Application, BME," McGraw-Hill, New York (1957).
4. Karassik, Igor J., and Carter, Roy, "Centrifugal Pumps," McGraw-Hill, New York (1960).
5. Neerken, Richard F., Pump Selection for the Chemical Process Industries, *Chem. Eng.*, Feb. 18, 1974, p. 104.
6. "Progressive Cavity Pumps," Robbins & Myers, Inc., brochures.
7. "Standards of the Hydraulic Institute," Hydraulic Institute, New York (1947).

The author

Jacoby A. Scher works for Fluor Engineers and Constructors, Inc. (Box 35000, Houston, TX 77035) as Principal Environmental Engineer, Process Dept., in wastewater treatment design projects for the petroleum, natural gas, chemical and coal conversion industries. Before, he worked as environmental chemist and engineer in wastewater treatment and air-pollution control in industry and as a consultant. He has a B.A. degree in chemistry from Rice University and an M.S. in environmental engineering from the University of Houston.

Section 2
WATER-POLLUTION INSTRUMENTATION

Water-pollution instrumentation

Here is a tour through a wastewater-treatment plant, with the information that you will need concerning the instrumentation used at each stage.

Robert G. Thompson, *Roy F. Weston, Inc.*

□ Current water-pollution-control legislation incorporates a discharge-permit system. Permits will require plants to monitor discharges, keep records of monitoring activities and report periodically on how they are meeting the discharge criteria. All this requires instrumentation.

As a help in the instrumentation and control design, two publications are useful. The first is one on design criteria for reliability [*1*], and the second is a handbook on monitoring industrial wastewater [*2*]. Both are from the Environmental Protection Agency (EPA).

Basic systems

The instrumentation on a wastewater treatment system serves two purposes. It provides information needed in running the plant, and it accumulates data required by the discharge permit.

Influent—Often it is desirable to measure influent parameters such as flow, pH, conductivity and the like to warn the operator that an unusual quantity of some undesirable material is approaching. (It may be diverted to a holding pond for treatment at a slower rate.)

Neutralization section—Here, pH is adjusted to aid in

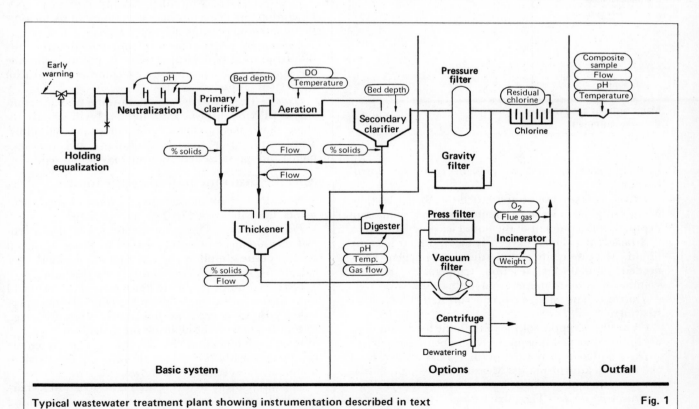

Typical wastewater treatment plant showing instrumentation described in text

Fig. 1

Originally published June 21, 1976.

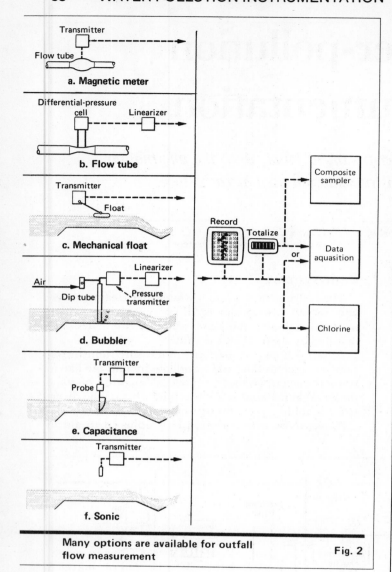

a. Magnetic meter

b. Flow tube

c. Mechanical float

d. Bubbler

e. Capacitance

f. Sonic

Many options are available for outfall flow measurement — Fig. 2

Parameters measured ↘ / Industry type ↓	Flow	pH	Temperature	Solids, suspended	Solids, settled	Solids, total	Color	BOD$_5$	DO	Oil	COD
Beverage	X	X	X					X	X		
Chemical	X	X	X					X		X	X
Flat glass	X	X	X	X							X
Cement and lime	X	X	X	X							X
Meat products	X	X		X	X			X		X	X
Metal finishing	X				X					X	X
Petroleum	X	X	X	X				X		X	X
Plastics	X	X	X					X		X	X
Pulp and paper	X	X	X					X		X	X
Steel	X	X	X	X				X			
Textile	X	X	X								X
How measured											
Direct	X	X	X						X		
Lab (composite sample)		X		X	X	X	X	X		X	X

Some monitoring parameters arranged by type of industry and by methods of measurement

further treatment and to prevent corrosion downstream.

Primary clarifier—Solids are settled-out in this section. Monitoring includes measurement of the solids bed depth, or solids content of the liquid leaving (or both).

Aeration basin—Here, oxygen is incorporated into the liquid. Measurements include dissolved oxygen (DO), needed for biological reaction; temperature; and air volume or power to the aeration mechanism.

Thickener—The sludge is concentrated for further treatment.

(With the exception of the outfall, the following steps are optional, depending on the wastes being treated.)

Digester—Used primarily in biological rather than chemical treatment plants. Measurements include pH, temperature and (in anaerobic systems) waste-gas flow.

Dewatering devices—The three common types of equipment—press-type filters, vacuum-drum filters and centrifuges—normally come equipped with instruments.

Incinerator—Dewatered sludge may be incinerated. The principal control is an O$_2$ exhaust-gas monitor on the system's discharge.

Filters—These may be either pressure or gravity types. Operation and efficiency may be monitored by measuring differential pressure across the filter.

Chlorination system—These are often used before the discharge of a biological system. A residual chlorine meter is used to determine excess chlorine and to trim the system properly.

Outfall—The discharge of a plant is the primary point where EPA requires monitoring. This will usually include flow measurement and composite sampling. Temperature (for thermal pollution), pH measurements and specialized monitoring may be needed.

Specific instrumentation applications

Now, let us look more closely at some of the instruments that will be used in these applications.

Flow monitoring—This is most common, and is a crucial measurement being often used as a pacing signal for composite sampling and chlorination, as well as for calculating amounts of pollutants in the discharge.

For piped flows, magnetic flowmeters (Fig. 2A) yield a linear output vs. flow; flow tubes (Fig. 2B) may provide a cost advantage, but linearizing is required, e.g., to coordinate with the chlorination system.

In open-channel flow (Fig. 2C–F), the various sensors can be used with any weir or flume—only the Parshall flume (the most common) is shown.

The mechanical float (Fig. 2C) is attached to a characterized cam to produce a linear output. A stilling well is used when measuring low flow (low head). The bubbler tube (Fig. 2D) requires an air source, and the output is nonlinear. Capacitance and sonic devices (Fig. 2E and 2F) are reliable and require little maintenance. Electronic linearization is a standard feature.

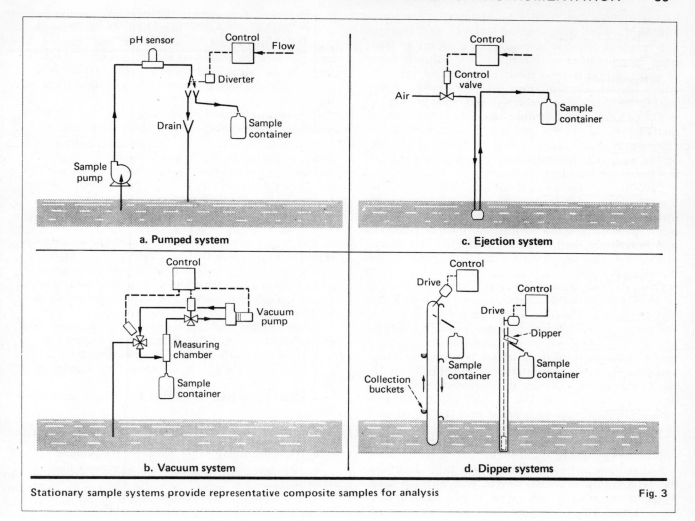

a. Pumped system

c. Ejection system

b. Vacuum system

d. Dipper systems

Stationary sample systems provide representative composite samples for analysis

Fig. 3

Flow totalizing is an integral part of flow monitoring systems, and is often used to activate samplers. Careful consideration must be given to external influences that may affect the calibration—e.g., the effect of two-way radios on sensor electronics, rainstorm wetting and foam on capacity probes, fouling of mechanical floats, and freezing of bubbler air supplies.

pH monitoring—Measurement employs standard pH hardware of an immersion type in the outfall stream or in a feedthrough-flow block configuration in a pump discharge line. The pumping system may be integrated with a composite sampler (Fig. 3A).

Temperature monitoring—If thermal pollution is a possibility, continuous recording will be needed. Otherwise, periodic readings (with high-temperature alarms) will usually be satisfactory.

Optional monitoring—Some parameters can provide early warning of problems or, with proper correlation, can satisfy legal monitoring requirements. For example, if TOC (total organic carbon) readings correlate with BOD_5, corrective action need not wait the five days for the BOD analysis. Dissolved oxygen (DO), residual chlorine, turbidity and conductivity can all be monitored directly and continuously if the need occurs.

Composite sampling—Almost every plant will have a sampling system in its outfall, and may have such sys-

tems at other critical locations. In the pumped system (Fig. 3A), a portion is diverted to the sample bottle at intervals. The vacuum sampler (Fig. 3B) uses a vacuum to suck a sample from the stream. The sampler line is purged with pressure from the pump discharge before and after sampling. In the ejection system (Fig. 3C), a submerged sample-cell is emptied at intervals into a container by using compressed air. Dipper systems (Fig. 3D) use mechanical sampling cups.

Sampling may be done at preset intervals, or based on the effluent flow. The sample may be refrigerated, and sample bottles of various sizes are available.

The discharge permit will require data based on sampling, so the system must be reliable. In choosing a system, consider the following:
- Pickup-point fouling (size of solids).
- Change of composition (due to pressure changes) of sample during transport from stream to bottle.
- Biological and mechanical interference buildup between samples.
- Sampling accuracy (sample size and frequency).
- Mechanical reliability.

Waste processing

Neutralization, which requires pH control, is a common first step in waste treatment. Whole texts [3] have been written on such control, but the systems are generally feedforward (measure influent pH and flowrate and compute amount of neutralizing agent) or feedback (measure after neutralization and adjust amount of agent to give desired result). The control valve requires special selection since the lime slurry usually used for neutralization is abrasive and tends to plug.

The pH electrode assemblies are a frequent cause of malfunction. Ultrasonic and mechanical cleaning devices are available. They should be tested, if possible, on the particular neutralization system to be used, before being purchased and installed.

Clarifiers and thickeners—Solids concentration is the most common measurement; this aids in delivering solids at a consistently high value. For 0 to 2% solids, optical-type sensors, sonic devices and nuclear radiation instruments are used. Between 2 and 5%, sonic and radiation devices are effective. Above 5%, radiation instruments are used.

Control of the drawoff from the bottom of a clarifier or thickener by solids-concentration monitoring can improve system efficiency, as can control of the sludge-bed depth. Bed-depth measurement is usually made by placing an optical or sonic-type solids-concentration sensor at the depth of the interface between the sludge and clear liquid, but the interface is ill-defined and difficult to sense. Two devices at different elevations may be used to give high- and low-level indications. Or, portable equipment can be lowered into the liquid in a "sounding" fashion.

All the devices suffer from the same problem—it is difficult to keep the equipment clean.

Sludge flowrate measurement will help determine the relationship between recycle and forward flow. Magnetic flowmeters are commonly used. To keep them clean, heated tubes and ultrasonic cleaners are available, but in some cases a high-velocity flow may be all that is needed to keep the flowtube clear.

Flumes are also used to measure forward and recycle flow where gravity-feed systems have been designed. The measurement systems of Fig. 2, with totalizer and recorder, may be used. When flumes are employed for measuring recycle flow, capacitance probes should be able to reject fouling by the considerable quantity of sludge, as the level rises and falls.

Aeration—Here, the primary measurements are DO (dissolved oxygen) and temperature. Depending on the operation, continuous monitoring, grab samples or portable instrument readings may be required. The location of the sensor for continuous monitoring will probably have to be experimentally determined during startup.

Maintenance of probe systems must be evaluated. Continuous-cleaning devices are available—they both reduce maintenance and provide turbulence that helps in obtaining uniform readings. If misapplied, a particular attachment may defeat its own purpose by mechanically failing when fouled or by forcing solid matter into the probe membrane, causing premature failure.

DO measurement may help reduce power costs on aeration equipment and improve process reliability.

Optional stages—Digesters, like thickeners, require flow measurement. For aerobic digesters, pH is the only other variable commonly measured. In an anaerobic digester, temperature measurement and control is needed, since the process is accelerated by a temperature increase. Measurements of pH and quantity of discharged gas (mostly methane) are also indications of the operation of this stage.

In general, dewatering, incineration and filtration equipment comes complete with instrumentation.

Outfall—Each outfall will have its own requirements, based on the discharge permit. However, some types of instruments are common to virtually every outfall system. The table shows some of the parameters that may have to be measured, depending on the industry.

Some general thoughts

The degree of sophistication of the instrumentation required in a wastewater treatment plant is primarily determined by two factors. First, the inflow, uniformity and composition of the waste to be treated and, second, the acceptable variation in operating range at any step in the process.

In specifying and applying instrumentation, one must never lose sight of the following basic questions:

■ How frequently does the process change?

■ Can we use a simple indicator and manual control?

■ How dependable is the hardware that we are selecting?

■ What are the data requirements of our discharge permit?

■ What is the class of service of the instrument:

a. Critical—one that may cause a permit violation if it fails?

b. Operational—where observation of the trend in the process change is essential to control?

c. Convenience—where automatic equipment and continuous monitoring replace manual methods?

References

1. "Design Criteria for Mechanical, Electric and Fluid System and Component Reliability," EPA 430-99-74-001, U. S. Govt. Printing Office, Washington, D.C.
2. "Monitoring Industrial Wastewater," Handbook #6002, Technology Transfer, U.S. Environmental Protection Agency, Washington, D.C.
3. Shinskey, F. G., "pH and pIon Control in Process & Waste Streams," Wiley, New York, 1973.
4. Lund, Herbert, F., ed., "Industrial Pollution Control Handbook," McGraw-Hill, New York, 1971.
5. Lipták, Béla G., ed., "Instrument Engineers' Handbook," Vol. I, Vol. II and Supplement, Chilton, Philadelphia, Pa., 1969, 1970, 1972.

The author

Robert G. Thompson is presently responsible for instrumentation and control application design with Roy F. Weston, Inc., Weston Way, West Chester, PA 19380. He has held positions as instrument and electrical maintenance superintendent for a large glass-fiber operation, and as control-design supervisor for a paper machinery manufacturer. He has also worked on the division engineering staff of a synthetic-fiber manufacturer as an instrument engineer, with primary emphasis on waste treatment and boiler applications. He is a member of the Philadelphia Section of Instrument Soc. of America.

Instruments for environmental monitoring

Plants are required to monitor many pollutants in both air and water. Here is a rundown of the instruments available for the job.

D. M. Ottmers, Jr., D. C. Jones, L. H. Keith and *R. C. Hall, Radian Corp.*

☐ Government regulations require that industry monitor for various gaseous pollutants, particulates, lead, and the so-called "priority pollutants." Hence, monitoring of air and water qualities has become increasingly important during the past decade. Thus, it is important to know:

■ The instrumental methods that are currently accepted for pollutant monitoring.

■ The instrumentation available for this purpose.

This article discusses instrumentation for monitoring both air and water quality. Emphasis is placed upon how these instruments function, their approximate costs, and what systems may be used or developed in the near future.

Air-quality monitoring

Air-quality monitoring consists primarily of measuring various gaseous pollutants, particulates, and lead. Instrumentation for monitoring these atmospheric pollutants are discussed in the sections that follow. Quality assurance considerations and research in air-quality monitoring are also discussed briefly.

Gaseous pollutants

Federal regulations have established allowable limits for sulfur dioxide, nitrogen dioxide, carbon monoxide, and ozone [1,2]. To aid in attaining compliance with the oxidant standard, a standard has also been established for non-methane hydrocarbons [3].

Monitors for these gases usually are continuous, i.e., the indicated pollutant concentration may vary continuously with time. These monitors are automated, and are capable of unattended operation for days or weeks.

Also, manual samplers are available that determine an average pollutant concentration. They do this by chemically scrubbing the pollutant from a known quantity of air collected at a constant rate during the sampling period, with subsequent analysis of the captured material. However, the servicing and analysis to support these samplers is manpower intensive, and they are generally not well suited to the observation of frequent short-term averages. With improvements in the reliability and accuracy of the continuous monitors, manual samplers have gradually lost favor with most users. Consequently, this article will be limited to a discussion of continuous analyses.

The basic elements of a pollutant monitoring system are shown in Fig. 1. Air is taken in by the sample manifold, which is made of an inert material such as glass or Teflon. The sample manifold brings a fast-moving stream of air to the immediate vicinity of the monitor in order to minimize the residence time of the air sample in that portion of the system upstream of the measurement cell. Even though the sample-handling system is inert, reactions may occur between gaseous species in the air, e.g., reaction of ozone and nitric oxide to produce nitrogen dioxide. Thus, the residence time in the air transport system should be minimized.

Monitors may include filters to remove particulates that might otherwise cloud optical windows, plug orifices, and the like. Some monitors may use a scrubber to remove a gas in the air that would interfere with measurement of the one of interest.

Support gases are required for some measurement methods. These are consumed during the measurement and thus may increase operational costs and logistics problems.

Measurement cells can be based on many different methods for the pollutants of interest. Basically, the pollutant is introduced into an environment where it will undergo a chemical or physical reaction whose output can be converted to an electrical signal. Examples include:

Flame photometry—The gas of interest is pyrolyzed by a flame, with degradation products reacting to produce a particular wavelength of light. This light is detected by a photomultiplier tube and converted to an electrical signal.

Infrared—The gas of interest absorbs infrared light, which is beamed through the sample cell. This absorption results in heating, which is measured as a pressure imbalance, and converted to an electrical signal.

Chemiluminescence—The gas of interest is mixed with a gas with which it undergoes a spontaneous reaction. The reaction gives off light of a particular wavelength, which is measured via a photomultiplier tube and converted to an electrical signal.

Electrochemical—The gas of interest is dissolved in a cell with electrodes where it will react with other dissolved species. This reaction, or auxiliary reactions, generates an electric current at the electrodes that is proportional to the concentration of the gas of interest.

Fluorescence—Ultraviolet light is beamed through the

Originally published October 15, 1979.

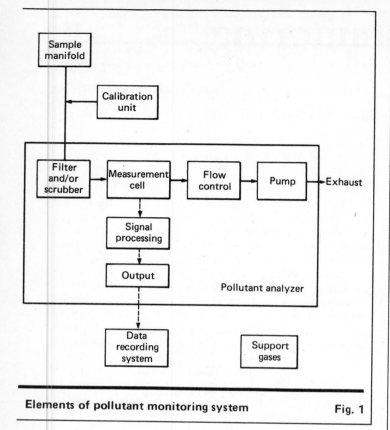

Elements of pollutant monitoring system   Fig. 1

sample each data channel at some frequency. Their output may be merely a voltage printed on a paper tape; more-sophisticated systems may provide a machine-readable output. Small computerized data-acquisition systems based on microprocessors can provide on-site reduction, with data storage on magnetic tape or disk. These units also may assume control functions such as automatic calibration of the instruments, calculation of zero and span drift, sequencing of high-volume samplers, and so forth.

Table I provides a listing of monitors that meet EPA's reference or equivalency specifications for the four gaseous criteria pollutants. The type of detection system used is specified for each monitor, along with remarks on support gases, scrubbers, etc.

Non-methane hydrocarbons are typically analyzed via a gas chromatograph with a flame ionization detector. No EPA-approved instruments are available for this measurement.

Particulates

Regulations also have been established for total suspended particulates [5]. Only one measurement technique is approved, the high-volume sampler. This technique uses a filter to trap particulates during a 24-h sampling period. The filter is contained in an all-weather housing, built according to a specific design. The air flowrate is measured by a chart recorder (continuously) or a rotameter (manually). This flowrate must be calibrated periodically. The filters are weighed (at a controlled humidity) before and after exposure to determine the particulate loading. The air-flow volume is converted to standard conditions (1 atm pressure and 25°C) and used to determine the particulate concentration in the sampled air.

High-volume samplers are typically run every sixth day, every third day, or daily. The sampling period is midnight to midnight.

Lead

A federal standard for ambient lead has recently been promulgated [6]. The measurement procedure involves chemical analysis of filters collected in the manner (described above) for particulates. The standard is based on a three-month average of lead on filters collected no less frequently than once every six days.

Quality-assurance considerations

Air-quality monitoring should be supported by an adequate quality-assurance program. Included are functions such as calibration method and frequency, zero/span checks, traceability of standards, analyzer audits, reports, and documentation. Quality-assurance guidelines are available from EPA [7-9]. However, EPA regional offices, or other permitting offices, may have considerable latitude in specifying particular procedures.

Research in air-quality monitoring

Research in air-quality monitoring is currently very active. For gaseous pollutants there is a major activity in remote sensing, i.e., measurement of the pollutant concentration in the air at some point far removed from

sample cell, either continuously or in pulses. Absorption of this light induces emission of fluorescent light by the species of interest. This is detected by a photo-multiplier tube and converted to an electrical signal.

Second-derivative spectroscopy—Ultraviolet light is absorbed by the gas of interest in the sample cell. Measurements are based on the shape characteristics of the absorption spectrum rather than on the change in intensity. The output signal is proportional to the second derivative of intensity with respect to wavelength.

Ultraviolet spectroscopy—Ultraviolet light passes through a sample and is measured by a photoelectric detector. The decrease in intensity—which is exponentially proportional to the concentration of the absorbing pollutant—is then converted to an electrical signal.

Generally, only one type of measurement cell is used in any particular monitor. As shown in Fig. 1, other items in the system include a flow-control system and a sample pump. Flow control may be achieved via a capillary orifice, needle valve, and so on, with suitable pressure control.

The calibration unit must supply zero air, and gases of known concentration to check or adjust instrument response. This unit may be manual or automated.

Various types of outputs may be supplied with air-quality instrumentation. Panel meters, digital displays, recorder jacks, and terminal strips are all used in various brands.

The data recording system provides a permanent record of the monitoring results. Strip-chart recorders provide a continuous record but are manpower intensive in terms of data reduction. Data loggers typically

the monitor. These systems are typically based on lasers or correlation spectroscopy. For particulates, much research has been done on size fractionating devices, which allow respirable and inhalable particulates to be quantified. New regulations will probably be required before these new techniques enjoy widespread application.

The analysis of specific organic compounds in ambient air is receiving much attention. This is currently achieved by collecting air samples and transporting them to a laboratory. However, the use of special gas-chromatograph–mass-spectrometer systems for real-time analysis shows considerable promise.

Water-quality monitoring

The chemical parameters for monitoring water pollutants are very extensive. They include traditional measurements such as pH, color, biochemical oxygen demand (BOD), chemical oxygen demand (COD), total solids (TS), nitrogen, oxygen, sulfate, sulfite, sulfide, chloride, ammonia, hardness, etc. The instruments for measuring and monitoring these parameters are common and have been the subject of many past articles, books, and reference manuals [10]. They will not be further discussed here. Instead, this discussion will be limited to a new list of chemical parameters that both industry and the EPA are required by law to begin monitoring in the near future.

These new chemical parameters are the so-called "priority pollutants," and are listed in Table II. They are distinguished from chemical parameters previously monitored for, by being specific organic compounds and isomers. (The elements are still measured as total copper, zinc, etc.)

The priority pollutants resulted from a 1976 court settlement after the EPA was sued by several environmental groups for failing to implement portions of the Federal Water Pollution Control Act. This history of the development of the list of 129 priority pollutants, and present methodology used to analyze for them is summarized in a recent review [11].

Regulations for promulgation of new-source performance standards and pretreatment standards started to be issued in June 1979, and are to be completed for 21 major industrial categories by June 1981. These regulations will require application of best available technology (BAT) economically achievable by June 1984. The EPA is spending about $90 million to gather analytical, engineering, and economic data on which to base these regulations. It is estimated that the economic impact to American industry to meet the treatment and monitoring requirements of these regulations will be between $40 and $60 billion.

Sample treatment

Before the instrumentation needed to monitor for the priority pollutants is discussed, a brief description of sample treatment is necessary to put the complexity of the monitoring requirements in perspective. There are eight different procedures required to prepare a single sample for analysis.

The organic compounds are divided into four fractions. The very-volatile organic compounds are col-

EPA-approved pollutant monitors [4]			Table I
Method and cost	**Brand and model**	**Range(s), ppm**	**Notes**
Sulfur dioxide			
Flame photometry $4,000-6,000	Meloy SA185-2A	0.5, 1.0	1,2
	Meloy SA285E	0.05, 0.1, 0.5, 1.0	
	Monitor Labs 8450	0.5	
	Bendix 8303	0.5, 1.0	
Fluorescence $4,000-6,000	Thermo Electron 43	0.5	3
	Beckman 953	0.5, 1.0	
	Meloy		
	Monitor Labs		
Electrochemical $7,000-8,000	Phillips PW 9755	0.5	3,2
	Phillips PW 9700	0.5	
Derivative spectroscopy $8,000	Lear-Siegler SM 1000	0.5	3
Other	Asarco 500	0.5	
Nitrogen dioxide			
Chemiluminescence $5,000-6,000	Monitor Labs 8440E	0.5	3
	Bendix 8101-C	0.5	
	CSI 1600	0.5	
	Meloy NA530R	0.1, 0.25, 0.5, 1.0	
	Beckman 952A		
	Thermo Electron 14 B/E	0.5	
	Thermo Electron 14 D/E	0.5	
Ozone			
Chemiluminescence $4,000-7,000	Meloy OA325-2R	0.5	4
	Meloy OA350-2R	0.5	
	Bendix 8002	0.5	
	McMillan 1100-1	0.5	
	McMillian 1100-2	0.5	
	McMillian 1100-3	0.5	
	Monitor Labs 8410E	0.5	
	Beckman 950A	0.5	
	CSI 2000	0.5	
	Phillips FP 9771	0.5	
Ultraviolet spectroscopy $3,000-4,000	Dasibi 1003-AH	0.5, 0.1	3
	Dasibi 1003-PC	0.5, 1.0	
Infrared $6,000-7,000	Bendix 8501-5CA	50	3
	Beckman 866	50	
	MSA 202S	50	
	Horiba AQM-10	50	
	Horiba AQM-11	50	
	Horiba AQM-12	50	

Notes:
1. Support gas (hydrogen) is needed.
2. Scrubber is required.
3. No support gases needed.
4. Support gas (ethylene) is needed (except Phillips 9771).

lected in a headspace-free 40-milliliter (mL) vial. They are stripped from solution by nitrogen or helium which is bubbled through a 5-mL aliquot. These compounds are trapped in a stainless-steel tube packed with Tenax and silica gel, or a similar solid adsorbent. When the purging is complete (12 min) the trap is rapidly heated to 180°C; the adsorbed compounds are flashed off the trap and retrapped on the head of a gas chromatographic column after which they are analyzed by one of the instrumental methods discussed later.

The remaining organic sample is collected in a one-gallon jug. A 1-L aliquot is extracted with a hexane/methylene-chloride mixture to remove the pesticides and polychlorinated biphenyls (PCBs). The extract is concentrated to 10 mL using a Kuderna-Danish concentrator on a steam bath. Florisil chromatography of this concentrate provides three fractions that contain all the pesticides and PCBs. Each of these fractions is concentrated again to 10 mL prior to instrumental analysis.

A 2-L aliquot of the water is made alkaline to pH 10 and extracted with methylene chloride. Then it is made strongly acidic with hydrochloric acid and reextracted with methylene chloride. The first (alkaline) extract contains both the basic and neutral priority pollutants and the second (acidic) extract contains the phenols. Each of these extracts is concentrated to about 10 mL using a Kuderna-Danish concentrator and then each is concentrated to exactly 1.0 mL using a micro-Snyder 3-ball condenser column. These concentrated extracts contain the majority of the priority pollutants.

Total cyanides are determined by reflexing 500 mL of an acidic water solution. Cyanide, as HCN, is vaporized and trapped in a sodium hydroxide solution. Buffer is added, followed by chloramine-T solution and pyridine-barbituric acid reagent. A red-blue dye is formed and the concentration is determined by using a spectrophotometer.

Total phenols are determined by distilling 500 mL of the water to remove most interferences (the phenols steam distill). Solutions of 4-aminoantipyridine and potassium ferricyanide are added to form a yellow dye. This, in turn, is extracted with chloroform and the concentration is determined by use of a spectrophotometer.

The metals are digested three different ways before they are analyzed by atomic absorption. In addition, three different techniques of atomic-absorption analysis are used: flame excitation, graphite furnace, and cold-vapor analysis. One group of metals (cadmium, chromium, copper, nickel, lead, and zinc) are reanalyzed using the graphite furnace if they are not first detected using flame excitation.

Finally, the asbestos samples are filtered through Nuclepore filters and the retained particles are carbon coated under vacuum. The organic filter is dissolved with chloroform, leaving the fibers embedded in a carbon film. A portion of the film is magnified 20,000 times with a transmission electron microscope, and the asbestos fibers are identified by selective-area electron diffraction. A representative area of the electron microscope grid is counted, and the concentration of asbestos, in millions of fibers per liter, can be calculated from the size of the water sample.

Analytical measurements

The instruments currently used to provide measurements of the priority pollutants are:

- Spectrophotometers, for total cyanides and total phenols.
- Atomic-absorption spectrophotometers, for the elements.
- Transmission electron microscopes, for asbestos.
- Gas chromatographs (GC) with electron capture detectors, for pesticides.
- Computerized gas-chromatograph–mass-spectrometers (GC-MS), for the purgeables, base/neutrals and phenols.

All of these instruments—except the spectrometer and gas chromatograph—are relatively expensive. The EPA is moving to substitute less expensive or more cost-effective instruments for monitoring purposes whenever possible. For instance, gas chromatographs with specific detectors and high-pressure liquid chromatographs will, in some cases, be able to substitute for gas-chromatograph–mass-spectrometers (GC-MS). These instruments are cheaper than the GC-MS and, if only a few of the organic priority pollutants need to be monitored, can offer significant savings. However, if many of the organic priority pollutants must be monitored, the cost of all these instruments plus the more labor-intensive sample preparation requirements may more than offset the investment in a GC-MS instrument.

Another substitution forthcoming will be inductively-coupled argon-plasma emission-spectroscopy (ICAP). Although an ICAP is many times more expensive than an atomic absorption spectrophotometer, it can analyze many of the elements at once and the labor savings, if thousands of samples are to be analyzed, will offset the extra investment in the instrument.

The instrumental methods presently in use, and those under consideration as alternative methods, are summarized in Table III.

Priority pollutant analysis

Instrumentation for measuring the priority pollutants is discussed in this section. Here, the discussions involve: the equipment types and their approximate costs, associated data systems, principles of operation, and sample detection.

Equipment types and approximate costs

Gas chromatography (GC) and gas chromatography coupled with mass spectroscopy (GC-MS) are the primary instrument systems used for analysis of the organic priority pollutants. High performance liquid chromatography (HPLC) is also used, but at present is recommended for only a few of the priority pollutants. These systems are extremely flexible and can be equipped with a variety of options to alter their characteristics and meet different analytical needs. The major features of these systems are compared in Table IV.

The prices of gas and liquid chromatographs are approximately the same. Basic instruments can be purchased for approximately $6,000. These instruments, however, include few options and cannot be used for the analysis of all the pollutants. More-flexible units

Priority pollutants as listed by EPA   Table II

Purgeable organics

Acrolein	1,1-Dichloroethylene	Bromoform
Acrylonitrile	1,1,2-Trichloroethane	Dichlorobromomethane
Benzene	1,1,2,2-Tetrachloroethane	Trichlorofluoromethane
Toluene	Chloroethane	Dichlorodifluoromethane
Ethylbenzene	2-Chloroethyl vinyl ether	Chlorodibromomethane
Carbon tetrachloride	Chloroform	Tetrachloroethylene
Chlorobenzene	1,2-Dichloropropane	Trichloroethylene
1,2-Dichloroethane	1,3-Dichloropropene	Vinyl chloride
1,1,1-Trichloroethane	Methylene chloride	1,2-*trans*-Dichloroethylene
1,1-Dichloroethane	Methyl chloride	*bis*(Chloromethyl) ether
	Methyl bromide	

Base/neutral-extractable organics

1,2-Dichlorobenzene	*bis*(2-Ethylhexyl) phthalate	Benzo(k)fluoranthene
1,3-Dichlorobenzene	Di-*n*-octyl phthalate	Benzo(a)pyrene
1,4-Dichlorobenzene	Dimethyl phthalate	Indeno(1,2,3-c,d)pyrene
Hexachloroethane	Diethyl phthalate	Dibenzo(a,h)anthracene
Hexachlorobutadiene	Di-*n*-butyl phthalate	Benzo(g,h,i)perylene
Hexachlorobenzene	Acenaphthylene	4-Chlorophenyl phenyl ether
1,2,4-Trichlorobenzene	Acenaphthene	3,3'-Dichlorobenzidine
bis(2-Chloroethoxy) methane	Butyl benzyl phthalate	Benzidine
Naphthalene	Fluorene	*bis*(2-Chloroethyl) ether
2-Chloronaphthalene	Fluoranthene	1,2-Diphenylhydrazine
Isophorone	Chrysene	Hexachlorocyclopentadiene
Nitrobenzene	Pyrene	N-Nitrosodiphenylamine
2,4-Dinitrotoluene	Phenanthrene	N-Nitrosodimethylamine
2,6-Dinitrotoluene	Anthracene	N-Nitrosodi-*n*-propylamine
4-Bromophenyl phenyl ether	Benzo(a)anthracene	*bis*(2-Chloroisopropyl) ether
	Benzo(b)fluoranthene	

Acid-extractable organics

Phenol	4,6-Dinitro-*o*-cresol	2-Chlorophenol
2-Nitrophenol	Pentachlorophenol	2,4-Dichlorophenol
4-Nitrophenol	p-Chloro-*m*-cresol	2,4,6-Trichlorophenol
2,4-Dinitrophenol		2,4-Dimethylphenol

Pesticides/PCBs

α-Endosulfan	4,4'-DDE	Aroclor 1016
β-Endosulfan	4,4'-DDD	Aroclor 1221
Endosulfan sulfate	4,4'-DDT	Aroclor 1232
α-BHC	Endrin	Aroclor 1242
β-BHC	Endrin aldehyde	Aroclor 1248
δ-BHC	Heptachlor	Aroclor 1254
γ-BHC	Heptachlor epoxide	Aroclor 1260
Aldrin	Chlordane	2,3,7,8-Tetrachlorodibenzo-
Dieldrin	Toxaphene	*p*-dioxin (TCDD)

Metals

Antimony	Chromium	Selenium
Arsenic	Copper	Silver
Beryllium	Lead	Thallium
Cadmium	Mercury	Zinc
	Nickel	

Miscellaneous

Asbestos (fibrous)	Total cyanides	Total phenols

Instrument for analyzing priority pollutants   Table III

Priority pollutants	Present instrument	Alternative instrument
Phthalate esters	GC-MS	GC with EC detector
Haloethers	GC-MS	GC with Hall detector
Chlorinated hydrocarbons	GC-MS	GC with EC detector
Nitrobenzenes	GC-MS	GC with EC detector
Nitrosamines	GC-MS	GC with alkali FI detector
Benzidines	GC-MS	HPLC with electrochemical detector
Phenols	GC-MS	GC with EC and FI detector
Polynuclear aromatics	GC-MS	HPLC with fluorescence detector
Pesticides	GC, GC-MS	– – –
Halogenated purgeables	GC-MS	GC with Hall or EC detector
Nonhalogenated purgeables	GC-MS	GC with FI detector
Nonhalogenated hydrocarbons	GC-MS	GC with FI detector
Elements	AA	ICAP
Asbestos	Electron microscope	– – –
Cyanide	Spectrophotometer	– – –
Total phenols	Spectrophotometer	– – –

GC-MS, gas-chromatograph-mass-spectrometer
GC, gas chromatograph
AA, atomic absorption spectrophotometer
EC, electron capture
FI, flame-ionization
HPLC, high-performance liquid chromatography
ICAP, inductively coupled argon-plasma emission-spectroscopy

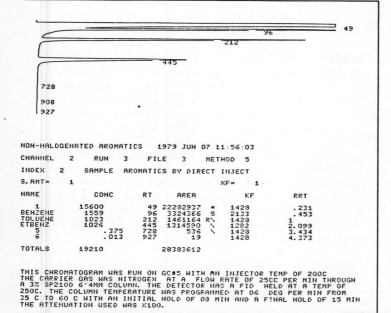

```
NON-HALOGENATED AROMATICS   1979 JUN 07 11:56:03
CHANNEL   2     RUN   3     FILE   3     METHOD   5
INDEX   2     SAMPLE   AROMATICS BY DIRECT INJECT
S. AMT=   1                          XF=   1
NAME          CONC      RT      AREA          KF        RRT
  1          15600       49  22282937   *    1428      .231
BENZENE       1559       96   3324366   S    2133      .453
TOLUENE       1023      212   1461164   R\   1428     1
ETBENZ        1026      445   1314590   \    1282     2.099
  5            .375     728       536   \    1428     3.434
  6            .013     927        19        1428     4.373

TOTALS       19210            28383612
```

THIS CHROMATOGRAM WAS RUN ON GC#5 WITH AN INJECTOR TEMP OF 200C
THE CARRIER GAS WAS NITROGEN AT A FLOW RATE OF 25CC PER MIN THROUGH
A 3% SP2100 6'4MM COLUMN. THE DETECTOR WAS A FID HELD AT A TEMP OF
250C. THE COLUMN TEMPERATURE WAS PROGRAMMED AT 06 DEG PER MIN FROM
35 C TO 60 C WITH AN INITIAL HOLD OF 03 MIN AND A FINAL HOLD OF 15 MIN
THE ATTENUATION USED WAS X100.

Printer/plotter output   Fig. 2

capable of multiple pollutant analysis generally cost from $10,000 to $15,000. The addition of a data system for automated data processing and reporting of results is an important consideration and adds another $5,000 to $10,000. If a data system is not purchased, a strip-chart recorder is required, which costs $750 to $1,500.

Gas-chromatograph–mass-spectrometers are considerably more expensive than the other chromatograph systems and cost approximately $80,000 to $120,000 for an instrument suitable for analyzing all the organic priority pollutants. The higher cost is primarily due to the complexity of the mass spectrometer. Not only is the mass spectrometer complicated in terms of components such as vacuum pumps, data system and the mass analyzer, but it is also quite complicated to operate and requires a skilled operator. In addition to their higher cost, GC-MS systems are usually floor-mounted and require a space two to four times that occupied by a gas or liquid chromatograph.

Although all of the present priority pollutants can be analyzed by either GC or GC-MS techniques, the choice between the two is not particularly easy. Such factors as the number and type of compounds to be analyzed, nature of the samples, type of personnel available for operating the equipment, and degree of confirmation required must be considered. After these factors have been considered, the chromatographic capability required for the analyses must be determined. In general, a minimum of two gas chromatographs equipped with a total of four different detectors is required to analyze all compounds (see Table II). The cost effectiveness of GC versus GS-MS techniques have been discussed in detail elsewhere [12,13] and will not be presented in this article.

Data systems

The data system is an essential component of all modern gas-chromatograph–mass-spectrograph instruments. Analytical capability and ease of use are directly related to the data system. Specifications related to specific ion-monitoring, background-subtraction capability, spectra searching and compound identification techniques, size of reference library, ability to use central spectra libraries via telephone communication, and computer instrument tuning ability should be thoroughly investigated.

In contrast to the gas-chromatograph–mass-spectrometer, a data system is not absolutely essential to gas chromatographs or high-performance liquid chromatographs. Nevertheless, a data system should be seriously considered as an integral part of any chromatograph system. Besides extending the analytical capability of the chromatograph, these systems can generate detailed final reports of analysis with accompanying chromatograms and be used to totally automate the analysis. In so doing they greatly reduce the amount of operator expertise required and enable most analyses to be routinely performed.

A variety of data systems are available for chromatography. Instruments range from simple digital integrators to complex computing integrators with printer/plotter outputs. The more-complex systems may be able to handle data from more than one input

Instruments available for priority pollutant analysis   Table IV

Instrument	Use	Price range*	Major U.S.† manufacturers	Comment
Gas chromatograph	All organic priority pollutants	$6,000-25,000	Hewlett-Packard Perkin-Elmer Tracor Varian	Instrument specificity depends upon the detector used
Liquid chromatograph	Benzidines and polynuclear aromatics	$6,000-28,000	Altex Du Pont Hewlett-Packard Micromeritics Perkin-Elmer Spectra-Physics Tracor Varian Waters	Not compatible with spraying techniques of analysis. Compounds must have a high ultraviolet absorbence or be fluorescent
Gas chromatograph-mass spectrometer	All organic priority pollutants	$50,000-150,000	Finnigan Hewlett-Packard	Requires a skilled operator

*Price depends on the options chosen.
†Listings are in alphabetical order and do not include manufacturers of speciality instruments.

(multichanneled). Level of data-processing sophistication is usually not related to the number of data channels that can be handled, however. Simple digital integrators usually do no more than print peak areas, retention times, and area percents or concentrations. The more sophisticated printer/plotter systems are capable of peak identifications, multilevel calibrations based on one or more internal or external standards, flexible integration and baseline-correction algorithms, and detailed reporting with custom headings. Most of these systems can be user-programmed for total control of integration and reporting. In addition, many instruments can be equipped with floppy-disk or magnetic-tape systems for retaining programs and raw chromatographic data. This allows the data to be reprocessed using different integration algorithms and reporting formats.

A typical printer/plotter output of a computing integrator is shown in Fig. 2. The first information presented is the chromatogram with retention times of the components printed at each peak. Printing the retention times on the chromatogram greatly facilitates locating integration results for a given peak. After the chromatogram, a heading is printed that may contain information regarding the type of analysis (pesticides, etc.), date and time of analysis, details of the instrument (channel, run, file, method), and the sample analyzed. Other information such as the analyst, correction factors, etc., can also be included. Results of the analysis can be presented in a variety of formats. Information presented usually includes peak names, concentrations, retention times, integration areas, method of baseline correction, and compound response factors. The report can be terminated with a statement that identifies the chromatograph used and all operating parameters of the chromatographic analysis.

Peaks are identified in the report by either name or number. They are identified by name according to information supplied by the user prior to the analysis. This information consists of the name to be used, the expected retention time, and the window of retention time for which a given name is to be applied to a peak. If a peak does not fall within one of these retention time windows, it is assigned a numerical value.

It should be understood that identification of a peak by name does not necessarily mean the peak is actually the compound identified, because retention times are only characteristic of a compound's identity. Usually several other compounds can be found that will have the same retention time as the compound of interest under a specific set of chromatographic conditions, and theoretically hundreds of compounds may have that same retention time simply because there are so many organic compounds (~3,000,000) known.

Information concerning component concentrations, retention times, peak areas, and method of baseline correction can be presented in various ways depending upon preference and the particular integrator used. Concentrations can be in parts per million, parts per billion, weight percent etc. Retention times of peaks may be output in units of time (seconds, minutes) or converted to other parameters such as relative retention times or Kovat's Retention Indices. Peak areas are normally output as total integrator counts. A symbol or code usually follows the peak area that tells how the peak was integrated and the method of baseline correction used. Peak heights may also be shown in addition to or in place of peak areas.

Principles of chromatography

Chromatography [14,15] is the physical method of separating a mixture into its components by an equilibrium process established between a stationary and a mobile phase. The mobile phase is a gas in gas chromatography and a liquid in liquid chromatography. As presented in Table V, gas chromatography consists of

					Table V

Forms of chromatography available for pollutant analyses

Chromatography	Mobile phase	Stationary phase	Classification	Major use
Gas	Gas	Solid	Gas-solid	Separation of gases and hydrocarbons
	Gas	Liquid	Gas-liquid	Various organic separations except gases
Liquid	Liquid	Solid	Normal phase (absorption)	Nonpolar to moderately polar compounds
	Liquid	Liquid	Reverse phase (liquid-liquid)	Most nonionic compounds
	Liquid	Ion exchange resin	Ion exchange	Ionic compounds
	Liquid	Polymer (gel)	Steric exclusion (gel permeation)	Macromolecules

gas-liquid and gas-solid chromatography; whereas liquid chromatography is subdivided into normal phase, reverse phase, ion exchange, and steric exclusion.

The basic process responsible for the separation can also be used to categorize the chromatography. For instance, partition chromatography is the separation process due to equilibria established between the mobile phase and a liquid, and adsorption chromatography involves equilibria between the mobile phase and a solid.

The separation process occurs during transport of the sample through the chromatographic column. Two major types of columns have been developed for gas chromatography, packed and capillary. Packed columns, as their name suggests, contain a packing that either acts as an adsorbent or supports a stationary liquid phase. These columns are commonly coiled tubes (2 to 4 mm I.D.) made of metal or glass. Capillary columns contain the liquid phase coated on the inside wall of the tube. They are considerably smaller in diameter (0.25 to 0.5 mm) than packed columns, and vary in length from approximately 10 to 150 m. In contrast to gas chromatographs, modern columns for high-performance liquid chromatographs are almost exclusively made of metal, packed with very small particles (5 to 10 μm) and only 15 to 25 cm in length.

As shown in the block diagram of Fig. 3, the components of a basic gas-chromatograph system consist of a carrier-gas supply with pressure and flow controls, injector (injection port), column, column oven, detector, signal processor, temperature-control electronics, and output device. The carrier gas serves as the mobile phase that transports the sample through the column and into the detector. The injector is normally a heated septum inlet and forms a gas-tight seal between the column and the ambient environment.

The column and detector form the "heart" of the chromatograph system. The column performs the analytical separation, and the detector transforms the separated constituents into electrical signals. These signals are usually very small currents that require amplification by the signal processor before they are capable of driving an output device. The temperature-control electronics maintain the injector and detector at fixed temperatures, and enable the column oven to be operated at a fixed temperature (isothermal operation) or to be temperature programmed at a fixed rate from an initial to a final temperature. The great advantage of temperature programming is that it allows compounds of significantly different volatilities to be analyzed in a single chromatographic run.

The components of a liquid chromatograph system are similar to these in gas chromatography with three main differences:

1) A pump is used for driving the liquid mobile-phase through the column.

2) A solvent programmer is used instead of a temperature programmer, which is used to change the mobile phase composition during an analysis.

3) A column oven is not essential.

Two main techniques are used for introducing priority pollutant samples into a chromatograph. These are syringe injection and the use of a sample valve. Syringe injection is used to introduce small aliquots (1 to 10 μL)

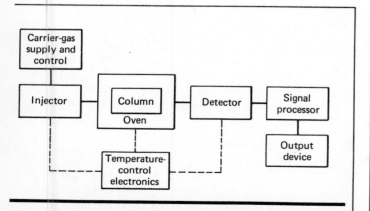

Block diagram of a basic gas-chromatograph system   Fig. 3

Characteristics of chromatograph detectors commonly used for priority pollutant analysis   Table VI

Detector	Type	Classification[1]	Use	Sensitivity[2]	Specificity	Complexity[3]
Flame ionization	GC	Universal	Hydrocarbons	Nanogram	None	2
Photoionization	GC	Universal/Selective	Aromatics	High picogram	Low	1
Electron capture	GC	Selective	Pesticides, PCBs[1] chlorinated hydrocarbons, phthalate esters	Picogram	Variable	2
Nitrogen-phosphorus	GC	Specific	Nitrogen-containing compounds	High picogram	High	3
Hall detector	GC	Specific	Halogen-containing compounds	High picogram	High	3
Mass spectrometer	GC	Specific	All volatile compounds	Nanogram	High	4
Absorbance	HPLC	Selective	Polynuclear aromatics	Nanogram	Variable[4]	1
Fluorescence	HPLC	Selective	Polynuclear aromatics	High picogram	Variable	2
Electrochemical	HPLC	Selective	Benzidines	High picogram	Variable	3

[1] Universal, response is approximately the same for all compounds; selective response is greater for certain compounds; specific, response is specific for compounds that contain a given element.

[2] Quantity of material that can normally be detected in a typical analysis.

[3] Complexity: 1, requires few adjustments and little maintenance; 2, requires several adjustments; 3, requires several adjustments and routine maintenance; 4, requires numerous adjustments and routine maintenance by a skilled operator.

[4] Specificity depends upon the nature of the compounds present.

of liquid samples, primarily into gas chromatographs. Sample valves are used to introduce samples into both gas and liquid chromatographs. In gas chromatography the valve forms part of a gas sparging apparatus and directs the thermally desorbed volatile pollutants onto the head of a gas chromatography column. On the other hand, in liquid chromatography a valve is used to inject a liquid sample, normally 10 to 100 microliters (μL) in volume.

Sample detection

A wide variety of detectors exist for chromatography. Characteristics of the nine detectors commonly used for analysis of priority pollutants are summarized in Table VI. As shown in this table, the detectors used for analyzing the various categories of pollutants (see Table I) range from non-specific to highly specific. They also range from those needing little maintenance or knowledge for use to those requiring a skilled operator. Although the mass spectrometer should not be considered as a standard chromatograph detector, it is included since it is often used as such in the analysis of pollutants. A brief description of each of these nine detectors is presented below.

The *flame ionization detector* (FID) is a universal detector that responds to all organic compounds, but it is primarily used for the detection of hydrocarbons. Sensitivity is in the low nanogram-per-peak range (1×10^{-11} g/s). The detector consists of an air/hydrogen flame burning in a polarized environment (Fig. 4). During normal combustion the flame produces very few ions. However, when an organic compound enters the flame, ions are created that are collected by a collector elec-

trode to produce an ion current. The ion current, which is directly proportional to the mass of the compound entering the flame, is then amplified by an electrometer. The detector response is linear over a very wide range (approximately 10^4). The FID requires hydrogen and air burner gases at flowrates of approximately 30 and 300 cm^3/min, respectively. Detector maintenance requires occasional optimization of burner gases and cleaning of the flame jet and collector electrode.

The *photoionization detector* (PID) is similar to the FID in response characteristics. It is primarily used for the detection of organic compounds. Response is approximately ten times greater for aromatic compounds than it is for other compounds and for this reason is recommended for the detection of aromatic pollutants. The PID like the FID is a simple device that is easy to operate and maintain. The detector consists of a high energy UV light source, polarizing electrode and collector electrodes which are contained in a heated housing.

The detector signal is generated from an ion current that is formed by the high energy UV ionization of the compound. The detector is slightly more linear and sensitive than the FID, and detector operation requires no additional gases.

The *electron-capture detector* (EC) is a selective detector that responds to compounds that have a high electron affinity. The detector contains a radioactive foil (usually Ni-63) that ionizes nitrogen or argon/methane carrier gas to produce an electron cloud. This electron cloud is sampled with a collector electrode to produce a standing current. Electron negative compounds, such as the chlorinated pesticides and PCBs, that enter the detector cavity, deplete the electron population that can

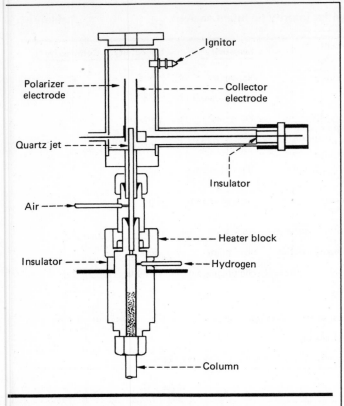

Polarizer electrode

Ignitor

Collector electrode

Quartz jet

Insulator

Air

Heater block

Hydrogen

Insulator

Column

Cross-section of the flame-ionization detector   Fig. 4

be sampled, thereby resulting in a detector signal. Modern electron-capture detectors are operated in a constant-current mode in order to extend their linear range to approximately 10^4. This is done by varying the sampling rate as a function of electron demand (compound concentration in the detector). The detector is extremely sensitive and will respond to sample quantities of sensitive materials as low as 10^{-12} or 10^{-13} grams. Due to its high sensitivity, the detector is also sensitive to contamination from GC column bleed and to dirty samples, but with reasonable caution is easy to use and requires little maintenance.

The *nitrogen-phosphorus detector* (N-P) is a FID that has been modified to contain an alkali source between the flame jet and the collector electrode. In recent designs the source contains a nonvolatile alkali such as rubidium silicate. The source is electrically heated to a dull red and held at a negative electrical potential with respect to the collector electrode. Hydrogen flowrates of only 1 to 5 cm^3/min are used, because such a low flow will not support a flame as classically perceived. Instead, a plasma is created around the source. The alkali acts as a catalyst and selectively ionizes certain nitrogen and phosphorus species that are created in the plasma. In this manner, an ion current similar to that in a FID is established. Selectivity to nitrogen compounds relative to hydrocarbons is usually 10^4. The N-P detector is very sensitive and can be used to detect subnanogram quantities of most nitrogen compounds. Response is approximately ten times greater for phosphorus than it is for nitrogen. Frequent adjustment of the source tempera-

ture and hydrogen flowrate may be required to maintain the same detector-response characteristics. In addition, the alkali source may need replacement every one to six months.

The *Hall electrolytic conductivity detector* (HECD) is a specific detector that can be used for the detection of halogens, sulfur, and nitrogen containing compounds. Detector components include a reactor, electrolytic conductivity cell, electrolyte system, and electronics for measuring electrolytic conductivity and heating the reactor. The relationship of these components is shown in Fig. 5. The detector operates by catalytically converting organic halogen, sulfur, or nitrogen to HX, SO_2, and NH_3, respectively. These molecules are then extracted into an electrolyte stream and the resulting change in electrolytic conductivity output is measured as the detector signal. Selectivity is achieved by either removing unwanted reaction products with a scrubber, converting them to substances that do not change the electrolyte conductivity (i.e., CH_4), or suppressing the conductivity of certain ions by using nonaqueous electrolytes. The detector can be operated in two main modes, reductive and oxidative. Halogen and nitrogen compounds are detected in the reductive mode, using hydrogen as the reaction gas. Sulfur compounds are detected in the oxidative mode with air as the reaction gas. The detector exhibits subnanogram sensitivity and selectivities of 10^5 or greater (relative to hydrocarbons). Linear range of operation is from 10^3 to 10^5, depending upon the element detected. Maintenance involves periodic replacement of the electrolyte, reaction tube, scrubber (not used in detecting halogen compounds), and the ion exchange resin.

The *quadrapole mass spectrometer* is the only system that is commonly employed as a chromatographic detector for priority pollutant analysis. This system can be considered as a mass-selective gas-chromatograph detector. Compounds entering the ion source (Fig. 6) are either directly ionized by electrons emitted from a heated filament (electron-impact mode), or indirectly by interaction of a reaction gas that has been previously ionized by electron bombardment (chemical-ionization mode). When the compound is ionized, it fragments into characteristic ions that are separated by the quadrapole filter. The separation process is very rapid and usually takes only a few milliseconds. This process results in the formation of an ion spectrum that can be used to confirm the identity of a compound. The data system can be used to record and compare spectra to those contained in a library (memory), or it can selectively monitor specific ions, resulting in a mass-specific chromatogram. Selection of characteristic ions or ratios of certain ions to be monitored results in high specificity for most compounds.

The ion information, separation, and detection must be performed under high vacuum with the aid of sophisticated electronics. Consequently, most mass spectrometers are fairly complex to operate and maintain. Sensitivity is normally in the nanogram range for most of the priority pollutants. Specificity is very high, but depends upon the ions chosen for monitoring the individual compounds.

The *ultraviolet absorption* (UV) detector is probably the

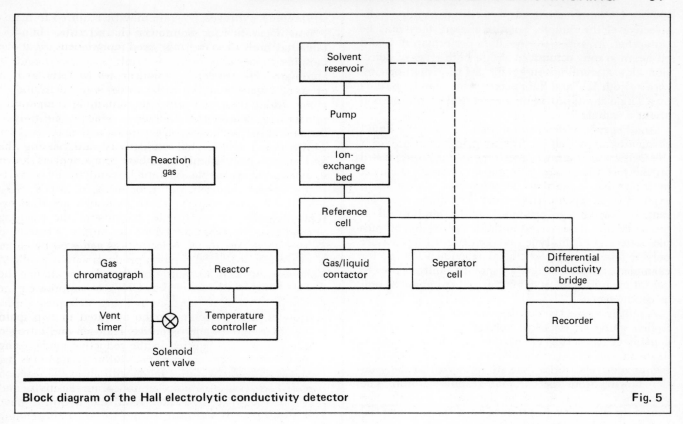

Block diagram of the Hall electrolytic conductivity detector   Fig. 5

most commonly used detector for high performance liquid chromatography (HPLC). UV detectors fall into two classes; namely, fixed wavelength and variable wavelength. Both types of UV detectors operate on the well known principles of photometry, which involve the absorption of light by certain types of organic and inorganic compounds. The amount of light absorbed by the sample compounds is directly proportional to the concentration of the compound in solution. The solvent used as the mobile phase in the chromatographic process must of course be transparent to the ultraviolet light.

Fixed-wavelength detectors operate at a single wavelength of light, which is provided by the emission of special lamps containing appropriate elements. For example, the most commonly used lamp in this type of detector is the mercury-vapor lamp, which has primary emission at 254 nm wavelength. In order to obtain other wavelengths, the lamp must be changed to another type.

Variable wavelength detectors incorporate a broad-spectrum lamp that emits a continuous band of wavelengths. The desired wavelength is then obtained by use of a grating or prism monochromator which, in effect, transmits only the selected wavelength to the sample cell.

Both types of UV detectors are moderately selective in the types of compounds they will detect. Fixed-wavelength detectors are best suited to the detection of aromatic compounds, and can be very sensitive for certain strong UV-absorbing compounds such as polynuclear aromatic hydrocarbons (PNAs). Variable wavelength detectors offer additional selectivity, and sometimes added sensitivity, for many other types of compounds.

Fluorescence detectors depend on the fact that certain types of compounds, when irradiated with light of a certain wavelength, emit light of a longer wavelength, that is, they fluoresce. Fluorescence detectors are somewhat similar to UV absorption detectors in that they require fixed or variable light sources and some type of monochromator. In fluorescence detectors, however, the amount of light *emitted* by the sample is measured rather than the amount of light absorbed.

Since relatively few types of compounds fluoresce, the degree of selectivity provided by fluorescence detectors is generally much higher than that of UV detectors. In addition, for certain types of compounds, fluorescence-detection-sensitivity is much better than by UV absorption.

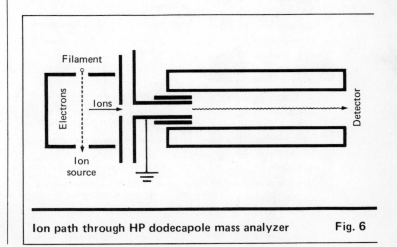

Ion path through HP dodecapole mass analyzer   Fig. 6

Both UV and fluorescence-type detectors are nondestructive of the sample. For this reason, they may be connected in series to the high-performance liquid chromatography column to obtain both types of detection almost simultaneously. By use of a dual-pen recorder, both UV and fluorescence peaks can be plotted on a single chart, providing a great deal of information about a sample.

Another type of detector for high-performance liquid chromatography that is currently gaining acceptance is the *electrochemical detector*. This type of detector is based on the fact that many compounds undergo chemical reactions that cause them to conduct a current when exposed to an electrical potential. The amount of current, or the voltage produced in the sample cell, is measured and then related to the concentration of sample. Detectors of this type may be designed to measure only a type of electrochemical reaction such as, for example, conductance or they may be multipurpose to permit measurement of oxidation and reduction as well as conductance.

A multipurpose type of detector will provide a controlled source of electrical voltage or current and be capable of measuring the current or voltage produced in the sample cell. The sample cell generally consists of two or more electrodes that are immersed in the solution flowing from the chromatographic column. The presence of sample solute in the carrier solvent then causes the electrodes to produce an electrical potential or current, which is measured.

Electrochemical detectors can be very sensitive and specific for certain types of compounds. Detection limits in the picogram range are sometimes obtainable. The use of these detectors, however, puts a number of restraints on the type of mobile phase that may be used for the chromatographic separation since the mobile phase plays an integral part in the electrochemical reactions taking place in the cell.

Summary

Instrumentation for monitoring air and water pollutants has been described. As noted, monitoring of atmospheric pollutants is more firmly established and automated. However, more-sophisticated measurement of trace organics in the atmosphere may be in the offing. Monitoring of water pollutants is in a transitional phase. Much effort is currently being expended to define suitable, yet less-expensive measurement systems. This article has described the currently used systems and presented a general description of the type of instrumentation available.

References

1. *Federal Register,* Vol. 36, Nov. 25, 1971.
2. *Federal Register,* Vol. 44, Feb. 8, 1979, for NAAQS, Ref. and Cal. Method—(Chemiluminescent was also reference method in *Fed. Reg.,* 11/25/71)
3. *Federal Register,* Vol. 36, Nov. 25, 1971.
4. *Federal Register,* Vol. 40, Apr. 25, 1975.
5. *Federal Register,* Vol. 36, Nov. 25, 1971.
6. *Federal Register,* Vol. 43, Oct. 5, 1978.
7. "Quality Assurance Handbook for Air Pollution Measurement Systems, Vol. I, Principles," EPA Environmental Monitoring and Support Laboratory, Research Triangle Park, NC 27711.
8. "Quality Assurance Handbook for Air Pollution Measurement Systems, Vol. II, Ambient Air Specific Methods," EPA-600/4-77-027a, May 1977, EPA Environmental Monitoring and Support Laboratory, Research Triangle Park, NC 27711.
9. "Guidelines for Development of a Quality Assurance Program: Reference Method for the Determination of Suspended Particulates in the Atmosphere (High Volume Method)," EPA-R4-73-0286, Office of Research and Monitoring, U.S. EPA, Washington, D.C., June 1973.
10. American Public Health Assn., American Water Works Assn., and Water Pollution Control Federation, "Standard Methods for the Examination of Water and Wastewater," American Public Health Assn., Washington, D.C. 20036, 1976, pp. 1193.
11. Keith, L. H., and Tellieard, W. A., "Priority Pollutants I—A Perspective View; *Environ. Sci. Technol.* Vol. 13, No. 4, pp. 416–423 (1979).
12. Brenner, N., others, "Gas Chromatography," Academic Press, New York, 1962.
13. Budde, W. L., and Eichelberger, N. W., *Anal. Chem.,* Vol. 51, p. 567A (1979).
14. Finnigan, R. E., others, *Environ. Sci. Technol.* Vol. 13, p. 534 (1979).
15. Snyder, L. R., and Kirkland, J. J., "Introduction to Modern Liquid Chromatography," John Wiley & Sons, New York, 1974.

The authors

Delbert M. Ottmers, Jr. is Assistant Vice-President, Engineering and Chemistry, Radian Corp., 8500 Shoal Creek Blvd., P.O. Box 9948, Austin, TX 78776, telephone (512) 454-4797. He has been involved, at Radian, in process design and development activities, with primary emphasis on air/water pollution problems. He holds B.S., M.S. and Ph.D. degrees in chemical engineering from the University of Texas. He is a member of AIChE, Tau Beta Pi, Phi Lambda Upsilon, Omega Chi Epsilon, Sigma Xi and Phi Kappa Phi.

David C. Jones is Assistant Vice-President, Technical Staff, Radian Corp. His major interest at Radian has been ambient air quality monitoring. He is an authority on instrument selection, calibration procedures, and quality assurance methods for air monitoring stations. He holds a B.S. in chemistry and a Ph.D. in physical chemistry from the University of Texas, and is a member of Alpha Chi Sigma and the National Assn. of Corrosion Engineers.

Lawrence H. Keith is Manager, Analytical Chemistry Div., Radian Corp., where he is responsible for maintaining the technical quality of the analytical staff and assuring that the laboratories keep pace with advances in instrumentation and technology. He holds a B.S. in chemistry from Stetson University, an M.S. in organic chemistry from Clemson University and a Ph.D. in natural products chemistry from the University of Georgia. He is a member of Sigma Xi, Gamma Sigma Epsilon, Kappa Kappa Psi, and ACS (Chairman, Executive Committee, Div. of Environmental Chemistry, and Past Chairman, Central Texas Section).

Randall C. Hall is Department Head/Senior Staff Scientist, Radian Corp., where he serves as head of the Organic Chemistry Dept. and as acting group leader of the Analytical Separation group within that department. He holds a B.S. in chemistry and a Ph.D. in analytical-organic and physical-organic chemistry from Texas A&M University, and is a member of ACS and Sigma Xi.

Gaging and Sampling Industrial Wastewaters

Before you can mount a water-pollution control program, you must first find wastewater volumes (gaging), and collect representative samples to determine pollutants.

JOSEPH G. RABOSKY and DONALD L. KORAIDO, Calgon Corp.

Wastewater sampling and gaging are the keystones of any water-pollution control program because they are vital to the collection of data needed for effective treatment. Analysis of samples identifies contaminants and pinpoints their concentration, while gaging determines total wastewater flow.

Sampling and gaging may be conducted entirely in-house or in cooperation with a qualified consulting firm. In either case, plant personnel should be thoroughly familiar with what is required. The purpose here is to discuss generally accepted techniques and to provide the plant engineer with the knowhow needed to conduct his own sampling and gaging surveys.

Plant familiarization is the first step. Sewer plans should be thoroughly reviewed and each sewer physically examined to determine its course and destination. Then, the exact source and the point at which each waste stream enters the sewer should be determined. (In many old plants, prints of sewer layouts do not exist, so dye-tracing may be the only practical way to confirm the layout.)

After this is done, sampling and gaging points can be established to ensure that both the flow data and the analytical data obtained from the samples will be representative and reliable.

Begin at the Outfalls

A logical place to start is at each outfall since samples collected there will characterize total plant effluent. However, it is also important to characterize the smaller

Originally published January 8, 1973.

Nomenclature

A	Water area, sq. ft.
a	Velocity of stream needed to overcome meter friction and start rotation, ft./sec.
b	Current-meter constant (differs for each machine)
C	Coefficient characteristic of stream bed
C_f	Steam concentration of salt at Point B, Fig. 5
C_i	Concentration of salt being added to Stream Q_s by Stream Q_i
C_s	Concentration of salt in stream Q_s before addition
c	Width of weir end contractions (should not be less than $\frac{3}{4} L$.
H	Head (height of water above weir bottom), ft.
H_a, H_b	Parshall flume gage readings, ft.
L	Width of weir notch at water level
L'	Distance from crown of pipe to water level, ft.
N	Revolutions of meter cups, rev./sec.
Q	Flowrate, cu. ft./min.
Q'	Flowrate, cu. ft./sec.
Q_i	Salt solution flowrate
Q_s	Stream flow rate at Point A. in Fig. 5
Q_t	$Q_i + Q_s$ at Point B in Fig. 5
R	Hydraulic radius of stream, ft.
S	Slope of water surface on stream bottom
V	Mean velocity of stream, ft./sec.
v	Mean velocity of stream, ft./min.
W	Parshall flume throat width, ft.
Z	Distance between channel bottom and weir-notch bottom, ft. Pipe diameter, ft.

tributaries to the main sewers that ultimately discharge to the plant outfalls. This is often necessary to determine what processes and plant operations contribute contaminants to the wastewater. More-specific analytical data can be obtained by sampling streams discharging to the tributary sewers.

Once sewer layout and sample points have been established, the next step is to determine the type of sample to be collected at each point. Generally, samples are either grab or composite. A grab sample is a single sample of the wastewater, while a composite sample is one composed of a series of samples collected over a period of time and then blended for analysis.

Grab samples represent only the conditions that exist at the instant the sample is collected. Since they do not typify average conditions, they generally should not be reported to a regulatory agency and should never be used as a basis for treatment. However, grab samples are invaluable in determining the effects of intermittent dumps on a continuous discharge. Grab samples should be collected whenever an unusual discharge is observed or suspected at sampling points during the time composite samples are being collected.

Composite samples provide more-meaningful data because they represent average plant conditions over a relatively long period of time and are invaluable in determining treatability and equipment design.

The sampling method used should be one that presents a representative picture of the quality of the waste stream. Some operations have a steady flow and steady production rate, while in others these may both vary. Hence, a sampling program must be tailored to meet the conditions within the plant.

Consideration should be given to the length of time of the collection period for composite samples. If the plant is in operation for only one shift per day, sampling must be done during this working shift. If the plant operates two or three shifts per day but the discharges of wastewater and plant production rates are relatively uniform, sampling during one shift is sometimes sufficient. However, to obtain an accurate representation of the total daily effluent, particularly if there is great variability in flow or in production rate, strong consideration should be given to collecting composite samples over the entire operating day. Also, if samples are to be truly representative, no less than three composite samples should be collected at each point, to effectively determine the concentrations and types of contaminants discharged.

Manual Composite Sampling

When flowrate is relatively uniform, samples can be manually composited by collecting a series of grab samples of fixed volume at given time intervals and mixing them. The sampling interval should never exceed one hour and should preferably be less.

If flowrate varies, however, grab samples making up the composite sample should be proportioned to flow, i.e., as the flow at the sampling point increases or decreases, the volume of the sample collected should be increased or decreased proportionately. The flowrate at the sample point should be determined with a flow measurement device each time a portion of the composite sample is collected. The sample can then be proportioned to the flow by means of a table or graph.

Before making a graph or table, it is important to determine the quantity of sample required for analysis since flow may be on the low end of the scale of the graph, resulting in samples too small to make up the quantity needed.

One major advantage of manual sampling is that personnel can observe visible changes such as color, suspended solids, floating oil, etc. When such changes occur, grab samples can be collected to determine the effect of these changes on the waste stream. A disadvantage is the greater manpower requirement. A minimum of three men would be needed for a plant operating around-the-clock. Of course, if sampling points are many and widely separated, more men per shift will be required.

Automatic Composite Sampling

Composite samples can also be collected automatically. A number of automatic samplers are on the market, ranging in cost from $250 to $1,500. Some of these devices continuously collect samples in one large container, while others collect portions in separate containers, thus providing a series of composite samples. Some automatic samplers operate continuously, while others work on an intermittent timed sequence. And, some devices are programmed to collect samples in proportion to flow, as illustrated in Fig. 1. This type of sampler uses a probe

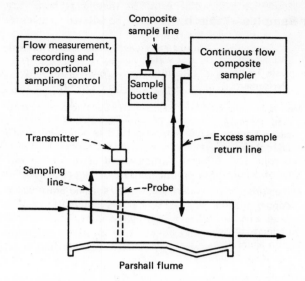

AUTOMATIC device proportions samples to flow—Fig. 1

Parameters	Analytical Tests Performed on Organic Wastewaters	Analytical Tests Performed on Inorganic Wastewaters
pH	X	X
Alkalinity or acidity	X	X
Suspended solids	X	X
Volatile suspended solids	X	
Total solids	X	X
Volatile total solids	X	
Settleable solids	X	X
Chloride		X
Sulfate		X
Calcium	X	X
Magnesium	X	X
Hardness	X	X
Aluminum		X
Iron	X	X
Copper	X	X
Fluoride		X
Manganese		X
Nickel		X
Lead		X
Cadmium		X
Mercury		X
Arsenic		X
Zinc	X	X
Hexavalent chromium		X
Total chromium		X
Total phosphate		X
Total nitrogen	X	
Ammonia nitrogen	X	
Oil and grease	X	X
Total organic carbon (TOC)	X	
Biochemical oxygen demand (BOD)	X	
Chemical oxygen demand (COD)	X	
Phenol	X	
Cyanide		X

that senses the flowrate through a weir or Parshall flume; the quantity of sample collected is based upon this flow.

Regardless of the type of automatic sampler used, the primary benefit is reduced labor requirement. Only occasional attention is necessary to maintain the unit and to pick up the collected sample.

Some automatic-sampler problems include: (1) batteries running down in battery-operated units; (2) plugging of pump tubing and orifices when waste streams contain high amounts of suspended matter; and (3) problems of obtaining representative samples from waste streams that may be stratified or contain high quantities of suspended solids, and oil or grease.

Floating oil and grease, suspended solids that settle very rapidly, and any substance that has a tendency to stratify in the waste stream will create problems whether the stream is sampled manually or automatically. However, these problems are more pronounced when using an automatic sampler. Since the sample-pickup line generally is placed below the surface of the water to obtain as representative a sample as possible, floating oil or grease may be entirely missed. Rapidly settling suspended solids also may get by the sample line. Conversely, if the sample line is placed near the bottom of the stream, a disproportionate amount of settling solids may be collected. If there is stratification in the stream, the sample pickup line will collect only from the stratum in which it is located, resulting in an unrepresentative sample.

Because these problems are sometimes impossible to overcome, allowances must be made when interpreting the analytical results from samples collected under such conditions. In most cases, however, these will not exist if samples can be collected at points where rapid water movement exists. The discharges over weirs and through Parshall flumes usually create sufficient turbulence to cause good mixing of the constituents in the waste stream. Sewers that flow rapidly or discharge in waterfall fashion are also good collection points. Sampling from stagnant

areas such as tanks, flooded manholes, and pools in a sewer should be avoided.

Onstream Monitors and Analyzers

The discussion so far has assumed that the samples collected will be analyzed in a laboratory. However, wastewater streams can be continuously and automatically monitored and analyzed by reliable onstream effluent-monitoring equipment currently on the market. Some onstream analyzers can be automated even to the point of self-calibration.

More than 40 important parameters can be monitored and analyzed by such equipment. However, most units are limited in the number of parameters they can si-

Handling of Wastewater Samples—Table II

Analytical Tests To Be Performed	Quantity of Water Required, Quarts	Special Handling Required	Analytical Tests To Be Performed	Quantity of Water Required, Quarts	Special Handling Required
TOC, COD, BOD	2	Generally, no preservation step is required.*	Metals (cadmium, lead, mercury, silver, barium, tin, cobalt, berylium, molybdenum)	1	No specific preservation is required.†
Cyanide	1	The pH of the sample should be raised to 11 or higher with sodium hydroxide.*			
Phenol	1	The pH of the sample should be lowered to less than 4 with concentrated phosphoric acid. Then add 1 gram of copper sulfate.*	Metals (iron, copper, manganese, zinc, aluminum, nickel, chromium, arsenic, magnesium, calcium, antimony)	1	No specific preservative is required.†
Nitrogen (Kjeldahl, organic, ammonia and nitrite)	1	Add 2 ml. of mercuric chloride per quart of sample (prepared by dissolving 27 g. of mercuric chloride per l.*	Selenium	1	No specific preservative is required.†
			Radioactivity	1	No specific preservative is required.
Oil and grease	1	Add 10 ml. of 1:1 hydrochloric acid.*	Routine mineral analyses	1	No specific preservative is required. However, the sample should be refrigerated.
Sulfide	1	Add 2 ml. of zinc acetate solution (prepared by dissolving 22 g. of zinc acetate per 100 ml. of distilled water) to the *empty* sample bottle. With a minimum of aeration, fill the bottle to the brim with the sample and cap bottle immediately.	Algaecides, pesticides and chlorinated hydrocarbons	1	A glass bottle is required. The bottle closure must be Teflon-lined. No specific preservative is required. However, the sample should be refrigerated.
Sulfite	1	Add 10 ml. of EDTA solution to the *empty* sample bottle. With a minimum of aeration, fill the bottle to the brim with the sample and cap bottle immediately.	Coliform bacteria	½	A *sterilized* glass bottle is required. No specific preservative is required. However, the sample should be refrigerated and analyzed within 36 hr. after collection.

*Sample should be refrigerated and the test performed as soon as possible.

†Sample should be acidified upon reaching the laboratory.

multaneously monitor and analyze. Many units can automatically analyze only one parameter while more-expensive units will analyze from three to ten simultaneously. Other units can automatically analyze different parameters by changing the sensing devices.

Generally, however, automatic onstream analyzers are not used in an initial or occasional plantwide wastewater sampling and gaging survey. But the results of such a survey may indicate that automatic equipment should be installed at critical points. Regulatory agencies sometimes insist that certain parameters be continuously monitored

to ensure that they are within the allowable discharge limits. This is particularly true of treated effluents discharging from the plant's wastewater treatment facilities. As always, the sample point should be selected and the equipment installed so that the sample collected is representative of the waste stream in question.

Other Sampling Considerations

In addition to the selection of sample points and the methods of sampling, there are three other important

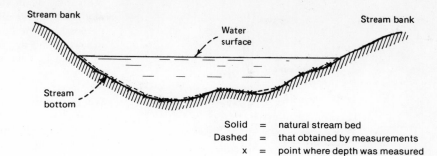

STREAM bottom as measured using a staff gage (dashed line) compared with actual natural contour (solid line)—Fig. 2

Solid   =   natural stream bed
Dashed   =   that obtained by measurements
x   =   point where depth was measured

considerations that must be taken into account in a sampling program. These are: volume of sample required, analyses required, and proper handling and preservation of the samples.

The volume of sample required depends on the analyses to be performed. These in turn depend on the manufacturing or processing operation involved and on the types of chemicals or materials added to the process water. Local, state and federal regulatory agencies often demand specific analyses on wastewater discharges from particular types of manufacturing operations.

Typical analyses often performed on organic and inorganic wastewater samples are shown in Table I.

When samples have been collected, proper handling is often of paramount importance to obtain valid test results. Many analyses require that samples be specially preserved immediately after collection. This prevents change in characteristics of particular constituents during shipment and storage.

Special handling requirements for particular constituents in wastewater samples, and the quantity of water required to determine the amount of any constituent in the sample, are shown in Table II. The procedures shown are based on criteria established in Ref. 1.

Finally, sampling safety precautions are important. One hazard frequently overlooked by plant managers is the presence of pathogenic bacteria in sanitary-waste streams. Personnel directly involved with sanitary-waste streams should be inoculated against typhus and typhoid (and must wash their hands before eating or smoking to prevent the ingestion of pathogenic bacteria that could make them ill even though they have been previously inoculated).

Gaging of Flows

Directly associated with a sampling program is the need to develop flow data for various wastewater streams. Such data may be necessary to obtain an approximation of inplant water consumption or to estimate the quantities of contaminants in the waste streams.

Flowing water may be classified into two convenient categories—open-channel flow or pipe flow. In open-channel flow, the water surface is exposed to the atmospheric pressure. In pipe flow, the water is subject only to hydraulic pressures, with no surface exposed to the atmosphere.

Numerical Methods for Flow Measurement

Any natural stream may show variations in stream depth, width and slope as well as other physical differences.

Numerical methods to calculate flows in streams or open channels are widely known. By using these formulas, the velocities of water in the stream bed can be determined. The general expression is:[2]

$$v = CR^x S^y \tag{1}$$

Formulas such as this assume a constant depth, area, velocity and discharge at every point along the stream. Although this is not true for natural streams, the methods give good approximations.

The first formula similar to the above form is the Chezy formula, developed by Antoine Chezy in 1769:[2]

$$v = C_1 R^{1/2} S^{1/2} \tag{2}$$

The coefficient C_1, called the Chezy coefficient, takes into account flow resistance.

The second numerical method was formulated in 1889 by Robert Manning:[3]

$$v = \frac{1.49}{n} R^{2/3} S^{1/2} \tag{3}$$

This equation establishes a more definitive flow-resistance coefficient which is equal to $1.49/n$. The "n" is called Manning's coefficient of roughness and can be obtained from hydraulics handbooks and some textbooks. Since it is both easy to use and relatively accurate, this formula is most widely employed for the calculation of open-channel flow.

The coefficient C_1 in the Chezy equation is equal to C in the general expression. The coefficients x and y in the general expression equal 1/2 and 1/2, respectively, in the Chezy equation. However, Manning's equation differs from the Chezy equation because exponents x and y are 2/3 and 1/2, respectively, and the coefficient C in the general expression is equal to $1.49/n$ in Manning's equation.

The relationship between the two formulas is shown in the following equation:[2]

$$C_1 = \frac{1.49}{n} R^{1/6} \tag{4}$$

If this expression is used in either the Manning or Chezy formula, they become equalities.

When evaluating these formulas, the variables that

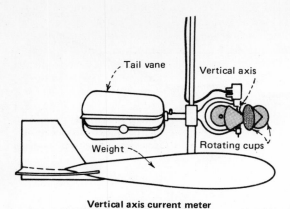

Vertical axis current meter

Horizontal axis current meter

CURRENT meters are made either with vertical axis (top) or horizontal axis (propeller type)—Fig. 3

have to be determined in addition to the flow-resistance-coefficient are the hydraulic radius and the stream bed slopes. The hydraulic radius can be calculated by dividing the water area by the wetted perimeter.[2] The slope, which is the decrease in elevation within a specified distance, can be determined from existing data or by surveying techniques.

In order to determine the water area and wetted perimeter to be used in hydraulic-radius calculations, a sufficiently long staff gage or other suitable device for

measuring depth should be used. At random intervals along a line extending from one bank to the other and perpendicular to flow, the device should be dropped into the water and the depth recorded. Care should be taken to ensure that the device is perpendicular. If a flexible line, e.g., rope or string is used, sufficient weight should be placed on the bottom of the line so that it becomes taut in the current and no deflection results. Fig. 2 shows the results of a study of this type. The ×'s indicate the points at which depth measurements were taken. Depth measurements taken at short intervals will give a more accurate stream-bed profile than will those taken at greater intervals. Using this profile, the wetted perimeter and cross-sectional area of stream flow can be approximated, and then used to determine the hydraulic radius and then the stream velocity.

Another method to determine this velocity is by the use of current meters. All of these devices operate on the same basic principle, i.e., a rotating member that turns with the least possible friction is mounted so that its speed of rotation is proportional to the velocity of the fluid. Several different meters exemplify this principle. One type has six conical cups that drive electric contacts and rotate about a vertical axis. These contacts close a battery circuit that causes a particular number of clicks for each revolution to be transmitted to headphones worn by the operator.[3] Included are tail vanes to keep the meter facing into the current and heavy weights to hold the meter as vertical as possible. Smaller versions of this meter are available to measure flow at shallow depths.

Another type of meter has a propeller, which turns about a horizontal axis, as the rotating element. The contacting mechanism and suspension system are similar to the vertical-axis type. One advantage of the propeller-type meter is that vertical currents will not be indicated as positive velocities as they are with vertical-axis meters. (A vertical velocity measurement overestimates or inflates the actual stream velocity.) Also, the horizontal type (propeller) meters frequently incur mechanical disadvantages. Despite these disadvantages, however, the stream velocity can be determined rather accurately. Fig. 3 illustrates the difference in principle between the two types of current meters.

The general equation used when determining velocity by current meters is:[3]

$$v = a + bN \qquad (5)$$

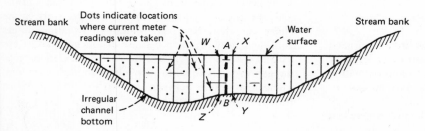

Note: Two dots in each section indicate velocities were measured at 0.2 D and 0.8 D where D equals total depth from water's surface. One dot indicates velocity measurement was taken at 0.6 D.

FLOW determination by means of current meter requires the integration of flows in a number of areas—Fig. 4

The constants a and b are known for particular meters, and N is the only variable that must be determined from the stream flow.

Although stream velocity can be calculated by formula or measured by current meter, flow or discharge of the stream must be determined. This can be done using the continuity equation as follows:

$$Q = VA \qquad (6)$$

Water area of the stream can be found by depth measurements as was explained above and illustrated in Fig. 2. If velocity was calculated mathematically, the discharge, Q, can be found by simply multiplying the calculated velocity by the water area. However, a more complicated procedure must be followed if velocity was determined by the current meter. The method of obtaining stream flow, Q, using the current meter is:[3]

1. Divide the water area of the stream into vertical sections with each section being less than 10% of the total flow (see Fig. 4).

2. Measure the total depth of each section.

3. Determine the mean velocity of each section by averaging the velocities measured by the current meter at 0.2 and 0.8 of the total depth from the water's surface. This should be done with the deeper sections; however, velocity measurements should be taken at 0.6 of the total depth from the water's surface in shallow sections since the mean velocity at this depth is nearly equal to the average of the velocities at 0.2 and 0.8 depth. The mean velocity at these depths can be equated as follows:

$$V_{avg.} \simeq V_{0.60} \simeq \left(\frac{V_{0.20} + V_{0.80}}{2} \right)$$

4. Measure the velocity when the meter is at the 0.8 depth, by starting a stopwatch on a click (or impulse) from the meter and stopping it on another click about 45 sec. later. (One click should be subtracted from the total since the first is considered to be zero.) Calculation of velocity from the meter calibration can be determined since the number of clicks and the time are known variables.

5. Raise the meter to the 0.2 depth level. Repeat Step 4. However, in shallow water near the shore, a single velocity determination at 0.6 depth may be used. (Note: If velocities are high, the meter will not be suspended vertically but will be carried downstream by the current. Appropriate corrections should be made to make sure the meter is near the desired depth where velocity readings are to be determined.)

6. Compute the total discharge as follows:

a. Average the velocity as mentioned in Step 2.

b. Determine individual sections by extending the boundary of each to the midpoint of the distance between verticals where meter readings were taken (WZ and XY in Fig. 4). The area of this section is found by multiplying the measured depth at the vertical AB by WX, which is the width of section $WXYZ$.

c. Add the increments of discharge for all the vertical sections.

Since $Q = VA$, the discharge is easily found by multiplying average velocity and area for each vertical section and summing the discharges for all the vertical sections.

The formula techniques and current-meter approach are limited primarily to large open channels such as streams and rivers, and cannot be used for other applications such as flows within pipelines.

Water meters on influent pipes will give the water consumption for that particular line. To determine total water usage, a meter can be installed on the plant influent lines (this should be done before a wastewater survey is conducted if a water balance for the plant is desired). Flow data of individual processing lines can be determined by installing a meter on lines feeding the particular operation.

Flow also can be determined with a bucket and stopwatch. This is a simple method and involves the time it takes to fill a bucket (or other container) of known volume.[4] This method is accurate if the time to fill the container is greater than 10 sec. If the time interval is less, the results would be questionable.[6]

Frequently, wastewater is pumped through pipes to a sewer or stream. If both the pump rate (from a pump curve plotting head vs. flow) and the length of operation are known, the pump discharge can be calculated. At times, pump curves are unreliable because the pump is old and worn. To insure accuracy, therefore, the pump should be recalibrated prior to checking flowrates.

Floating Objects, Dyes and Salt

Floating objects, dyes and salt solutions are sometimes used to give the velocity of the water flowing in the sewer while other devices such as weirs and Parshall flumes give volumetric measurements of flow. If velocity determinations are made, the cross-sectional area of flow within the pipe that is occupied by the water must be determined also. An approximation of water velocity within the sewer can be obtained using floating objects and dyes. If the distance between manholes or points of observation for the dyes or floating objects is known, the time for the object or dye to appear at these observation points can be measured. Dividing the distance between the two points by the time it takes for the dye or soluble salt to appear yields average velocity that can be expressed as follows:[4]

$$V = d/t \qquad (7)$$

If floating objects are used, the above equation must be modified. Since floating objects in open channels travel at a velocity about 1.2 times the mean velocity, the above equation now becomes:

$$V = (d/t)/1.2 \qquad (8)$$

It is assumed that the channel has its water surface exposed to the atmosphere. Whichever technique is used, sufficient time should be allowed for the dye, salt or object to appear. Inherent inaccuracies in this method are the assumptions that cross-sectional area and velocity are constant and that nothing in the channel creates turbulence or disturbances in flow.

Another technique used for flow measurement is the dilution method. A salt brine, such as lithium chloride, LiCl, is added to the stream at known concentrations.

Samples are then taken for analysis after the two streams are well mixed. Downstream measurements are monitored continuously by conductivity measurements or periodically sampled for laboratory analysis. Fig. 6 illustrates this technique.

Since Q_s is desired, the equations can be solved for this variable:

$$C_iQ_i + C_sQ_s = C_fQ_t = C_f(Q_i + Q_s) \qquad (9)$$

$$Q_s = \frac{(C_f - C_i)Q_i}{C_s - C_f} \qquad (10)$$

Other Instruments for Flow Measurement

Different types of devices can be used for flow measurement depending on the channel features and flow volumes. For instance, if the channel has regular or uniform features, e.g., a culvert, one may need a device different from that required to measure a natural stream that has very irregular characteristics. Also, distinct devices will be required for small flows versus large flows. If continuous monitoring is desired, however, still other devices may be necessary.

One device for flow measurement that has widespread use is the weir. A weir is nothing more than a channel constriction that requires the water to flow through a predetermined opening. By simply measuring the head (i.e., the height of the water above the bottom of the weir opening), the flow of the water can be determined since each type of weir has been calibrated, and such data appear in many hydraulics handbooks. Several types of weirs exist, many of which are named after the type of opening through which the water flows.

Certain considerations are necessary for accurate flow measurements using weirs. Sharp, square-edged cuts should be made in the material used for weir construction.[4] Along with this consideration, the distance from the bottom of the notch to the channel bottom should be at least 2.5 times the head (H) over the weir,[4] i.e., $Z \geq 2.5 H$. This reduces the velocity of approach to a negligible value. The second consideration is that the nappe, or both top and underside of the falling water overflow, should be aerated to prevent a vacuum from forming that would increase the discharge over the weir for a given head.[2] Obviously, leaks must be sealed around the weir plate so that all flow to be measured passes through the notch.[4] An inert cement or asphalt-type material is preferable for this purpose.

In order to ensure a uniform depth of flow, the weir should be level, and debris should not be allowed to accumulate near the weir plate.[6] The channel upstream from the weir should be straight in order to assure quiescent or quiet flow.[6] And the weir should be selected after the flow rate is estimated by other techniques; this may be done by various methods already mentioned such as floats, dyes, etc.

The first type of weir to be considered is the V-notch. The 30-deg. V-notch is rarely used; the 60-deg. and 90-deg. V-notch weirs are common. The 90-deg. V-notch is based on the Barnes flow formula:[7]

$$Q' = 2.48 H^{2.48} \qquad (11)$$

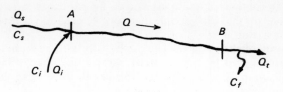

FLOW measurement by brine dilution—Fig. 5

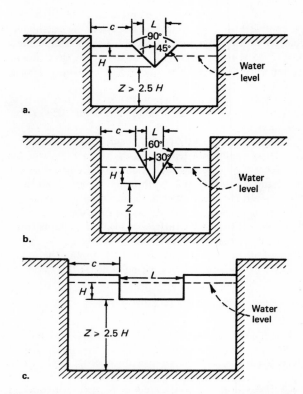

WEIRS, 60- and 90-deg. V-notch, and rectangular—Fig. 6

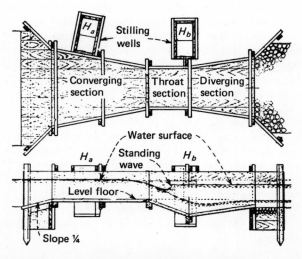

PARSHALL flumes give minimum head loss—Fig. 7

Methods of Flow Measurement and Their Application to Various Types of Problems—Table III

Device or Method	Flow Range Measurement	Cost	Ease of Installation*	Accuracy of Data	Application
Mathematical formulas	Small to large	Low	N/A	Fair	Open channels and pipe flow.
Water meters	Small to large	Low	Fair	Excellent	Pipe flow.
Bucket and stopwatch	Small	Low	N/A	Good	Small pipes with ends or where joints can be disconnected.
Pump capacity and operation	Small to large	Low	N/A	Good	Lines where water is being pumped.
Floating objects	Small to medium	Low	N/A	Fair	Open channels.
Dyes	Small to medium	Low	N/A	Fair to average	Pipe flow and open channels.
Salt dilution	Small to medium	Low	N/A	Fair	Pipe flow and open channels.
Orifice meter	Small to large	Medium	Fair	Excellent	Pipe flow.
Weirs and flow recorders	Small to large	Medium	Difficult	Good to excellent	Open channels.
Parshall Flume	Small to large	High	Difficult	Excellent	Open channels.
Venturi meter	Small to large	High	Fair	Excellent	Pipe flow.
Magnetic flow meter	Small to large	High	Fair	Excellent	Pipe flow.
Flow nozzles	Small to large	Medium	Fair	Excellent	Pipe flow.
Pitot tube	Small to medium	Medium	Fair	Good	Pipe flow.
Rotameter	Small to medium	Medium	Fair	Excellent	Pipe flow.

*N/A = Not Applicable

Fig. 6a illustrates the 90-deg. V-notch weir.

For a 60-deg. V-notch weir (Fig. 6b), the flow is based on the formula:[7]

$$Q' = 1.43 \, H^{5/2} \tag{12}$$

The advantages of the V-notch weir include greater sensitivity, which makes it useful for measuring low flowrates, and greater heads for a given discharge, which allow for more accurate measurements. A maximum of about 90 gpm. limits the practical use of V-notch weirs.

The second type of weir uses a rectangular notch. Two basic types of rectangular weirs exist, those with end contractions, which are walls at the sides of the weirs that make the weir narrower than the channel of the stream, and those without them. The rectangular weir is illustrated in Fig. 6c.

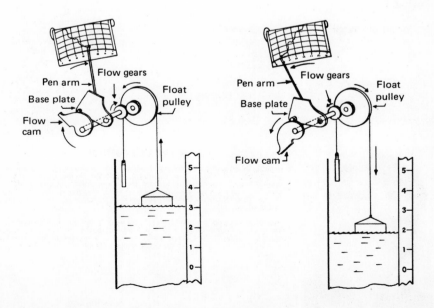

WATER-LEVEL recorders can be used with weirs to provide a record of stream flow—Fig. 8

The discharge is based on the Francis formula:[6]

$$Q' = 3.33 \, LH^{3/2} \tag{13}$$

The Gourley and Crimp formula similarly describes weir discharge by the formula:[7]

$$Q' = 3.1 \, L^{1.02} \, H^{1.47} \tag{14}$$

Only minor differences exist between the exponents of these two formulas. Schoder also formulates the discharge over a weir with modified end contractions by the equation:[6]

$$Q' = 3.00 \, LH^{3/2} \tag{15}$$

Rectangular weirs with modified end contractions are used when there is insufficient channel width to insert a rectangular weir with standard end contractions, or the available height is too small to permit the use of a triangular weir.

The rectangular weir is normally used where flow capacity for triangular V-notch weirs is exceeded, since it is capable of conveying greater quantities of water. Other weirs are sometimes used—their descriptions may be found in hydraulics handbooks or in Ref. 11.

One of the most widely used devices for flow measurement is the Parshall flume. The instrument can be built when there is need for continuous measurement of stream or sewer flow at or near the ground's surface. The flume has the ability to clean itself, thus requiring low maintenance; and accuracy is achieved with both free-flow discharge or submerged flows. Free flow conditions exist when the critical depth exists in the throat section and a hydraulic jump occurs on the diverging section. If the hydraulic jump becomes submerged, submerged flow conditions occur. Since only a small hydraulic-head loss occurs with flow through the flume, this device becomes most useful when there is a minimum allowable head loss in a sewer or channel. Fig. 7 illustrates the Parshall flume:

The equations used to describe Parshall flumes are dependent upon throat width as can be seen by the following:[2]

Width of Throat	Equation
3 in.	$Q' = 0.992 \, H_a^{1.347}$
6 in.	$Q' = 2.06 \, H_a^{1.58}$
9 in.	$Q' = 3.07 \, H_a^{1.53}$
12 in.–8 ft.	$Q' = 4WH_a^{1.522W^{0.026}}$
10 ft.–50 ft.	$Q' = (3.6875 \, W + 2.5)H_a^{1.6}$

The previous equations apply for free flow conditions.

If the flow becomes submerged, the actual discharge is reduced, and the above equations yield inflated results. Submerged flow is indicated by a small wave or ripple which forms near the upstream end of the diverging section. A hydraulics handbook should be consulted to determine what factors should be applied to give correct discharge rates if submerged flow conditions exist.

Flow recorders are generally used in conjunction with weirs for continuous measurement. Most types of continuous recording devices use a float and counterweight arrangement. The float rests on the water and is attached to a wire that circumscribes the upper half of a wheel. The counterweight is attached to the other end of the wire (see Fig. 8). The wheel is attached to a drum with a recording chart affixed along the drum's perimeter. As the float moves with the rise and fall of water surface elevation, the drum moves accordingly. A pen that travels in a longitudinal direction along the chart records the movement of the float on the chart. Most short-term flow recorders operate for 24 hr. while others are designed to function independently for a 30-day period, i.e., it takes the pen 24 hr. or 30 days to travel the length of the chart.

Other methods of measuring flow in pipes include the orifice meter, venturi meter, pitot tube, flow nozzles and magnetic flow meters. These methods are familiar to most chemical engineers and can be found in standard texts and handbooks (e.g., Ref. 8 and 12). Hence, they will not be further discussed here.

Bibliography

1. "Standard Methods for the Examination of Water and Wastewater." 13th ed., AWWA, APHA, WPCF, 1971.
2. Chow, V. T., "Open Channel Hydraulics," McGraw-Hill, New York, 1959.
3. Linsley, Ray K. Jr., others, "Hydrology for Engineers," McGraw-Hill, New York, 1958.
4. Paulson, E. G., "Water Pollution Control Programs and Systems," in "Industrial Pollution Control Handbook," H. F. Lund, ed., McGraw-Hill, New York, 1971.
5. Weber, R. L., others, "Physics for Science and Engineering," McGraw-Hill, New York, 1959.
6. "Planning and Making Industrial Waste Surveys," Metal-Finishing Industry Action Committee of the Ohio River Valley Water Sanitation Commission, 1952.
7. "Hydraulics and Useful Information," Chicago Pump Co., Div. of FMC Corp., 1965.
8. McCabe, W. L., Smith, J. C., "Unit Operations of Chemical Engineering," McGraw-Hill, New York, 1956.
9. Metcalf and Eddy, Inc., "Wastewater Engineering," McGraw-Hill, New York, 1972.
10. Ripley, K. D., Monitoring Industrial Effluents, *Chem. Eng., Deskbook Issue*, May 8, 1972.
11. Urquhart, Leonard, C., ed., "Civil Engineering Handbook," McGraw-Hill, New York, 1959.
12. Perry, Robert H., others, eds., "Chemical Engineers' Handbook," 4th ed., McGraw-Hill, New York, 1963.

Meet the Authors

Joseph G. Rabosky is a project manager of the environmental engineering of Calgon Corp., P.O. Box 1346, Pittsburgh, PA 15230. He received his bachelor's degree in chemical engineering from Penn State University, and a master's in civil engineering from the University of West Virginia. He is a member of AIChe, Water Pollution Control Federation of Pennsylvania and the American Water Works Assn.

Donald L. Koraido is a senior project engineer in the environmental engineering dept. of Calgon Corp., P.O. Box 1346, Pittsburgh, PA 15230. He received his B.S. degree at Penn State University and is a member of the Water Pollution Control Federation and the Water Pollution Control Assn. of Pennsylvania.

How To Measure Industrial Wastewater Flow

Effluent flows from industrial plants are often most easily measured in open channels or at pipe-ends. Weirs and flumes are the simplest devices for charting such flows; here is how to use them.

THOR THORSEN and ROLF OEN, SINTEF*

Environmental factors, resource utilization, and government requirements for cleaner effluents have led to a new technology called wastewater engineering.

Originally a part of sanitary engineering, with increasing awareness of industrial effluents, wastewater engineering grew more and more like chemical engineering. Especially since many water-pollution problems in the chemical process industries require a change in the production process itself.

To control any industrial chemical operation, we require information about the process parameters, including those for flow and composition. With wastewaters, the most important parameters are those associated with quality. That is, do we have clean water in the waste stream? Or, is the discharge of chemicals in the effluent beyond acceptable limits?

To answer these questions, we must perform both chemical and physical quantitative analyses, and flow measurement. Furthermore, we must continuously record the necessary parameters. In practice, such analyses may be very difficult to do and expensive; but as for flow measurements, the tools can be simple and inexpensive.

Analysis Techniques

The most common methods for analyzing effluents continuously are potentiometry and photometry. In Table I, we have put together feasible techniques and relative costs for some important parameters in wastewaters. Actual costs, of course, will vary with equipment design and required tolerances. As shown in Table I, continuous

* The Foundation of Scientific and Industrial Research at the University of Trondheim

Originally published February 17, 1975.

analysis is either expensive or unreliable. The very rapid development of potentiometric analysis will lead to reliable electrodes for use in wastewaters. Because of low cost, we expect that most inorganic parameters in the future will be measured by this method.

In our opinion, the best parameter for organics is the TOC (Total Organic Carbon). Suitable equipment for continuously doing TOC analysis should become avail-

Continuous Methods for Effluent Analysis—Table I

Parameter	Method	Cost	Reliability
pH	Potentiometry	Low	Very good
Cadium	Potentiometry	Medium	Fair
Chromium (VI)	Potentiometry	Medium	Bad
Copper	Potentiometry	Medium	Fair
Cyanide	Potentiometry	Medium	Not good
Fluoride	Potentiometry	Medium	Good
Lead	Potentiometry	Medium	Fair
Nitrate	Photometry	High	Good
Phenol	Photometry	High	Good
Phosphate	Photometry	High	Good
Ionic strength	Conductiometry	Low	Good
Suspended solids	Photometry	Medium	Good
COD[1]	Photometry	High	Good
BOD[2]	—	—	—
TOC[3]	—	—	—

[1] Chemical Oxygen Demand.
[2] Biochemical Oxygen Demand. Impossible to perform continuously.
[3] Total Organic Carbon. Expected to be widely used in future.

Methods and Their Limitations for Measuring Effluent Flow—Table II

Method	Capacity	Continuous Recording	Cost	Ease	Accuracy	Application
Formulas	Large	No	Low	Good	Low	All
Pump capacity	Large	No	Low	Good	Low	All
Water meter	Large	Occasionally	Low	Good	Medium	Pipe
Bucket/watch	Small	No	Low	Good	Medium	o.e.*
Tracer	Large	Yes	Medium	Fair	Medium	All
Orifice	Large	Yes	Medium	Difficult	Excellent	Pipe
Pitot tube	Large	Yes	Medium	Fair	Medium	Pipe
Rotameter	Medium	Occasionally	Medium	Difficult	Excellent	Pipe
Weirs	Large	Yes	Medium	Good	Excellent	o. ch.†
Flumes	Large	Yes	High	Good	Excellent	o. ch.†

*o.e. = open end of pipe or channel.
†o. ch. = open channel

able in the near future. For reference purposes, the BOD (Biochemical Oxygen Demand) test may be performed in the laboratory of each industrial effluent from time to time. At present, the control of effluent is best performed in the laboratory on samples taken at intervals—

especially for small plants where continuous equipment is not economically feasible. However, this procedure does not apply to pH and flow. These may be measured and recorded continuously, at a reasonable cost.

Flow Measurement

Quantitative analysis of an effluent will yield only the relative concentrations of the components, and whether the concentrations of any of them are too high. But we know nothing about the quantities of pollutants until flow is measured.

Methods for measuring effluent flows are given by Rabosky and Koraido [1]. Thorsen [2] lists some alternative methods, as given in Table II. Unfortunately, waste pipes in many process plants are in a very poor condition and, often, nobody knows where they are. Fouling and corrosion of waste pipes and channels are other problems.

Effluents are most generally accessible in open channels or at the ends of pipes. Hence, the methods and equipment for measuring flow in existing plants, or for

WEIR FLOW is function of level above restriction—Fig. 1

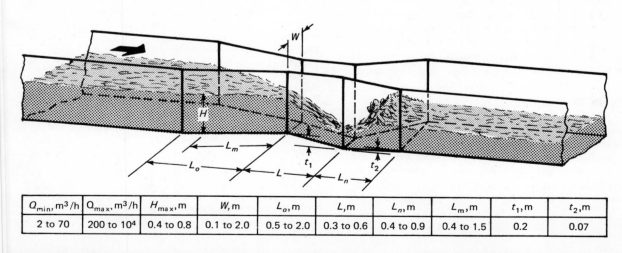

Q_{min}, m³/h	Q_{max}, m³/h	H_{max}, m	W, m	L_o, m	L, m	L_n, m	L_m, m	t_1, m	t_2, m
2 to 70	200 to 10⁴	0.4 to 0.8	0.1 to 2.0	0.5 to 2.0	0.3 to 0.6	0.4 to 0.9	0.4 to 1.5	0.2	0.07

PARSHALL FLUME has a bottom that declines in the throat section and inclines in the diverging section—Fig. 2

temporary installations, are to a great extent the same. Such flow-measuring devices must:

1. Withstand corrosion.
2. Have minor sensitivity to fouling.
3. Serve primarily for use in open channels and/or pipe-ends.
4. Be simple and inexpensive.
5. Maintain an accuracy within 5%.

Point 4 is important here because these investments for pollution control seldom give any payback. Point 5 stresses an accuracy of 5%. This should be good enough because the flow measurement is most often just informative or for rough control purposes.

Operating Principles for Weirs and Flumes

Weirs and flumes seem to be good choices for flow measurements in open channels. Their operating principle is very simple, and they use the same method for sensing flow, which is measurement of the effluent level. Once calibrated, the measuring devices are very responsive to change in flow, and are almost insensitive to wear.

For a weir, the measuring principle is that water level above the bottom of a vertical, abrupt restriction of the flow channel is a function of flow. Level is ideally measured just upstream from the point where the liquid surface begins to curve downwards, that is about 3 to 4 times the maximum height of liquid above the restriction. Usually, the channel is also restricted in width, as shown in Fig. 1.

For a flume, the restriction of the channel cross-sectional area is horizontal. The restriction is not abrupt, but consists of a converging section, a throat, and a diverging section. The bottom of the flume may be flat, but it may also decline in the throat, and incline in the diverging section, as shown for the Parshall flume in Fig. 2. The aim of the throat is to cause free-flow. Then, the total water level in the first part of the diverging section is a function of flow and flume dimensions. We will not go into additional details of construction for weirs and flumes because recent articles have described these details [1,3].

Both methods are well suited for direct measurements in open channels. With pipe-ends, a weir is easy to use. We construct an open box with a weir in one of the walls, and let the effluent fall into the opposite end of the box. Though weirs are easy to construct, flumes have the advantage of being self-cleaning, i.e., solids do not accumulate.

Flow Calculations for Liquids

Under ideal conditions, we can easily calculate flow from the upstream level, but this situation does not exist in effluent measurements. The general equation for flow is:

$$Q = KWH_n \qquad (1)$$

where Q = flow, m³/h; K = constant for a given flume or weir; W = width of narrowest part, m; H = liquid level, m; and n = exponent, which ideally equals 1.5.

The factors affecting Eq. (1) are: weirs do not have

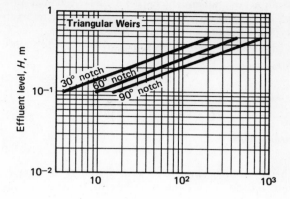

GRAPH gives flows for different weir notches—Fig. 3

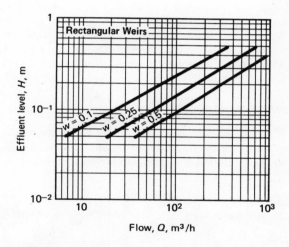

WEIR FLOW is correlated to width given in meters—Fig. 4

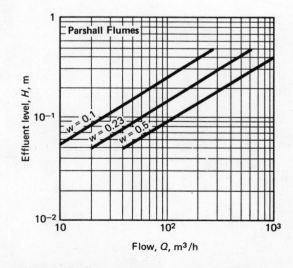

THROAT WIDTH of flume is also measured in meters—Fig. 5

infinite width; streamlines are so curved that Bernoulli's equation does not entirely apply; there is friction at walls and edges; and high velocity upstream, i.e., the reservoir not being infinite and calm.

From various sources, we have derived these formulas:

Triangular weir:

$$Q = 5 \times 10^3 \tan (\phi/2) H^{2.5} \tag{2}$$

Rectangular weir:

$$Q = 850 \left(7.63 + \frac{H}{H_\phi}\right) W H^{1.5} \tag{3}$$

Parshall flume:

$$Q = 8.52 \times 10^3 (11.4^{\log W}) H^{(1.57 + 0.09 \log W)} \tag{4}$$

where ϕ = vertical angle in weir, °; and H_o = distance from the bottom of the weir box to the crest of the weir, m.

In weirs, the edges of the weir chest should be 2 mm thick, or less, on the upstream side. The vertical channel walls and bottom of channel or weir box should be at least $2.5 H_{max}$ from nearest weir edge.

In Parshall flumes, the dimensions given in Fig. 2 may be used, with intermediate values derived by linear interpolation.

With careful measurements, it is possible to get an accuracy of ±3% by these procedures. Fig. 3, 4 and 5 yield the flow for various levels of liquid for different weirs and flumes.

Calibration Method

If with reasonable effort, we cannot attain the conditions on which the formulas are based, we have to calibrate the measuring devices. In our experience, there may easily be a 20% error if the conditions are wrong. But if free flow exists, or nearly so, one can achieve good accuracy and reproducibility for continuous measurements once the device is calibrated. Calibration may be done in the laboratory, or by using methods such as bucket and stopwatch; or tracer techniques, depending on magnitude of flow.

The upstream velocity energy causes many of the major deviations from flow as calculated from Eq. (2), (3) or (4). With actual effluent channels and pipes, this deviation yields a lower value for the exponent n. In our experience, the deviation in flow often has a magnitude of 5% to 20%—giving a value for the exponent that ranges from 1.25 to 1.45.

In our work on flow measurement of industrial effluents, we have become very practical instead of trying to work out empirical formulas for any given form, declination and material of construction of waste pipes and channels. Using formulas and all sorts of corrections would make us feel we were steadily losing contact with correct measurements. This has led us to improvise in field work, which can be of tremendous help under dirty, wet, cold and otherwise unattractive and space-limited working conditions.

Therefore, we always reconstruct our field measurements in the laboratory in a calibrated circulation stand, using a precision orifice (100 to 700 m³/h) and Woltman-counter (0 to 150 m³/h) for calibration. Such calibrations give us ±2% accuracy on single measurements; and well below ±1% accuracy for a calibration curve. The apparatus is shown in Fig. 6.

One of the major problems is to create the same upstream velocity and turbulence in the laboratory as in

CIRCULATION STAND, calibrated for close accuracy, reconstructs field measurements in the laboratory—Fig. 6

RHEOSTAT is operated by float behind notch weir—Fig. 7

the field. This is much easier with weir boxes where we can let liquid fall free into the end of the box opposite the weir. The effects of waves and turbulence are almost removed by inserting a perforated metal plate between the inlet and weir. With flumes, and weir baffles in channels, this is not as easy to do. We also have to measure the slope of effluent pipes or channels, and reconstruct it in the laboratory at a distance at least five diameters upstream. Therefore, we try to use weirs as much as possible if the problem of sludge buildup is not great.

Level Measurement

Level measurement is best performed in a small well located outside the box or flume, with a subsurface connection to the main stream. With calibrated devices, this procedure may not be necessary. Putting the level probe directly into the main stream generates more noise in the measurements due to waves, turbulence and floating matter.

Several measuring principles are possible: ultrasonic, static subsurface pressure (bubble pipe or membrane probe), conductivity (submerged long electrodes), and float-operated sensors. One simple, cheap and reliable device is a rheostat operated by a float on a rod. With few electronics, this device can give a proper signal to most recorders. Fig. 7 shows such an outfit, including a NiCd-pack for field work.

Sampling Industrial Effluents

The practical part of defining an industrial effluent consists of three main tasks: (1) recording process parameters and how production takes place, (2) taking samples for analysis, and (3) measuring flow.

If we cannot directly analyze the effluent, we must take samples. These must be representative over the entire cross-section of the channel. And this is almost impossible to do if emulsions, sludge, particles, or non-watersoluble liquids concentrate in different parts of the cross-sectional area, because the linear velocity of the stream will vary from the center toward the channel walls.

Samples must also be representative over time. But then we have to withdraw liquid continuously, and we will get either very big volumes, or very slow withdrawal, which usually means that we must use a small sample tube. Taking samples discontinuously means we run the risk of losing short concentration peaks that may contain a greater amount of pollutants. So we must take into account the relation between the rhythm in the production process, the time to take one sample, t_s, and the interval between samples, t_i, to get a reasonable chance to hit concentration peaks. Then, we multiply every concentration-peak hit with the relation t_i/t_s to get the actual number of peaks.

Continuous measurement of the effluent flow is essential if the data are to be used for projecting a treatment plant.

Recording the process parameters is very important in every effluent-defining job. And this should be done before taking samples and measuring flow. Information about the polluting process should guide the engineer when selecting sample intervals, in determining where and how to take samples, and in establishing the total time for taking samples and measuring flow. Recording the process parameters should also be done during the whole period of sample-taking and gaging. This provides a check on whether the process is working normally, and gives the engineer a better chance to explain the analysis and flow-data.

References

1. Rabosky, J. B. and Koraido, D. L., Gaging and Sampling Industrial Wastewaters, *Chem. Eng.*, Jan. 8, 1973, pp. 111–120.
2. Thorsen, T., Measurement of Effluent Flow in Industry, Intl. Scandinavian Congr. Chem. Eng., KEM TEK 3, Copenhagen, 1974.
3. Reitel, W. M. and Sobel, R., Waste Control Management, *Chem. Eng.*, June 18, 1973, pp. 59–63.

Meet the Authors

Thor Gustav Thorsen is a group leader in the water pollution group (a part of the Section for Environmental Technology) for Sintef, The Foundation of Scientific and Industrial Research at the University of Trondheim (7034 Trondheim—NTH, Norway). Previously, he was a scientist at Sintef, doing research in absorption processes, the treatment and protection of metal surfaces, water pollution, and other fields. He received an M.Sc. degree in chemical engineering from the Norwegian Institute of Technology, and is a member of the Soc. of Norwegian Engineers and of the Chemical Soc. of Norway.

Rolf Helge Oen is an engineer with Sintef's water-pollution group, which is concerned primarily with water-pollution-control methods and effluent measurement for a wide variety of industries. Previously, he was an engineer in product development (chewing gum, puddings, sauces) at the Nidar-Erbe A/S Chocolate Factory, and a technician at the University of Oslo. He received his degree in engineering from the Chemical Section of the NKI Technical School.

Measurement uncertainties in the pollutant-discharge permit system

A reported discharge having an associated uncertainty greater than 50% of its permit authorization is not unusual. Industry should determine its ability to measure effluent data, so as to recognize whether it is reporting an actual value or an accumulated error. Here are statistical procedures for doing this.

David L. Russell, Allied Chemical Corp. and James J. Tiede, Hazelton Laboratories America, Inc.

☐ No matter how carefully we make a measurement, there will be some inaccuracy in it. The magnitude of accumulated inaccuracies associated with sampling, flow measurement and chemical analysis in determining the amount of a pollutant discharged by a plant is of great significance when the measurements are used to judge compliance with a regulation or a permit.

The National Pollutant Discharge Elimination System (NPDES) of the U.S. Environmental Protection Agency does not address the inherent inaccuracies in data measurement. During development of the "Effluent Guidelines," industrial efforts were directed toward the statistical evaluation of plant performance data, with little attention given to the overall accuracy of measurements.

Guideline estimates of plant performance have become "absolute" rules in NPDES permit negotiations, causing hardship to many industries. In one specific case where the Effluent Guidelines were strictly followed, a plant was required to limit its net discharge of a particular pollutant to 0.5 ppm, when the analytical accuracy of the chemical testing method employed was about 0.25 ppm.

To comply with existing and proposed permit limitations, industry needs to: (1) determine and state the limits on its ability to measure effluent data; (2) use these limits to set internal performance standards for pollution control; and (3) attempt to gain recognition of the magnitude of measurement uncertainties in current and future permit negotiations.

Part of a typical NPDES permit for a chemical plant is shown in Fig. 1. It describes effluent characteristics, discharge limitations and monitoring requirements.

There are two discharge limits: the "daily average," defined as the daily average for the month (based upon a specified number of samples per month); and the daily maximum. Monitoring requirements specify the frequency of sampling and the type of sample.

An example demonstrating the magnitude of uncertainty associated with measuring the discharge of a

1. During the period beginning . . . . and lasting until . . . . , the permittee is authorized to discharge from . . . .

Such discharges shall be limited and monitored by the permittee as specified below:

Effluent characteristic	Discharge limitations kg/d or lb/d		Monitoring requirements	
	Daily avg	Daily max	Measurement frequency	Sample type
Flow-million gal/d	—	—	daily	continuous
Total suspended solids	—	—	1/wk	24-h composite
Ammonia (as N)	—	—	1/wk	24-h composite
Organic nitrogen	—	—	1/wk	24-h composite
Oil-grease	—	—	1/wk	grab
Temperature	—	—	daily max.	continuous
BODs	—	—	1/wk	24-h composite

2. The pH shall not be less than 6.0 nor greater than 9.0 and shall be monitored daily by means of a grab sample.
3. There shall be no discharge of floating solids or visible foam in other than trace amounts.
4. Samples taken in compliance with the monitoring requirements specified above shall be taken at the following location(s):

Typical NPDES effluent limitations and monitoring requirements　　　　　　　　**Fig. 1**

Originally published October 9, 1978.

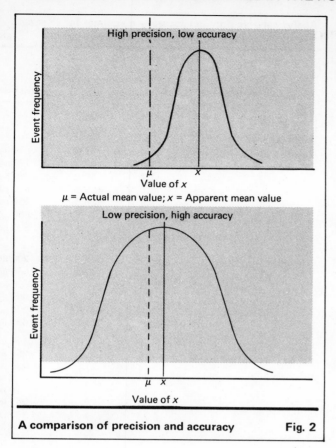

High precision, low accuracy

μ = Actual mean value; x = Apparent mean value

Low precision, high accuracy

A comparison of precision and accuracy   Fig. 2

Nomenclature

Var (A)	analytical variance
Var (B)	between-day variance
Var (W)	within-day variance
C	concentration, mg/L
D	discharge, lb/d
K	constant, lb/mg/million gal/L
Q	flowrate, million gal/d
S	standard deviation
e	number of days of sampling/mo
g	number of analyses on individual day
m	d/mo
n	analyses/sample
r	samples/d
u	subsamples/d
v	number of days subsamples are collected
x	number of days samples are collected

pollutant will follow a brief introduction to the required statistical procedures.

Precision and accuracy

Precision is the ability to exactly duplicate the results of a single measurement. Accuracy is the ability to determine the exact or true value. The difference between the two is illustrated in Fig. 2. The precision and accuracy of various analytical instruments and test methods vary widely, depending on factors that are beyond the analyst's control.

Calculation of variance

Variance is the amount of spread or variation in a set of measurements. Small variance indicates precision in data measurement.

Mathematically, variance is the average of the sum of the squares of the differences from the average value in a set of observations.

Let the variance of X be denoted by Var (X), and let the covariance between two values Y and Z be denoted by Covar (Y, Z). For,

$$X = Y - Z \qquad (1)$$

$$\text{Var } (X) = \text{Var } (Y) + \text{Var } (Z) - 2 \text{ Covar } (Y, Z) \qquad (2)$$

If Y and Z are independent, Covar $(Y, Z) = 0$ [1]. Now, if

$$X = (YZ)$$

and Y and Z are independent, then:

$$\text{Var } (X) = \text{Var } (YZ)$$

This can be approximated by a Taylor Series expansion of X as:

$$\text{Var } (X) = \left(\frac{dX}{dY}\right)^2_{\mu Z} \text{Var } (Y) + \left(\frac{dX}{dZ}\right)^2_{\mu Y} \text{Var } (Z) \qquad (3)$$

where μ refers to the mean value of the variable given in the subscript.

Net basis calculations

Most NPDES permits are based on net values, an allowance over influent values. Representing Q as flowrate, C as concentration of a particular pollutant, and E and I as effluent and influent values, then the net discharge value, D_{NET}, of the pollutant is:

$$D_{\text{NET}} = Q_E C_E - Q_I C_I \qquad (4)$$

If $\bar{X}_K$ and S^2_K denote the estimates of the mean and variance of variable K, then S^2_D, the estimate of the variance of D_{NET}, can be expressed by combining Eq. (2) and (3) as:

$$\text{Var } (D_{\text{NET}}) = S^2_D = \bar{X}^2_{C_E} \cdot S^2_{Q_E} + \bar{X}^2_{Q_E} \cdot S^2_{C_E} + \\ \bar{X}^2_{C_I} \cdot S^2_{Q_I} + \bar{X}^2_{Q_I} \cdot S^2_{C_I} \qquad (5)$$

All covariances are assumed to be equal to zero.

S_D, the square root of the variance, is an estimate of the standard deviation of D_{NET}.

Determination of the daily average

Now let us determine the magnitude of uncertainty in the calculation of the daily average.

Typically, the total daily variance is made up of three components: (1) analytical variance, (2) within-day variance, and (3) between-day variance.

A composite sample of r sub-samples/d taken at fixed time intervals is collected for analysis.

The stream being sampled is a once-through cooling water, with a relatively constant flowrate. A small amount of iron from plant production enters the cooling water from leaking heat exchangers and other minor sources. The feed water also contains some iron. The discharge permit is based on the *net* contribution of iron, as may be the case for a plant that has quantities of a priority pollutant in its intake or raw water.

Since the plant's cooling-water losses are very small with respect to the flow, the net discharge, D, is equal to the flowrate, Q, times a constant, K, times the difference in concentration, C.

$$D = KQ(C_E - C_I) \qquad (6)$$

If Var (A) is the analytical variance, Var (W) the variance within the day, and Var (B) the variance between the days, then the variance in D due to differences in measuring concentration, Var (C), is:

$$\text{Var }(C) = \frac{1}{2}\text{ Var }(A) + \text{Var }(W) + \frac{(m-e)}{m}\text{Var }(B) \quad (7)$$

The analytical variance is halved because each sample is analyzed twice, and the multiplier on the between-day variance term is a correction for e days of sampling/mo of m days [5].

If a continuous, exactly flow-proportional, sampler were employed, the within-day variance, Var (W), would be deleted. However, if the sampler is not truly flow-proportional, the term must be retained, and the variance calculated.

The variance between the days, Var (B), is deleted for calculation of a particular-day estimate of variance, but must be retained for calculation of the variance of any nonparticular day [2].

Once Var (C_E) and Var (C_I) are calculated, and the variance in D due to inaccuracies in flow measurement, Var (D_Q), has been determined (as shown later), the total variance in D, Var (D_T), can be expressed as:

$$\text{Var }(D_T) = S_D^2$$
$$= K^2\, Q^2\, [\text{Var }(C_E) + \text{Var }(C_I)] + K^2\, [\text{Var }(Q) \cdot (\bar{C}_E^2 + \bar{C}_I^2)] \quad (8)$$

This is similar to Eq. (5), where $\bar{C}_E$ and $\bar{C}_I$ represent average values for the respective variables.

The standard deviation in the daily average is:

$$S_D \text{ avg.} = \sqrt{\frac{\text{Var }(D_T)}{e}} \qquad (9)$$

Analytical variance—Var (A)

For a sample taken on day j, multiple analyses are made to determine the concentration C_{ji}, where i is the ith sample on day j. If n analyses are made on the sample, then:

$$\bar{C}_j = \frac{\sum\limits_{i=1}^{n} C_{ji}}{n} = \frac{C_{ji} + C_{j2} + \cdots C_{jn}}{n} \qquad (10)$$

If samples are collected for x days, then:

$$\text{Var }(A) = \frac{\sum\limits_{j=1}^{x}\sum\limits_{i=1}^{n}}{x}(C_{ji} - \bar{C}_j)^2 \qquad (11)$$

This is the estimate of analytical variance.

Estimates of the standard deviation associated with

particular analysis methods have been published in "Standard Methods" [3], and "Methods for Chemical Analysis of Water and Wastes" [2]. Accuracy and precision data for some water and wastewater methods are shown in Table I.

If the tabular value for relative precision is P_M, and the average value of the measured parameter from analysis is $\bar{X}_M$, then the analytical variance is:

$$\text{Var }(A) = \left(\frac{P_M}{100\ \bar{X}_M}\right)^2 \qquad (12)$$

Analytical precision and accuracy guidelines for selected water and wastewater parameter concentrations [3]   Table I

Parameter and method	Relative precision, % (P_M)	Relative accuracy, %	Concentration
Hardness—EDTA	2.9	99.2	
Metals—atomic absorption			
Cadmium	21.6	91.8	50 µg/L
Chromium	47.8	83.7	110 µg/L
Copper	11.2	96.6	1,000 µg/L
Iron	16.5	99.4	300 µg/L
Lead	23.5	84.0	50 µg/L
Silver	17.5	89.4	550 µg/L
Zinc	8.2	99.6	500 µg/L
Nitrogen			
Ammonia (Nessler)	46.3	89.8	200 µg/L
	18.0	96.0	1,500 µg/L
Nitrate (Brucine)	66.7	92.4	50 µg/L
	5.5	94.0	1,000 µg/L
Organic (Nessler)	94.8	45.0	200 µg/L
	43.1	90.7	1,500 µg/L
Oxygen demand—biochemical	15	N/A	175 mg/L
—chemical	6.5	N/A	200 mg/L
Residue—total	52	N/A	
—suspended	33		15 mg/L
	10		242 mg/L
	0.76		1,707 mg/L

Iron concentration data for 2-day sampling period   Table II

Sample number	Concentration (mg/L)			
	1st day influent	1st day effluent	2nd day influent	2nd day effluent
1	0.40 / 0.25 >0.32	0.35 / 0.38 >0.37	0.22 / 0.25 >0.23	0.60 / 0.52 >0.56
2	0.25 / 0.30 >0.27	0.84 / 0.91 >0.88	0.30 / 0.36 >0.33	1.34 / 1.30 >1.32
3	0.34 / 0.38 >0.36	0.56 / 0.62 >0.59	0.34 / 0.35 >0.35	1.00 / 1.10 >1.05
4	0.45 / 0.44 >0.45	0.77 / 0.91 >0.84	2.60 / 2.70 >2.65	4.45 / 4.40 >4.42
5	0.30 / 0.36 >0.33	0.52 / 0.53 >0.53	1.00 / 1.00 >1.00	0.45 / 0.49 >0.47
6	0.56 / 0.56 >0.56	0.57 / 0.57 >0.57	0.88 / 0.88 >0.88	0.46 / 0.42 >0.44

$C_I = 0.644$ mg/L, $C_E = 1.003$ mg/L, $n = 2$

Summary of calculation of uncertainty for daily average   Table III

Step 1 — Calculate from influent and/or effluent data (Table II) Var (A), using Eq. (11): Var (A) = 0.002

Step 2 — Calculate from influent and effluent data:
W_1 and Var (W) using Eq. (13) (14) and (15)
W_1 influent = 0.4201 Var (W_1) = 0.0348
W_1 effluent = 1.1959 Var (W_E) = 0.0995

Step 3 — Calculate between-day variances from Eq. (16) and (17) (data base for calculations not shown)
B_1 influent = 0.1374 Var (B_1) = 0.0329
B effluent = 0.2820 Var (B_E) = 0.0405

Step 4 — Calculate Var (C_1) and Var (C_E) from Eq. (7) for appropriate sampling frequencies

Samples/mo, (e)	Var (C_1)	Var (C_E)	Var (C_1) + Var (C_E)
1	0.0676	0.1397	0.2073
4	0.0679	0.1360	0.2039
8	0.0647	0.1316	0.1963

Step 5 — Calculate standard deviation in the daily average from Eq. (8) and (9), where Q = 4.46, Var (Q) = 0.1563 and K = 8.34

$e = 1$ $S_D = [8.34^2 (4.46^2 (0.2073) + 0.1563 (0.644^2 + 1.003^2))]^{1/2}$ = 17.38
$e = 4$ $S_D = [8.34^2 (4.46^2 (0.2039) + 0.1563 (1.4208))/4]^{1/2}$ = 8.62
$e = 8$ $S_D = [8.34^2 (4.46^2 (0.1936) + 0.1563 (1.4208))/8]^{1/2}$ = 5.99

Step 6 — Calculate confidence intervals for the daily average:

e	S	Degrees of freedom	Student's t	95% confidence interval
1	17.38	—	—	34.76*
4	8.62	3	3.182	27.43
8	5.99	7	2.365	14.17

*Assuming normal distribution

Variation within the day—Var (W)

Our sampler takes r individual (grab) samples/d and collects them as a single composite sample. To obtain an estimate of the within-day variance, Var (W), the single composite must be divided into a number of subsamples. To obtain a final result, two intermediate statistics, W_1 and W_2, must be calculated.

Suppose we collect u subsamples per day for v days, and C_{ab} is the bth subsample collected on the ath day, then $\bar{C}_a$ is calculated according to Eq. (10), and

$$W_1 = \frac{\sum_{a=1}^{v} \sum_{b=1}^{u} (C_{ab} - \bar{C}_a)^2}{v(u-1^*)} \tag{13}$$

*Degrees of freedom

W_1 contains Var (A), which must be removed. Therefore, W_2 is defined as:

$$W_2 = \frac{(W_1 - \text{Var}(A))}{n} \tag{14}$$

Finally, since a 24-h composite sample is composed of r individual samples, and since we have taken r/u individual samples for each subsample, then:

$$\text{Var}(W) = \frac{W_2}{u} \tag{15}$$

Variation between days—Var (B)

If the average concentration of a particular pollutant on an individual day is C_i, and $\bar{C}$ is the mean value of g

---

Calculation of uncertainty for influent and effluent daily values for a single day   Table IV

Using previously developed data from the first day of sampling, estimate the standard deviation of the discharge.

C_1 = 0.383 mg/L for the first day of sampling
C_E = 0.638 mg/L for the first day of sampling
Var (C_1) = Var (W_1) + Var (A)/2 = 0.0358
Var (C_E) = Var (W_E) + Var (A)/2 = 0.1005

where:
Q = 4.46 million gal/d, and Var (Q) = 0.1563,
$S^2 = 8.34^2 (4.46^2 (0.0358 + 0.1005) + 0.1563$
$(0.383^2 + 0.638^2))$
S^2 = 194.60,
S = 13.95 lb/d

Assuming a normal probability distribution, the true mean net discharge for this particular day has a 95% probability of being in the range of 9.49 lb/d ± 2 (13.95), or between −18.41 and +37.39 lb/d.

analysis determinations on individual days, then B_1, the variance of the concentration between those days, is expressed as:

$$B_1 = \frac{\sum_{i=1}^{g}(C_i - \bar{C})^2}{g - 1^*} \qquad (16)$$

*Degrees of freedom

B_1 contains analytical and within-day variances, which must be removed, so Var (B) is:

$$\text{Var } (B) = \frac{B_1 - \text{Var } (A) - n \text{ Var } (W)}{n} \qquad (17)$$

Variance of flow measurement—Var (D_Q)

A carefully installed flowmeter may have an inaccuracy of about 1.5% of the flow. In an open channel, the inaccuracy may approach 20%. The most frequently overlooked source of open-channel flowmetering error is failure to observe energy and boundary conditions for the installation. Such oversights can greatly change the head-discharge relationship upon which most open-channel flow measurement is based.

A recent example points this out: A surface-level measuring device that relied on a weighted bobbing line to determine depth of flow was installed in an open channel. A theoretical calibration curve, relating depth of flow to discharge, was computed. However, the upstream flow conditions, energy relationships and water velocity in the channel were not considered when installing the meter.

When the flowmeter was found incapable of measuring less than a 25% change in flow, conditions of installation were examined. A steep upstream slope permitted the channel water velocity to reach 20 ft/s. At that speed, the depth-discharge relationship for the channel was found substantially different from the theoretical calibration curve.

The following is a calculation of Var (D_Q):

A particular flowmeter installation on a 90°V notch weir has a head discharge relationship of [4]:

$$Q = 1.62 \, H^{2.5} \qquad (18)$$

where Q is in million gal/d, and H is in ft. The uncertainty in the reported discharge, Var (D_Q), is actually a combination of the uncertainty in measurement of H and the precision of the flowmetering system, and is expressed as:

$$\text{Var } (D_Q) = \left(\frac{dQ}{dH}\right)^2 S_H^2 + \text{Var (System)} \qquad (19)$$

When $H = 1.5$, $S_H = \frac{5}{8}$ in. (0.05 ft) and $Q = 4.46$ million gal/d, then:

$$\text{Var } (D_Q) = 1.62^2 (2.5 \, H^{1.5})^2 (0.05)^2 + \text{Var (System)}$$

The variance in the system is generally stated in terms of the precision of mechanical and electrical components, which in this case is estimated as 3% of the flow. Thus:

$$\text{Var } (D_Q) = 0.1384 + [(0.03)(4.46)]^2 = 0.1563 \qquad (20)$$

$$\text{and: } S_Q = [\text{Var } (D_Q)]^{1/2} = 0.395 \text{ or about 8.9\%}$$

The estimate of the standard deviation in the measurement of Q is 0.395 million gal/d.

Practical example of analysis of variance

Table II shows concentration data for a two-day sampling period. Table III summarizes the calculation of uncertainty for the daily average, based on the data in Table II and the example just given.

If the NPDES permit-limit daily average is 30 lb/d, and compliance monitoring is done once per month, the standard deviation is ±17.38 lb/d, more than half the daily average. The 95% confidence interval for the daily average is 30 ± 34.8 lb/d, assuming normal distribution.

When monitoring is done more frequently, the uncertainty in the average decreases. At 4 samples/mo, the standard deviation is ±8.62 lb/d; at 8 samples/mo, it is 5.99 lb/d. The 95% confidence intervals for 4 and 8 samples/mo are 30 ± 27.43 and 30 ± 14.17 lb/d, respectively.

For our example, the amount of iron discharged during a particular day (a number required by the NPDES permit daily maximum) has a measurement variance of 13.95 lb/d, as calculated in Table IV. For a net discharge of 9.49 lb/d as calculated from Eq. (6), the 95% confidence interval is 9.49 ± 27.9 lb/d, or between −18.41 and 37.29 lb/d. The negative value is possible because this is a net discharge calculation.

References

1. Ostle, B. and Mensing, R. W., "Statistics In Research," Iowa State Univ. Press, Ames, Iowa, 1975.
2. Methods for Chemical Analysis of Water and Wastes, EPA, USEPA-1625/6-74-003a, 1974.
3. "Standard Methods," 14th ed., American Public Health Assn., New York, 1975.
4. Chow, V. T., "Open Channel Hydraulics," McGraw-Hill, New York, 1959.
5. Cochrane, W. G., "Sampling Techniques," 3rd ed., John Wiley & Sons, New York, 1977.

The authors

David L. Russell is a senior environmental engineer for the Specialty Chemicals Div. of Allied Chemical Corp. at B and A Works, Wilmington Turnpike, Marcus Hook, PA 19061. Prior to joining Allied, he was a private consultant for nine years. He received a B.S. in Civil Engineering from the University of Illinois (Urbana) and holds an M.S. in Environmental Engineering from West Virginia University. Mr. Russell is a registered engineer in Ill. and Pa.

James J. Tiede is a senior biostatistician for Hazelton Laboratories America, Inc., 9200 Leesburg Turnpike, Vienna, VA 22180. At the time this article was written, he was a statistical scientist for the Specialty Chemicals Division of Allied Chemicals Corp., Buffalo, N.Y. Dr. Tiede received his Ph.D. in statistical science from State University of New York at Buffalo and a B.S. in mathematics from St. John Fisher College, Rochester, N.Y.

Optical instruments monitor liquid-borne solids

By passing a beam of light through a process or effluent stream and measuring changes in the liquid's optical properties, these instruments can detect the count and size distribution of suspended particles.

Alvin Lieberman, Royco Instruments, Inc.

☐ Various optical instruments have been developed for measuring the number and size of particles suspended in various fluids. Such instruments have been devised for analyzing both gas and liquid streams; however, this article will be limited to those units used to determine the characteristics of particles suspended in liquids. Since many design variations exist, this discussion will not attempt to cover all of them; instead, attention will be given to several optical devices that illustrate general principles.

Optical instruments offer several advantages. They can produce real-time data; they can handle a broad range of liquids; they can operate online (within specified concentration limits); they can measure over a fairly wide dynamic size range; they can gather a large amount of reproducible data quickly; and they can be operated by relatively unskilled personnel.

However, certain limitations on optical instruments should also be kept in mind. They take size data based on optical properties that must be referred to a calibration base; they yield a size distribution based on particle population, rather than particle mass; their concentration restrictions may necessitate sample dilution; and they are subject to bias if air bubbles or immiscible liquids are present.

How they work

Commercially available optical instruments fall into two general classes. In the first, an assemblage of particles is scanned to reveal information on overall concentration. With more-sophisticated instruments of this type, spatial filtering of scattered light permits determination of mean diameter and variance of particle size distributions.

In the second class of instruments, the optical system views a volume so small that individual particles in a flow are observed and each one provides an indication of size. Both concentration and size data can be derived, since the instrument reports the number of particles over several size ranges measured in a given volume of liquid.

Originally published December 18, 1978.

Photometers

The operating principle of these devices is quite simple. A light beam is directed through the sample. The total amount of light scattered at specific angles, or absorbed by the opaque particles, can be related directly to the quantity of particulate material in the liquid. Empirical calibration to a turbidity unit scale is normally used to define the quantity of suspended solids, although the degree of scattering is dependent on the size, shape and concentration of the particles.

Online photometric analyzers are used primarily to measure turbidity and low concentrations of solids in process liquids or water lines. Back-scattering systems are used to determine concentrations in slurries con-

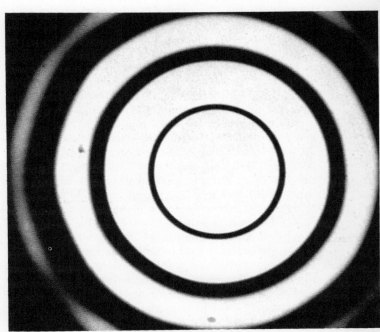

Laser beam diffracted by suspended particles produces typical Fraunhofer pattern

taining 10–15% solids. Angular scattering systems can be used to monitor solids contents up to 50,000 ppm. Absorption analyzers, which handle lower concentrations (to approximately 1,000 ppm) are used for monitoring water treatment systems and process lines.

Depolarization measurements

Under development by the Precision Products Div. of Badger Meter, Inc. (Tulsa, Okla.) is a method for determining suspended particle concentration, based on the depolarization of scattered light. A depolarization ratio is determined by measuring the intensity of scattered light passing through a polarizing prism. Measurements are made along the optical axes both perpendicular and parallel to the plane of incident polarized light. The ratio of intensities represents the depolarization ratio.

This ratio can be related to concentration, since multiple scattering increases with the number of particles present. Primary scattered light becomes increasingly polarized with multiple scattering.

Reports indicate that this method gives good response to solids concentrations in waste liquids from a few ppm to 5,000 ppm, with no undue bias from particle size or dissolved materials. Solids buildup on the optical windows is not believed to be a problem.

Fraunhofer diffraction analyzers

The French company Industrielle des Lasers Cilas (Marcoussis) offers a particle-size analyzer (Model 227) that measures the intensity of Fraunhofer diffraction patterns produced by passing He-Ne laser light through a suspension. Although the pattern diffracted through polydisperse material is monotonic, smaller particles diffract the laser beam over a larger angle than do larger particles. The overall energy level at any point varies with concentration.

Diffracted light passes through a rotating mask, whose window diameters increase toward the periphery. A photodetector then determines the levels of light energies at various radii from the laser beam's axis. From this information, the particle distribution can be deduced using a preprogrammed matrix calculation.

Size measurements can be made from 1 to 128 microns. Cumulative weights are calculated at 1, 2, 4, 8, 16, 32, 64 and 128 microns. Data reported by the manufacturer indicate good results with cement and alumina samples. Sample concentrations should be in the range of 0.1 to 1 g in 500 mL of liquid. Measurement time of approximately 5 min is required.

The Leeds & Northrup Co. (North Wales, Pa.) has developed a line of low-angle light-scattering instruments (Microtrac) that operate on the principle of Fraunhofer diffraction, using laser light sources, optical filters, and microprocessors. These instruments measure particles in either liquid or air suspension, and may also be used online for process monitoring.

Although the intensity of the total light scattered (photo) by particles is proportional to the square of their diameters, the angle of diffraction is inversely proportional to their diameters. By placing a specially shaped spatial filter in the diffraction plane of the collecting lens, one can relate light fluxes to scattering angles; these fluxes are related to particle volume.

The spatial filter is rotated to extract signals serially. (Collected light can be related to the second, third or fourth power of the particle diameter.) Once these signals are detected and converted to digital form, a microprocessor computes the distribution of particle volume and surface area. Output data are presented in the form of the particle volume in each of 13 size bands. Information is also given on the percentage of particles passing each size interval, as well as their mean diameters, surface areas and relative total volume. Measurements are made over two size ranges: 1.9 to 176 microns and 3:3 to 300 microns.

The product line has recently been extended to include a low-cost monitor that provides a continuous measure of solids concentration in wastewater and other industrial streams. This instrument measures true volumetric concentration, which can be converted to mass concentration with a one-time, single-point gravimetric calibration.

Similar instruments are being used in refining operations and chemical processes, including the manufacture of powdered resins, catalysts and a variety of powdered and slurried chemicals.

Lab methods detect particle groups

This brief summary of some operating laboratory methods is restricted to those used to characterize particle assemblages in liquids. Holographic analyzers and automated image analyzers are treated later as single-particle devices.

1. Bagchi and Vold [1] describe a method for determining the average particle size of coarse suspensions from measurements of apparent specific turbidity. They show that the turbidity is inversely proportional to the average radius from about 12 to 50 μm for materials whose turbidity is independent of the wavelengths of light. The method has been used with materials in concentrations of 1–10 g/L.

2. Dobbins and Jizmagian [2] describe two methods for finding mean size by measurement of optical cross-section. In the first method, when concentration is known, the relationship between mean scattering cross-section and volume-surface mean diameter is fixed. In the second method, both the volume-surface mean diameter and concentration can be measured if transmittances are measured at two wavelengths. Data were obtained for mean diameters ranging from 1–5 μm.

3. Groves, Yalabik and Tempel [3] discuss the operation of a centrifugal photosedimentometer using laser light. A transparent hollow disk is rotated, causing particles to settle under the influence of centrifugal force. The relationship between sedimentation time and optical density at any point along the radius is a characteristic of the particle size distribution.

A data logger is used to record light-transmission level, location along the radius, and settling time. This enables computer treatment of information to yield relatively rapid descriptions of particle size data. The method can size particles from 5–10 μm to less than 0.1 μm, and can handle sample concentrations up to one percent.

4. Jordan, Fryer and Hemmen [4] use a hydro-

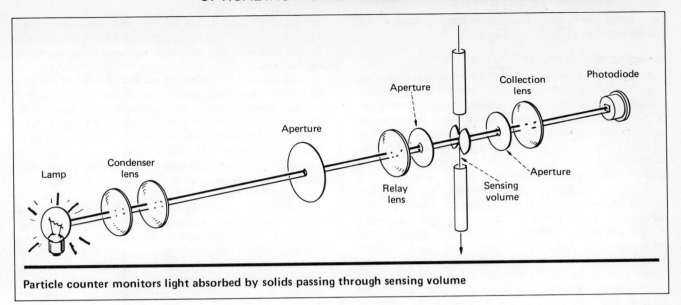

Particle counter monitors light absorbed by solids passing through sensing volume

photometer to find particle distribution in the size range from 2–50 μm via a sedimentation method. Optical density at a fixed point in a gravity settling column is measured as a function of time and is correlated to mass loading.

5. Robillard, Patitsas and Kaye [5] describe the use of Mie scattering measurements at two wavelengths to determine the mean diameter and refractive index of particulate material. The method requires comparison of experimental data with computer-generated curves derived from Mie theory, followed by selection of best-fit parameters. Size distribution is inferred from the intensity ratio of the scattering extrema. This work follows and extends earlier work by T. P. Wallace [14]. It is not applicable to multimodal distribution.

Nonimaging, single-particle counters

Among these are three optical single-particle counters commercially available in the U.S. All pass a liquid sample through an illuminated sensing volume; the amount of light detected depends on the degree of absorption or scattering as the particles traverse the sensing volume. The sensing volume is small enough so that single particles are detected, even when they occur in relatively high concentrations. From the amplitude of the light pulse, particle size information can be derived. By sorting and counting the light pulses, the instrument can ascertain the distribution and concentration of particles.

Climet Instrument Co. (Redlands, Calif.) manufactures a liquid particle analyzer (Model Cl 220), which detects the white light scattered by individual particles as they traverse the sensing zone. Scattered light is collected at scattering angles from 15 to 105 deg. Suspended particles—sized in terms of their equivalent optical diameter as referred to latex spheres—are detected and can be resolved between 2 and 200 μm. A sample flowrate of up to 500mL/min can be used, and by varying the ratio of sample fluid to filtered recirculating fluid, one can measure particle populations between 1 and 10^8 per milliliter.

HIAC Instruments Div. of Pacific Scientific Co. (Montclair, Calif.) manufactures a design that observes a telecentric white-light beam and detects reduction in light level caused by particle passage. The firm makes a series of sensors that can detect particles from 1 μm to a maximum of 9 mm. Each sensor has a dynamic range reported to be 60:1, based on the smallest detectable particle (as referred to an equivalent optical diameter for a latex sphere) and the minimum dimension of the internal passageway. The smallest sensor handles particles 1 to 60 μm, in concentrations up to some 30,000/mL at a flowrate of 4–6 mL/min. The largest one handles particles to 9 mm in concentrations up to 0.2/mL at a flowrate of 100 L/min.

Royco Instruments, Inc. (Menlo Park, Calif.) manufactures two instrument packages. One observes a focused beam and detects reduction in white-light level caused by particle passage. Two flow cells are available with this package, both of which can detect particles of less than 2 μm with a dynamic range of at least 100 to 1. Sizing is based on calibration with either latex spheres or the AC Fine-Test Dust [15]. The smallest cell has a concentration capability of 12,000/mL at flowrates to 35 mL/min; the larger cell can handle concentrations of 3,000/mL at flowrates to 150 mL/min.

The second package detects near-forward scattering produced by particles passing through a laser beam. The same flow cells can be used with this package, permitting identical concentration and flowrate capabilities. Minimum size sensitivity is 0.5 μm, and a dynamic range of 50:1 is achieved.

Spectrex Corp. (Redwood City, Calif.) markets a device that uses light scattered from a laser beam to detect particles larger than approximately 5 μm to 100 μm in concentrations to 1,000/mL within a glass container. The laser beam can scan a volume of 10 cc in 15 s. Since out-of-focus particles and the bottle walls are not detected by the collection optics, only particles within the in-focus point for both the illumination and collection optics are reported.

A microprocessor attachment automates the complete size analysis, prints and plots size and mass distribution, and provides a settling scan.

Imaging single-particle counters

Image analysis systems. Many manufacturers produce image analysis systems. These are basically video-camera devices that are focused on a microscope stage or on a photograph. Signals from the video tube are transmitted to a computer processor for data reduction. A wide range of information can be developed by the computer, including length and area size data, shape factors, and statistical analyses.

Differential scattering systems. Science Spectrum, Inc. (Santa Barbara, Calif.) has developed instruments that can locate a single particle and retain it electrostatically in position so that differential scattering intensity can be recorded as a function of scattering angle. In this way, particle composition and shape information can be retrieved by matching measured data with those in a computer file. Particles as small as 1 to 2 μm can be studied, but the measurement time is on the order of seconds per particle. Data produced include average size, size distribution, shape, and other particle suspension characteristics.

Lab methods for analyzing single particles

Numerous techniques for characterizing single particles have been reported in the literature. Included among these are:

1. Bartholdi et al. [6] describe a scanning system in which a particle is illuminated by an argon-ion laser beam as it passes through a flow chamber. Scattering between 7.5 and 21.5 deg is measured by a 128-element photodiode array. Measurement is initiated by a delay signal once the particle has passed through a Coulter orifice. Approximately 500 microseconds per particle are required to record a complete scatter pattern for particles larger than approximately 10 μm.

2. Breitmeyer and Sambandam [7] detail an inline holography analysis, where holograms are made of blood cells in saline solutions to determine orientation and cell shape. An argon-ion laser was used and photographs were observed for data reduction.

3. Eisert et al. [8] discuss a microphotometer that uses a flow-through system for orienting nonspherical cells along the flowaxis. A focused laser beam at the axis of the hydrodynamically focused flowpath is arranged so that its diameter is smaller than that of the cell. If constant flow velocity is maintained, the pulse width will indicate the cell length in the flow system. Sizing from approximately 5 μm to 300-μm lengths at rates up to 50,000/s is reported.

4. Kaye [9] describes the use of a low-angle laser-light-scattering instrument. Scattered light is measured at angles as low as 1.5 deg from volumes of approximately 10 microliters, illuminated with a 5-mW He-Ne laser. Particles as small as 0.1 μm can be detected. The sample flowrate is extremely low, but concentration capabilities are high.

5. Mullaney et al. [10] have employed a forward angle photometer. Hydrodynamic focusing maintains a sample stream of 50-μm dia. that is illuminated by a He-Ne laser beam of 100-μm dia. Light scattered at between 0.5 and 2 deg is collected. Sensitivity from approximately 3 μm up to 20 μm has been obtained with a flowrate of 1 mL/min at concentrations up to 50,000/mL.

6. Ricci and Cooper [11] outline the use of a flying spot laser that scans a flowing slurry stream with a two-dimensional raster scan. Particles interrupt the focused beam that would otherwise illuminate a photodiode. Particles from 5 μm to 1,000 μm are measured in concentrations up to 2%, in streams flowing at rates up to 5 gal/min. The laser beam, reduced to 10-μm dia., scans a 3-cm distance at a frequency of 500 Hz and a velocity of 50 m/s.

7. Salzman et al. [12] describe a flow system, multi-angle light-scattering instrument. Particles are characterized by their light-scatter patterns. A He-Ne laser is focused at the center of the sample stream. As each particle is illuminated by the beam, a 250-μsec pulse of scattered light is collected by a 32-photodetector array laid out in a concentric pattern that encompasses scattering angles from 0.3 to 20 deg. The scatter pattern is then stored in a computer for analysis by a mathematical clustering algorithm.

8. Uzgiris and Kaplan [13] describe application of a laser Doppler velocimeter to determination of electrophoretic mobility distributions. A standard LDV system was used to observe the frequency shift in scattered laser light due to applied voltage on the suspension of particles in a sample. Mobilities ranging from 1–5 μm/s/V/cm were recorded.

Material handling requirements

Batch systems. Since essentially all of the optical devices use flow systems, it is necessary to transfer material from a container through the instrument. One may assume that the necessary preliminary steps of adequate particle dispersion and uniform mixing have been completed. Since the instruments measure particles as local nonhomogeneities in the liquid, it is necessary to avoid artifact introduction. This requires elimination of bubbles, inclusions of immiscible liquids, and contamination from airborne dust or surface debris on the container. In addition, if concentration data are required, then the need for careful volumetric control is obvious.

For the single-particle counting devices, if the maximum recommended concentration is exceeded, coincidence errors can occur. In this case, it may be necessary to dilute the sample. If dilution is required, then aside from the normal care in maintaining good volumetric control, dilution ratio selection must be chosen high enough to avoid coincidence errors, but not so high as to reduce concentration of large particles in typical distributions to a level that does not permit accumulation of statistically significant data.

If a suitable dilution ratio cannot be found for all size ranges, then it may be necessary to measure small particles at a high dilution ratio, and large particles at a low ratio. For the most part, dilution is a manual procedure; however, there are many commercially available automatic dilutors manufactured for clinical chemistry procedures. These typically dilute by a factor ranging from 10 to 500, with high precision and accuracy. Diluted samples of 10–20 ml are typically produced; the concentrated material is aspirated in microliter quantities and then dispensed along with clean diluent in milliliter quantities.

Inline systems. If an optical particle counter is connected inline, then the problems of solids dispersion and mixing, sedimentation during sample storage, careless handling, etc., associated with container preparations are eliminated. However, a number of other problems are present. First, it is necessary that sample stream flowrate be controlled; an isokinetic sample probe may be required; a means of ensuring gas bubble elimination may be needed; an inline dilution system may be required.

Sample stream flowrates can be controlled by using pressure-compensated flow control valves with suitable protection to avoid orifice clogging, or by using a metering pump; both are downstream of the particle counter. Bubble control can be attained to some extent either by maintaining system pressure so high as to ensure that all gases remain in solution in the liquid or by incorporating a low-pressure vacuum deaerator immediately upstream of the particle counter inlet.

At present, no commercially available inline dilution system is available for dispersion sampling. Some one-of-a-kind developments have been made. These are basically modifications of batch dilutors. Clean diluent is pumped at a fixed flowrate into a mixing chamber, where a smaller sample flow is mixed with it. The diluted sample is then fed to the particle chamber. These systems are limited by the available quantity of clean diluent, particle losses in the mixing chamber, and flow measurement stability.

Some preliminary work has been reported on fixed-ratio sample-diluent flow systems feeding into motionless mixers for minimum particle loss before presentation to the optical instrument. Clean diluent is supplied by recirculating the diluted stream through a suitable filter.

Data development

Data production. Optical instruments have a number of common features in terms of the data they produce. In each device, the measurements are dependent upon the optical properties of the particulate material being different from those of the substrate; the primary data output is related to a projected area for the particles; a truncated size measurement is always made, with the low end varying with instrument sensitivity. These comments apply to both multiple- and single-particle analyzers in which size information is produced.

Multiple-particle optical instruments will usually produce data describing a particle-size number distribution. With judicious data-processing circuitry, either differential or cumulative distributions can be defined. Single-particle optical instruments will produce data based on the number of particles in several size ranges.

Depending on the resolution of the instrument, the width of each range can be made small enough so that essentially all particles in the range can be represented as having the mean diameter of the range. In this way, mathematical manipulation can be performed to convert from a number base to a volume or an area base with minimum error.

Data processing. The conversion from number base to volume or area base includes certain assumptions that must be kept in mind if comparison with data obtained by other means is desired. First, the size description is based on calibration to an idealized basis; next, a limited range of shape factors is assumed; next, uniform particle composition in the sample is assumed. Once these assumptions are accepted, conversion from optical diameters to aerodynamic diameters or sieve data can be accomplished for a range of materials. In some cases, empirical calibration may be required.

All of the optical instruments produce data that can be easily transmitted to a computer for storage and/or processing. The data can be a series of d.c. signal levels, or parallel or serial output digital pulses. Part or all of the data processing system can be included with the instrument.

Data can be provided that indicate cumulative or differential particle-size distribution curves, mean diameters with standard deviations, total quantity of particles per unit volume of liquid, and number of particles in several particle size ranges. With this easily acquired and easily processed data, application of the instrument output through a computer or microprocessor to a control function for process lines can be accomplished with relative ease. It is necessary to select a particular size parameter or ratio of sizes as the dependent variable in a control function, define an optimum operating range, and choose the process parameter that controls the variable of concern.

References

1. Bagchi, P., and Vold, R.D., *J. Colloid Science,* Vol. 53, No. 2, 1975, p. 194.
2. Dobbins, R.A., and Jizmagian, G.S., *J. Opt. Soc. of America,* Vol. 56, No. 10, 1966, p. 1,351.
3. Groves, M.J., Yalabik, H., and Tempel, J.A., *Powder Technology,* Vol. 11, No. 3, 1975, p. 245.
4. Jordan, C.F., Fryer, G.E., and Hemmen, E.H., *J. Sed. Petrology,* Vol. 41, No. 2, 1971, p. 489.
5. Robillard, P., Patitsas, A.J., and Kaye, B.H., *Powder Technology,* Vol. 10, No. 6, 1974, p. 307.
6. Bartholdi, M., et al., Optics Letters, Vol. 1, No. 6, 1977, p. 223.
7. Breitmeyer, M., and Sambandam, M.K., *J. Assn. Adv. Med. Instrum.,* Vol. 6, No. 6, 1972, p. 365.
8. Eisert, W.G., et al., *Rev. Sci. Instr.,* Vol. 48, No. 8, 1975, p. 1,021.
9. Kaye, W., *J. Colloid Int. Sci.,* Vol. 44, No. 2, 1973, p. 384.
10. Mullaney, P.F., et al., *Rev. Sci. Instr.,* Vol. 40, No. 8, 1969, p. 1,029.
11. Ricci, R.J., and Cooper, H.R., *ISA Trans.,* Vol. 9, No. 1, 1970, p. 28.
12. Salzman, G.C., et al., *Clin. Chem.,* Vol. 21, No. 9, 1975, p. 1,297.
13. Uzgiris, E.E., and Kaplan, J.H., *J. Colloid Interface Sci.,* Vol. 55, No. 1, 1976, p. 148.
14. Wallace, T.P., and Kratohirl, *J. Polymer Sci.,* Vol. 8, Pt. A-2, 1970, p. 1,425.
15. American National Standards Inst. Calibration Method, ANSI B93.28, 1973.

The author

Alvin Lieberman is vice-president of Royco Instruments, Inc., 141 Jefferson Drive, Menlo Park, CA 94025, which makes optical instruments for particle measurement in gases and liquids, and electrical-resistance blood cell counters. He holds B.S. and M.S. degrees in chemical engineering from Illinois Institute of Technology. He has served on the Technical Advisory Committee for the Illinois Air Pollution Control Board, and on the National Academy of Engineering Ad Hoc Panel on Abatement of Particulate Emissions from Stationary Sources.

Section 3
PRIMARY TREATMENT METHODS

Primary-waste-treatment methods

Removal of dissolved solids, suspended solids and oily matter from waste waters involves procedures based on the physical and chemical properties of the stream. Disposal of undesirable materials ultimately requires some type of mechanical separation.

Leslie D. Lash and *Edward G. Kominek,* Envirotech Corp.

☐ Primary treatment covers a wide range of equipment and applications for handling the almost infinite combination of waste compositions. A classification of treatment methods is shown in Table I (T/I), and of waste compositions in T/II.

The requirements for primary treatment will vary, depending upon whether the effluent will be subjected to secondary treatment, or discharged to a municipal sewer system, or to a river or lake with an effluent quality that meets federal, state and/or local standards.

As a general rule, several of the treatment methods listed in T/I, used alone or in combination, will reduce a pollution load to the required level. Selection of the optimum treatment for a particular plant should include an evaluation of these alternatives:

Should the waste be discharged to a municipal sewer for partial or complete treatment? This will depend upon an evaluation of capital and operating costs against sewer charges. A major factor may be company policy regarding return on investment.

If a treatment plant is to be installed, answers to the following questions are important:

■ How consistently must required effluent quality be obtained? (Consider EPA daily maximum and monthly average requirements.)

■ Can plant operations be easily modified to handle increased waste loads—either volume or quantity of pollutant, or both?

■ Can the plant be easily modified to produce internally a higher quality effluent as, for example, the EPA's BATEA standards that go into effect in 1983?

■ Will plant operations create any odor- or air-pollution problems?

■ What methods are available for disposal of sludge produced? Will the method chosen be acceptable for the life of the plant?

■ Can the plant produce a reusable effluent?

■ Can a maintenance problem affect effluent quality sufficiently to cause curtailment of plant operations? For how long?

■ What may be the effects of poor effluent quality from the standpoints of bad publicity, civil liability or legal liability? Is the plant required to handle storm water? If so, is the design adequate?

Answers to these questions require study by a consulting engineer, unless a company has an engineering staff qualified to evaluate waste-treatment projects.

The layout of a waste-treatment plant should be based upon the worst combination of flow and pollutant-load variation that may be encountered. It is

Classification of waste-treatment methods		T/I
Primary		
Screening	Coagulation	
Centrifugal separation	Gravity separation	
Neutralization	Induced-air flotation	
Aeration	Dissolved-air flotation	
Chemical oxidation	Media filtration	
Chemical reduction	Diatomite filtration	
Precipitation	Ultrafiltration	
Secondary		
Aerated lagoons	Activated sludge	
Contact media	High rate activated sludge	
Trickling filters	Extended aeration	
Disk contactors		
Advanced		
Media filtration	Reverse osmosis	
Granular activated carbon	Ion exchange	
Powdered activated carbon	Electrodialysis	

Originally published October 6, 1975.

Possible compositions of wastes	

Acids and alkalis (pH from 1.0 to 12.0

Suspended solids
  Coarse to submicron
  Particulate to colloidal
  Density heavier, equal to, or lighter than water

Fats, oils and greases
  Free, or physically or chemically emulsified
  Wide variation in density or specific gravity
  Wide variation in globule or particle size

Dissolved solids
  Mineral
    Oxidizing or reducing chemicals
    Cationic or anionic metal ions
    Toxic materials
  Organic
    Biodegradable
    Nonbiodegradable
    Toxic materials

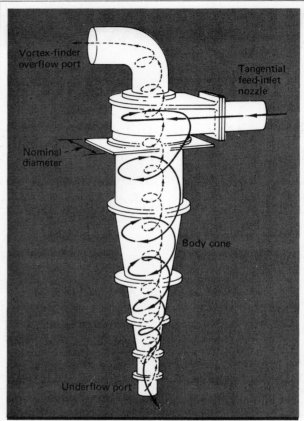

Hydroclone removes dense, suspended solids

almost inevitable that pipelines will break and tanks will inadvertently be dumped. Emergency operating conditions may increase waste volume even if storm-water treatment is not required. A decision must then be reached whether or not to install surge or equalization basins at the head end of a plant, to remove gross suspended solids before equalization, or to adjust pH as a corrosion preventative before any other treatment. The discussion of primary treatment methods that follows in the various sections may aid in answering these questions.

Screening

Screening is a mechanical process that separates particles on the basis of size. There are several types, which have static, vibrating or rotating screens [1]. Openings in the screening surfaces range from several inches down to 35 micron (μm).

The most commonly used screens in industrial processes are classified as:
- Bar screens for removing large suspended solids.
- Rotating drums having screen openings ranging down to 250 μm, for removing suspended solids on the outer surface of the drum.
- Vibrating screens with screen cloth down to 100 μm.
- Inclined stationary screens down to about 250 μm.
- Microstrainers with rotating drums that retain suspended solids on the inside of the drum, and that have screen apertures down to 23 μm.

Screens are used for treating raw water, storm water, and wastewaters from food-process plants, textile mills, and pulp and paper mills. The amount of solids removal depends upon screen size and characteristics of the water or wastewater being filtered. Microstrainers are claimed to be equivalent to sand filters in their capacity to remove suspended solids, including microscopic solids such as various forms of plankton and general microscopic debris [2].

Centrifugal separation

Hydrocyclones remove relatively dense suspended solids from a liquid stream. Frequently, they are used for separating grit from storm water or wastewater. Fig. 1 (F/1) shows the operation. Fluid-pressure energy is used to create rotational fluid motion in a cyclone. This motion causes relative movement of materials suspended in the fluid, and permits separation of these materials from the fluid. The bulk of the fluid is removed through a top outlet, and the separated solids are discharged through a second outlet located in an axial position at the bottom.

Neutralization

Alkaline wastes can be neutralized with sulfuric or hydrochloric acid as an emergency measure. If a highly alkaline waste is normally discharged, neutralization with flue gas either by bubbling it through diffusers in a holding tank or by releasing it countercurrent to the alkaline waste in a spray tower may be the best solution.

If strong alkali and strong acid wastes are regularly discharged, segregating the wastes and controlling their release to provide self-neutralization should be the cheapest alternative. Eluates from large demineralization units are an example.

Acid wastes can be neutralized with sodium hydroxide, calcium hydroxide or limestone. The selection should consider the products of neutralization and

```
┌─────────────────────────────────────────┐
│  Reduction of hexavalent chromium    (T) │
│                                      (III)│
│  ─────────────────────────────────────── │
│                                           │
│                           Time*           │
│       pH                  min             │
│       ──                                  │
│       1.0  . . . . . . . . . . . .   0.1  │
│       1.5  . . . . . . . . . . . .   0.5  │
│       2.0  . . . . . . . . . . . .   5.0  │
│       3.0  . . . . . . . . . . . .  20.0  │
│       4.0  . . . . . . . . . . . .  60.0  │
│       5.0  . . . . . . . . . . . . 200.0  │
│                                           │
│       *Time to reduce 99% of hexavalent   │
│       chromium to trivalent form.         │
│       Source: Ref. [3].                   │
│                                           │
└─────────────────────────────────────────┘
```

whether or not the waste contains metallic cations that must be removed by precipitation at high pH.

The objection to sodium hydroxide is its cost, which generally precludes its use. Granular or powdered limestone can be used effectively with hydrochloric acid wastes. However, precipitation of calcium sulfate on the limestone particles reduces the latter's efficiency for the treating of sulfuric acid wastes. Neutralization with limestone practically stops at a pH of about 5.0.

If it is necessary to increase the pH to 9.0, or higher, to precipitate dissolved metals, an amount of lime (equivalent to the bicarbonates and free carbon dioxide in solution) is added. This may turn the situation in favor of using lime for total neutralization. An exception would be the neutralization of strong acids only to a pH of about 4.0, followed by aeration to remove free carbon dioxide. Since no bicarbonates would be produced, the pH could then be increased with a minimum amount of lime.

If lime or sodium hydroxide were used for neutralization, pH control could be effective—provided that alkali demand did not exceed the capacity or the rate of response of the chemical feeders. Neutralization of a strong acid is difficult to control accurately over short time periods because of the rapid change in pH in the range from about 3.5 to 9.0, caused by the addition of a small amount of alkali. This can be controlled by stepwise pH control, or by equalization in a holding tank after neutralization.

Chemical reduction

In waste-treatment operations, chemical reduction is used principally for treating plating wastes that contain hexavalent chromium salts. Reduction changes the hexavalent chromium ions (chromate) into reduced trivalent chromium ions. These can then be precipitated as the hydroxides. Common reducing agents are sulfur dioxide, sodium metabisulfite or ferrous sulfate. The reduction of chromate ions with sulfite or metabisulfite is pH dependent, as listed in T/III.

Ferrous ion reacts with hexavalent chromium to reduce the chromium, thereby oxidizing the ferrous ion to the ferric form. This reaction occurs rapidly at pH levels below 3.0. Approximately 2.5 times the stoichiometric amount of ferrous sulfate is necessary to obtain complete reaction.

Adding iron to chromate wastes increases the quantity of sludge that is precipitated. Considering the 2.5 times theoretical dosage of ferrous sulfate required, approximately 15.2 parts of ferric hydroxide will be precipitated per part of chromium reduced. One part of chromium is equivalent to 1.98 parts of chromic hydroxide—so the total quantity of sludge produced will be 17.2 parts per part of chromium.

On the other hand, when a sulfite is used as the reducing agent, the quantity of sludge produced is only 1.98 parts per part of chromium (when the pH is raised to about 9.0 to precipitate chromic hydroxide), unless the waste contains sufficient sulfate to cause calcium sulfate precipitation.

Many plants that do chromium plating or anodizing operations also have waste pickling acid from descaling operations. The waste pickling acid will contain ferrous sulfate and free sulfuric acid. This can be used advantageously for chromate reduction.

Chemical oxidation

Oxidation is frequently carried out in waste-treatment operations, and is generally accomplished by aeration. Chemical oxidation, using chlorine or hypochlorite, is often limited to cyanide oxidation. Cyanide wastes are produced from plating operations and from the hardening of steel.

Free cyanide ions can be oxidized to cyanate. The reaction requires 2.73 parts of chlorine and 1.13 parts of caustic per part of cyanide [3]. The rate of reaction is pH dependent, but goes to completion in about 30 min, at a pH above 8.5. At a lower pH, cyanogen chloride may form, which is volatile and somewhat poisonous. Hence, this oxidation should always be carried out at a pH of 8.5, or higher.

Cyanates are further oxidized by the addition of chlorine and caustic to produce carbon dioxide and nitrogen. This requires 4.09 parts of chlorine and 3.08 parts of caustic per part of cyanide. In the majority of cases, sufficient chlorine and caustic are added to complete the oxidation of cyanide to carbon dioxide and nitrogen in one stage. At a pH of 8.4, the reaction is complete in 10 min; at a pH of 9.6, it takes 40 min for completion.

The addition of chlorine, equivalent to the chemical oxygen demand (COD) of the wastewater, is necessary to ensure oxidation of cyanides. If a system has simple cyanides, the presence of excess chlorine assures complete oxidation of the cyanides. This can readily be controlled by monitoring the oxidation-reduction potential (ORP) of the treated wastes.

If the wastewater contains cyanide complexes such as ferrocyanide or ferricyanide, oxidation will proceed at a slower rate. An oxidation system must, therefore, be designed to handle the complex cyanides if present. The required reaction time is determined by tests.

Aeration

Aeration may be used for odor control and partial removal of COD as in oxidation of ferrous iron, sulfides or sulfites.

Fixed aerators such as spray aerators, multiple-tray aerators (similar to cooling towers), cascade aerators and diffused-air units have been used [4]. Mechanical-surface aerators are now widely used. The latter include motor-speed aerators or slow-speed aerators with gear reducers, in floating or fixed configurations. Submerged turbine aerators may also be used.

Waterfall-type aerators include spray aerators, which consist of fixed nozzles that spray a fountain of water into the air. Cascade aerators flow water over obstructions to produce turbulence and expose new surfaces for absorption of air. Multiple-tray aerators use perforated, wire-mesh or slat trays that are stacked vertically.

Diffused-air units have perforated pipes, porous diffuser tubes or various patented blocks or impingement devices. Compressed air generates fine bubbles that provide longer contact time with the liquid, resulting in increased air-absorption efficiency compared with waterfall aerators.

Mechanical-surface aerators generally provide the most efficient means of absorbing air, based on pounds of oxygen absorbed per horsepower-hour. The energy is divided into shear, which creates the solids-liquid interface; and pumping, which removes the oxygenated liquid.

A typical fixed low-speed aerator is shown in F/2. The nonclog open impeller permits operation even with debris in the pond. Conventional operating speeds are 37 to 78 rpm. Low-speed aerators are available in sizes from 5 to 200 hp.

A floating motor-speed aerator with doughnut-float assembly uses a propeller to pump liquid up through a draft tube and across a deflector plate, intimately contacting the water stream with air. Alternately, fiberglass-reinforced plastic pontoons, foam-filled to prevent sinking, may be used to float the aerator mechanism and drive motor. Motor-speed aerators are available in sizes from 5 to 100 hp.

Acid-mine wastes are treated by lime neutralization and aeration, followed by clarification/thickening. The ferrous sulfate present in acid mine waters is formed by leaching iron pyrite as shown by:

$$FeS_2 + 3\tfrac{1}{2}O_2 + H_2O \rightarrow FeSO_4 + H_2SO_4$$

Adding oxygen through aeration, and pH control by lime, the ferrous iron may be oxidized and precipitated as ferric hydroxide, according to:

$$2Ca(OH)_2 + 2FeSO_4 + \tfrac{1}{2}O_2 + H_2O \rightarrow 2CaSO_4 + 2Fe(OH)_3$$

Precipitation of iron hydroxide with lime will produce a hydroxide floc that settles slowly, compacts poorly, and is difficult to dewater. However, premixing the iron hydroxide sludge with lime prior to neutralization produces a denser and more rapid settling sludge than described above.

Manganese and other elements similar to iron may also by oxidized and removed in a similar manner.

Because an aerator is a gas-liquid contactor, absorption or stripping of gases may be performed by aerators—for example, stripping of excess carbon dioxide.

With hydrogen sulfide, a dual action is obtained. Under the proper pH range, oxidation of the sulfide and/or stripping of the gas can occur. Because of the poisonous nature of hydrogen sulfide, it is often advantageous to achieve oxidation rather than stripping, by

maintaining a pH above the acid range. However, colloidal sulfur will precipitate and the free sulfur may be difficult to remove.

Precipitation

Precipitation reactions are frequently carried out in water and waste treatment. Lime softening using lime to precipitate calcium carbonate and magnesium hydroxide is frequently practiced in water treatment. In some cases, sodium carbonate is added to react with the sulfates and chlorides of calcium and magnesium, thereby reducing whatever noncarbonate hardness that may exist.

In waste-treatment operations, the precipitation of metallic hydroxides from plating, metallurgical and steel-mill wastes and also from acid mine drainage is commonplace. Wastes that contain phosphates must also be treated to precipitate calcium or aluminum phosphate. Lime or limestone is used for the precipitation of calcium sulfite or calcium sulfate from stack-gas scrubber-water.

Essentially, every reaction is carried out with the stoichiometric amount of precipitant and at the optimum pH. If the solubility of a compound exceeds the required effluent concentration, an excess of the precipitant may be added to reduce the solubility, due to common-ion effect.

Precipitation reactions are usually carried out in the presence of previously formed sludge. This minimizes supersaturation. It also results in more rapidly settling precipitates because precipitation occurs on existing slurry particles.

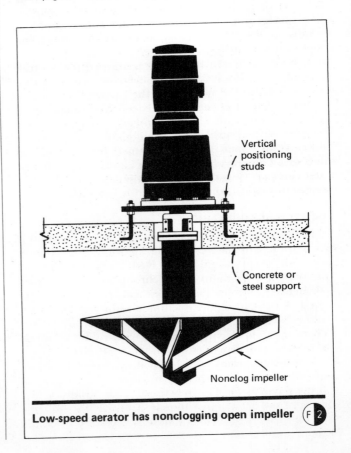

Vertical positioning studs

Concrete or steel support

Nonclog impeller

Low-speed aerator has nonclogging open impeller   (F)(2)

Slurry recirculation may be internal, as in a solids-contact unit; or it may be external, with underflow being recycled. When low-pH wastes are treated for metal precipitation, a more rapid settling and denser sludge will be produced if underflow from the clarifier or thickener is premixed with lime, before being added to the wastewater. It may be advantageous to recycle up to 30 lb of sludge per lb of newly precipitated sludge.

Flocculation and coagulation

Flocculation is the process of bringing together fine particles so that they agglomerate. Coagulation is the method by which chemicals are dispersed to change the characteristics of the suspended particles so that agglomeration will occur. The coagulant causes suspended solids to agglomerate by reducing zeta potential (commonly defined as the isoelectric pH of the waste) to zero. When a metallic cation such as aluminum or iron is used, metallic hydroxides are initially formed as micron-size particles, which adsorb colloids.

Polymers are also used, either as coagulants or as coagulant aids. Polymers may be cationic, and adsorb on a negative flow particle; anionic, which permit bonding between a colloid and polymer; or nonionic, which adsorb and flocculate by hydrogen bonding between the solid surfaces and the polar groups in a polymer.

Laboratory tests are necessary to establish optimum conditions for coagulation. These can be accomplished by determining zeta potential or by jar tests in which coagulant and coagulant-aid dosages and pH are varied. Jar tests establish the subsidence rate of the flocculated particles and the volume and sludge concentration that will result from the optimum treatment. They can also be used to determine whether or not solids contact is beneficial. If the jar tests are carried out at the minimum water or waste temperature at which the plant must operate, the flocculation time and settling rate data can be used, with appropriate scaleup factors, for a full-size plant.

Gravity separation

Gravity separation refers to the removal of suspended solids or oils when the specific-gravity difference from water causes settling or rising of the solids or oils during passage through a tank under suitably quiescent conditions.

Static conditions that affect gravity separation are:

1. Water temperature, which affects both viscosity and density of the water.

2. Specific gravity of the oil or suspended solids.

3. Size and shape of suspended oil droplets, and size and shape of particles.

4. Solids concentration, which results in free or hindered settling.

Settling velocities of particles will change with time and depth, as particles agglomerate and form larger floc sizes. Separation characteristics can be determined by static tests.

Since it is virtually impossible to duplicate in static tests the conditions that will be encountered in plant-scale operation, area scaleup factors ranging from 1.25 to 1.75 [4], and volume scaleup factors ranging from 1.5 to 2.0, have been suggested. For solids-contact units, it has been recommended that a factor of at least 2 to 1 should be used. On the other hand, a scaleup factor of 1.3 has been recommended for an inclined-plate settler [5].

The dynamic factors involved in gravity-separator design depend primarily upon hydraulic factors. Conventional sedimentation tanks may be:

Tank	Flow
Rectangular	Rectilinear
Center feed	Radial
Peripheral feed	Spiral
Peripheral feed	Radial
Square	Center-feed radial

Settling tanks [4] are divided into four zones:

1. Outlet zone, to provide smooth transition from the influent flow to a uniform steady flow desired in the settling zone.

2. Outlet zone, to provide smooth transition from settling zone to the effluent flow.

3. Sludge zone, to receive and remove settled material and prevent it from interfering with the sedimentation of particles in the settling zone.

4. Settling zone, to provide tank volume for settling, free of interference from the other three zones.

A primary clarifier-sedimentation unit is shown in F/3, where a beam-support structure is used. F/4 shows a center-support mechanism with top feed.

If a sedimentation unit is being used for removing flocculated solids, it is important that the velocity in the influent channel be kept low, not exceeding about 2 ft/s to prevent breakup of floc [F/5]. A top-feed clarifier is not recommended because there will be a

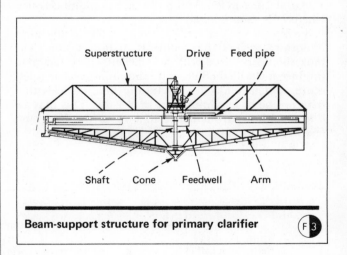

Beam-support structure for primary clarifier   F 3

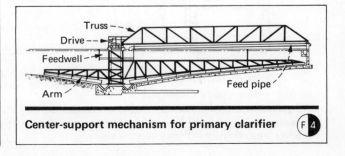

Center-support mechanism for primary clarifier   F 4

3 to 4-ft difference between the water level in the influent pipe of flume and the level in the feedwell. The velocity resulting from this change in elevation will cause considerable turbulence in the inlet line and also in the feedwell.

Size of the feedwell is also important. It is recommended that it not be less than 20% of the tank dia., with a depth of 55 to 65% of the side-water depth of the tank. The preferred exit velocity from the feedwell is in the range of 1 to 2 ft/min, to avoid currents that will upset the settling zone.

Outlet-zone weir-loading rates have been recommended in the range of 10,000 to 30,000 (gal/day)/ft, depending upon the type of floc.

The sludge zone should be designed for moving sludge to its point of withdrawal with a minimum disturbance. A floor slope of 1 to 12 is recommended because it is difficult to move sludge if the tank has a lesser slope or is flat.

If the sludge contains organic solids that might become septic, it is desirable to screed-in the tank floor with the scraper mechanism and to use squeegies on the scraper blades. For secondary clarifiers, where the sludge may be nitrified, it is desirable to use vacuum suction pipes instead of squeegies for removing the settled solids with each rotation of the scraper. The velocity of the scrapers should be kept below 1 ft/min, so as not to disrupt the settling process or to resuspend the settled sludge.

The quantity and settled density of the sludge will determine the desired torque capability of the scraper drive. As a rule of thumb, this is generally rated according to the square of the diameter of the tank. Torque in ft-lb will range from $3D^2$ for a secondary activated-sludge clarifier to $25D^2$ for a primary pulpmill clarifier, where D is tank diameter.

Surface area of a settling tank is one of the most important factors that influence sedimentation. The tank should be sized with a sufficient safety factor to produce a clarified effluent at minimum water temperature, and to allow for separating floc particles that may not be of maximum size or density. This may be due to either partial deflocculation in the inlet and feedwell, or to partially ineffective coagulation.

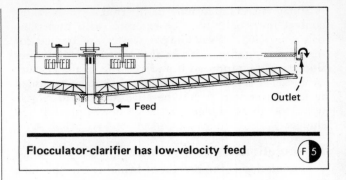

Flocculator-clarifier has low-velocity feed   F/5

If the release of entrained air might cause floating floc, or if floating solids may be present at times, a scum baffle plus a scum skimmer that is rotated by the scraper mechanism should be installed.

Tank performance and removal of flocculent particles will depend upon depth. The efficiency of removal, however, is not linearly related to the detention time. Excessively deep tanks are not considered desirable because of increased cost. However, a minimum depth of 11 ft is generally recommended for clarifiers ranging from 30 to 65 ft dia., with the depth increasing with diameter. A minimum depth of 15 ft is recommended for tanks 200 ft dia. or larger.

The American Petroleum Institute manual recommends that "The primary function of an oil/water separator is to separate free oil from refinery wastewater. Such a unit will not separate soluble substances nor will it break emulsions and therefore should never be specified for these purposes" [6]. API separators, as diagrammed in F/6, are recommended to be designed within the following limits:

■ Horizontal velocity: maximum 3 ft/min, or 15 times the rise rate in ft/min of oil globules 0.015 cm. dia., whichever is smaller.

■ Depth: 3 ft minimum to 8 ft maximum.

■ Depth-to-width ratio: 0.3 minimum to 0.5 maximum.

■ Width: 6 ft minimum to 20 ft maximum.

A corrugated-plate interceptor (CPI) was developed by the Royal Dutch/Shell group, as shown in F/7.

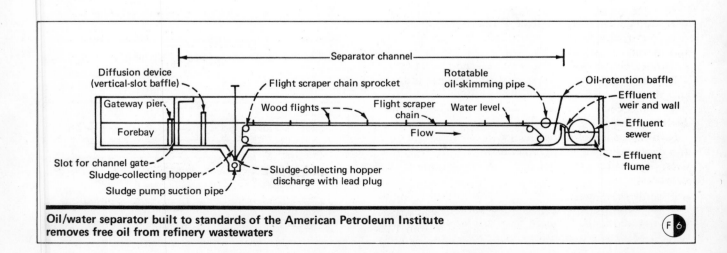

Oil/water separator built to standards of the American Petroleum Institute removes free oil from refinery wastewaters   F/6

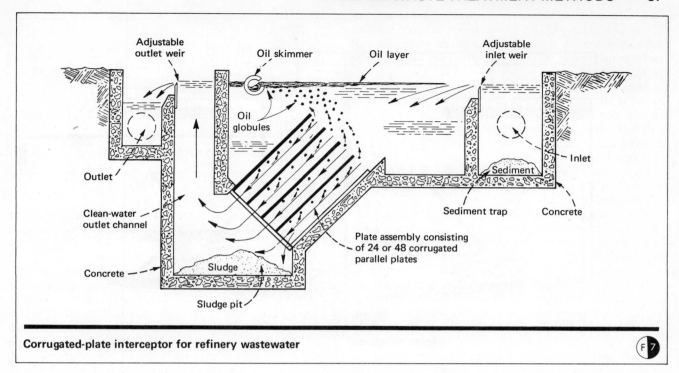

Corrugated-plate interceptor for refinery wastewater

F 7

Wastewater enters a separator bay and flows downward through corrugated plates arranged at an angle of 45° to the horizontal. Oil collects on the underside of the plates and rises to the surface where it is skimmed. Solids settle into a sludge compartment, with the clarified waste discharging into an outlet channel [7].

It has been reported that the parallel-plate separators achieve higher oil removal than API separators having the same area. For the same amount of oil removal, the parallel-plate separator should permit the use of a smaller installation.

Tube-and-plate settlers are also being used in sedimentation-type units. It has also been reported that the presence of inclined tubes will increase settling rates by a factor of up to three, compared with settling in a regular vertical vessel under the same conditions. Also, for most new applications, utilization of tube modules may reduce sedimentation volume by one-third to one-half of that required by conventional basins.

Installation of tube or plate modules in existing clarifiers will increase the treatment capacity accordingly, if they can handle the higher hydraulic load. For new installations, space requirements and the installed cost of the two types of units will be the major factors for evaluating tube or plate type settlers against conventional sedimentation units.

Solids-contact units

Solids-contact units, sludge-blanket units, and flocculating clarifiers are characterized by combining in a single basin the processes of mixing, coagulation and flocculation, floc conditioning, liquid-solids separation, and sludge removal.

F/8 illustrates the basic features of a solids-contact unit. Influent water or waste is discharged into a draft tube, where it is mixed with four or five volumes of recirculating slurry. Chemicals for treatment may be added to the influent line or to the draft tube. A slow-speed turbine, operating with a variable-speed drive, pumps the draft-tube contents upward into the cone-shaped reaction chamber. Approximately 25 min of reaction time is allowed for precipitation and/or coagulation and floc formation. An optimum concentration of slurry is maintained in circulation—up to 5% by weight when calcium carbonate or calcium sulfate is precipitated.

A volume of the circulating slurry equal to the influent is displaced into the annular solids-separation zone outside of the reaction chamber. The hydraulic design permits upward separation of the clarified water or waste from the solids that settle to the tank bottom. The clarified water or waste is removed through radial and peripheral launders at the surface of the separation zone.

A rotating scraper mechanism moves the settled solids to the sludge-concentration sump from which the sludge is intermittently discharged through a timer-controlled, pilot-operated valve.

Circulating slurry is continuously pumped back through the bottom of the draft tube, and mixed with the influent and the treating chemicals. The circulation rate can be varied, together with the sludge-discharge rate.

These adjustments are necessary to maintain the optimum slurry concentration for various treatment applications that may produce solids ranging from fragile amorphous flocs to dense crystalline precipitates.

Characteristics of these units are:

1. Slurry recirculation, using a large slow-speed turbine for recirculating settled solids.

2. Agitated central mixing of chemicals and recycled sludge, with solids separation through a sludge

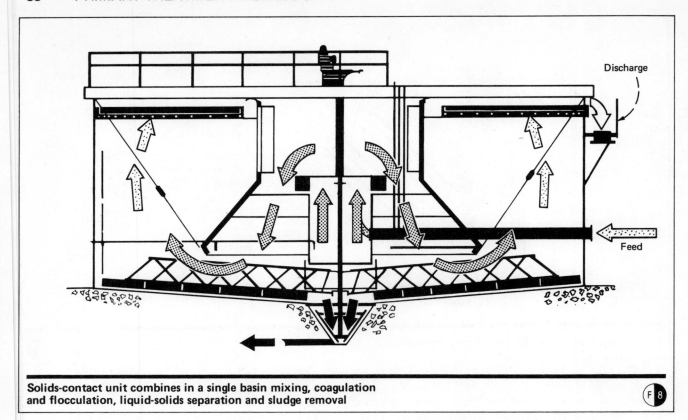

Solids-contact unit combines in a single basin mixing, coagulation and flocculation, liquid-solids separation and sludge removal   (F 8)

blanket kept in suspension by the upward flow.

3. Steady discharge, with pulsating feed of chemically treated influent to a suspended fluidized blanket.

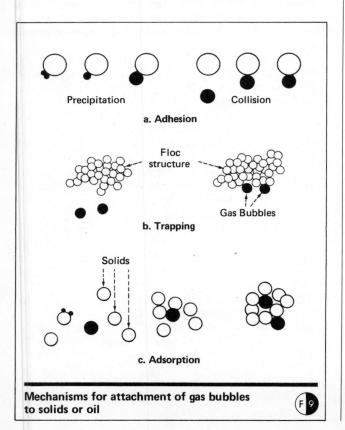

Mechanisms for attachment of gas bubbles to solids or oil   (F 9)

4. Chemically treated influent and recycled sludge flocculated with slow-speed paddle agitator or turbine. Minimum fluidization of sludge blanket due to low overflow rates. This flocculating action with low sludge density is achieved in flocculating reactor-clarifier units.

Design features of solids-contact clarifiers are [9]:

1. Rapid and complete mixing of chemicals, feed-water and slurry solids must be provided. This should be comparable to conventional flash-mixing capability and should incorporate variable control of $\bar{G}t$ values, usually by adjustment of recirculator speed. $\bar{G}$ is the velocity gradient that is a measure of shear intensity, $(ft/s)/ft$, and t is time, s.

2. Mechanical means for controlled circulation of the solids slurry must be provided with at least a 3 : 1 range of speeds. The maximum peripheral speed of mixer blades should not exceed 6 ft/s. The pumping capacity of mixers can be estimated [10].

3. Means for measuring and varying the slurry concentration in the contacting zone up to 50% by volume.

4. Sludge-discharge systems should allow for easy automation and variation of volume discharge. Mechanical-scraper tip speed should be less than 1 ft/min, with speed variation of 3 : 1.

5. Sludge-blanket levels must be kept a minimum of 5 ft below the water surface.

6. Effluent launders should be spaced so as to minimize the horizontal movement of clarified water.

Recommended standards [11] for solids-contact units include: not less than 30 min for flocculation; not less than 2 h retention for clarifiers or 1 h for softeners; weir loadings not exceeding 28,800 (gal/d)/ft for softeners, and 14,400 (gal/d)/ft for clarifiers.

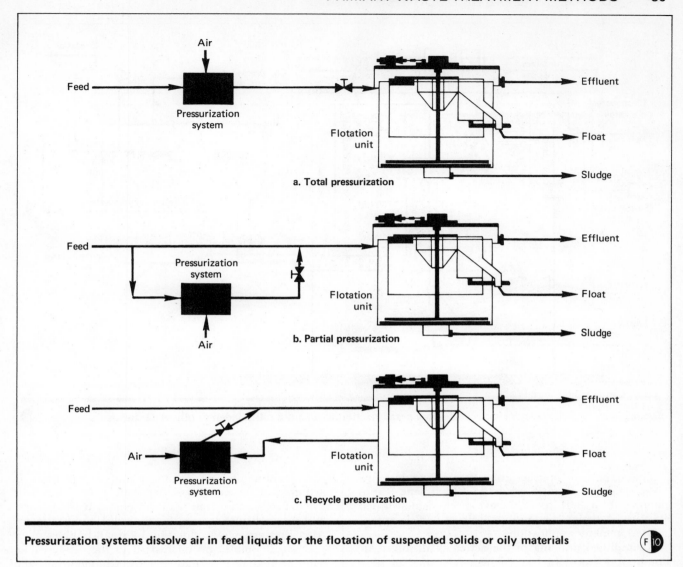

a. Total pressurization

b. Partial pressurization

c. Recycle pressurization

Pressurization systems dissolve air in feed liquids for the flotation of suspended solids or oily materials

F 10

Dissolved-air flotation

In dissolved air flotation (DAF), air is intimately contacted with an aqueous stream at high pressure, dissolving the air. The pressure on the liquid is reduced through a backpressure valve, thereby releasing micron-sized bubbles that sweep suspended solids and oil from the polluted stream to the surface of the air-flotation unit. Applications include treating effluents from refinery API separators, metal finishing, pulp and paper, cold-rolling mill, poultry processing, grease recovery in meat-packing plants, and cooking-oil separation from french-fry processing. An increasingly important application is the thickening of sludge.

Attachment of gas bubbles to suspended solids or oily materials occurs by several methods, as illustrated in F/9. The suspended-solids/gas mixture is carried to the vessel surface after: precipitation of air on the particle; collision of a rising bubble with a suspended particle; trapping of gas bubbles, as they rise, under a floc particle; and adsorption of the gas by a floc formed or precipitated around the air bubble.

To dissolve air for flotation, three types of pressuriza-tion systems are used. These are illustrated in F/10.

Full-flow or total pressurization is used when the wastewater contains large amounts of oily material. The intense mixing occurring in the pressurization system does not affect the treatment results.

Partial-flow pressurization is used where moderate to low concentrations of oily material are present. Again, intense mixing by passage through the pressurization systems does not affect treatment efficiency significantly.

The recycle-flow pressurization system is for treatment of solids or oily materials that would degrade by the intense mixing in the other pressurization systems. This approach is used following chemical treatment of oil emulsions, or for clarification and thickening of flocculent suspensions. As shown in F/11, the pressurized system is the clarified effluent. Air is dissolved in this stream at an elevated pressure, and mixed with the feed stream at the point of pressure release.

Mixing of streams prior to entering the flotation zone results in intimate contact of the pressurized air/water mixture with the suspended solids to effect efficient flotation.

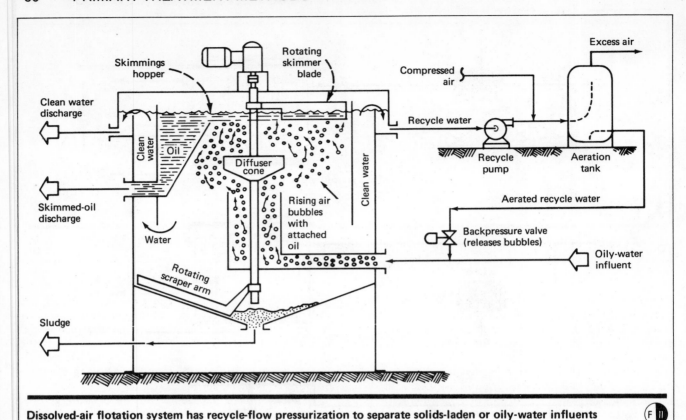

Dissolved-air flotation system has recycle-flow pressurization to separate solids-laden or oily-water influents (F 11)

The amount of pressurization flow is based on the air-to-solids weight ratio required for treatment. Ratios ranging from 0.01 to 0.06 have been demonstrated effective. A design air-to-solids ratio of 0.02 is appropriate for many applications.

A simplified drawing of the action of dissolved-air flotation is shown in F/11. The solids-laden or oily-water influent mixture enters the flotation vessel, and the air-solids mixture rises to the liquid surface. The air-solids mixture has a specific gravity less than water. Solids having a specific gravity greater than water tend to settle to the bottom and are removed by a rotating scraper arm. Attached to the same shaft is a rotating skimmer blade that removes the floating matter from the surface of the vessel into a skimmings hopper. Clean water passes underneath a skirt and then must leave the vessel through a launder, which is located in the peripheral region.

A portion of the effluent water is recycled for pressurization. Compressed air is introduced into the discharge of the recycle pump, and intimate contact with the water is achieved in the aeration tank. Maximum solubilization efficiency is important at this point. The aerated recycle water is then returned through a backpressure valve, where the pressurized air is released, and mixed with the influent for flotation.

Flocculants such as synthetic polymers may be used to improve the effectiveness of dissolved air flotation. Also coagulants such as filter alum may be used to break emulsified oils and to coagulate materials for improved flotation recovery.

A combination of flotation and clarification is achieved in the same vessel by the clarifier, shown in F/12. Feed, pressurized with air, enters at the bottom and disperses into the top flotation-disk compartment. After pressure release, flotation by small bubbles takes place, with solids and oil carried to the surface. The concentrated float is gathered by skimmers into the scum box. Removal to discharge is by a screw-flight conveyor.

The partially settled effluent from the flotation-disk region passes under the baffle and is then separated in the annular settling zone. Fine particles, after release of the flotation bubbles, settle in this clarification region and are plowed to the center for discharge as a sludge out the bottom. Also, the dense particles that settle initially move to the bottom of the equipment as part of the sludge.

Induced-air flotation

Flotation by induced air uses devices similar to minerals beneficiation-flotation machines. Air is drawn into the cell by action of the rotor, is mixed with the water, and transformed into millions of minute bubbles. Oil particles and suspended solids attach themselves to the gas bubbles and are borne to the surface of the water. Skimmer paddles push the contaminated froth from the top of the cell into collection launders.

Properly conditioned oily wastes leave nearly 100% oil-free, and in most cases suspended solids are reduced significantly.

A typical design is a four-cell unit, each cell having

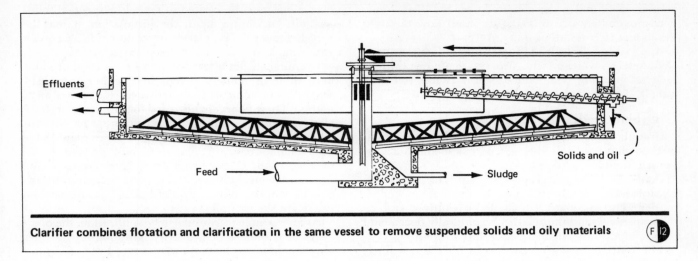

Effluents

Feed

Sludge

Solids and oil

Clarifier combines flotation and clarification in the same vessel to remove suspended solids and oily materials   F 12

a retention time of 1 min. Capacity may be determined by running bench tests on the water under consideration. These tests also determine the amount and type of chemical reagent best suited for a particular application.

A diagram of the operating mechanism is illustrated in F/13. The shaft, driven through a V-belt drive, extends into the tank, at the end of which is mounted a star rotor. A steel standpipe extends down from the bearing-assembly plate and has attached to it the disperser and disperser hood. Rotor speed decreases as the size of machine increases, in order to maintain a constant tip-speed. Typical rotor speed varies from 437 to 186 rpm.

The mechanism is designed to self-induce air into the water at atmospheric pressure. No auxiliary equipment is required for mixing the air with the liquid. A rapidly spinning rotor pumps the water through the disperser, where the induced air intimately mixes with the water. The rotor and disperser configuration produces exceedingly small gas bubbles that greatly enhance flotation.

As the rotor spins, it acts as a pump to force the water through the disperser, and also creates a vacuum in the standpipe. This vacuum forces air into the standpipe through an opening leading to the rotor. As the air and water travel through the disperser at high velocity, a shearing force is created that causes the air to form minute bubbles. As these bubbles continue outward and upward to the surface, oil particles and suspended solids are attached to the bubbles, interface. When the air breaks the liquid surface, the oil and suspended solids are left on the surface.

The sloping side directs the action of rising currents containing oil and suspended solids toward the direction of the overflow lip. At this point, the skimmer paddles skim off the floating contaminants. The tank is designed with sloping sides at the lower part to recirculate the water for continuous purification. The conical shape of the disperser hood, with its perforated surface, acts to quiet the liquid surface.

Wastewater enters the feed box, where its turbulence and energy are dissipated. Due to the differences in hydrostatic head between feed box and discharge, liquid moves into the first cell, where the flotation cleanup

begins. The water moves progressively through the other three cells, in series, where further cleanup is done by air flotation. Any contaminant floated in one cell is prevented from moving to the next cell by baffle plates. The contaminant-float launders permit recirculation of the float from the last cell, reducing the total amount of float to be handled.

From the fourth cell, water enters the discharge box, which removes any residual bubbles, free oil and suspended solids with another skimmer paddle. The discharge box contains a pneumatic level control that keeps the proper water level in the flotation cells and discharge box for froth-skimming by the paddles. The level controller actuates a discharge valve, which is sized to accommodate inadvertent surges in the feedrate.

Efficiency is often improved with chemical aids injected in the water, upstream in the flotation cell. In addition to breaking oil-in-water emulsions, they also make the air bubbles more stable and promote formation of froth on the surface of the water. The total

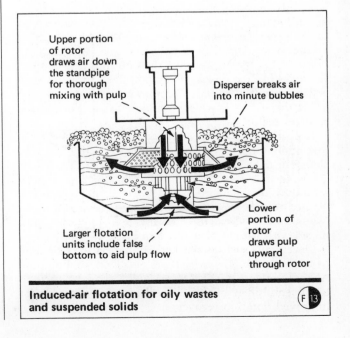

Upper portion of rotor draws air down the standpipe for thorough mixing with pulp

Disperser breaks air into minute bubbles

Larger flotation units include false bottom to aid pulp flow

Lower portion of rotor draws pulp upward through rotor

Induced-air flotation for oily wastes and suspended solids   F 13

volume of oil and water collected over the float flume is approximately 0.5 to 3% of the total flow through the induced-air flotation unit.

Applications include API-separator-effluent waters, olefin quench-water systems, separation of fine powders such as grinding compounds, storm runoff (oil terminals, etc.), and cleanup of benzol wastewater.

Granular-media filtration

Granular-media filtration is a liquid-solids separation method that uses flow through porous media, such as sand, to remove particulates.

Generally speaking, granular-media filtration is applicable for the removal of 5 to 250 mg/l of suspended solids and up to 200 mg/l of oil. To be readily filterable, the suspended solids must either be naturally flocculent, or be made so by chemical coagulation.

The heart of any granular-media filter is the filter bed. The size and depth of filter media are the most important design parameters. Dual-media shallow-bed filters frequently will be designed with 0.5 mm sand and 0.9 mm anthracite. Generally, for shallow-bed applications, the filtration rate will be between 2 and 6 gpm/ft^2. For deep-bed filter applications, design rates of 10 to 15 gpm/ft^2 are achieved with sand media of 1 to 2 mm dia. These values are for gravity filters. Increased rates (and occasionally, increased problems) may be derived with pressure filters.

Terminal headloss for gravity shallow-bed filters is normally between 5 and 7 ft of water. For gravity deep-bed filters operating at considerably higher filtration rates, terminal headlosses in the same range are common. Pressure filters often operate in the range of 15 to 25 ft of head, with rates of 4 to 15 gpm/ft^2, depending on the application.

Major granular-media filter outputs are (a) filter effluent, and (b) wastewater from backwashing operations. Backwash water generally amounts to about 1 to 5% of the total flow. The waste from backwashing

operations must be treated to remove suspended solids and oil. Depending upon the installation, it may be recycled to a primary clarifier or collected in an equalization tank and treated in a separate clarifier.

Granular-media filters are used for removal of solids and oil from refinery wastewaters, either from API separators or effluents from activated-sludge treatment plants. (These filters are also used for municipal tertiary waste-treatment, and for clarification prior to activated-carbon adsorption.) Pulp and paper plants have used them for removing fibers and solids prior to recycle of wastewater. Heavy-metal precipitates are recovered in granular-media filters, and process streams are clarified by similar units. Applications of granular-media filters are shown in T/IV.

As an example of granular-media filtration, the operation of a shallow-bed, gravity, granular-media filter may be visualized from the diagram of F/14. The filter is completely automatic on a controlled cycle. Several filters are customarily used in parallel to avoid interruption of the flow stream when the filter backwashes. During filtration, wastewater is passed through the media, and solids are removed. As the cycle progresses, the pressure drop across the filter increases to a preset maximum, indicating that the bed is full of suspended solids, and a backwash-cycle controller is energized. Alternately, a timed cycle or monitoring of the suspended solids may be used to terminate filtration.

A two-stage backwashing process is used: (1) air scour, and (2) water backwash.

Initially, liquid is drained down to the surface level of the media. At this point, air at approximately 4 (std. ft^3/min)/ft^2 is introduced through the underdrain nozzles to thoroughly scrub off adherent solids from the media. Then, a three-way valve is repositioned to allow entrance of backwash water from the storage compartment. Backwash water flows by gravity through the collection chamber and the filter compartment to remove the suspended solids that were fluidized by the air wash. The backwash discharges by gravity to settling basins, where all of the suspended solids settle out and water is recovered. The addition of polymers to the backwash water accelerates the settling rate of the solids.

Granular-media filter operates on a controlled cycle   F/14

Applications for granular-media gravity filters

Shallow-bed filters
(single medium, or multiple media)
1. Polishing clarified water.
2. Filtering flocculated low-turbidity waters.
3. Filtering sidestream from cooling towers.
4. Filtering waste solids.
5. Filtering tertiary effluents
6. Filtering after physical/chemical treatment.
7. Clarifying chemical processing streams
8. Recovering valuable suspended-solids products.

Deep-bed coarse-media filters
(single medium, or multiple media)
1. Clarifying steel-mill wastes.
2. Filtering tertiary effluents.

Hydraulic backwash begins at a high rate (approximately 25 gpm/ft²). However, as liquid level in the storage compartment lowers, the rate falls off to approximately 10 gpm/ft² at the end. This allows restratification of the dense sand and the light coal fractions. Such action reestablishes the desirable coarse-to-fine filter path for downflow filtration.

Filter performance is based on removal of suspended solids plus terminal headloss across the filter, from which cycle times are chosen. Cycle time on industrial waste normally ranges from 8 to 24 h of operation. However, when a heavy load of suspended solids passes into the filter, the runs are shortened considerably. Equalization of flow and concentration minimizes the effect of this type of shock.

Removal of suspended solids by granular-media filters generally ranges from 30 to 70% without pretreatment. However, with flocculation and/or coagulation, the removal range normally spans from 80 to 98%. Polymers tend to flocculate the solids present and assist removal of solids in the filter. Coagulants such as alum break emulsions and improve filterability of the solids present. Essentially complete removal of suspended solids may often be achieved when pretreatment is used.

Deep-media filters have a bed greater than 4 ft deep. Generally, coarse media of 1 to 2 mm sand are used. Coarse coal may also be placed on the top to remove solids and increase the storage volume. A typical bed would be 6 to 8 ft of sand, 1 mm in dia. Coal is also used, for example, with 5 ft of sand (1 to 2 mm in dia) topped by 3 ft of coal (3 to 6 mm dia.).

Increased filtration rates and additional storage volume are advantages of deep-media filters. But because of the greater porosity, bleeding of solids into the effluent is more likely to occur. To prevent this, flow through the column must be carefully regulated to avoid surges. Also, it is wise to divert the initial filtrate (after backwash) to avoid breakthrough of high suspended solids at the beginning of a filter cycle. These may be solids from the bed, or new solids that break through prior to setting up of an effective filter pattern. Diversion for 10 to 20 min is typically adequate.

With increased solids in the feed, the voids between the media fill faster, and shorter cycles result. This is often encountered with storm flow, such as in a petroleum refinery, where suspended solids increase, flowrates increase, and cycle times terminate prematurely. With automatic filters, backwash takes place at ultimate pressure loss, and the filter is regenerated for continued operation.

A key operating aspect for obtaining long and trouble-free service from a granular-media filter station is using adequate backwash procedures. This implies scouring all filtered materials from sand- and coal-grain surfaces and flushing this released material from the filtering chamber. For multiple media, proper restratification of the filter bed is then necessary. Experience indicates that air-scour followed by hydraulic fluidization of a filter bed—first at high rates, then at decreasing rates—is an effective backwashing procedure.

With organic material (such as from biological treatment) or oily material, vigorous backwash is necessary to clean the media. In certain instances, it has been necessary to go to a double backwash to achieve adequate cleansing. Proper backwash should be carefully considered in the design phase.

Ultrafiltration

A relatively new development, closely related to reverse osmosis, uses semipermeable membranes for separating emulsified oils, free oils, and suspended solids from wastes [12]. The process involves pumping dilute waste at a pressure of 30 to 50 psig through an ultrafiltration membrane. The membrane porosity prevents passage of the oils and suspended solids so that the permeate contains only materials in true solution.

It is claimed that ultrafiltration is generally effective. However, no data are available to indicate the susceptibility of the membranes to plugging or to deterioration by solvents or chemicals that may be present in the wastewaters.

References

1. Matthews, C. W., Screening, *Chem. Eng.,* Feb. 15, 1971, pp. 99–104.
2. Givans, G. R. and Boucher, P. L., Microstraining: Description and Application, *Water Sewage Works,* Reference Number 1962.
3. Eckenfelder, W. W., Jr., "Industrial Water Pollution Control," McGraw-Hill, New York, 1966.
4. "Water Treatment Plant Design," American Water Works Assn., New York, 1969.
5. Rubin, E. and Zahave, E., Enhanced Settling Rates of Solid Suspensions in the Presence of Inclined Planes, presented at AIChE meeting, Houston, Mar. 16–20, 1975.
6. "Manual on Disposal of Refinery Wastes," American Petroleum Institute, Washington, 1969.
7. Corrugated Plate Separators, Bulletin B 931, Heil Process Equipment Co., Cleveland, Ohio.
8. Beychok, M. R., Waste Water Treatment Processes for Refinery and Petroleum Plants, presented at Instituto de Ingenieros Quimicos de Puerto Rico, San Juan, July 1971.
9. "Process Design Manual for Suspended Solids Removal," Environmental Protection Agency, Washington, Jan. 1975.
10. Rushton, J. H. and Mahoney, L. H., Mixing Power and Pumpage Capacity, presented at annual meeting of American Institute of Mining, Metallurgical and Petroleum Engineers, New York, Feb. 15, 1954.
11. "Recommended Standards for Water Works," Great Lakes-Upper Mississippi River Board of State Sanitary Engineers, Health Education Service, Albany, N.Y., 1968.
12. Lim, Y. H. and Lawson, J. R., Treatment of Oily and Metal Containing Waste Water, *Pollution Engineering,* Nov. 1973.

The authors

Leslie D. Lash is marketing specialist of industrial water and waste equipment for Eimco BSP Div. of Envirotech Corp., Salt Lake City, UT 84110. His major research interests are the chemical and biological treatment of sewage and industrial wastes, and countercurrent separation processes. He has a B.S. and M.S. in chemical engineering from the University of Idaho and University of Utah, respectively. Mr. Lash is currently secretary-treasurer, Environmental Div. of AIChE, and is a registered professional engineer in Utah.

Edward G. Kominek is market manager of industrial water and waste equipment for Eimco BSP Div. of Envirotech Corp. He has had many years of experience in water- and waste-treatment activities, including process design of systems. He has a B.S. and M.B.A. from the University of Chicago, and is a registered professional engineer in Arizona, Illinois and Ohio. Among his memberships are ACS, AIChE, the Water Pollution Control Federation and the American Petroleum Institute.

Designing parallel-plates separators

Here is a way to design parallel-plates interceptors to separate oil globules from water. According to the author, in addition to being smaller, these interceptors are in many ways superior to widely used separators for this type of service.

Julio G. Miranda, Empresa Nacional del Petróleo

☐ Traditionally, oil-water separation in effluents from petroleum refineries has been achieved by means of the American Petroleum Institute (API) type separators as the first separation step. Even though a properly designed API separator is very efficient and easy to operate, it has disadvantages that in some cases are very difficult to overcome. Typical problems are: construction cost, space requirements, evaporation losses, fire hazard, high steam-consumption to avoid freezing of heavier products, etc.

An API separator has been designed to allow oil globules 0.015 cm dia. and larger to rise from the separator bottom (at the entrance of the unit) to the liquid surface, just before the outlet of the separator. In other words, the globules have to cross the entire separator depth to rise to the surface.

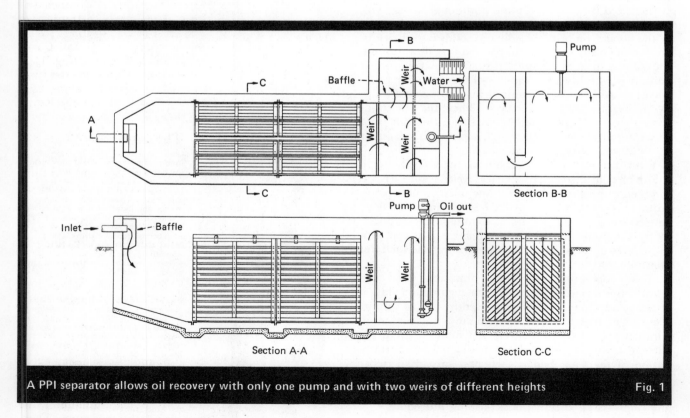

A PPI separator allows oil recovery with only one pump and with two weirs of different heights   Fig. 1

Originally published January 31, 1977.

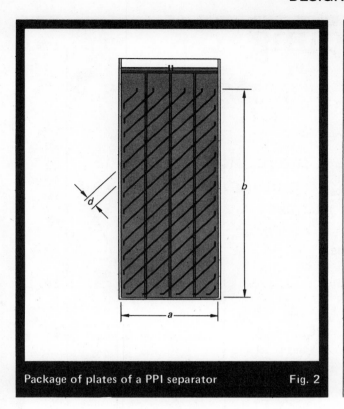

Package of plates of a PPI separator   Fig. 2

Parallel-plates-separator operation

The difficulties mentioned above have been minimized by the parallel-plates interceptor (PPI), a different type of separator first introduced by Shell Oil Co. in 1950, and whose use has received wide acceptance during the last years.

By means of a PPI, the oil path has been reduced to a slight distance by a set of parallel plates inclined at 45 deg. Oil coagulates at the undersurface of each plate and slides upward to the liquid surface where it can be skimmed off. Furthermore, solid particles collect on top of each plate and slide down to the bottom. These properties make a PPI much smaller than its equivalent API separator, which allows automatic oil recovery without skimmers, with only one pump (see Fig. 1), and by means of two weirs of different heights [1].

Although the PPI has been increasing in popularity, very little has been written about criteria to get a sound design of this separator. This article deals with a theoretical design approach that agrees fairly well with at least one existing PPI [2]. The development of this approach will be based on one package of plates such as the one shown in Fig. 2.

Retention time

The retention time (t_r) is provided by two equations:

$$t_r = d\sqrt{2}/V_t \qquad (1)$$
$$t_r = AL/Q_A \qquad (2)$$

where V_t = rising velocity of the bubbles, cm/s; d = distance between plates, cm; A = transversal area, cm^2, (a) (b) in Fig. 2; L = separator length, cm; and Q_A = flow through A, cm^3/s (Fig. 1).

If a and b are much greater than the distance between plates "d", then $R_H = d/2$. Hence, the Reynolds number (N_{Re}) is given by:

$$N_{Re} = 2dQ_A\rho/Au \qquad (3)$$

where ρ = fluid density (usually the same as water); and u = fluid viscosity (usually the same as water). Therefore:

$$A = 2dQ_A\rho/uN_{Re} \qquad (4)$$

Combining Eq. (1), (2) and (4):

$$L = uN_{Re}\sqrt{2}/2\rho V_t \qquad (5)$$

But since $u/\rho = v$, where v = kinematic viscosity, finally:

$$A = 2dQ_A/vN_{Re} \qquad (6)$$
$$L = vN_{Re}\sqrt{2}/2V_t \qquad (7)$$

Eq. (6) and (7) are design equations; the terms Q_A, ρ and u are known for each case, and V_t is a function of the minimum globe diameter to be separated. Then, by Stokes' Law:

$$V_t = (g/18u)(\rho_w - \rho_o)D^2 \qquad (8)$$

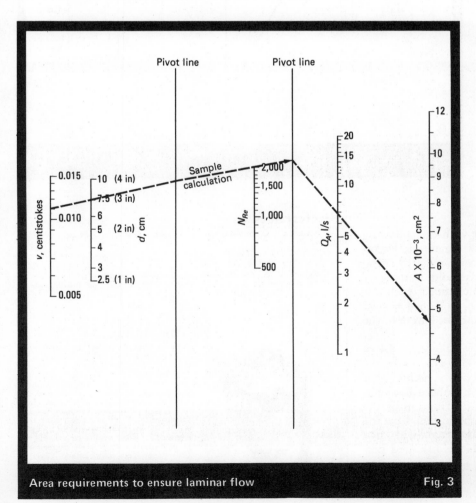

Area requirements to ensure laminar flow   Fig. 3

where g = acceleration due to gravity; u = absolute viscosity of water; ρ_w = density of the water; ρ_o = density of the oil; and D = diameter of the oil particle. (All of the above equations must be used with consistent units.)

Since there are two equations—(6) and (7)—and four unknown variables—A, L, d and N_{Re}—two variables must be arbitrarily fixed, and unless space restrictions are critical, it is convenient to assign values to d and N_{Re}.

Because PPI dimensions are much smaller than the equivalent API separator, it is possible to design with laminar flow—i.e., a Reynolds number equal to or less than 2,000. In addition, the smaller the distance between plates (d), the smaller the transversal area required for a given flowrate and a given Reynolds number; therefore, the greater the efficiency [3]. On the other hand, a small distance could mean clogging and diminished flow area by floating debris and materials not retained by inlet trash racks. This would increase maintenance costs.

Taking the above into consideration, it seems convenient to design in the following ranges:

$$500 \leqslant N_{Re} \leqslant 2,000 \text{ and}$$
$$1 \text{ in} \leqslant d \leqslant 4 \text{ in}$$

Eq. (6) and (7) may be represented graphically to cover these ranges, as shown by the nomographs in Fig. 3 and 4.

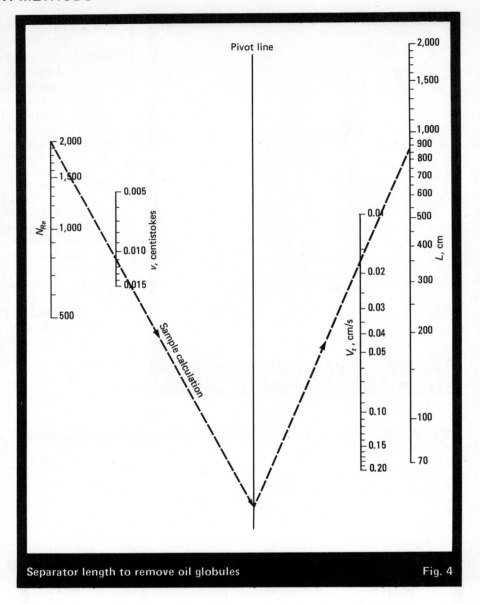

Separator length to remove oil globules   Fig. 4

Sample calculation

It is desired to treat an oily flow with the following characteristics: total flow = 27.8 l/s; kinematic viscosity = 0.011 centistokes*; rising velocity = 0.018 cm/s; distance between plates = 7.5 cm; Reynolds number = 2,000. The separator is to have two channels, each one with two packages of plates, i.e., Q_A = 27.8/4 = 6.95 l/s. What should be the total area and the length of the separator?

From Fig. 3, A = 4,740 cm^2 (total area = $4A$ = 18,960 cm^2), and from Fig. 4, L = 865 cm.

For a given type of oil (an assigned ρ_o value) at a fixed temperature, V_t depends only on globule diameter. Therefore, if the globule-diameter distribution of the oily flow is available, it may be superimposed on the rising velocity in Fig. 4. The nomograph will then provide the separator length to achieve a desired percentage of oil removal.

*1 stoke = (g)/[(s)(cm)(ρ)]; 1 centistoke = 0.01 stoke.

References

1. Miranda, J. G., Sump design for oil/water separators, *Chem. Eng.*, Nov. 24, 1975, p. 85.
2. Kirby, A. W. W., The Separation of Petroleum Oils from Aqueous Effluents, *The Chem. Engineer*, Apr. 1964.
3. Ramírez Z., A., and Moreno L., O., Estudio Experimental de Separación de Aceite-Agua por Gravedad en Equipo Piloto, Revista del Instituto Mexicano del Petróleo, July 1973.

The author

Julio G. Miranda is staff engineer of the technical department of the Concón refinery of Empresa Nacional del Petróleo, Casilla 242, Concón, Chile, where he has been in charge of design, project supervision and starting up of the refinery's pollution-control facilities. He has worked for the same company for 10 years, previously holding positions in the R&D laboratory and as substitute chief engineer for R&D quality control. He holds a chemical engineering degree from Universidad Técnica del Estado (Chile) and is a registered chemical engineer.

Single waste-treatment vessel both flocculates and clarifies

The principles of flocculation and clarification for wastewater treatment are reviewed here as separate topics; then we discuss how those two operations are economically wedded in modern treatment plants.

Robert N. Kovalcik, Dorr-Oliver, Inc.

☐ Flocculation during wastewater treatment converts finely divided suspended solids into larger particles so that efficient, rapid settling can occur. The two operations, flocculation and settling (clarification),* work especially well together when combined into a single vessel.

In such a single-tank setup, flocculation can take place in an inner, circular chamber, and clarification in an outer, annular compartment. Good solids-removal is achieved by the transfer of well-formed flocs to the clarification stage without disintegrating. There the flocs settle out quickly. The overflow effluent can be significantly more clear than when the operations are not combined. And, much less of flocculating agents is needed.

Secondary-treatment clarifiers equipped with flocculation zones now are being used to process as much as 45 million gal/d of industrial wastewater.

Before reviewing the operation of combined flocculation-clarification equipment, it is well to take more-basic looks at flocculation and clarification separately.

Forming the floc

Flocculation is often employed as a prelude to clarification (settling) when the suspended solids in a wastewater are light. In some of these cases, the solids tend to coalesce naturally into flocculant particles. In either event, chemicals can be added to promote this floc formation.

Flocculation is often preceded by coagulation. Solids in a colloidal suspension normally exhibit a net electri-

*Nomenclature in this field is not uniform. Some sources use "clarification" as the name of the entire process of coagulation, flocculation and settling; others use it as a synonym for settling. This article takes the latter approach.

cal surface charge within the hydrodynamic boundary layer around each particle. The charge is the same for all particles in a suspension, which causes them to repel each other and remain in a stable and unsettleable suspension. Coagulation is the process of neutralizing the particle charge, increasing the probability of constructive collisions between the particles. Flocculation is the subsequent physical process of particle cohesion that leads to formation of large, dense particles.

The pH of the wastewater also affects the charge state of suspended particles. Hydrogen or hydroxyl ions surround each particle according to its net charge. This orientation also must be overcome before flocculation can take place. And, for optimum flocculation, the zeta potential—a measure of the intrinsic particle-surface charge density—should be minimized to allow successful collisions.

Two kinds of flocculants

Chemical flocculants added to the suspension can be inorganic or organic; different mechanisms of coagulation and flocculation are associated with the two. Inorganic flocculants, such as aluminum sulfate (alum), sodium aluminate, ferrous sulfate and ferric chloride, dissociate in water to form ions (provided that the process conditions are suitable). These ions can neutralize the repelling charges on suspended particles, allowing increased collisions and the subsequent agglomeration necessary to produce a floc. If an excess of flocculant beyond that required for electrostatic particle neutrality is used, metal hydroxide flocs will form. These large, fluffy flocs "sweep" and entrap suspended solids, further enhancing wastewater clarification.

Organic flocculants, also known as polyelectrolytes,

Originally published June 19, 1978.

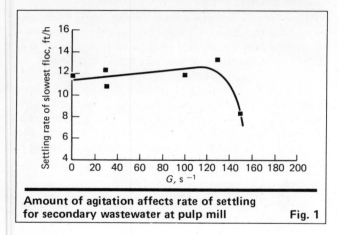

Amount of agitation affects rate of settling for secondary wastewater at pulp mill   Fig. 1

combine electrostatic, chemical and physical mechanisms to bring about coagulation and flocculation. These long-chain, high-molecular-weight, water-soluble compounds have active sites along the chain with varying degrees of anionic, cationic, or nonionic characteristics. Electrostatic attraction causes some sites to adsorb some of the colloidal particles; chemical reactions at other sites bind additional particles; and bridging between the polymers can cause still other particles to become mechanically trapped in the molecular chain and incorporated into the structure. The choice of polyelectrolyte, its dosage, and the optimum pH condition will depend on the wastewater being treated, temperature, and other process factors.

In some cases, the use of one flocculant alone will produce significant improvements. In other situations, multiple additives with different chemical characteristics may be needed.

The ultimate use to which recovered solids or clarified liquids are put after treatment also affects the selection of flocculant. In many cases, the pH requirements for landfill applications or the characteristics of allowable liquid effluents for discharge or recycle will restrict the kinds of additives that can be used.

Designing for flocculation

Chemical flocculants added to a wastewater feed stream must be uniformly distributed throughout it. Flocculants normally act in relatively dilute concentrations: inorganic flocculants range from 20 to 200 ppm in the stream, and polyelectrolytes from 0.02 to 5 ppm.

The point at which flocculants are introduced into the feed stream is critical. Turbulence should be sufficient to assure complete dispersion, but not at so high a level as to cause floc shear. In many cases, the flocculant is added upstream of the flocculating chamber, so that thorough mixing occurs prior to the flocculation stage. Adjustment of pH levels, if needed, should be prior to flocculant addition and feed-stream introduction into the clarifier.

For effective flocculation, the number of collisions between particles must be maximized. Even though some random collisions will occur, added energy in the form of mechanical agitation greatly aids the process. Conventionally, this has consisted of gentle mixing in a basin or tank.

However, destructive turbulence must be avoided, to keep the formed flocs from breaking up. Floc strength varies with the material but, in general, flocs are easy to shear and destroy. When flocs form naturally without chemical addition, or originate as a chemical precipitate, they tend to reform after shear. Those that are formed by the addition of organic polyelectrolytes rarely reform; and if reformation occurs, the floc exhibits significant differences from the original structure.

The amount of energy added to the system by mechanical agitation can be defined in terms of G, the mean velocity gradient in the fluid. The value of G is proportional to the power input, the absolute viscosity of the fluid, and the volume of the container where flocculation is occuring. The dimensionless parameter, Gt, incorporates a detention-period term expressing the time during which energy addition takes place. Typical values of G and Gt, for wastewater treatment, range from 20 to 75 s^{-1} (reciprocal seconds) and 10^4 to 10^5, respectively.

On to clarification

Clarification generally involves the removal of turbidity and suspended solids by settling. Discrete, non-flocculating particles such as coarse mineral materials follow rather closely the laws of fluid-particle dynamics formulated by Newton and Stokes. On the other hand, the settling of flocculant suspensions cannot be analyzed so straightforwardly. The characteristic feature of flocculant suspensions is that they are agglomerations of particles. The main factors affecting agglomerate settling are the available area and the detention time.

A determination of the velocity at which solids settle requires knowledge of the particle density, particle volume, fluid density, interaction forces governed by solids concentration, and drag forces. The effect of agitation emerges more explicitly in Fig. 1, which illustrates the relationship between clarification settling rate and the mean velocity-gradient parameter, G, for typical secondary wastewater treatment at a pulp and paper mill. In this particular example, the optimum, or fastest, settling rate was achieved after an amount of mechanical agitation that corresponds to a G of approximately 120 s^{-1}.

By algebraic manipulation of equations involving settling velocity and system flowrates, designers can determine an area requirement for the settling equipment. In addition to clarification area requirements, the size and configuration of the system must allow the fluid to stay in a quiescent or a controlled low-velocity condition for long enough to allow continued flocculation and settling.

In 1953, a scientific basis from which to model the above phenomenon was developed by Dorr-Oliver. Researchers found that the rate of flocculation corresponds kinetically to a second-order chemical reaction. In differential form,

$$-\frac{dC}{dt} = KC^2 \qquad (1)$$

where C is concentration of suspended solids and K is a constant. Thus, the settling rate is proportional to the square of the concentration. Taking into account the time lag prior to flocculation, as well as material that will remain in suspension regardless of the treatment, an operating expression has been defined:

$$K(t - t_o) = \frac{1}{C - C_\infty} - \frac{1}{C_o - C_\infty} \qquad (2)$$

where C is the concentration of suspended solids at time t, C_o is the concentration at time t_o when the untreated feed gets its dose of flocculant (if flocculant is used), and C_∞ is the residual concentration of suspended solids once the fluid has reached its terminal clarity.

Clarifer tanks can be square, rectangular, or circular, as long as the area and detention-period requirements are met. Differences in clarifier designs most often concern the methods used for feeding, solid-underflow removal, and withdrawal of clear-liquid overflow. Feed and withdrawal must be accomplished in such a way that flocculation and clarification take place unhindered. If concentrated underflow is desired, sludge thickening must not be disturbed.

Internal configurations of clarifiers vary widely. The method of feed may be peripheral or central, and at the surface or submerged; underflow configurations may provide for central or distributed withdrawal; and withdrawal of the overflow may be accomplished using a peripheral or radial launder. There are advantages claimed for each variation. In the design of clarifiers, the overflow rate (converted into velocity terms) must not exceed the settling velocity of the slowest floc or floc group. Flow patterns should not cause turbulence, or short-circuiting of flow from feed to withdrawal.

Some amount of floating scum is almost a certainty in operating clarifiers, and it must be removed to prevent fouling of the overflow system. This is done by a skimming mechanism. A baffle, concentric with the overflow weir, retains the scum in position, allowing it to be swept into a trough.

Combining flocculation and clarification

The combination of mixing, flocculation and clarification in one tank is more economical and efficient than conventional separate treatment units. Such a combined system can operate as follows:

First, a dilute solution of flocculating chemicals is uniformly distributed in the feed slurry, which then enters a gently agitated compartment in the center of the tank, where the floc forms and grows. The water drifts radially outward to a quiescent sedimentation zone where solids separate from the liquid by settling. Solids removal works effectively, because well-formed flocs move to this zone without disintegration and then settle rapidly.

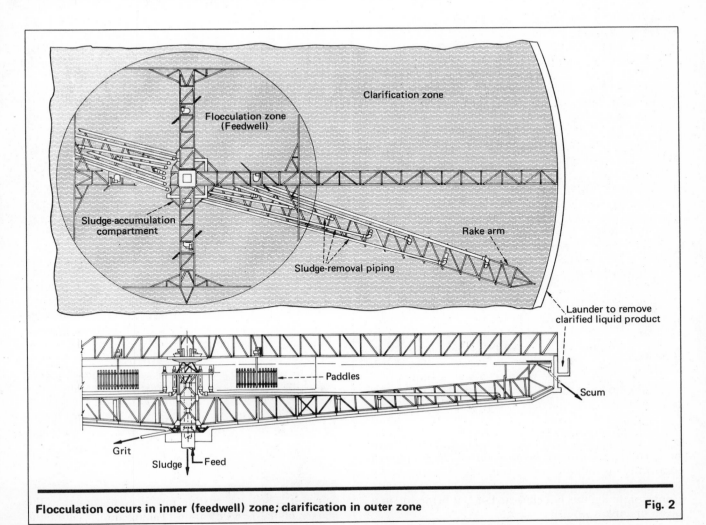

Flocculation occurs in inner (feedwell) zone; clarification in outer zone　　　　　　　　　　　　　**Fig. 2**

The main benefit is usually a lessened requirement for flocculant. Also, since there is no need for equipment to transfer fluid between two tanks, the user saves in equipment costs, and the (internal) transition from the flocculation stage to the clarification stage is smooth, with minimal floc disruption. Combination units require less monitoring, thanks to their all-in-one configuration. And the resultant ease of operation lends itself to better control of process conditions.

Equipment for flocculation/clarification

One version of combined flocculation/clarification equipment is shown in Fig. 2. Its basic design structure is a circular tank.

The feed stream enters a large concentric feedwell (the flocculation zone) and is gently agitated by gate-type paddles that superficially resemble picket-fence gates. The number and size of paddles depends on the desired G parameter and the unit size, which is determined by the flowrate. This feedwell detains the incoming fluid for a period of time so that flocculation can occur. In this version, feed is introduced radially from an annulus at the center column. Other designs provide different means for feeding. A key aim is to assure control of both the velocity and the dispersion of material into the flocculation zone.

By selecting the proper paddle size and speed, G values from 7.2 to $45.5 s^{-1}$ can be obtained in this model. Feedwell detention periods from 3 to 54 min. are achieved by choice of feedwell size and feed rate.

Although the agitating paddles are constantly in motion, the feedwell in the version shown remains stationary. In some designs, rotating feedwells are employed, to aid flocculation. Care must be taken with a rotating feedwell so that gross fluid rotation does not occur in the clarification zone close to the feedwell, which would effectively destroy the quiescent conditions desired for clarification.

Appropriate piping to remove sludge rapidly can be incorporated into flocculating clarifiers. In the unit shown, a series of pipes located along the clarifier rake-arms, and positioned a few inches from the tank floor, continuously withdraw sludge over the entire tank floor area, as soon as the material settles. These pipes discharge into a central sludge-accumulation compartment, from which the sludge overflows into an axial discharge pipe at the center column. Clarified liquid overflows into a launder at the periphery.

When the system is applied for activated-sludge treatment, fresh biological sludge is removed before anoxic conditions occur. It can be quickly returned to seed further growth in aeration basins, or discarded.

Actually, rapid sludge removal allows a three-phase separation to be made. In wastewater installations with large aeration or holding lagoons preceeding secondary treatment (as often found in, for instance, the pulp and paper industry), wind carries airborne dirt and grit into the system. If not removed, the accumulation of grit will cause operating problems. Heavier grit particles fall directly to the tank floor and can be removed via a centrally located underflow pipe (see Fig. 2). The three phases, thus, are the clarified overflow, a sludge underflow, and centrally collected grit.

While existing clarifiers can be modified to perform as combination flocculation/clarification units, there are problems in this approach. Adding the larger feedwell and the agitation mechanism needed for flocculation will reduce the tank space available for clarification. In a unit previously sized for specific application, this modification might result in an undersized clarification stage, giving unsatisfactory performance. The alteration of an existing clarifier also will be limited by structural considerations. For instance, feedwell and agitator support may be difficult. Adding the center-column feed system and the rapid-sludge-removal piping would be even more difficult.

Two typical applications

A pulp-and-paper mill in Maine has installed two 255-ft-dia. flocculating clarifiers that together are treating up to 45 million gal/d of mill wastes, in an activated-sludge circuit.

After initial processing, combined plant wastes go to an aerated lagoon that promotes the growth of aerobic bacteria, which breaks down the organics present. The wastewater then is directed to the two flocculation/clarification units, from which activated sludge can be recycled back to the process at rates up to 70% of the wastewater feed. The underwater parts of both units feature the rapid-sludge-removal approach. The rake arms are equipped not only with sludge uptake pipes but also with alternate rake and deflector blades. Settled activated sludge is removed through the sludge uptake pipes for return to the aeration lagoon, while heavier solids raked to the tank center go to waste.

Two 85-ft-dia. flocculation/clarification units are employed in a chemical-plant treatment facility in Plaquemine, La., that handles wastewater from the manufacture of organic chemicals. Wastes are held in a stabilized basin and then enter an oxygenation chamber for BOD reduction. Effluent from the oxygenation chamber is split into two streams, feeding two 130-ft-dia. clarifiers that are followed by the combination flocculation/clarification units. Clarification overflow contains only 30 to 50 ppm. of suspended solids.

References

1. Camp, T. R., "Flocculation and Flocculation Basins," *Proc. Am. Soc. Civil Engineers,* Paper No. 2722, 1955.
2. Grutsch, J. F., and Mallatt, R. C., "Filtration and Separation: Optimizing Granular Media Filtration," *Chemical Engineering Progress,* April 1977, p. 57.
3. Tenney, M. W., and Verhoff, F. H., "Chemical and Autoflocculation of Microorganisms in Biological Wastewater Treatment," *Biotechnology and Bioengineering,* Vol. 15, 1973, p. 1045.

The author

Robert N. Kovalcik, application engineer for sedimentation technology at Dorr-Oliver, Inc., Stamford, CT 06904, is currently doing research and development in that field. A graduate of Rensselaer Polytechnic Institute with a B.S. in chemical engineering, he is working toward an M.B.A. degree at the University of Connecticut. He is active in AIChE.

Estimating Hydrocyclone Efficiency

The authors have developed a graphical method for estimating the effects of changes in operating variables on the performance of a hydrocyclone.

K. NAGESWARA RAO and T. C. RAO, Julius Kruttschnitt Mineral Research Centre (Australia)

The general performance of a hydrocyclone is influenced by design variables (e.g., dimensions of the vortex finder and spigot) as well as operating variables (feed pressure and physical properties of the feed solids and feed pulp).

Hydrocyclone performance is best represented by an actual efficiency curve, which is a plot of the particle size versus the percentage of the feed of that size entering the underflow discharge. Such a curve does not pass through the origin of the graph. Kelsall [1] explained this by suggesting that if R_f is the fraction of feed fluid entering the underflow, then R_f percent of all sizes of particles are discharged in the underflow. This led to the concept of corrected efficiency:

$$E_c(d) = \frac{E_a(d) - R_f}{100 - R_f} \times 100 \qquad (1)$$

Yoshioka and Hotta [2] developed a method of reducing similarly shaped performance curves to a single "reduced efficiency curve" by plotting $E_c(d)$ vs. d/d_{50}. From an analysis of carefully controlled experimental data, Rao et al. [3,4] have developed the following expressions to define the performance of industrial-size hydrocyclones.

$$WOF = 1.1\, WF - K_1 - 10\, Spig \qquad (2)$$

$$\log d_{50} \text{ (corrected)} + K_2 = \frac{V.F.}{2.6} - \frac{Spig}{3.5} + \frac{P}{10.7} - \frac{WOF}{52} \qquad (3)$$

The centrifugal efficiency of a hydrocyclone operation is defined by its reduced efficiency curve. This is represented by the general equation:

$$E_c(d) = \frac{\exp(\alpha x) - 1}{\exp(\alpha x) + \exp(\alpha) - 2} \times 100 \qquad (4)$$

The factor α is the variable parameter that describes completely a change in the shape of the curve. Studies showed that K_1, K_2 and α are specific to a particular ore and hydrocyclone. They can be calculated by determining the performance of the hydrocyclone under one set of operating conditions (that is, from the data of a test on the system). The usefulness of these equations for designing, optimizing and controlling hydrocyclone systems has been detailed [5–11].

From Eq. (2) and (4), it follows that to determine the corrected and actual performance curves of the hydrocyclone for a particular set of operating conditions, the following are needed:

1. d_{50C} and R_f for the operation.

2. The reduced efficiency curve for the ore and the hydrocyclone under consideration.

We should also point out that Rao and Deb Kanungo [12] generalized Eq. (3) to eliminate K_2. The new expression is:

$$WOF = 0.97\, WF - 4.3\, Spig - 0.53 \qquad (5)$$

This can be rewritten to directly calculate R_f in percent:

$$R_f = 3 + 430\, Spig/WF + 53/WF \qquad (6)$$

More recently, Lynch and Rao [13] carried out detailed

Nomenclature

Q	Throughput, ton/h
VF	Diameter of vortex finder, in
$Spig$	Diameter of spigot, in
PW	Percent water in the feed pulp
WOF	Water rate of overflow stream, ton/h
P	Feed pressure, psi
d	Particle size
d_{50C}	Size of the particle having a corrected efficiency of 50%
WF	Water rate of feed stream, ton/h d/d_{50C}
x	d/d_{50C}
R_f	Percent feed water to underflow
$E_a(d)$	Actual gross efficiency of size d
$E_c(d)$	Corrected efficiency of size d
α, K_1, K_2	Constants

Originally published May 26, 1975

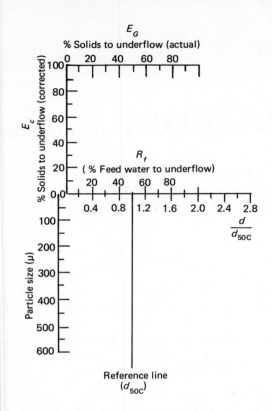

NOMOGRAPH for estimating hydrocyclone efficiency—Fig. 1

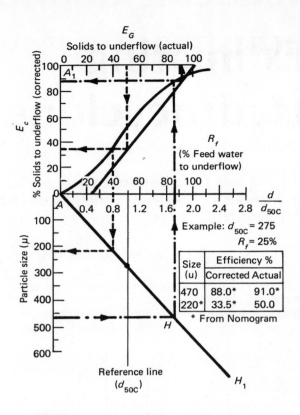

TYPICAL EXAMPLE of applying nomograph—Fig. 2

test work on 4-, 6-, 10- and 15-in hydrocyclones to extend the model developed earlier. These results confirmed that the following arguments are valid:

1. Equations of the form (2) and (4) completely define the performance of hydrocyclones.

2. The constants are dependent on the material being classified.

After establishing the constants and reduced efficiency curve from one test, the engineer can calculate the effects of any of the variables. These calculations, however, are lengthy and time consuming, and thus we have developed a nomograph to simplify the process (Fig. 1).

Fig. 2 shows a hypothetical reduced efficiency curve (ABC) superimposed on the nomograph. The steps involved in predicting the performance curve for any conditions of operation are:

1. Estimate d_{50C} and R_f for the operating conditions from Eq. (2) and (3).

2. Mark point D corresponding to R_f on the R_f scale. Join point D to E, which is 100 on the EG scale.

3. Mark point H on FG (reference line) so that $FH = d_{50C}$. Then draw line AHH_1.

The paths to be followed in predicting the efficiency of particles of a given size, and the size of particles having a given efficiency, are illustrated in Fig. 2.

Lynch et al. [14] have shown the reduced efficiency curves also exist for a rake classifier. They have developed empirical expressions for d_{50C}, R_f and the reduced efficiency curve, so it should be possible to prepare nomographs to describe rake-classifier operations. #

References

1. Kelsall, D. F., *Chem. Eng. Sci.*, **2**, 1953, p. 254.
2. Yoshioka, H. and Hotta, Y., *Chem. Eng.*, Japan, **19**, 1955, p. 632.
3. Rao, T. C., Ph.D. thesis, University of Queensland, Brisbane, Australia, 1966.
4. Lynch, A. J. and Rao, T. C., *Ind. J. of Technology*, **6**, 1968, p. 106.
5. Draper, N. and Lynch, A. J., *Proc. Australas. Inst. Min. Metall.*, **213**, 1965, p. 89.
6. Lynch, A. J., Whiten, W. J. and Draper, N., *Trans. I.M.M.*, **76**, 1967, p. C169.
7. Draper, N., Dredge, K. H. and Lynch, A. J., paper 22, 9th Commonwealth Mining and Metallurgical Congress, 1969.
8. Lynch, A. J., Rao, T. C. and Whiten, W. J., *Proc. Australas. Inst. Min. Metal.*, **223**, 1967, 71.
9. Putman, R. E. J., "Decision Making in the Mineral Industry," Special Vol. 12, The Canadian Inst. of Min. & Met., 1971, pp. 461–472.
10. Mular, A. L., *Can. Min. Met. Bull.*, **64**, 1971, p. 34.
11. Pitts, J. D., Holsinger, S. R., Johnson, N. W. and Crowell, D. E., AMIRA Symposium on Optimization and Control, Brisbane, July 1974.
12. Rao, T. C. and Deb Kanungo, P. S., *Trans I.M.M.*, **82**, 1973, p. C105.
13. Lynch, A. J. and Rao, T. C., Modelling and Scale-Up of Hydrocyclone Classifiers, 11th International Mineral Processing Congress, Sardinia, May 1975.
14. Lynch, A. J., Rao, T. C., Whiten, W. J. and Kelly, *Proc. Australas. Inst. Min. Metall.*, **224**, 1967.

Meet the Authors

Dr. T. C. Rao is an assistant professor in the department of metallurgical engineering, Indian Institute of Technology, Kanpur, India. Between May 1973 and May 1975 he was on study leave at the Julius Kruttschnitt Mineral Research Centre, University of Queensland, Brisbane 4067, Australia, working on classification characteristics of hydrocyclones. He worked from 1961 to 1966 as a teaching fellow and then a research fellow in the department of mining and metallurgical engineering at the University of Queensland and was awarded his Ph.D. degree from that institution in 1966. He has thirty-three publications on topics of mineral beneficiation processes and their computer control, ceramic research and beneficiation of ores and industrial minerals.

K. Nageswararao holds the degree of M.Sc. in Metallurgy from the Indian Institute of Technology, Kanpur, India and is presently working for his Ph.D. degree at the Julius Kruttschnitt Mineral Research Centre, University of Queensland, Australia.

Reducing fluoride in industrial wastewater

The most widely used method for reducing fluoride-ion concentrations in industrial wastewaters involves fluoride precipitation as the calcium salt. Techniques employed for potable-water systems— alum precipitation and alumina adsorption—may "polish" effluent from the initial treatment to achieve still-lower fluoride levels.

Edgar G. Paulson, Industrial Pollution Control, Inc.

☐ Fluoride contamination occurs in a wide range of industrial effluents—wastewater from aluminum and steel production, metal finishing and electroplating, glass and semiconductor manufacturing, ore-beneficiation, and fertilizer operations.

Concentration limits for fluoride of 1.0 to 1.5 mg/L have been established from time to time for surface waters, based upon two premises: (1) that these waters will serve as sources of potable water, and (2) that processes used in water-treatment plants would normally have no effect on the fluoride concentration.

Various studies on the toxicity of fluoride to aquatic life have produced conflicting results, but apparently this toxicity decreases with an increase in dissolved solids and in hardness of the water. The requirement to meet drinking-water standards, and not the toxic effect on aquatic life, becomes the limiting factor. It would, for any localized site, call for establishment of fluoride levels in a receiving stream based upon the impact of fluoride on local fish and other aquatic life.

Standard analytical methods are available for determining fluoride content of water. The most commonly used technique involves distillation of the sample, followed by colorimetric determination of the fluoride using SPADNS reagent on the distillate. Development of the specific-ion electrode started a trend toward direct determination of fluoride in the waste sample, by use of a TISAB II (total ionic strength adjustment buffer) to eliminate the effects of complexing ions such as aluminum. Recently, some analysts have been using the specific-ion electrode instead of the SPADNS reagent to minimize the possibility of interference from entrainment of sulfate during the distillation step.

Most of the development work on these procedures has involved relatively low fluoride concentrations. Concentrated spent process solutions, containing very high fluoride levels, require special precautions to ensure that distillation is being used and that all the fluoride is recovered in the distillation step.

In some cases, the sample should be diluted prior to distillation (if necessary) to reduce contents to between 1 and 3 mg/L for complete fluoride recovery and to minimize entrainment by interfering ions. Also, elapsed time between collection and analysis can affect results in the case of spent process solutions.

The standard analytical methods, based upon distilled samples, measure ionized fluorides, as well as

Originally published October 17, 1977.

complexed fluorides which might or might not respond to a given treatment. Fluoride values determined on undistilled samples with a specific-ion electrode have been evaluated against results obtained on distilled samples, in an attempt to differentiate between the forms of fluoride present in a wastewater. These comparisons indicate the presence of complexed fluorides in a waste stream but do not correlate directly with treatability and with potential removal capability for a particular treatment.

In considering treatment to reduce fluoride concentration, the following factors should be considered:

1. The most widely used fluoride-removal treatment is precipitation as a calcium salt.

2. This treatment will produce an effluent containing 12 to 30 mg/L of fluoride.

3. For further reduction, defluoridation techniques employed to treat potable water can be used as additional treatment for industrial wastewaters, but more study is needed on each specific application.

Calcium precipitation

The most commonly used treatment for practical removal of fluoride involves adding a soluble calcium salt. This forms an insoluble calcium fluoride:

$$Ca^{++} + 2F^- \rightarrow CaF_2\downarrow$$

This will lower the fluoride concentration to less than 30 mg/L but above the theoretical value of 7.8 mg/L,

the actual result depending upon the quality of the water undergoing treatment. (The theoretical solubility of calcium fluoride is 16 mg/L, which corresponds to the 7.8 mg/L fluoride-ion concentration, determined on pure compounds in distilled water.)

In a real wastewater situation, with various interfering and complexing materials present, excess calcium will shift the solubility reaction and may even produce lower fluoride levels, due to the common-ion effect. The solubility product is a constant, equal to the product of the molar concentration of the calcium ion and the square of the molar concentration of the fluoride ion. Any increase in the excess concentration of calcium ion consequently reduces the solubility of fluoride ion to satisfy the solubility-product constant.

However, any fluoride present in any complexed form does not readily react with, nor is it precipitated by, addition of calcium salt. In this instance, the level attainable by this treatment will remain above the theoretical solubility of the fluoride ion, and additional calcium may have little impact on the final fluoride-ion content.

Sludge can cause other effects. Recirculation of sludge may noticeably improve fluoride removal and give lower residual concentration.

To determine attainable fluoride levels when treating a specific waste stream, studies must be undertaken to duplicate planned treatment procedures and conditions, based on agreed analytical methods.

Calcium requirement

The first requirement in removing fluoride by precipitation is to assure the presence of enough calcium to react with the fluoride. To precipitate a pound of fluoride ion theoretically requires 1.06 lb of calcium. This corresponds to 2.2 lb of 90%-active lime or 2.9 lb of calcium chloride. Automatic pH control is often used for addition of lime for fluoride precipitation.

This treatment has proven successful because many fluoride streams have low pH and may contain hydrofluoric and other strong acids. Just satisfying the alkali requirement, however, does not by itself guarantee that sufficient calcium has been added to complete the precipitation reaction. A simple stoichiometric balance conducted on the calcium ion will show whether the primary requirement of adequate calcium ions has been satisfied.

Feeding both calcium hydroxide and calcium chloride for fluoride precipitation will help meet the calcium requirement, while operating within a predetermined pH range and at the same time proportioning chemical addition on the basis of need. The principal instrumentation for this purpose, a pH controller, will vary the lime requirement with the acidity of the waste being treated. Interlocking the calcium chloride feed system with the lime feed system, to maintain a definite ratio of $Ca(OH)_2$ to $CaCl_2$, will accurately proportion the $CaCl_2$.

Another alternative: proportion the $CaCl_2$ to the quantity of wastewater being treated and vary the lime feed according to pH. This system requires periodic testing to ensure adequate calcium addition, but not too much of an excess. The relationship between acidity

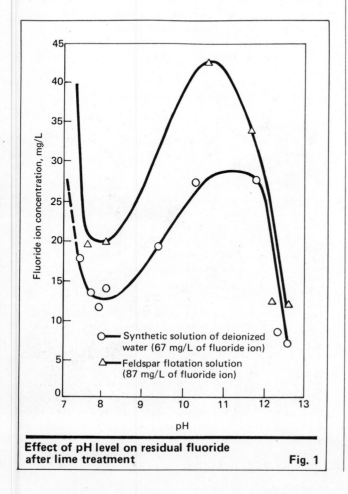

Effect of pH level on residual fluoride after lime treatment   **Fig. 1**

Synthetic solution of deionized water (67 mg/L of fluoride ion)

Feldspar flotation solution (87 mg/L of fluoride ion)

and fluoride content of the incoming waste can dictate which of these systems to use for a given situation.

Required pH

The next requirement: establish the desired pH range to obtain minimum fluoride levels. Field experience indicates that a pH between 8 and 9, or above 12, will assure minimum solubility, as shown in Fig. 1. Since the pH was raised by adding considerably more calcium, as lime, the lower fluoride concentration may result from higher calcium values as well as elevated pH.

Additional factors enter into the selection of proper pH for maximum fluoride precipitation: (1) what other constituents are present in the waste stream, (2) the desired final concentration for these constituents, and (3) how the fluoride-removal process integrates into the total treatment process.

Quite often, for instance, the fluoride-bearing stream also contains heavy metals. These are normally removed from the water, by precipitation as the hydroxide, at pH of 8.5 to 9.0. Certain metals (chromium, aluminum, zinc) are more soluble at pH 12 than at the less-alkaline values. To remove fluoride and heavy metals simultaneously, the overall results (maximum removal of the several constituents involved) may become the guiding factor in selecting the optimum pH, as compared to the best pH to get rid of fluoride only.

Sometimes, fluoride is being removed as a pretreatment on a segregated stream; the supernatant can then be blended with other waste streams for heavy-metal precipitation. In this case, the pH can be chosen primarily on the basis of fluoride removal.

Also, final pH values for discharge to a receiving stream generally fall in the range of 6.5 to 9.0. So with precipitation as the final step of the treatment prior to discharge, the necessary high pH would then require neutralization with acid prior to final discharge.

On the other hand, the supernatant may be incorporated into another stream for additional treatment. Then the high alkalinity needed previously for heavy-metal removal may be useful in this stage. This will avoid having to add more alkali where the subsequent step calls for an elevated pH.

Reaction time

The third requirement is reaction time. This can vary widely. For some waste streams, a 30-minute reaction time (followed by liquid/solids separation) gives maximum fluoride removal. With other wastes, long-term retention, as much as 24 hours, can increase fluoride removal.

Generally, with pure compounds, a reaction time of 30 minutes is more than enough. Hypothetically, at least, complexed fluorides (1) hydrolyze, (2) react with calcium, and (3) precipitate from solution. This removes more fluoride and lowers its residual concentration. The only way to determine the required reaction time exactly for a specific waste is to conduct treatability studies on that particular waste stream.

Treatment dynamics

Fluoride can be removed in either a batch operation or a continuous flow-through process. Batch treatment is usually employed when volumes are small and/or the generation rate is low (on the order of 25,000 gpd, or with flowrates below 50 gpm). Continuous treatment is generally preferred for larger volumes (100,000 gpd, or 70 gpm). Each situation, however, calls for individual consideration.

Batch treatment

Batch processing requires several holding and treatment tanks, each sized on the basis of (1) quantity of waste to be treated, (2) frequency of treatment, and (3) volume of treatment-chemicals added. While one tank, or set of tanks, is receiving waste, the contents of the other tanks are being treated and disposed of.

The usual treatment practice involves collecting a predetermined volume of effluent in a tank. After agitation of tank contents and addition of treatment chemicals, pH must be adjusted to the proper range.

Tank contents should undergo agitation for at least 30 minutes. Depending upon characteristics of the precipitate, addition of ionic and/or polymeric coagulants may help to promote liquid/solids separation. A multispeed or variable speed agitator will facilitate mixing and improve precipitate growth.

After the precipitate has formed and agglomerated, the mixer is shut off, allowing solids to settle to the bottom of the tank. The clear supernatant is then decanted from the tank, either to discharge or to a subsequent treatment step; the settled solids are transferred into a container for disposal. Total removal of settled solids from each batch is not essential, but a conical or sloped bottom in the tank does make it easier and faster to get rid of most of the solid material.

Designing auxiliary equipment associated with a batch-treatment operation should take into account the capacity of the chemical-addition system, the decant system, and the sludge-removal system. Assuming an 8-h cycle, it should take no longer than 1 h to introduce the treatment chemicals into the tank, no longer than 3 h to discharge the supernatant, and at most 1.5 h to empty the sludge.

Continuous treatment

A typical flow-through system consists of equalization, reaction tank, liquid/solids separation, chemical-feed systems, sludge-handling systems, and instruments and controls—all size-based on the following criteria:

1. Equalization will even out fluctuations in concentration and hydraulic loading on the treatment facility.

Proper equalization represents one of the most critical features in successful flow-through treatment. This equalization tank should have mechanical agitation to produce a uniform concentration.

The sizing of this system depends upon operating practice within the plant where the wastewater originates. With reasonably uniform quantity and concentration of waste, retention times of one to two hours will provide adequate equalization. However, processing operations involving wide fluctuations in hydraulic rates as well as concentrations can require equalization time of as much as 8 h.

Regarding disposal of spent process solutions, two principal alternatives exist: (a) Collecting spent process solutions separately and then blending them into the dilute stream at a controlled rate. (This method can reduce the size of the required equalization facilities.) (b) Dumping these solutions into the same collection system as the dilute fluoride-containing waters. (This calls for considerably larger equalization facilities and may require two tanks in series to minimize fluctuations in concentration.)

2. Waste is transferred, either by pumping or by gravity, from the equalization tank to a reaction tank. Treatment chemicals, usually lime, are added at this point, with feed based on pH control and with a minimum retention time of 30 minutes. As noted above, calcium chloride can be added simultaneously, proportioned to the lime requirement or the flow, to optimize precipitation.

3. The waste then flows by gravity to a liquid/solids separation unit. From here, the treated water overflows either into a surface stream or to subsequent treatment steps—generally sized at a rise rate of about 0.3 to 0.5 gpm/ft². A full bottom rake in the separator promotes sludge removal. Also, solids-contact units with internal sludge recirculation may increase chemical efficiency.

Automatic control of sludge blowdown should depend upon volume of waste treated. This involves taking a signal from the totalizer, or the flowmeter, to a field-adjustable counter; this actuates a field-adjustable timer that controls the duration of sludge blowdown.

Addition of coagulants and polymers to the stream—either prior to entering, or into the center mixing well of the solids-contact unit—should be proportioned to the waste-stream flow by use of a counter/timer device. Recently, plants have begun using variable-speed drives on chemical-feed systems, based upon the flowrate of the effluent.

4. To reduce fluoride concentration to solubility levels calls for polishing the clarifier effluent to remove trace quantities of suspended solids—through the use of polishing lagoons or by filtration. Polishing lagoons with retention time of 8 to 24 h can reduce suspended solids to levels of 2 mg/L or less.

Filtration can reduce suspended solids to less than 1 mg/L. Actually, few plants filter their final effluent; this limits the amount of field experience for this step. Effluent-filtration involves a water containing precipitated calcium salts. Consequently, the usual practice of stabilizing this water by adding acid would redissolve some of the suspended material which the filter is trying to remove.

Dual, or multi-media, filters would probably work best for this type of situation, and they should have surface-wash or air-scour capabilities. Periodic chemical cleaning of the filters will minimize incrustation and keep the media in good operating condition.

5. Sludge from the solids-contact unit, containing between 1 and 2% solids by weight, then transfers to a

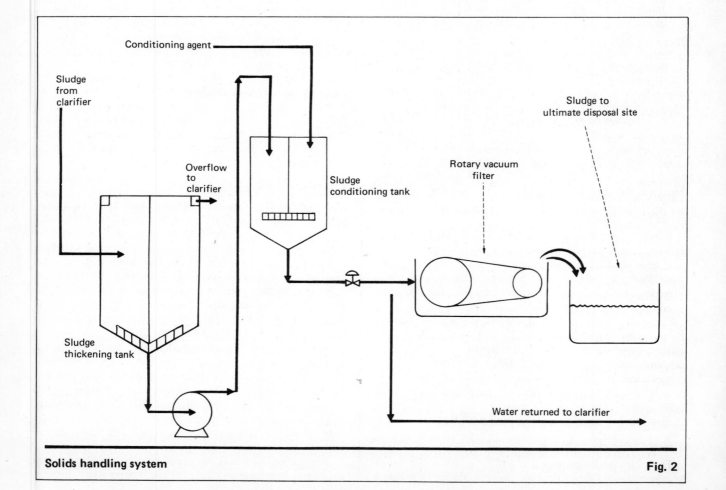

Solids handling system   **Fig. 2**

holding tank—possibly equipped with internals to promote further thickening. This equipment combination can boost sludge concentrations to the range of 2 to 10% solids. Other mechanical means (such as vacuum filters) can dewater the sludge even further, to make ultimate disposal more convenient or economical.

Fig. 2 depicts a typical sludge-handling system.

Neutralization procedures

When treating spent hydrofluoric acid solutions separately, the acid must be blended into the contents of the reaction vessel having enough lime in it to keep the charge on the alkaline side at all times. Neutralization of hydrofluoric acid with lime is an exothermic reaction. But when neutralization and precipitation take place simultaneously, the overall heat of reaction does not appear to become excessive. However, if the system goes over to the acid side, the neutralization reaction generates so much heat that it must be dissipated. Also, the reaction may liberate fumes which require wet-scrubbing to control.

The usual procedure involves collecting the spent hydrofluoric acid in a separate container and metering it into an agitated mixing vessel. Lime metered into the reaction vessel at the same time maintains pH of the solution at 9 or above. With continuous monitoring of pH as treatment proceeds, an interlock arrangement will cut off the acid-metering pumps if pH falls below a preset level (in this case, pH 9).

One way to produce even lower fluoride concentrations involves segregation of the fluoride-containing waters to assure maximum calcium precipitation. Blending the resultant supernatant with other process streams will reduce the overall fluoride level. This may meet the limiting criteria with no additional treatment needed.

For still further fluoride reduction, the two most generally promising treatments are *alum precipitation* and *adsorption on activated alumina*. In evaluating these processes for wastewater treatment, users must consider the following factors:

1. Fluoride-removal techniques have been developed primarily to reduce fluoride in potable waters free of the variety of contaminants present in wastewater effluent. At the present state of the art, each individual application requires a trial evaluation program.

2. Screening will be needed to see whether fluoride levels are reduced enough by these methods.

3. Pilot-plant studies can tell whether the removal processes work dependably under expected variations in influent characteristics. Even though fluoride content may stay in a relatively narrow range, as a result of processing consistency prior to precipitation, amounts of presently unidentified interfering substances may vary widely.

4. Pilot tests will be required also to develop engineering data vital for scaling up the process.

5. The impact of these fluoride-reducing processes on final water quality, such as increases in total dissolved solids (TDS), must receive full consideration.

6. Fluoride-containing streams from these processes may require further treatment to convert the fluoride to a form suitable for ultimate disposal.

7. It will take some lead-time to implement this program fully and in a logical, stepwise manner.

Alum precipitation

Alum (aluminum sulfate) has been used to remove still more fluoride from potable waters. The efficiency of alum flocculation affects the degree of fluoride removal. Optimum pH can range between 6.5 and 7.5 [2]. The alum requirement is a function of the flocculating pH [3]. Taking the alum requirement at pH 7 as a baseline, it will require about twice that amount of alum at pH 8 and 3.5 times as much at pH 9, when treating the same influent concentration of fluoride and reducing it to the same level.

The quantity of alum consumed depends upon both the initial fluoride concentration and the desired final fluoride level. This alum input rises as the residual fluoride decreases. To get a residual below 4 mg/L, it takes significantly more alum [3].

For example, with a solution containing 15 mg/L of fluoride, and flocculating it at pH 7, it takes about 30 lb of alum/lb of fluoride to lower the residual to 4 mg/L. To reduce the residual to 1.5 mg/L would take some 90 lb of alum. These order-of-magnitude estimates—for preliminary planning purposes only—may change significantly when treating a specific waste stream.

Since efficient removal depends upon good alum flocculation, an effective process utilizing this approach would involve flash mixing, flocculation, and liquid/solids separation. The criteria for water-clarification equipment should generally apply to other major items of process equipment. Sludge recirculation may not notably improve chemical efficiency, but it should improve floc formation and settling rates. Polymer addition might also help in this regard.

Alum increases sludge

Alum produces extensive amounts of sludge. Estimates show that the settled sludge will amount to 18 to 40% of the volume treated. Handling this sludge usually involves thickening, mechanical dewatering, and ultimate disposal.

To point up the significance of this treatment step, every pound of fluoride precipitated (as CaF_2) produces about 2 lb of calcium fluoride sludge (on a dry-weight basis). Using 28 lb of alum per lb of fluoride—to reduce fluoride level from 14.5 mg/L down to 4 mg/L—will yield 11.2 lb of dried sludge per lb of fluoride removed. To reduce the same fluoride concentration to a level of 1.5 mg/L will take 90 lb of alum, and will produce about 27.4 lb of sludge, per lb of fluoride removed.

How this sludge will be handled becomes a serious consideration in deciding whether this will be a viable solution to the problem of achieving lower fluoride levels. Disposal methods for metal-hydroxide sludges have become increasingly restricted in recent years. This means that costs associated with sludge disposal have climbed significantly. The current status of regulations involving disposal of metal-based sludges does not provide any good guidelines for even short-term alternatives. This further confuses the treatment picture.

Potentially, alum can be recovered, for reuse, from the fluoride-containing sludge. Technical feasibility and relative economics would have to be determined for each situation, and process technology developed and refined, before the potential can be considered valid.

Activated alumina

Activated alumina can reduce fluoride-ion levels to less than 2 mg/L. Initial studies on defluoridation of potable waters have found that, at a regeneration level of 12 lb of alum/ft^3 of alumina adsorbent, the activated alumina has a capacity of some 2,000 gr/ft^3.

Additional studies [3] on this process have defined a number of other significant factors:

1. Flowrate has an impact on the attainable final fluoride concentration. A flowrate of 1.25 gpm/ft^3 can reduce fluoride level to below 1.5 mg/L—and a definite exhaustion point appears in the run cycle. At a flowrate of 5.5 gpm/ft^3, a continuous breakthrough occurs, and the fluoride concentration steadily increases as the run proceeds.

2. Regeneration with sulfuric acid shows a capacity of 1,700/ft^3 for the first regeneration and 900/ft^3 after two regenerations. It seems to make no difference in regeneration results whether the H_2SO_4 is used as a 1% or a 4% solution.

3. The rate at which the acid regenerant passes through the bed affects the capacity of the bed as well as the amount of activated alumina depleted during each regeneration cycle. Decreasing the regenerant flowrate by 75% will increase the amount of activated alumina depleted per cycle by a factor of 7, and will reduce regeneration capacity.

4. Regeneration with alum also lowers the capacity of alumina after the first regeneration.

5. The acidity of an alumina column affects its capacity. When treating an influent having pH of 11 to 12, the capacity is only 25% of the capacity at influent pH of 7 to 8. Not surprisingly, the use of an alkaline wash after acid regeneration will reduce the capacity of the particular column.

6. A water rinse after acid regeneration produces a rinse effluent on the acid side. This will require post-rinse adjustment of pH before the rinse water can be discharged.

7. Sulfuric acid regenerant causes plugging and solidification in the alumina bed.

8. Hydrochloric acid regeneration produces satisfactory results and avoids bed-plugging. Reusing the HCl regenerant three times does not affect capacity; it does reduce the amount of alumina dissolved in each regeneration cycle.

9. Some complexed forms of fluoride are not removed by activated alumina.

This process, too, presents some problems, chiefly (1) handling the regenerant, and (2) the impact this regenerant will have on overall effluent quality. Collecting spent regenerant and displacement rinse, and bleeding them into a calcium fluoride precipitation stage, will require more lime for neutralizing the acid. (The lime will also precipitate alumina dissolved from the adsorption bed.)

Sulfuric acid as a regenerant potentially increases the sulfate concentration by about 1,500 mg/L just to remove 15 mg/L of fluoride. Separate treatment of the H_2SO_4 regenerant would reduce this effect, since both CaF_2 and $CaSO_4$ precipitate out. The resulting increase in dissolved sulfate would amount to only 100 mg/L or so while removing 15 mg/L of fluoride.

Using hydrochloric acid as the regenerant, based upon four regeneration cycles before discarding the HCl, would add 1,200 mg/L of chloride ion just to remove 10 mg/L of fluoride.

Summary

Fluoride control has become increasingly necessary as antipollution regulations grow more stringent. Several practical control measures are available, but fluoride-removal techniques today require cautious consideration, because several of the basic factors can limit their applicability:

1. The process in most-general use now is precipitation of fluoride as a calcium salt.

2. This process generally produces effluents with fluoride concentrations as low as 12 to 30 mg/L.

3. For still lower concentrations, first segregate the fluoride-containing streams and treat them separately, to find out whether this will give an overall effluent with the desired final fluoride concentration.

4. Processes developed for defluoridation of potable-water supplies may apply also to the treatment of industrial wastewaters.

5. These processes would most likely be used, as a polishing stage, on effluent from a calcium-precipitation step.

6. These fluoride-reduction techniques substantially increase dissolved solids, which may not be desirable.

7. Alum flocculation can produce substantial quantities of sludge.

8. The various defluoridation operations require extensive investigation to determine their applicability to any given effluent situation.

9. It will take a long lead-time to put any of these defluoridation processes to work.

References

1. Parker, C., and Fong, C. C., Fluoride Removal Technology and Cost Estimas, *Industrial Wastes*, Vol. 21, No. 6, Nov–Dec 1975, pp. 23–27.
2. Culp, R. L., and Stoltenberg, H. A., Fluoride Reduction at LaCrosse, Kansas, *J. American Water Works Assn.*, Vol. 50, pp. 423–431, 1958.
3. Zabban, W., and Helwick, R., "Defluoridation of Wastewater," Proceedings of the 30th Annual Purdue Industrial Waste Conference, Purdue University, 1975.

The author

Edgar G. Paulson is vice-president and technical director of Industrial Pollution Control, Inc., 45 Riverside Ave., Westport, CT 06880, where he is in charge of the design of industrial waste-treatment plants and startup assistance. Previously, he was with International Telephone & Telegraph, as assistant technical director of environmental technology, and with Calgon Corp. as manager, Environmental Engineering Dept. He holds B.S. and M.S. degrees in chemical engineering from the University of Pittsburgh, and belongs to AIChE, Pennsylvania Water Pollution Control Assn., and Air Pollution Control Assn.

Techniques for Removing Metals From Process Wastewater

Concern for the potential hazards of trace metals in effluent streams has only recently become widespread. The task of meeting ever-more-stringent effluent standards is now being imposed on the chemical process industries.

THEODORE W. CADMAN, University of Maryland, and ROBERT W. DELLINGER, U.S. Environmental Protection Agency

Numerous techniques have been proposed or are being used to meet the environmental problem of trace metals in effluent streams. This survey presents summaries of heavy-metal removal techniques, along with pertinent references, and results available to date.

Removal by Chelation and Extraction

Many heavy-metal ions are commonly removed and recovered by chelation with ammonium pyrrolidine dithiocarbamate, which is added to a solution in stoichiometric quantities. The chelate is then removed by extraction [23].

Lead, cobalt and nickel will chelate in this way at a pH of 2.8. Methyl isobutyl ketone will extract the metal chelates, the ions being in the 40 μg/l range [12]. Copper, lead, nickel, zinc and cobalt may be so removed at a pH between 4 and 5 (adjusted with HCl) in the several-ppb range, with a solution-to-methyl-isobutyl-ketone ratio of 750/20 [8]. Pyrrolidine dithiocarbamic acid and its ammonium salt will chelate cadmium, cobalt, copper, lead, manganese, molybdenum, vanadium, nickel and zinc at a pH of 4.5 to 6, with the metal ions extracted with chloroform [23].

Other agents for chelating heavy-metal ions include potassium thiocyanate, which will chelate molybdenum in the presence of stannous chloride (reducing agent). Molybdenum dissolved in 500 ml of water at a concentration of 10 μg/l requires 3 ml of 10% solutions of potassium thiocyanate and stannous chloride. The thiocyanate complex is extracted by 16 ml of isopropyl ether [11].

Foam-Separation Recovery

Valuable metal solutes at low concentration in process streams may be effectively recovered by foam separation (fractionation), a technique applicable to strontium, chromium, cobalt and nickel. Forced by nitrogen gas to rise in a column, the foam (carrying adsorbed nickel complex) is withdrawn to a foam collector. The foamate

product is completely recycled to the bottom of the column, steady state being reached in 1 h. In experiments, enrichment ratios of 16 were achieved at a pH of 5.5 to 6.5, nitrogen flows of 500 ml/min or less, nickel ion concentrations of less than 0.06 g/l, and a surfactant concentration of 0.4% of the nickel ion concentration [4].

Enrichment ratios of 12 were obtained for strontium at a pH of 8, a strontium ion concentration of 1.6 g/l, and a nitrogen flow of 100 ml/min. Sodium o-hydroxyphenyl butyl benzene sulfonate was the surface-active agent, and a steady state was achieved in 20 min [33].

Autoclave Recovery

Marine microorganisms are cultivated in sterile seawater, commonly prepared in an autoclave in which a liquid is heated above its boiling point under pressure with steam at 15 psi and 250°F.

The process forms a precipitate in sample flasks. When a sample was autoclaved at 121°C (236°F), returned to a pressure of 1 atm, then passed through 0.3-micron-porosity filters, the precipitate was recovered by scraping the flask. After drying for four days at 50°C, all but 2 to 3 mg of the precipitate was recovered from the filters. An analysis showed 66% of the titanium recovered, as well as major portions of the chromium, copper, lead, manganese and tin [15].

Adsorption of Strontium

Strontium-90 is a bone-seeker and therefore a very hazardous isotope in water. Activated alumina, Al_2O_3, will selectively adsorb strontium solutions containing high concentrations of certain competing cations. In the cationic exchange, one strontium ion replaces one sodium ion on the activated aluminum-oxide surface. Spherical activated alumina is ground to pass through 80-mesh screen and heated 16 h at 250°C for maximum activity.

The process is very pH dependent; for alumina concentrations of 0.2 moles/l and strontium concentrations

Originally published April 15, 1974.

of 1×10^{-5} moles/l, only 10% of the strontium was removed at a pH of 5, but 98% at a pH above 8. With strontium concentrations of 1×10^{-3} moles/l, removal fell to 55% at an alumina concentration of 0.2 moles/l and a pH above 8 [6].

Manganese Removal

Spray aeration, iron fixation, manganese fixation, filtration, and contact with granular activated carbon is a traditional means of removing manganese from water.

Manganese and organic-matter levels were sharply lowered in a pilot-plant study in which spray aeration and filtration with filter aid (with potassium permanganate the oxidizing agent) reduced the load on granular activated carbon by 75%.

The cost of chemicals and filter aid was 1.5¢/1,000 gal of water treated. Activated carbon cost 4.0¢/1,000 gal but regeneration by burning or chemical means reduced the cost to 0.8¢/1,000 gal [3].

If lime-soda treatment facilities are lacking, manganese is best removed by an oxidation process. The oxidation was once mostly done by means of an aeration system, which is now being replaced by either a regeneration-type manganese zeolite bed or continuous permanganate feed with subsequent filtration.

Manganese zeolite is the oxidizing source and filter medium in the regeneration-type zeolite-bed approach. The alternate treatment of process green sand (sodium zeolite) with manganese sulfide and potassium permanganate produces manganese zeolite, a bed of which oxidizes soluble manganese to insoluble oxides and filters the latter out of water at 0.09 lb of manganese/ft³ of 16–20-mesh manganese zeolite [47]. After the oxidizing capacity of the bed has been used up, soluble manganese breaks through. Breakthrough is prevented by batch regeneration with a ½% solution of potassium permanganate ($KMnO_4$). However, $KMnO_4$ is an uneconomical regenerating agent at high flows or high manganese concentrations, or both. This has led to the replacement of batch feeding with continuous permanganate feeding.

The continuous-feeding process involves adding $KMnO_4$ continuously ahead of a filter bed of manganese zeolite, on top of which is a layer of anthracite. The bed need not be regenerated because soluble manganese is oxidized by the $KMnO_4$ before it reaches the bed. After deposits have built up, the bed is backwashed as an ordinary sand filter would be.

The reactions of the manganese ion with $KMnO_4$ are [1]:

$$3MnSO_4 + 2KMnO_4 + 2H_2O = \\ 5MnO_2 + K_2SO_4 + 2H_2SO_4 \quad (1)$$

$$3Mn(HCO_3)_2 + 2KMnO_4 + 2H_2O = \\ 5MnO_2 + 2KHCO_3 + 4H_2CO_3 \quad (2)$$

$$8MnSO_4 + 2KMnO_4 + 7H_2O = \\ 5Mn_2O_3 + K_2SO_4 + 7H_2SO_4 \quad (3)$$

In one instance, 1.67 mg/l of $KMnO_4$ is said to be necessary to remove 1 mg/l of dissolved manganese [1]; in another, a 2 to 1 ratio is recommended [47]. Using the more conservative figure, 33 lb of $KMnO_4$ would treat 1 million gal of water having a manganese concentration of 0.2 mg/l. With $KMnO_4$ at 30¢/lb, 1 million gal of treated water would cost $10. Continuous feeding requires 50% less capital than the standard aeration process, and can be expanded by simply adding manganese zeolite filters in parallel [47].

Radiation and Magnetic Field Treatment

A method and apparatus developed in 1970 for making seawater potable by extracting ions from it at low cost also removes dissolved materials.

The invention, which is not limited to separating materials from seawater, uses electromagnetic radiation, preferably in the ultraviolet or X-ray spectrum, to change the charge on ions in the water. Radiation of the proper wavelength strips electrons from the ions, reducing them to a higher ionization potential.

An electromagnet extending around the circumference of the pipe concentrates the ions in localized areas of the stream cross-section. Water at central regions of the flow is withdrawn through an outlet pipe, which deflects the water of lower ion concentration to outside the pipe. Water of higher ion concentration continues flowing through the pipe. A bar magnet with one pole located close to the side of the pipe cuts down on Brownian motion of the ions in water [45].

Paper Chromatographic Separation

Paper chromatography affords a means of separating inorganic cations. Separation depends on the ability of a solvent to cause a differential migration of cations on thin layers of cellulose. A mixture of chloroform, methanol, acetone, isopentanol and formic acid in ratios of 1:1:1:1:0.5 was found to effectively separate mercury, copper, lead, bismuth and cadmium [27].

In a study to separate manganese, iron, cobalt, nickel, copper and zinc, cellulose was first impregnated with the liquid ion-exchanger tri-N-octylamine (TNOA) as a hydrochloride salt. The mechanism of separation was an ion exchange between chloride and metal ions. Hydrochloric acid at a concentration of 4.5 molar separated zinc, copper, nickel, cobalt and manganese but not zinc from iron [24]. (Zinc was later separated from iron with a 0.05 molar solution of HCl.)

Ion Removal by Rotating Electrodes

Useful for desalting water, a method developed in 1969 for removing ions from an ionized liquid electrically and mechanically may also be applied to many process systems. An ionized solution is passed into a central chamber, where it is contacted with two rotating bipolar electrodes that adsorb ions on the central chamber side. Cations in the liquid are attracted toward the surface of the negatively charged side of a cylindrical electrode and follow it as a thin stream into a side-space as the electrode rotates. The reverse side of the rotating electrode, being positive, repels cations into the liquid in the side-space. The cations move toward an outlet, where they are attracted by a stationary negative electrode.

At the same time, anions may be treated similarly on

the opposite side of the central chamber and also removed in a side-stream. Carrying out this process a series of times makes possible the removal of liquid from the last central chamber, where the ion concentration meets required standards [42].

Biological Ion Removal

Experiments have shown that some strains of yeast recovered from acid mine-waters can precipitate dissolved copper in the form of copper sulfide, in the presence of H_2S generated by reducing elemental sulfur with glucose [10].

The following bacteria have been reported possibly useful in abating pollution: *Leptothrix ochracea, Clonothrix putealis, Siderococcus sp.* and *Toxothrix trichogenes.* These organisms can eliminate iron and manganese from water when added to a slow sand filter by oxidizing the metals to precipitable forms. The filter then removes the precipitate [18].

Removal and Recovery by Ion Exchange

The ion-exchange resin Chelex-100 (below), a deriva-

tive of iminodiacetic acid, affords a rapid and simple means of simultaneously concentrating a number of trace elements in water, the order of selectivity of the cations being Cu > Pb > Cr > Ni > Zn > Co > Cd > Mn, with a high preference for these cations over Na, K, Mg, and Ca [20]. The resin is suited equally to seawater and fresh-water applications [9].

Supported on sintered glass, the resin is suspended in an ion-exchange column, through which acidified water (pH 5–6) is passed. Cations may be eluted from the column with 4-normal HCl [20].

The resin reportedly will remove essentially 100% of the cobalt in seawater [20], and 100 g of it will remove all the cobalt, zinc, manganese, nickel, copper, cadmium and lead from 2,000 l of seawater [9]. A resin of the form of Chelex-100 affords an inexpensive means of recovering cobalt and zinc from a waste stream of the oxo-alkylation process [43].

Removal by a Chelating Polymer

Chitosan, a widely used ion-exchange resin, exhibits essentially the same high capacity for collecting of zinc and mercury ions in water solutions as Chelex-100. All the cobalt (0.5 μg/l), copper (3 μg/l) and zinc (10 μg/l) of the indicated concentrations was removed in a 10-cm × 1-cm chitosan column. Chitosan was also reported effective in removing manganese, titanium, chromium and lead. All these ions can be totally eluted from the column with 0.1-normal ethylenediaminetetraacetic acid (EDTA) [25].

Amberlite Ion-Exchange Resins

Amberlite polystyrene ion-exchange resins (impregnated on a paper filter or supported in an ion-exchange column) have been found effective for removing and recovering cadmium, cobalt, copper, nickel, lead, vanadium and zinc [35]. The resins have also been recommended for recovering zinc from such processes as that for viscose-rayon production [41].

Inorganic Ion Exchanger

Titanium arsenate was found to have high affinity for the following metal ions in the order listed: Pb > Cu > Cd > Sr > Zn > Mn > Ni > Co. These ions are effectively eluted from a column packed with titanium arsenate by a solution of 1.0 molar NH_4NO_3 and 0.1 molar HNO_3. The exchange capacity at room temperature is 0.95 milli-equivalents per gram, with a 0.01% dissolution upon elution [29].

Recovery by Another Ion-Exchange Resin

A comparison of 10–20-mesh Permutit-S1005 chelating resin with 50–100-mesh Chelex-100 ion-exchange resin in recovering metal ions from seawater revealed the Permutit resin to be 100% effective for cadmium, cobalt, copper, lead, manganese, molybdenum, nickel, vanadium and zinc, but only 10% for chromium. The results for Chelex-100 were the same except that 100% of the chromium was recovered but only 60% of the manganese [30].

Anion-Exchange Resin

Experiments with dibromo-oxine supported on the anion-exchange resin DeAcidite FF and tested on 1-liter and 5-liter seawater samples revealed that both cobalt and zinc were totally retained by the resin at their seawater concentration ranges. One hundred ml of 0.2 normal HCl recovers all the cobalt and 60% of the zinc, and 100 ml of 0.2 normal H_2SO_4 also recovers all the cobalt and 75% of the zinc [37].

A method devised to remove chromate from cooling-tower blowdown before it is discharged, and to recover the chromate in a reusable form, involves an anion-exchange resin of the same form as DeAcidite FF. The reactions governing this recovery process are:

Exhaustion:

$$HCrO_4^- + RCl \rightarrow RHCrO_4 + Cl^-$$

Recovery:

$$RHCrO_4 + OH^-/Cl^- = RCl + CrO_4^= + H_2O$$

RCl is the DeAcidite FF and OH^-/Cl^- the caustic-chloride regenerant. The chromate anion is recovered as a 3.5% solution. The lower the pH, the more effective the chromate recovery.

The cost for the chemicals involved is $0.51/lb of chromate recovered, an improvement over the standard reduction-followed-by-chromic-hydroxide-precipitation

Summary of Techniques for Removing Metals

Method	Source	Reagents	Concentration	pH Range	Ions	Reference
Extraction	Water	Ammonium pyrrolidine dithiocarbamate, methyl iso butyl ketone, chloroform	Several ppb	4.5-6	Cu,Pb,Ni, Zn,Cd,Co, Mn,Mo,V	8, 11, 12 23
Foam separation	Industrial processes	N_2, surfactant	0.06 g/l & lower	5.5-6	Ni,Cr,Co, Sr	4, 33
Autoclaving	Biological samples	Steam	ppb	—	Ti,Cr,Cu, Pb,Mn,Sn	16
Adsorption on alumina	Radioactive rinse water	—	Up to 1 x 10^{-3} moles/l	8.0	Sr	6
Aeration	Municipal & industrial water supplies	$KMnO_4$, activated carbon	0.2 mg/l	—	Mn,Fe	3
Manganese zeolite bed	Municipal & Industrial water	$KMnO_4$, anthracite	0.2 mg/l	—	Mn,Fe	1, 47
Ultraviolet radiation & magnetic field	Seawater, freshwater, industrial streams	—	—	—	All	45
Paper chromatographic separation	—	Mixture of chloroform, methanol, acetone, isopentanol & formic acid, TNOA-HCl	ppb	—	Cu,Pb,Cd, Bi,Hg,Mn, Co,Ni,Cu, Zn,Fe	24,27
Rotating electrodes	Process streams, seawater	—	—	—	All	42
Biological	a. Acid mine waters	Yeasts, sulfur, glucose	10s of mg/l	3.5	Cu	10
	b. Municipal & industrial water streams	Bacteria			Fe,Mn	18
Ion Exchange						
Chelex-100	Sea & fresh water, process waters	HCl	Varies	5-6	Cu,Pb,Cr, Ni,Zn,Co, Cd,Mn	9, 20, 43
Chitosan	Salt water	EDTA	ppb(varies)	—	Mn,Ti,Cr, Pb	25
Amberlite	Process liquid	Eluant	Varies	—	Cd,Co,Cu Pb,V,Zn, Ni	35,41
Titanium arsenate		NH_4NO_3+HNO_3	Varies	—	Pb,Cu,Cd, Sr,Zn,Mn, Ni,Co	29
Permutit-S1005	Seawater	Eluant	ppb	7.6 5.0(Mo,Cr) 6.0 (V) 9.0 (Mn)	Cd,Co,Cu, Pb,Ni,Zn Cr	30
DeAcidite FF + dibromo-oxine	Seawater, cooling tower blowdown	HCl or H_2SO_4	ppb 200 ppm	—	Co,Zn,Cr	16, 37
Zeo-Karb 225	Process streams, rain water	H_2SO_4	ppb	—	Sr	5, 13
Dowex	Water, process streams	Eluant	100 ppm	Varies	Ti,Sr,Mo Cd,Cr,Cu, Zn,Pb	7, 11, 34, 44
Precipitation						
Thioacetamide		HCl	ppb (varies)	0.75 8.0	Cd,Cu,Pb Cr,Ti,Zn, Mo	22
Metal sulfides	Cooling tower blowdown	HCl	15-200 ppm	2.3	Cr,Mo,Sr, Cu,Ni,Co	36, 38
Dibromo-oxine	Seawater	Acetone	ppb	8.0	Cu,Zn,Co, Mn,Pb,Cr	11, 31
Potassium Ferrocyanide	Electroplating solutions	Activated carbon	—	—	Pb,Sn,Cd, Zn,Fe,Ni, Co	2
Dialkyldithiocarbamates	Process streams	—	70 ppm	4.2	Zn,Cu,Fe	46
Oxalate or sulfate	Radioactive rinse water	—	40 ppb	—	Sr	28
Polyelectrolytes	Process solvents	Polygalacturonic & alginic acids	mg/l	4.0-4.5	Cu,Cd,Zn, Ni, Cr	14
Aluminum sulfates	Industrial & municipal water	—	0.20 mg/l	6.8-7.0	Pb,Cu,Cr, Cd,Zn,Ni	26
Lime	Industrial & municipal water	—	0.26 mg/l	9.6	Mn & others	32, 48
Carbonates	Industrial & municipal water	CaO or NaOH	0.2 mg/l	9.5	Mn & others	21, 32, 39
Hydroxides	a. Industrial & municipal water	$Ca(OH)_2$	100s of mg/l	>9.5	Pb,Cu,Cr, Ni,Co,Mn, Zn	26,39
	b. Seawater	$Th(OH)_4$	24 μg/l	6.0	Mo	17
	c. Seawater	$Mg(OH)_2$	60 μg/l	—	Co	19
	d. Process solutions	$Mg(OH)_2$+$Ca(OH)_2$	g/l	—	Sn,Pb	40
	e. Municipal & industrial water	NaOH	100s of mg/l	>9.5	Mn & others	32

method, for which the cost is \$1.93/lb of recovered chromate [16].

Other Ion-Exchange Resins

Cation-exchange resins have been found effective for removing zinc from viscous liquors of rayon plants at ion concentrations of 0.3% and less. Zeo-Karb 225 (14–52 mesh) was tested in a 1.9×25-cm glass column with a liquor flowrate of 100 ml/min. The zinc was recovered with a 25% (volume to volume) solution of sulfuric acid [5].

A study to find a means of monitoring rain water—involving Zeo-Karb 225 and DeAcidite FF supported on cotton wool—reported over 90% retention of strontium, with many ions collected that could be analyzed [13].

Experiments with Dowex 50W-X8 to investigate ion-exchange processes involving titanium cation adsorption from sulfate solutions removed 99.5% of the titanium. Contact times were 10 min and the solutions 0.5 normal titanium [7].

An investigation into removing metals from dry-cleaning solvents with Dowex 50 resin resulted in the following removal percentages: cadmium (40%), chromium (20%), copper (28%), zinc (40%), titanium (16.6%) and lead (26%). A 55-ml sample of mature dry-cleaning solvent was agitated for 16 h with 10 g of sulfonated styrene-divinyl-benzene (Dowex 50) [44].

Experiments on the adsorption of strontium on Dowex 50W-X8 recovered 95% of the strontium in rain water. A solution of 1.0 molar ammonium glycolate and 0.3 molar sodium chloride was used to elute the resin column [34].

Dowex 50W-X4 may be used to produce stable polyethylene lattice structures that are substantially free of metallic ions. In a study, 1,500 g of 50–100-mesh resin were used to treat 4,500 ml of polyethylene at a pH of 10.9. Having initially a metal ion concentration of 2.9 wt % (as metallic aristate), the polyethylene was made virtually free of metal ions [44].

Dowex A-1 chelating resin was found to adsorb essentially all molybdenum from water at a concentration of 10 μg/l and a pH of 1.1. The molybdenum was recovered by eluting the resin column with sodium hydroxide [11].

Recovery by Precipitation

Thioacetamide may be used as a precipitant for metal ions. Highly soluble in water, it hydrolyzes to give hydrogen sulfide by the reaction:

$$CH_3CSNH_2 + 2H_2O = CH_3COONH_4 + H_2S$$

The precipitates are coarsely crystalline, and easily filtered and washed.

At a pH of 0.75, adjusted with HCl, thioacetamide will effectively precipitate cadmium, copper and lead. At a pH of 8.0, the recovery is maximum for most other elements, although cobalt, nickel and manganese were not recovered well (thioacetamide at this pH is not considered satisfactory for these elements). The recovery of chromium, titanium and zinc is complete, and that of molybdenum partial, at a pH of 8.0 [22].

Precipitation With Metal Sulfides

Chromic acid or chromate salts are added to cooling water as oxygen inhibitors in the 15–200-ppm range, forming a passive protective film on metal parts. During cooling operations, their concentrations increase due to evaporation. Although they are insoluble and cannot react ionically with dissolved chromium and exchange electrons, metal sulfides can be used to condition cooling water before it is discharged as waste.

Iron pyrite, added as hard granular particles, will effectively reduce chromic acid. (The particles must be scrubbed to remove a film that will render them inactive.) At a pH to 2.3 (adjusted with HCl), the rate of reduction of chromium is such to enable the treatment of 100 gal/min of water containing 60-ppm chromic acid. The solution is next treated with magnesium oxide to increase the pH, and the following reaction takes place:

$$3FeS_2 + 12H_2O + 3H_2CrO_4 = 3Cr(OH)_3 + 3Fe(OH)_3 + 9/2\ H_2SO_4 + 3/2\ H_2S$$

The insoluble chromium and iron hydroxides are collected as sludge [38].

Metal sulfides (specifically, of cadmium) are excellent for removing chromium, molybdenum, strontium, copper, nickel, cobalt and vanadium, forming easily filtered precipitates [36].

Precipitation With Dibromo-Oxine

Dibromo-oxine (5,7-dibromo-8-hydroxyquinoline), which forms insoluble complexes with many metals, is an efficient coprecipitator. A procedure useful in trace-metal analysis involves adding, while stirring, a saturated solution of dibromo-oxine in acetone to seawater (10 ml/l of seawater). Metal-ion recoveries upon treatment of water solutions with dibromo-oxine at a pH of 8 are indicated [31]:

	Cu	Zn	Co	Mn	Pb	Cr
Concentration (μg/l)	5	5	0.025	5	5	0.5
Recovery in water, %	100	100	99	95	—	100
Recovery in seawater, %	100	100	99	85	42	100

Molybdenum and other trace elements may be effectively precipitated from fresh-water samples at a pH of 5.2 with dibromo-oxine [11].

With Potassium Ferrocyanide

The high cost of replacing rhodium plating solutions has stimulated interest in a technique for removing metallic contamination. Potassium ferrocyanide has been reported useful in restoring rhodium plating solution to production standards, being capable of completely removing lead, antimony, cadmium, zinc, iron, nickel and cobalt from solution.

From 95 to 190 l (25–50 gal) of plating solution can be treated with a 500-ml saturated solution of potassium ferrocyanide. In 24 h, the metal complexes will precipitate and settle out [2].

Removal by Thiocarbamates

Soluble zinc, copper and iron impurities can be removed from nickel-plating solutions by mixing with dimethyldithiocarbamate, dibutyldithiocarbamate, or diethyldithiocarbamate of nickel, corresponding to the general formula:

These compounds complex with soluble metallic impurities to form sparingly soluble materials in acidic nickel-plating solutions. The insoluble materials are removed by filtration [46].

In an experiment, a 4,200-gal nickel-plating solution was treated with 1,000 g of nickel dimethyldithiocarbamate, with the solution maintained at a pH of 4.2. Three samples were withdrawn from the electroplating bath: the first (Sample A) was taken before treatment, the second (Sample B) after the addition of 500 g of nickel salt, and the third (Sample C) after 8 h of electrodeposition followed by another 500 g of nickel salt:

	Impurity, ppm		
	Sample A	**Sample B**	**Sample C**
Copper	40	20	1.0
Zinc	30	26	25
Iron	0.5	—	—

Copper and iron were effectively removed, whereas the zinc concentration was only slightly decreased. However, during 8 h of operation, a sizable quantity of impurities were continuously deposited into solution by the zinc die-casting. These additional zinc ions were removed [46].

Removal of Radioactive Strontium

In an isotopic-exchange method proposed for removing radioactive strontium from solution, strontium sulfate or strontium oxalate precipitate is agitated in a solution containing radioactive strontium ions, initiating a rapid ion-exchange of strontium.

The precipitate, which is easily filtered from solution, acquires the radioactivity. In a study, 5-ml samples having strontium concentrations in the 40 μg/l range were used in a batch-type process, with 300 g of precipitate added: 96% of the radioactive strontium was in the sulfate precipitate after 10 min, 95% in the oxalate after 12 min. Equilibrium was reached after 25 min, with a maximum exchange of 99% [28].

Precipitation With Polyelectrolytes

Polygalacturonic acid, a natural polyelectrolyte derived from pectins, has been reported to form sparingly soluble metal complexes. It precipitates a complex that is virtually solvent-free. (Purifying the acid improves its performance, and it is best used as a powder.) The following reactions govern the precipitation of the metal complex:

$$RCOOH + M^{++} = RCOOM^{+} + H^{+}$$

$$RCOOM^{+} + RCOOH = (RCOO)_2M + H^{+}$$

The following shows the retention and recovery of metal ions by the acid at a pH of 4.5, and after treatment of the precipitated complex with 1.5 normal HCl:

Ion	Retained, mg/l	Recovered, mg/l	Recovery, %
Zn	49	46.5	95
Cd	50.4	48.0	95.4
Ni	88.0	79.4	90.2
Cu	95.3	93.5	98.2

Only 5% of the 2.0174 g of powdered acid was lost during the recovery process; the remainder may be used again [14].

With Aluminum Sulfate

An experiment on removing metals from solution by chemical treatment with aluminum sulfate effectively removed lead, copper, chromium and zinc. Adding 100 mg/l of aluminum sulfate to water containing dissolved metal resulted in a pH of 6.8–7.0 [26].

Removal of Manganese

Manganese can be removed when precipitated by the two-stage addition of lime. The lime is added to water passing into a reservoir. This increases the pH and causes sedimentation across the reservoir. More lime is added before the solution is filtered in a sand-bed, and pH is adjusted to 9.6. In a study, manganese concentrations were reduced from 0.26 to 0.02 mg/l. Manganese is also likely to coprecipitate, removing chemicals that form insoluble hydroxides and carbonates at a pH of 9.6 [48].

Manganese removal is reported to be effective only at a pH of 9.5 or higher. Lime is recommended for removal because it will raise the pH to this level when it is added in reasonable quantities; however, it hardens water, and water softening may be necessary [32].

Soda ash can be added to water to remove manganese by precipitation and pressure filtration [39]. The ideal filter-aid (low cost and resistance) was found to be Celite at concentrations of from 2–4 g/l [21].

A drawback is that soda ash is more expensive than other precipitants, such as lime, and it will not raise the solution pH to 9.5 upon the addition of reasonable amounts [32]. Therefore, such a chemical as lime or caustic soda must be added to obtain the desired pH.

Precipitation With Hydroxides

Many metals can be recovered from solution by chemical treatment with calcium hydroxide. The addition of 100 mg/l of hydroxide was found effective for removing lead, copper, chromium, zinc and cobalt at a pH of 9.5 [26].

Molybdenum was reported to be easily precipitated from solution upon treatment with thorium hydroxide. From seawater samples containing 24 μg/l of molybdenum—to which 3–4 ml of 0.1 molar $Th(NO_3)_4$, forming $Th(OH)_4$, was added—recovery at the stated pH values was:

pH	6.0	7.5	8.5	9.5
Recovery, %	99.5	81.5	61.6	46.4

Molybdenum precipitates as Th(MoO$_4$)$_2$, which has a solubility of 2×10^{-6} g/ml [17].

Calcium hydroxide and pressure filtration has been recommended for removing manganese from industrial and municipal water supplies [39], and so has caustic soda [32]. The cost of caustic soda is approximately the same as that of soda ash and will easily raise the pH of a solution to the 9.5 necessary to precipitate manganese. Although more expensive than lime, caustic soda does not harden water, indicating that water-softening equipment is not needed [32].

Cobalt can be precipitated from seawater by Mg(OH)$_2$. At a cobalt concentration range of from 0.14 to 63 μg/l, 93% of it was removed in experiments [19].

A study to find a means of removing lead and tin impurities from an ammoniacal ammonium carbonate solution containing copper, nickel or cobalt resulted in a method that removes the two metals from solution by use of dead burnt dolomitic limestone, Ca(OH)$_2$ and Mg(OH)$_2$. A mixture of from 40%–70% Ca(OH)$_2$ and from 30%–60% Mg(OH)$_2$ is preferred.

After addition of the hydroxide mixture in weight ratios of hydroxide to contaminant in the range of 5 to 1 or 10 to 1, the solution is pressure filtered and the impurities removed. A solution of 125 g/l of copper having 0.41 g/l of lead and 0.47 g/l of tin was treated with 8 g of 59% Ca(OH)$_2$ and 41% Mg(OH)$_2$ for 16 min. After filtration, 93% of the lead and 99% of the tin were found removed. When 100% Ca(OH)$_2$ was used, 64.4% of the lead but only 1% of the tin were removed [40]. #

Acknowledgment

The partial support of this survey by the Chesapeake Research Consortium, Inc., through RANN-NSF (Research Applied to National Needs–National Science Foundation) funds is gratefully acknowledged by the authors.

References

1. Adams, J., *Water and Sewage Works*, **116**, July 1969, p. 259.
2. Baldwin, P. C., *Plating*, **55**, *6*, 1968, pp. 612–613.
3. Bell, G. R., *Research and Development Progress Report No. 201*, U.S. Office of Saline Water, August 1966.
4. Bhandarkar, P. G. and Rao, M. G., *Indian J. of Technol.*, **7**, *2*, February, 1969, pp. 65–66.
5. Blake, W. E. and Randle, J., *J. of Applied Chem.*, **17**, December 1967, pp. 358–360.
6. Bonner, W. P., Beirs, H. A. and Morgan, J. J., *Health Physics*, **12**, *12*, 1966, pp. 1691–1703.
7. Bonsack, J. P., *Ind. and Eng. Chem.*, **9**, *2*, 1970, pp. 254–259.
8. Brooks, R. R., Presley, B. J. and Kaplan, I. R., *Talanta*, **14**, *7*, July 1967, pp. 809–816.
9. Callahan, C. M., Pascual, J. M. and Lai, M. G., *U.S. Government Research and Development Report AD648485*, December 1966.
10. Ehrlich, H. L. and Fox, S. I., *Applied Microbiology*, **115**, *1*, 1967, pp. 135–139.
11. Fishman, M. J. and Mallory, Jr., E. C., *J. of the Water Pollution Control Federation*, **40**, *2*, 1968, pp. R67–R71.
12. Fishman, M. J. and Midgett, M. R., *Advances in Chemistry Series No. 73, Trace Inorganics in Water*, 1968, pp. 253–264.
13. Fletcher, W., Gibbs, M. J., Moroney, T. R., Stevens, D. J. and Titterton, E. W., *Australian J. of Science*, **28**, May 1966, pp. 417–424.
14. Jellinek, H. H. G. and Sangal, S. P., *Water Research*, **5**, 1971, pp. 51–60.
15. Jones, G. E., *Limnology and Oceanography*, **12**, *1*, 1967, pp. 165–167.
16. Kelly, B. J., *Materials Protection*, **8**, March 1969, pp. 23–25.
17. Kim, Y. S. and Zeitlin, H., *Analytica Chimica Acta*, **51**, *3*, 1970, pp. 516–519.
18. Kojima, S., *Kagaku Kogaku*, **33**, *10*, 1969, pp. 921–927 (From *Chemical Abstracts*, **72**, *35633X*, 1970).
19. Krishnamoorthy, T. M. and Visivanathan, R., *Indian J. of Chemistry*, **6**, March 1968, pp. 169–170.
20. Lai, M. G. and Goya, H. A., *U.S. Government Research and Development Report AD647661*, November 1966.
21. Levi, H. W. and Melzer, H., Low and Intermediate Level Radioactive Wastes Symposium, Vienna, 1965, pp. 335–370.
22. Mallory, Jr., E. C., "Advances in Chemistry Series No. 73, Trace Inorganics in Water," 1968, pp. 281–295.
23. Marcie, F. J., *Environmental Sci. and Technol.*, **1**, *2*, February 1967, pp. 164–166.
24. McCormick, D., Graham, R. J. T. and Bark, L. S., 4th International Symposium on Chromatography and Electrophoresis, Brussels, 1966, pp. 199–206.
25. Muzzarelli, R. A. A., Raith, G. and Lubertini, O., *J. of Chromatography*, **47**, *3*, 1970, pp. 314–320.
26. Nilsson, R., *Water Research*, **5**, 1971, pp. 51–60.
27. Poonia, N. S., *Indian J. of Chemistry*, **4**, *12*, 1966, pp. 511–513.
28. Qureshi, I. H., Shahid, M. S. and Hasany, S. M., *Talanta*, **14**, *7*, pp. 951–956.
29. Qureshi, M. and Nabi, S. A., *J. of Inorganic and Nuclear Chemistry*, **32**, *6*, 1970, pp. 2059–2068.
30. Riley, J. P. and Taylor, D., *Analytica Chimica Acta*, **40**, *3*, 1968, pp. 479–85.
31. Riley, J. P. and Topping, G., *Analytica Chimica Acta*, **44**, 1969, pp. 234–236.
32. Robinson, Jr., L. R. and Dixon, R. I., *Water and Sewage Works*, **115**, November 1968, pp. 514–518.
33. Schnepf, R. W., Gaden, E. L., Mirocynik, E. Y. and Schonfeld, E., *Chem. Eng. Prog.*, **55**, *5*, pp. 42–46.
34. Senegacnik, M. and Paljk, S., *Fresenius' Zeitschrift Für Analytische Chemie*, **232**, November 1967, pp. 409–426.
35. Sherma, J., *Separation Science*, **2**, *2*, 1967, pp. 177–185.
36. Shevchik, A. D., *Geol. Zh.* (Russian ed.), **29**, *3*, 1969, pp. 102–104 (from *Chemical Abstracts*, **72**, *93241d*, 1970).
37. Topping, G., *Limnology and Oceanography*, **14**, *5*, 1969, pp. 798–799.
38. U.S. Patent No. 3,325,401.
39. U.S. Patent No. 3,340,187.
40. U.S. Patent No. 3,374,091.
41. U.S. Patent No. 3,380,804.
42. U.S. Patent No. 3,448,026.
43. U.S. Patent No. 3,488,184.
44. U.S. Patent No. 3,509,084.
45. U.S. Patent No. 3,511,776.
46. U.S. Patent No. 3,518,171.
47. Wilmarth, W. A., *Water and Wastes Engineering*, **5**, *8*, 1968, pp. 52–54.
48. Wood, R. and Burden, B. A., *Nature*, **213**, 1967, p. 637, and *Institution of Water Engineers J.*, **21**, August 1967, pp. 677–681.

Meet the Authors

◀ **Theodore W. Cadman** is a professor of chemical engineering and Director of the Laboratory for Process Analysis and Simulation at the University of Maryland (College Park, MD 20742). His primary teaching and research interests have been in process analysis, control, design and simulation, areas in which he has published and served as a consultant. He received his B.S., M.S. and Ph.D. degrees in chemical engineering from Carnegie-Mellon University. A member of AAAS, ACS, AIChE, ISA and NSPE, he is listed in "Who's Who in the East" and "American Men of Science."

Robert W. Dellinger is a chemical engineer with the Effluent Guidelines Div. of the ▶ U.S. Environmental Protection Agency (Waterside Mall, 923 East Tower, 401 M St., S.W., Washington, DC 20460). Currently, he is responsible for the development of effluent limitations, guidelines and performance standards for the cane-sugar and glass industries. His chemical engineering degrees include a B.S. from Virginia Polytechnic Institute and an M.S. from the University of Maryland.

Heavy metals removal

Most heavy metals found in waste-treatment processes are in the inorganic form, but there are certain industries where they appear as organic compounds. Here, the technical methods that are in general use are discussed, and the optimum treatment systems for specific metals are described.

Kenneth H. Lanouette, Industrial Pollution Control, Inc.

☐ Heavy metals is the classification that is generally applied to those metals of particular concern in the treatment of industrial wastewaters: copper, silver, zinc, cadmium, mercury, lead, chromium, iron and nickel. Other metals that may be considered part of this category are tin, arsenic, selenium, molybdenum, cobalt, manganese and aluminum.

Most heavy metals found in waste-treatment processes are in the inorganic form. However, in some industries such as textiles and dyeing, heavy metals are in the organic form. Treatment technology is different for these two forms.

Inorganic heavy metals

The most common method for removal of inorganic heavy metals is chemical precipitation. Typical estimates of final concentration levels obtainable by chemical precipitation are shown in Table I [1].

Metals precipitate at various pH levels, depending on such factors as the metal itself, the insoluble salt that has been formed (e.g., hydroxide, carbonate, sulfide, etc.), presence of complexing agents such as ammonia, citric acid, ethylenediaminetetra-acetic acid (EDTA), etc. Theoretical precipitation curves for various metals as hydroxides are shown in Fig. 1 [2].

It is apparent that when two or more heavy metals are found in the same waste stream, the optimum pH for precipitation may be different for each ion. The question then becomes whether it is possible and practical to precipitate one or more of the metals separately at the source at one pH, and treat the remaining stream at another pH.

Alternatively, it must be determined if one pH can be found which will produce satisfactory—though not optimum—insolubility for each of the metal ions present in the wastewater.

The presence of complexing agents such as ammonia must also be carefully investigated. Ammonia may be controlled by careful pH adjustment, the use of sulfides,

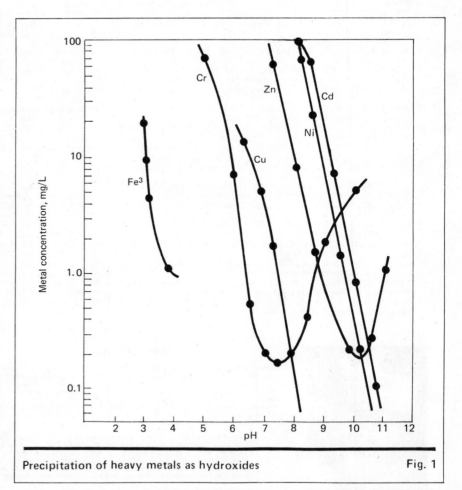

Precipitation of heavy metals as hydroxides

Fig. 1

Originally published October 17, 1977.

Typical concentration levels obtainable through chemical precipitation		Table I
Heavy metal	Achievable concentration, mg/L	Probable precipitating agent
Cadmium	0.3	Soda ash
Chromium^{+6}	0.05	–
Chromium, total	0.5	Caustic, lime
Copper	0.5	Caustic, lime
Iron	1.0	Caustic, lime
Nickel	0.5	Soda ash
Zinc	0.5	Caustic, lime

or removal by air stripping. Other complexing or chelating agents should also be investigated to determine their effect on the precipitation process.

In order to determine the optimum treatment process, a bench-scale testing program should be carried out using samples of the waste to be treated. Precipitation tests should use various likely chemicals to determine which produces the most acceptable effluent. The pH should be adjusted over a suitable range; use of polymers to aid in coagulation of the precipitates should also be investigated. The times required for proper mixing and sedimentation can also be determined by a bench-scale testing program.

Scale-up from bench-scale tests is usually reliable, provided that normal safety factors are applied.

Treatment methods

Chemical precipitation of heavy metals may be accomplished by either batch or continuous treatment systems. Batch treatment is usually preferred when the volumes to be treated are small, or where the waste may be variable from day to day and require modification of the treatment as characteristics change.

Batch treatment systems can be economically designed for flows as high as 50,000 gpd. A batch system is usually designed with two tanks, each one of sufficient volume to handle the waste generated in a specific time period—e.g., an 8-hour shift, a 24-hour day, etc. When one tank is full, a mixer is used to provide a homogeneous mixture, and a sample is taken and analyzed to determine the amount of heavy-metal contaminants present. Chemical addition based on contaminants present and pH of the waste is then calculated and the required amounts of chemicals are added. The tank contents are then mixed—using pH for controlling purposes—and, for heavy-metals removal, allowed to settle for 2 to 4 hours.

When treatment is complete, a second sample can be taken and analyzed to ensure that all contaminants have been removed. If, for any reason, contaminants are still present, treatment can be repeated or alternative treatment applied. When the operator is satisfied that the treated waste is suitable for discharge, the clear liquor is decanted. The settled sludge is drawn off periodically for disposal. The advantages of a batch treatment system are that nothing is discharged from the plant until the operator is satisfied that it meets effluent requirements. The system is also simple in its design and easy to operate (Fig. 2).

When wastewater characteristics are uniform or when volumes are large, a continuous treatment system is applicable. A usual continuous treatment system has an equalization tank of several hours to a day of detention time to even out any fluctuation in the wastewater characteristics and provide a uniform feed to the treatment system.

The first process step is the adjustment of the pH by addition of acid or alkali to the proper level for optimum precipitation. This chemical addition is controlled by a pH probe in the reaction tank, which activates the speed control of the chemical feed pump. A polymer is usually added to aid coagulation. Reaction times are in the range of 15 to 60 minutes.

The waste stream then flows to a sedimentation basin where the heavy-metal precipitate settles out of solution leaving a clear treated overflow for discharge to the receiving body of water. Detention times in the sedimentation basin usually range from 2 to 4 hours at overflow rates from 300 to 700 gal/ft^2 foot of surface area per day (Fig. 3).

Sludge recirculation

Recirculation of precipitated sludge to be mixed with the raw waste at the time of chemical addition can have beneficial effects. The presence of the

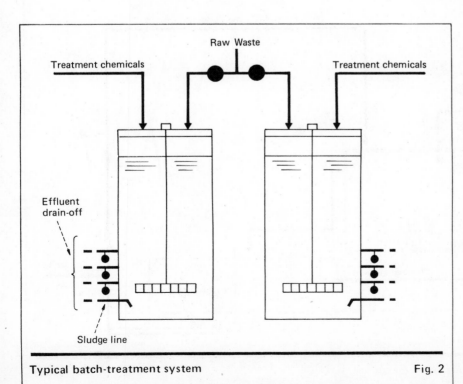

Treatment chemicals   Raw Waste   Treatment chemicals

Effluent drain-off

Sludge line

Typical batch-treatment system   Fig. 2

precipitated particle provides a seed for the newly formed precipitate to agglomerate.

In batch treatment systems, the settled sludge is left on the bottom of the tank. When a new batch is put in the tank, the mixer is turned on to resuspend the sludge and mix it with the tank contents.

In continuous treatment systems, the sludge can be recycled either externally around the clarifier or internally within the clarifier. With external recirculation, the sludge is pumped out of the sludge hopper in the clarifier and introduced to the raw feed in the chemical mixing tanks. With internal recirculation, the clarifier is designed with an internal mixer and baffles that provide recirculation within the clarifier.

Effluent filtration

A properly designed clarification system will remove the bulk of precipitated solids. However, the effluent may contain 5 to 20 mg/L of suspended solids including some of the heavy metal. Therefore it may be advisable to filter this effluent to ensure maximum removal of the metals.

The type of filtration system frequently used is a gravity filter with a sand bed or multimedia bed, similar to filters used for water treatment. Flowrates are 2 to 5 gpm per square foot of surface area. As the media entrap more solids, the pressure loss through the filter increases. When the loss of head is 8 to 12 ft, it is necessary to clean the filter by backwashing to ready it for reuse. Because the filter is out of use during the backwash cycle, which takes about one hour, two or more filters are usually required. Valveless gravity filters, mixed-media filters and pressure filters can be applied to this operation.

Sludge handling

The settled sludge from a clarification basin is frequently in the range of 1 to 2% solids. Hydroxide precipitation of the metals produces sludge that is usually gelatinous in character, thereby increasing the difficulty of dewatering. Lime will produce considerably greater quantity of sludge than caustic but is ordinarily easier to dewater.

The sludge can be further dewatered in centrifuges, vacuum or plate-and-frame filters or on sludge-drying beds. Centrifuges of the solid-bowl type usually dewater the sludge a little less than vacuum filters. The units are usually compact and require minimal floor space. Centrifuges operate with varying success depending on the nature of the sludge. Field trials should be run before selecting a unit.

Rotary vacuum filters might be expected to produce a cake of 20 to 25% solids. They require a vacuum pump and a filtrate pump as well as other auxiliary equipment. They also require considerable floor space, but are quite reliable in operation.

Pressure filters of the plate-and-frame type will produce sludges a bit more concentrated than vacuum filters. These filters occupy less floor space than vacuum filters, but require more labor and operator-attention.

Sludge-drying beds can, with sufficient retention, produce a fairly dry cake. However, they require con-

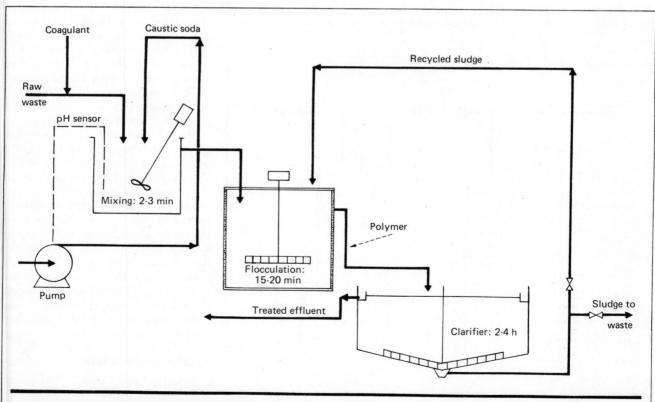

Typical continuous-treatment system for heavy metals

Fig. 3

siderable land area, and can only be used in locations where the incursion of leachate into the aquifer will not have a detrimental effect.

Sludge disposal

One of the biggest problems facing environmental authorities today is the question of how to handle and dispose of the sludge that has been generated in heavy-metal precipitation systems. There is always a possibility that the metals will solubilize if the pH falls below the optimum range. Therefore, these sludges should not be disposed of with municipal waste because the low pH generated by decomposition of the garbage can resolubilize the precipitate.

The final sludge-disposal lagoon, or dump pit, should be located so that it will have minimal impact on surface and ground waters. Best conditions are areas where natural (clay, rock) or artificial means (plastic liner) can prevent excessive amounts of leachate from getting into the ground water. Where these conditions do not exist, it may be necessary to install a collection system at the bottom of the pond so that leachate can be collected and piped to a treatment system for re-removal of soluble metals salts.

Chemical treatments have been developed to reduce the leaching possibilities of heavy metal sludges and may be used in conjunction with a landfill program to further minimize potential leachate contamination.

Metal values from sludge

Recovery of metal values from the sludge has been studied. Digesting the sludge in an acid medium, neutralization and electrolytic recovery have been investigated [3]. The cost estimate for recovery of copper,

nickel, and chromium in a small plant was $13.23/kg, which was quite high compared to current market prices. However, since these metal values are likely to increase as the years go by, one approach is to stockpile these metal-bearing sludges, either separately or in a regional disposal site, so that they are available for recovery in the future when the value of the metals and recovery technology make it economically attractive.

Even at the present time, while individual plant recovery may not be economical, regional recovery systems treating metal sludges from several plants in an area may be feasible. As the cost of sludge disposal increases, recovery becomes more attractive.

Chemical usage

The chemicals most frequently used for precipitation of metals are lime, caustic and sodium carbonate. Comparative costs of these chemicals are as follows:

Precipitating chemical	Theoretical requirements, mg/L	Approximate cost, $/ton
Quicklime, CaO	0.76	25
Hydrated lime, $Ca(OH)_2$	1.00	25
Caustic, 50% $NaOH$	1.08	140
Soda ash, Na_2CO_3	1.42	50
Sodium sulfide, Na_2S	—	250

Because of the lower cost, lime is preferred particularly in installations which use more than one-half ton per day. Caustic is used in smaller installations where daily chemical costs will not be significant. Soda ash is used in particular cases (e.g., cadmium and nickel precipitation) where it provides a better chemical reaction.

Lime—There are several types of lime available:

Material	Theoretical requirements compared to equivalent weight $Ca(OH)_2$
Quicklime (high calcium), CaO	0.76
Quicklime (dolomitic), $CaO \cdot MgO$	0.65
Hydrated lime (high calcium), $Ca(OH)_2$	1.00

Quicklime is cheaper than hydrated lime per unit of neutralizing capacity, but it requires the use of slaking equipment (about $10,000–15,000) and requires protective clothing for workers since it can cause severe burns. Quicklime is usually used when demand is over 5 tons/day. It must be stored in air-tight bins to prevent air-slaking, lime coats, and eventual plugging of pipes. Dolomitic lime (high magnesium) reacts more slowly than high-calcium lime. Magnesium in dolomitic lime may have side effects, either good or bad, on reaction.

Hydrated lime is more expensive but does not require slaking equipment and can be fed directly to the reaction system. It is used where demand is less than 5 tons/day, and its dust is less irritating than that of quicklime. Hydrated lime tends to bulk and bridge in

---

Theoretical solubilities of hydroxides and sulfides of heavy metals in pure water   Table II

Metal	Solubility of metal ion, mg/L	
	As hydroxide	As sulfide
Cadmium (Cd^{++})	2.3×10^{-5}	6.7×10^{-10}
Chromium (Cr^{+++})	$*8.4 \times 10^{-4}$	No precipitate
Cobalt (Co^{++})	$*2.2 \times 10^{-1}$	1.0×10^{-8}
Copper (Cu^{++})	$**2.2 \times 10^{-2}$	5.8×10^{-18}
Iron (Fe^{++})	8.9×10^{-1}	3.4×10^{-5}
Lead (Pb^{++})	$**2.1$	3.8×10^{-9}
Manganese (Mn^{++})	1.2	2.1×10^{-3}
Mercury (Hg^{++})	$*3.9 \times 10^{-4}$	9.0×10^{-20}
Nickel (Ni^{++})	$*6.9 \times 10^{-3}$	6.9×10^{-8}
Silver (Ag^+)	13.3	7.4×10^{-12}
Tin (Sn^{++})	$**1.1 \times 10^{-4}$	$*3.8 \times 10^{-8}$
Zinc (Zn^{++})	1.1	2.3×10^{-7}

Sources:

"Handbook of Chemistry and Physics," 50th ed., Weast, R.C., ed., Chemical Rubber Co., Cleveland, Ohio, p. B252.

*"Handbook of Analytical Chemistry," Meites, L., ed., McGraw-Hill, New York, pp. 1-15 to 1-19, 1963.

**"Ionic Equilibrium As Applied to Qualitative Analyses," Hogness and Johnson, Holt Rinehart & Winston Co., New York, pp. 360-362, 1954.

storage tanks, and therefore requires special agitating systems on cone bottoms. Dusting can also be a problem, and special dust collectors are required. Hydrated lime can be obtained in car lots, truck lots or 100-lb bags in completely dry form.

Caustic soda—Caustic soda, or sodium hydroxide (NaOH), is available in either liquid or dry form. Liquid caustic soda is usually preferred when smaller quantities are required because of ease of handling. It is usually obtained as 50% NaOH. Since freezing (crystallization) occurs at 53°F, indoor storage, or heating if stored outdoors, is required. Caustic is often diluted at the plant site because the freezing temperature at 20% concentration is −20°F.

Soda ash—Soda ash, or sodium carbonate (Na_2CO_3), has a theoretical equivalent weight requirement of 1.42, as compared to hydrated lime. It is not particularly corrosive by itself, and it has superior handling characteristics—little dust, good flow, and no arching in feeder. But soda ash is difficult to dissolve; it is stored in steel bins, and can be obtained in bulk by railcar or truck or in 100-lb bags.

Sometimes it provides superior precipitation, i.e. with cadmium it produces cadmium carbonate which is preferred to cadmium hydroxide for recovery purposes. Also with nickel, precipitation with carbonate gives lower final levels than precipitation with hydroxide.

Sulfides—Present technology for heavy metals removal is generally based on hydroxide precipitation, as previously described. However, when heavy metals are complexed with such agents as ammonia, cyanide, etc., hydroxide precipitation will not be sufficient to attain adequate levels. Also, with new emphasis on even-stricter effluent standards in the future, increased attention is being given to other methods of achieving even lower concentrations. Sulfide precipitation is a method of achieving these lower effluent levels.

Table II gives the theoretical solubilities of hydroxides and sulfides of various metals in pure water. In addition to having lower solubilities than hydroxides in the alkaline pH ranges, sulfides also tend to have low solubilities in the pH ranges below 7. Also, certain metal hydroxides tend to redissolve as the pH increases beyond the level of maximum insolubility [4].

One of the problems with the use of sulfides is that, with the use of sodium sulfide or similar compounds, excess sulfide in solution will form H_2S, which in itself is a pollution problem and requires removal. Adding exactly the right amount of sulfide is difficult to do—too little will leave some metal in solution, and too much will cause an additional pollution problem. Another problem is that metal sulfide sludges when exposed to air oxidize to the sulfate, and the metal ions can become resolubilized.

A process utilizing ferrous sulfide as the principle source of sulfide ion has been developed and appears to overcome the problem of generating H_2S from excess sulfide. The sulfide is released from the FeS only when other heavy metals with lower equilibrium constants for their sulfide form are present in solution. If the pH can be maintained at 8.5 to 9, the liberated iron will form a hydroxide and precipitate out as well.

Other removal methods

Activated carbon adsorption—Activated carbon will absorb hexavalent chromium, mercury and many metal compounds that have been complexed in the organic form as for dyes and pigments. The adsorptive capacity of any carbon depends on the pore size, size of molecule, pH of the solution and the initial and final concentration. Adsorption capacity generally increases as the pH decreases and, normally, adsorption efficiency increases as the concentration increases.

In order to determine the effectiveness of activated carbon for a particular metal-bearing waste, adsorption isotherms should be developed that indicate the amount of material adsorbed at a particular concentration. These tests are usually run on several different carbons at various concentrations in order to determine which provides the most effective treatment.

Granular carbon is usually the preferred type. It is two to three times as expensive as powdered carbon, but it can be chemically regenerated and reused. Powdered carbon is less expensive but usually can only be used on a once-through basis. It is difficult to handle due to a tendency to dust, and it must be removed from the waste stream, usually by coagulation or filtration. It is important to make an economic evaluation of the two types of carbon—granular or powdered—before a final choice is made.

Ion exchange—Ion exchange is an effective means of removing heavy metals from waste streams. There are a variety of resins for specific applications with various metals. When the resins are saturated, they must be regenerated with an acid or alkaline medium to remove the metal ions from the resin bed. The regenerant solution is smaller in volume and higher in concentration than the wastewater, but these metal values must then be adequately disposed of or recovered. Ion exchange is particularly applicable for waste streams which, for one reason or another, will not respond to more conventional treatment, or where the metals can be recovered from the regenerant solution and reused.

Reverse osmosis—Reverse osmosis is a treatment system utilizing semi-permeable membranes to produce a clear permeate and a concentrate containing the metal values. The system operates at pressures up to 600 psi and has been utilized effectively in the plating industry for the recovery and reuse of nickel in plating-bath waters. Applications of reverse osmosis are still in the development stage for the recovery of other metals. The system is essentially a concentration technique and is most effective if the concentrate can be reused.

Cementation—This is a metal-replacement process that occurs when a solution containing a dissolved metallic ion comes into contact with a more active metal, such as iron. This process is particularly applicable as a means of removing copper and silver from solutions. Scrap iron or steel wool is frequently used, and the less active metal is cemented onto the iron as the iron replaces it in solution. The weaker metal—copper, silver, etc.—is recoverable as a pure metal and the iron can be precipitated as an iron hydroxide if necessary. The reaction is carried out in an acid solution.

Other active metals, such as zinc dust, can be used for the cementation of metals such as gold, cadmium, mercury and lead [5].

Removal techniques for various metals

Cadmium—This metal is generated in the waste streams from pigment works, textiles, electroplating, chemical plants, etc. The U.S. Public Health Service (USPHS) has set a limit of 10 μg/L of cadmium in drinking water. Metal finishing guidelines show an attainable limit of 0.3 mg/L of cadmium in the effluent.

Cadmium is toxic to human beings when injected or inhaled. It is toxic to fish in concentrations as low as 10 μg/L. Cadmium acts synergistically with other elements such as copper and zinc to increase toxicity [6,7].

Cadmium as hydroxide, carbonate and sulfide forms an insoluble precipitate.

Precipitation in the hydroxide form is dependent on high pH. Pourbaix has determined that theoretical solubility of cadmium hydroxide is approximately:

pH	Solubility, mg/L
8	3,000
9	30
10	0.03
11	0.003

Therefore, with lime or caustic as the precipitating agent, pH should be maintained at about 11 for maximum insolubility. Operating at such a high pH requires final treatment of the effluent discharge to lower the pH to the allowable limit of 9. Since this pH adjustment will redissolve any remaining cadmium, it is important that all precipitate be removed, preferably by filtration, before this final pH adjustment is made.

If other metals are present that precipitate at lower pH levels (8 to 9), it is often preferred to have a treatment system operating at one pH for all metals. In this case, other chemical agents must be considered.

Treatment of cadmium with sodium carbonate (soda ash) will give good levels of removal at a slightly lower pH, in the range of 9.5–10. Due to the value of cadmium, it is often desirable to send the precipitated sludge to a reprocessor for recovery of the cadmium, or to reuse it. Whether the cadmium is in the hydroxide or carbonate form may be important to the reprocessor or plant user.

Sulfide precipitation can be carried out at low pH with good removals. For example, it is reported that sulfide precipitation at a pH of 6.5 reduced a raw waste ranging from 0.5 to 1.0 mg/L of cadmium to a final effluent of 0.008 mg/L. In this particular instance, BaS was used but Na_2S, NaHS or H_2S could have been applied as well [8].

Chromium—Chromium is found in two forms in waste streams—hexavalent and trivalent. Hexavalent chromium is found in the waste streams of plating operations, aluminum anodizing, paint and dye operations, and other industries. Trivalent chromium is found in the waste streams of textile dyeing, the ceramic and glass industry and photography.

Limitations on domestic water suppliers of 50 μg/L total chromium have been recommended by the EPA [6]. The usual chromium limits for most effluents are less than 0.05 mg/L for hexavalent chromium and 0.5 mg/L for trivalent chromium.

Treatment for chromium usually consists of a two-stage process: first, the reduction of hexavalent chromium to the trivalent form and, second, precipitation of the trivalent chromium. For this reason, chrome-bearing streams are segregated and treated separately, particularly in the reduction of the hexavalent to the trivalent form. The reduced chrome-bearing stream can then be blended with other metal-bearing streams for further treatment. Waste can be treated either batch-wise or continuously. For a continuous flow-through system, equalization of the raw waste flow, by means of a holding tank with several hours detention as a first step in the treatment process, is most important.

The reduction of hexavalent chromium is accomplished at a pH of 2 to 3. The pH can be adjusted either by manual measurement or by an automatic pH control unit with the addition of sulfuric or hydrochloric acid. The level of pH controls the rate of reaction. At pH 2, the reduction reaction is almost instantaneous, while at a pH of 3, the reaction takes about 30 minutes.

After the pH is adjusted a reducing agent is added. Reducing agents most commonly used are sodium dioxide and sodium metabisulfite. The amount of reducing agent added may be controlled based on oxidation-reduction-potential (ORP) readings or by analysis of the hexavalent chromium remaining in solution. Sulfur dioxide gas is toxic, so adequate venting and safety precautions must be provided.

Typical relationship between ORP and Cr^{+6} is [9]:

ORP	Cr^{+6}
590	40 ppm
570	10 ppm
540	5 ppm
330	1 ppm
300	0

This relationship should be confirmed by field tests.

Trivalent chrome is removed by precipitation. Usually lime or caustic is added to increase the pH to between 7.5 and 8.5 for minimum solubility of chromium hydroxide. The solubility starts to increase as the pH goes below 7 on the low side and as it goes above 9 on the high side. Therefore, careful pH control in the range of 7.5 to 8.5 is important. As pointed out above, other metals in the waste stream will also precipitate in this step.

Theoretical chemical requirements per lb of Cr are:
1. For reducing with sodium metabisulfite, 3 lb. $Na_2S_2O_5$ and 1-1$\frac{1}{2}$ lb. H_2SO_4.
2. For reducing with sulfur dioxide, 2 lb. SO_2 plus 35 mg for each liter of water being treated.
3. For precipitation, 2.2 lb. $Ca(OH)_2$ or 2.5 lb. NaOH or 3 lb. Na_2CO_3.

Ferrous sulfate will reduce hexavalent chrome and jointly precipitate the trivalent chrome and ferric hydroxide at pH 8.5. It requires 17 parts of ferrous sulfate for each part of hexavalent chrome. This process is more expensive and produces considerably more sludge than the conventional process of pH reduction and use of SO_2. However, the overall economics are worth considering when the hexavalent chrome content is in the range of less than 10 mg/L and probably it will be the system of choice when the content is less than 5 mg/L.

Under alkaline conditions, sodium hydrosulfite can be used for the reduction of hexavalent chrome and requires 5 mg/L for each mg/L of chrome + 5 mg/L for each mg/L of dissolved oxygen present in the water. This is the most expensive method for the reduction of hexavalent chrome. It is generally used in alkaline cleaning baths with a build-up of hexavalent chrome that has been carried over from previous baths, or as an emergency treatment when hexavalent chrome is present in precipitation reactors and/or clarifiers.

Other means of removing chrome from wastewater are adsorption on activated carbon, ion exchange and biological treatment.

Chromium can be removed by activated carbon. Tests by EPA showed 98% removal of chromium by a combination of lime precipitation and activated carbon. Starting with an initial concentration of 5 mg/L, final concentration was 90 μg/L [10]. The activated carbon can be regenerated with sodium hydroxide if it is desired to recover the chrome. If the chrome is to be disposed of, sulfuric acid will reduce the hexavalent chrome to the trivalent state.

Ion exchange systems are used in those instances where the chrome is being recovered for reuse in the process. Chromium removals of 40 to 80% have been reported in activated sludge systems based on low initial levels of chromium. When chrome is removed in the process, it ends up in the sludge and may become a problem there [10].

Lead—Lead is found in waste streams from battery manufacture, printing, painting and dyeing, and other industries. Lead is a cumulative poison and concentrates primarily in the bones. It is generally felt that 0.1 mg/L can cause lead poisoning if ingested regularly [11]. USPHS drinking water standards recommended 50 μg/L as a maximum level [7].

Lead can be precipitated with lime (CaOH) or caustic (NaOH) to form lead hydroxide, $Pb(OH)_2$, with soda ash, $NaCO_3$, to form lead carbonate, $PbCO_3$, or with trisodium phosphate, Na_3PO_4, to form lead phosphate, $Pb_3(PO_4)_2$. A combination of soda ash and caustic has also been applied.

Low levels of residual lead have been obtained in bench-scale tests, but many plants are finding it difficult to attain effluent levels below 0.5 mg/L in actual waste treatment plant operations.

Bench-scale tests conducted by Nassau Smelting and Refining Co. [12] studied precipitation by caustic, lime and caustic soda/soda ash. It was found that both lime and caustic soda/soda ash gave good results. The optimum pH was 9.0 to 9.5. Influent lead was 5 mg/L and final lead was 0.01 to 0.04 mg/L.

Fig. 4 shows solubility levels of lead with different alkali agents. As can be seen, the soda ash/caustic soda systems produced slightly better results than the straight-lime system. Separan AP 30 was used as a coagulant aid [12].

A study by Westinghouse Electric Corp. compared the effectiveness of all of the above precipitation systems and their treatment costs. They found that, for their waste stream, a combination of trisodium phosphate and caustic soda (at a ratio of 1 : 2.5) was the most effective precipitant system. Residual lead concentrations were below 0.02 mg/L at a pH of approximately 8.5. The total chemical cost of the treatment was $2.84/thousand gallons [13].

Other EPA tests using lime and activated carbon showed 99.4% removal. Starting with an initial concentration of 5.0 mg/L lead, the final concentration was 30 μg/L [10]. Lead removal in a biological plant is 50 to 90% and it is removed in sludge at concentrations of 830 mg/kg [10].

Mercury—The presence of mercury in water supplies is a major concern today. The recommended drinking water standard is 2 μg/L and quality criteria for aquatic life is 0.05 μg/L.

The standard method of removing mercury is to adjust the pH to 5 to 6 with sulfuric acid and then add sodium sulfide to an excess of 1 to 3 mg/L. This forms an insoluble mercury sulfide. After filtration on a rotary vacuum-precoat or pressure-precoat filter, the effluent contains approximately 10 to 125 μg/L (average 50 μg/L) mercury from an initial waste concentration of 0.3 to 6 mg/L. Polishing of the effluent with activated carbon or synthetic resins has been reported to further reduce the residual levels [14].

An effluent level of 10 to 50 μg/L of mercury may not be sufficient to meet the stringent stream standards. Therefore, it may be necessary to consider supplementary treatment systems.

A table of alternative treatment methods and levels that can be accomplished is given below [15]. Some of these mercury processes have not been used in full-scale commercial operation:

Technology	Lower limit of treatment capability, μg/L
Sulfide precipitation	10–20
Ion exchange	1–5
Alum coagulation	1–10
Iron coagulation	0.5–5
Activated carbon	
High initial Hg	20
Moderate initial Hg	2.0
Low initial Hg	0.25

Ion exchange seems to be an effective means of removing mercury down to levels of 1 to 5 mg/L. The best of all these treatment methods is reported to be a two-stage system. The pH should be on the slightly acidic side.

Use of zinc dust for reduction of mercury has been reported as effective in removing 99% of the mercury. This process is based on the reduction of the mercury to the elemental state and the formation of a stable complex with the excess of zinc. This compound is subsequently removed by conventional solids-liquid separation methods [16].

Silver—The principle sources of silver in wastewaters are the electroplating and photographic industries. Because of the value of silver, extensive efforts are made to recover as much of it as possible.

The allowable limit of silver in drinking water has been set by the USPHS as 50 μg/L. Fish and lower organisms are susceptable to silver poisoning, and concentrations of from 4 μg/L to 0.44 mg/L have proven toxic for salmon fry, eels, stickleback and lower organisms [6,7].

The two principal recovery methods are electrolytic recovery and cementation (metal replacement). In electrolytic recovery, the solution is passed through two electrodes operating on direct current. Silver deposited on the cathode is stripped off periodically for recovery.

The cementation process passes the silver-bearing liquor through a bed of iron in the form of steel wool, iron shavings, etc. The silver ion replaces the ferrous ion, with the ferrous ion going into solution and passing out with the effluent. The pH of this system must be kept between 4 and 6.5 [17].

Either of these systems will reduce silver content to 50 to 100 ppm. Ion exchange has also been reported as a method of recovering silver from spent photographic solutions or plating rinse waters.

To obtain levels below 1 ppm, chemical precipitation is used. Silver can be precipitated as a chloride or a sulfide, both of which are insoluble compounds. Addition of sodium chloride, hypochlorite or chlorine will form insoluble silver chloride. In the plating industry, wastes are in the form of silver cyanide. In order to free the silver, as well as to destroy the cyanide, chlorine is added to oxidize the cyanide to cyanate and then to carbon and nitrogen. The available chlorine, furnished as chlorine gas or a hypochlorite solution, also reacts with the silver to form a silver chloride precipitate. The sequence of operations for a batch treatment system is as follows:

1. Measure cyanide with field test kit.
2. Add caustic to pH 11.
3. Add chlorine as required by the specific amount of cyanide required.
4. Mix for 15 to 30 minutes.

5. Add sulfuric acid to pH 8.
6. Add chlorine as required to maintain the residual chlorine as measured by a field test kit.
7. Mix for 15 to 30 minutes.
8. Add polyelectrolyte.
9. Settle for 4 hours and decant.

Sulfide precipitation can also be quite effective and may give lower final-effluent levels of silver. Bench-scale testing should be performed to determine whether chloride or sulfide is the preferable means of precipitation.

Sodium sulfide under acidic conditions can form hydrogen sulfide, which is very toxic. Therefore, it is important that the precipitation be carried out under alkaline conditions with the addition of amounts of caustic soda to raise the pH. Adequate ventilation should also be provided.

References

1. Lanouette, K. H., and Paulson, E. G., Treatment of Heavy Metals in Wastewater, *Pollution Eng.*, Oct. 1976.
2. Feigenbaum, H. N., Removing Heavy Metals in Textile Waste, *Industrial Waste,* Mar.–Apr. 1977.
3. "Development Document for the Metal Finishing Industry," EPA Report No. 440/1-75/040-a, p. 166.
4. "Proceedings of Seminar on Metal Wastes Treatment Featuring the Sulfex Process," The Permutit Co., Paramus, N.J.
5. Dean, J. G., Bosqui, F. L., and Lanouette, K. H., Removing Heavy Metals from Wastewater, *Environmental Science & Technology,* June 1972.
6. "Quality Criteria for Water," EPA Report, preliminary draft.
7. "Water Quality Criteria," California State Water Resources Control Board, Publication 3-a, Reprint, Dec. 1971.
8. Larsen, H. P., Shou, J. K. P., and Ross, L. W., Chemical Treatment of Metal-Bearing Mine Drainage, *J. Water Pollution Control Federation,* Aug. 1973, p. 1682.
9. Russell, L. V., Treatment of CCA-, FCAP, and FR-Type Wastewaters, from "Pollution Abatement in the Wood Preserving Industry," Forest Products Utilization Laboratory, Mississippi State University, 1970.
10. Cohen, J. M., "Trace Metal Removal by Wastewater Treatment," EPA Technology Transfer, Jan. 1977.
11. "Development Documents for the Metal Finishing Industry," EPA Report No. 440/1-75/040-a, Apr. 1975, p. 76.
12. Day, R. V., Lee, E. T., and Hochuli, E. S., Bell System's Metals Recovery Plant, *Industrial Waste,* Jul.–Aug. 1974, pp. 26–29.
13. Rohrer, K. L., "Chemical Precipitants for Lead Bearing Wastewaters," Westinghouse Electric Corp.
14. Iammartino, N. R., Mercury Cleanup Routes—II, *Chem. Eng.,* Feb. 3, 1975, pp. 36–37.
15. Paterson, J. W., "Wastewater Treatment Technology," Ann Arbor Science Publishers, Oct. 1975, p. 157.
16. Rickard, M. D., and Brookman, G., "The Removal of Mercury from Industrial Wastewaters by Metal Reductions," Proceedings of the 26th Purdue Industrial Waste Conference, Purdue University, 1971, pp. 713–721.
17. "Recovering Silver from Photographic Materials," Eastman Kodak Co., Book No. 0-87985-041-8, Rochester, N.Y.

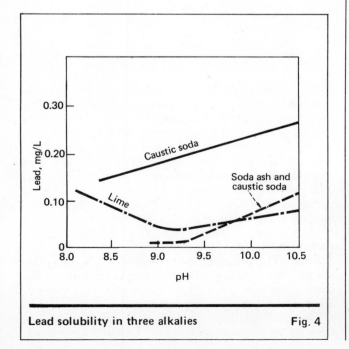

Lead solubility in three alkalies   Fig. 4

The author

Kenneth H. Lanouette is President of Industrial Pollution Control, Inc., 45 Riverside Ave., Westport, CT 06880, where he is in charge of supervising the design of industrial-treatment plants. Before, he was assistant to the president of Dorr-Oliver, Inc., where he was engaged in fabrication of waste treatment plants. Previously, he was responsible for waste-treatment facilities and chemical plants in Latin America. He holds a B.E. degree in civil engineering from the University of Southern California, and belongs to the American Soc. of Civil Engineers, AIChE, Water Pollution Control Federation. He is also a licensed professional engineer in Connecticut.

Section 4
SECONDARY TREATMENT METHODS

Selecting and specifying activated-carbon-adsorption systems

The type of polymer and adsorbent treatment for removing hazardous and toxic chemicals from waste streams should be tailored to the specific waste through a series of laboratory and pilot tests. Here is a summary of the choices available and the tests necessary to make them.

F. E. Bernardin, Jr., Calgon Corp.

☐ Adsorption on granular activated carbon is one of the best commercially proven methods for removing toxic organic chemicals from waste streams. It has demonstrated, in full-scale plants, a strong affinity for some of the best known toxic materials, such as DDT, PCBs, Dieldrin, Aldrin and many others. Furthermore, the technology and equipment for applying activated carbon to such a service are well proven and available.

However, a given application to a specific wastewater still requires careful examination. An engineer must evaluate the objective of the treatment system and peripheral problems that may affect that system. Many variables affect the optimization of a granular activated-carbon system (see box: "Removing mixed toxic chemicals").

The components of an adsorption system, in order of increasing process-engineering involvement, consist of: (1) carbon-handling and storage facilities, (2) reactivation facilities, and (3) the adsorption equipment.

The carbon handling and storage is an essential part of any adsorption system. Granular activated carbon is shipped either dry or wet; it is transported and stored on-site in slurry form, since hydraulic transportation of carbon slurries is clean, rapid and economical. Such slurries can be handled by water-jet eductors, centrifugal pumps, and pneumatic or hydraulic blowcases.

Similarly, reactivation in a furnace is an essential part of economical carbon adsorption systems. Both multiple-hearth and rotary-kiln furnaces are used, depending on the exhaustion rate, economics, space limitations, flexibility requirements, and plant preference.

The furnace size is dictated by the relative ease of activation, as well as by the amount of carbon to be reactivated and the desired excess capacity. The excess

Originally published October 18, 1976.

Removing mixed toxic chemicals

This chemical plant manufactures phenol, sulfuric acid, formaldehyde, pentaerythritol, sodium sulfite, sodium sulfate, and a number of synthetic resins and plastics. Its effluent, which averaged about 15 million gal/d, contained approximately 16,000 lb of BOD, 27,000 lb of COD, and 1,500 lb of phenol.

A target 90% reduction of these effluents was established in an agreement with the state regulatory agency; and a decision was made to install a clarification/adsorption system at a capital cost of about $2 million.

Prior to installing this system, the plant segregated its waste streams to reduce the flow of effluent for treatment to 400,000 gal/d. This collected effluent is adjusted for pH in an equalization basin and gravity-fed to a flocculation basin, where a nonionic polymer is added to enhance the formation of large floc particles. The flocculated wastewater then flows to a 40-ft² concrete clarifier, where essentially all suspended solids and floating material are removed.

The clarified waste is then pumped to either of two moving-bed adsorbers, each 12 ft dia. by 36 ft sidewall, made of steel, containing 124,000 lb of Filtrasorb granular activated carbon, and designed to accomodate a wastewater flow of 175 gpm.

This treatment has reduced waste to 1,450 lb/d of BOD, 2,675 lb/d of COD, 22 lb/d of suspended solids, and 0.5 lb/d of phenol. Approximately 70 gpm of the treated water is reused in the plant, and the rest is discharged to a river. The treatment plant has been operating over two years and is performing as designed.

capacity should be so limited as to have the furnace operating 80–90 percent of the time, since frequent shutdowns reduce refractory and arm life.

A hearth loading of 70–80 lb/(ft^2)(day) is generally required for multiple-hearth furnaces regenerating carbon for industrial wastewater applications. Depending on the application, this loading rate can range down to 50 and up to 120 lb/((ft^2)(day).

A 6% volumetric loading with 45 min at the reactivation temperature is generally used in countercurrent direct-fired rotary kilns for industrial wastewater carbons. Kilns can be designed with lifting flights to promote carbon drying and improve reactivation.

The adsorption equipment is the heart of the system. Three types of adsorbers are generally used in wastewater applications: downflow fixed-bed, packed moving-bed, and upflow expanded-bed.

Downflow fixed-bed adsorbers

These adsorbers offer the advantages of simple operation plus the ability to serve as a filter for simultaneous suspended-solids removal. Wastewater enters the top of the vessel and passes downward through the carbon bed, where the organics are adsorbed. The treated wastewater is collected in a header-lateral or tile underdrain at the bottom of the bed, and discharged to subsequent adsorbers or to the stream. The vessels can be operated either under pressure or by gravity flow.

If the bed is to serve as a simultaneous filter, 50% freeboard should be provided above the carbon to allow for expansion during backwash. Surface washers and an air scour are recommended for filtration applications. Backwash rates are generally 15–20 gpm/ft^2, and air scour rates 3–5 scfm/ft^2. As a rule of thumb, a carbon filter running at 5 gpm/ft^2 with a 50–75 mg/l suspended-solids loading will require backwashing for 10–20 min each day. Loadings are generally less than 5 gpm per ft^2 of cross-sectional area, if suspended solids are present, compared to rates up to 10 gpm/ft^2 in pressure systems with little or no suspended solids.

Fixed-bed adsorbers can be arranged for single-stage, multi-stage, or parallel operation. Single-stage adsorbers are used when the mass transfer zone is short (saturation of the carbon occurs shortly after initial breakthrough); two or more beds in series are used when an increased efficiency of carbon utilization is desired (see box: "Phenolic wastewater treatment"); two or more beds in parallel are used when the high flowrate would require excessively large vessel diameters.

Single-stage adsorbers are also used when the carbon exhaustion rate is low, and the cost of replacing or reactivating the carbon is a minor operating expense. Generally, when this occurs, the investment required for a multiple-column system cannot be justified by the lower carbon usage rate that would result.

When two or more stages are used, only one bed—that upstream—is removed at a time for reactivation; and the fresh carbon bed is put in the downstream, or polishing, position. When two or more beds are used in parallel, the effluents are blended, which is advantageous for reducing the carbon-exhaustion rate. In all cases, the entire vessel containing the exhausted carbon is emptied and refilled with fresh carbon.

Packed moving-bed adsorbers

The packed moving bed provides countercurrent operation, in that the fluid flows upward through a bed that is moved downward; portions of the bed are periodically removed from the bottom of the vessel, and replaced by reactivated carbon that is added at the top.

Capital investment for a moving-bed system is generally less than the investment for an equivalent fixed-bed system for treatments requiring large amounts of carbon. However, this type of bed generally requires the suspended solids to be less than 10 mg/l, which means that pretreatment is usually necessary.

Flowrates usually range from 3 to 7 gpm/ft^2. A height-to-diameter ratio of 3:1 is desirable to facilitate plug flow during carbon removal (pulsing). Pulse volumes range 2–10% of the total bed, so that the total bed is equivalent to 10–50 pulse-beds in series.

Adsorbers with total bed depths as great as 45 ft have been used without difficulty. Wastewater is fed peripherally through a ring header located on the bottom cone, while the treated water is collected in a ring header on the top cone. Screens or septa are submerged in the top of the carbon bed to retain the granules while passing the treated water.

Upflow expanded-bed adsorbers

If suspended solids in the effluent are not a concern, the upflow expanded-bed design should be considered. In this design, the wastewater enters the bottom of the beds at approximately 8–10 gpm/ft^2 and exits through the top of the vessel. The carbon near the top of the bed is expanded approximately 10%. A high rate of back-

---

Phenolic wastewater treatment

The plant is a producer of phenolic resins. Its wastewater contains floating oils, suspended solids, phenolics, and other organics. Flow is approximately 100,000 gal/d. Typical analyses are:

	Raw	Clarified	Carbon effluent
Suspended solids	220 mg/l	45 mg/l	<10.0 mg/l
Oil/grease	50 mg/l	25 mg/l	<5.0 mg/l
TOC	1,200 mg/l	650 mg/l	25.0 mg/l
Phenol	160 mg/l	130 mg/l	<0.1 mg/l

The treatment consists of separators for oil removal, coagulation utilizing a coagulant-aid and a polyelectrolyte, sedimentation, mixed-media filtration, and adsorption on activated carbon, which is supplied via an adsorption service that handles shipping, reactivation, and transfer of the material.

Adsorption is carried out in two fixed-bed adsorbers operated in series. Each vessel is 10 ft dia. by 11 ft sidewall and contains 20,000 lb of activated carbon.

The toxicity of this wastewater was evaluated by static tests in which bluegill sunfish were subjected to raw and final effluent samples. In the raw sample, all eight test fish died within the first 15 min of exposure. The fish exposed to carbon adsorber effluent showed no adverse reaction during the 10-day duration of the test.

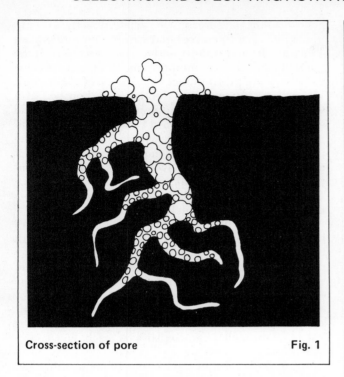

Cross-section of pore   **Fig. 1**

wash with air scouring may be needed occasionally, depending on the nature of the solids in the wastewater and the extent of biological activity in the bed.

The upflow expanded bed can be arranged for single-stage, series, or parallel operation. Design of the inlet distribution system is critical, since the distributors must be able to pass suspended solids, but retain the granular activated carbon. The concept of upflow expanded beds is relatively new, and as a result the engineering development has not progressed as far as fixed-bed and moving-bed technology.

Wet carbon in static contact with carbon-steel incurs a galvanic type of corrosion. Consequently, adsorber vessels are lined carbon-steel, stainless steel, fiberglass-reinforced plastic (FRP), and concrete. The final choice among these materials depends on the corrosivity of the wastewater in the presence of granulated carbon.

When toxic chemicals can be adsorbed

Activated carbon can be made from any carbon-containing raw material, e.g. coal, wood, nut shells, sewage sludge, or petroleum residues. Many raw materials can produce activated carbon with an extensive internal surface area. The differences between these products are their residual ash content, their hardness, and their activity. Activated carbon derived from bituminous coal is preferred for wastewater treatment due to its high hardness—a characteristic needed to keep down handling losses during reactivation.

The activation process creates within the individual pieces of carbon selectively sized pores, which provide the internal surface area where adsorption takes place. Calculations show that one pound of high-quality activated carbon has an internal surface area equivalent to approximately 125 acres. The major part of this area exists on the walls of micropores toward the interior of

the carbon particle as illustrated in Fig. 1.

A number of mechanisms govern the adsorption of organic molecules. First, molecules in the bulk of the solution must migrate to the carbon particle. Then they must migrate across a liquid film surrounding the particle, and thus into a pore. Thereafter they must migrate through the pore to finally come to rest on the ultimate adsorption site. When all of the available sites have been used up, the carbon particle is in equilibrium with the surrounding solution; and its capacity for further adsorption is exhausted.

Three principal factors affect the ease with which organic compounds are adsorbed: polarity, structure and molecular weight. Highly polar molecules are generally highly soluble; and vice versa. Since highly soluble molecules are adsorbed with difficulty (and vice versa), high polarity therefore reduces the ease of adsorption. Molecular structure affects the ease with which the molecules attach to the surface of the carbon particle; aromatic rings being conducive to adsorption.

Molecular weight affects ease of adsorption through two effects—solubility and surface attraction. The higher-molecular-weight compounds are generally less soluble, and consequently generally more easily adsorbable. Similarly, the surface attraction is generally greater for larger molecules, so that they are more easily adsorbed. However, this general rule holds only while the organic molecule is smaller than the pore size of the carbon. The major part of the pore surface area exists in pores ranging 10 Å to 1,000 Å, so that molecules larger than 1,000 Å are confronted with a severely limited surface-area capacity.

While these principles may afford some understanding of why a toxic chemical is more or less easily removed through adsorption, they do not help appreciably in estimating the equipment requirements of a specific treatment system. Such estimates require an adsorption isotherm and column-test results.

Adsorption isotherms

The equilibrium relationship between the concentration of an organic material adsorbed on the surface of carbon and the concentration of that material in the surrounding solution is given by the expression:

$$(x/m) = k\, C^{1/n}$$

where: (x/m) = concentration of the organic on the adsorbent, mg/g

C = concentration of the organic in the solution, mg/l

k, n = constants

Thus, if k, n and C are known, (x/m) can be calculated, and from that the theoretical equilibrium quantity of carbon. Since k and n vary with temperature, plots of the equation are known as adsorption isotherms. These are determined in the laboratory through tests on individual portions of activated carbon. A number of samples of weights ranging 0.1–20.0 g, for example, are placed in individual flasks. A given volume of the wastewater to be tested, e.g. 100 cm³, is added to each of the flasks; and the flasks are then put into a shaker and left there long enough for the contents to reach equilibrium. Average time is about 2 h, since

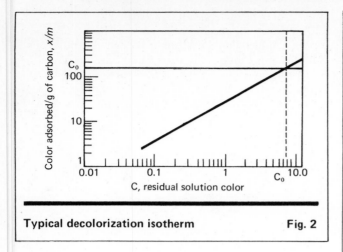

Typical decolorization isotherm   **Fig. 2**

wastewater contaminants of primary concern are usually adsorbed relatively rapidly.

Subsequently, the carbon is separated from the treated solutions on 0.45-micron filters, and the purified solutions analyzed for a particular characteristic, such as total organic carbon (TOC) or chemical oxygen demand (COD). Since the conditions in each of the flasks represent a different relationship between the concentration on the adsorbent and the concentration in solution, the data can be used to determine k and n.

In actual practice, this is done through a plot on log-log paper (Fig. 2). The plot, when extrapolated a small distance in either direction, tells the maximum amount of a particular contaminant that can be adsorbed from any concentration of that contaminant—and consequently the minimum amount of carbon necessary to treat a particular concentration of wastewater. More specifically, the plot tells us:

- The relative affinity of a component for the carbon.
- The maximum concentration of the contaminant on the carbon.
- The minimum concentration that can be attained in the solution.
- The sensitivity of adsorption to concentration in the solution (indicated by the slope of the plot).
- The effect of pH (by controlling the pH of the tested solutions).

Note that all this information is qualitative, since it relates to equilibrium conditions. Since equilibrium rarely represents the most economical conditions, additional tests are necessary for sizing actual adsorption facilities. These tests are known as the "column tests."

Column testing

The three major parameters for designing adsorption systems are: (1) contact time, (2) carbon usage rate, and (3) pretreatment requirements.

The contact time indicates the amount of activated carbon required onstream at any given moment—thus the size of the equipment, and the capital cost.

The carbon usage rate indicates the rate at which the carbon must be replaced and reactivated, and this defines the operating costs of the system. The pretreatment requirements, which are dependent on pH, sus-

pended solids, oil, grease, etc., relate directly to additional unit operations required for optimum treatment.

A wave-front or mass-transfer zone characterizes the gradual change in concentration of organic on the carbon during the continuous adsorption of the organic. When the solution moves through a carbon bed at a very rapid rate, this wave-front becomes long and drawn out. As a result, the bed begins to leak contaminant into the effluent very early, gradually increasing the concentration until complete breakthrough is achieved. If the solution moves through slowly, by contrast, the bed does not leak contaminant; and there is a sharp final breakthrough indicating a more complete utilization of the activated carbon (Fig. 3).

Contact time may be determined in the laboratory by a number of small carbon columns in series, with each column in the series representing a fixed contact time. Thus, if four columns each with 15 min contact time are put in series, it is possible to determine the effects of 15, 30, 45 and 60 minutes of contact time for any given volumetric throughput. By taking effluent samples from each column, data is collected to describe the breakthrough characteristics at each given contact time.

Fig. 4 shows breakthrough curves of an actual wastewater passed through five columns with contact times of 15, 40, 75, 105 and 240 min. The TOC in mg/l is plotted against the total volume throughput, in ml. As is shown, increasing the contact time from 15 to 40 min has virtually no effect on the breakthrough. However, an increase from 40 to 75 min results in the beginning of an effluent plateau at approximately 5,500 mg/l of TOC. A further increase to 105 min lowers this plateau and gives a significantly longer run before the average column has passed 4,000 mg/l.

Increasing the contact time to 240 min gives a more dramatic change in effluent quality. At this point, the additional requirement for equipment and carbon to achieve 240 min of contact should be worth the effort to achieve 4,000 mg/l in the effluent, because the amount of water that can be processed is much larger, considerably lowering operating costs. This is shown by a comparison with the curve for 75 min of contact.

At 75 min, about 400 cm³ were passed prior to exceeding 4,000 mg/l in the effluent, compared with over

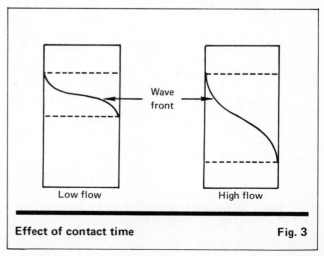

Effect of contact time   **Fig. 3**

2,100 passed at 240 min. This represents a 5.3-fold increase in wastewater handled by a column that is by volume of carbon only 3.2-fold larger. This example is a rather extreme case of a contact-time-dependent adsorbate. In the majority of cases, a contact time of 60 min or less is adequate for good utilization of the carbon.

Once the contact time has been determined, the second of the three major parameters, carbon usage rate, can also be determined through a column test. Data can be obtained by operating two columns in series until the effluent of the second bed exceeds the effluent limitations. At this point, the first bed in the series is removed and a fresh bed of carbon placed in the final position. This is known as staging. By repeating the staging procedure a number of times and analyzing effluent concentrations, a series of curves can be developed.

These curves should begin to approach an equilibrium condition, as exhibited by a duplication of the shape of the breakthrough curve. At this point of duplication, the volume throughput per unit weight of carbon can be derived and should permit an accurate estimate of carbon usage for treating the wastewater.

Data from an actual wastewater passed through four operations illustrate this principle (Fig. 5). Columns A and B were placed in series service. When the effluent from B exceeded 4,000 mg/l TOC, column A was removed, and column C placed in the polishing position after column B. When the effluent from C reached 4,000 mg/l, B was removed from service and D installed after C—and so on through column E.

In reviewing this data (Fig. 5), the shapes of the breakthrough curves for columns C, D and E take on a characteristic shape. Accordingly, the volume throughput between equivalent points on the curves determines the volume throughput per unit of carbon. It can be seen from the figure that this was about 2½ liters for columns C and D and just over 2½ liters for column E.

The operating performance of this system is thus approximately 2½ liters capacity for the quantity of carbon contained in each column (in this case 25 g). It is this figure—2.5/25 = 0.1 liter/g, or 1,000 gal of wastewater per 80 lb of carbon—that determines the rate at which carbon must be replaced and reactivated.

Wastewater pretreatment

Although granular carbon can be used to filter out suspended solids and oil, as well as to adsorb soluble compounds, it is sometimes more economical to pretreat a wastewater to remove such solids prior to adsorption.

In general, solid particles are held suspended in water because of surface charges, so that it is necessary to remove or reduce the surface charges in order to separate the solids. This can be done in a number of ways through the use of polymers. Polymers, which can be either positively or negatively charged, or have an overall net charge equal to zero, can be used (1) to link suspended particles, (2) to neutralize the charged particles, or (3) to both link together and neutralize the suspended particles. The objective in any case is to reduce the stability of the suspension, so as to allow separation of the particles by subsequent treatment.

In some cases (i.e., suspended solids and oil) dissolved-air flotation may be the most applicable separation technique. Polymers added to such a wastewater will produce a floc particle consisting of both solid and oil, and with a density nearly equal to or perhaps slightly less than that of water. Such a floc is quite amenable to dissolved-air flotation.

In other cases, neutralization plus linking the suspended particles may result in a sediment that is best removed through some settling technique.

In either case, the maximum removal of particles calls for an additional step, such as filtration, to polish the effluent. Filtration may also be applied directly to a wastewater in cases where the particle concentrations are suitable, as when colloidal particles can be destabilized inline and removed directly by filtration.

Sometimes a primary coagulant may be needed to begin the capture of colloidal particles and prepare them for treatment by a polymer. An inorganic primary coagulant such as alum, lime or ferric chloride in combination with a polymer may produce separations superior to the mere addition of the effects of using the two treatments separately. Also, clay can be used to add weight to the floc, so as to help settling, or to provide a point on which the materials in suspension can aggregate, thereby making coagulation with a polymer somewhat easier. The chemical structure of the polymers can be designed to provide the required charges, whether positive, negative or neutral (Fig. 6).

In the laboratory, the application of polymers to a wastewater is usually evaluated in a small-scale batch

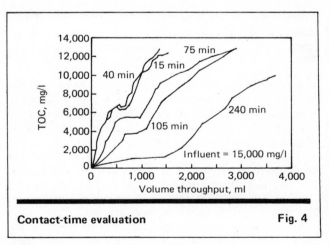

Contact-time evaluation   **Fig. 4**

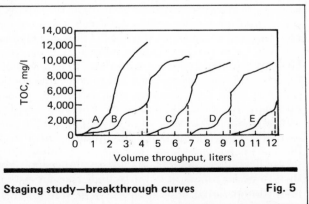

Staging study—breakthrough curves   **Fig. 5**

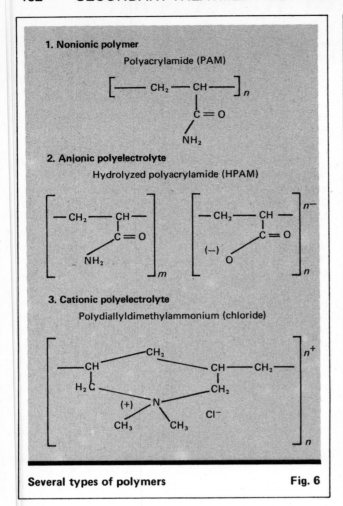

1. Nonionic polymer

   Polyacrylamide (PAM)

2. Anionic polyelectrolyte

   Hydrolyzed polyacrylamide (HPAM)

3. Cationic polyelectrolyte

   Polydiallyldimethylammonium (chloride)

Several types of polymers   Fig. 6

test termed the jar test. Various dosages of polymer and/or primary coagulants are added to volumes of wastewater, after which the solids are evaluated for flocculation time, size of floc particle, and speed of settling (in the case of sedimentation).

These tests will usually give an indication of the type of polymer to be tried. Depending on the application, dosages may range from a few tenths of a milligram per liter (in the case of water) to hundreds of milligrams per liter for sludge-dewatering applications.

Since the size of a floc particle and its speed of settling are somewhat dependent on the equipment used, it is not uncommon for dosages in plant operations to differ from the jar-test results. Full-scale trials are usually required to optimize the application of polymers.

One of the primary considerations of using polyelectrolyte polymers, or any coagulant, to remove solids from water is that the coagulant must come into contact with the solid. This becomes most evident when conducting a full-scale trial treatment. The direct application of a highly viscous high-molecular-weight polymer will usually not result in optimum treatment, because the polymers must be significantly diluted prior to application to achieve maximum dispersion. Also, they must be mixed at a fairly high rate for a short period immediately after application, so as to provide a more intimate contact with the solids.

Usually a separate mixing chamber is used, although in plants constructed without such mixing facilities Parshall flumes or weirs may be used to promote mixing. These latter are of particular interest when upgrading treatment plants that were not necessarily designed for the addition of polymers.

With the application of good coagulation and subsequent solid/liquid separation, the wastewater is ready for treatment by adsorption through contact with granular activated carbon.

Field pilot studies

In many cases it is necessary to run field pilot studies to verify the laboratory conclusions. Such tests demonstrate the process on a continuous basis under actual plant conditions of flow and concentration variations. They are generally desirable if suspended-solids removal is required or biological activity is expected. Granular carbon can be used as a filter for suspended solids and oil removal as well as an adsorber. Also, in the case of dilute, biodegradeable wastewaters, biological degradation of some of the organics can occur due to the growth of microorganisms on the surface of the carbon.

Generally, a pilot plant consisting of four $5\frac{1}{2}$-in. by 6-ft Plexiglas columns connected in series is used. This unit is designed to operate at a downflow rate of $\frac{1}{4}$–$\frac{1}{2}$ gpm.

The first column in the series will contain 3 ft of granular carbon, and will serve as a filter as well as an adsorber, so that series operation permits the study of bed depths and contact time provided by 3, 8, 13 and 18 ft.

The pressure drop across the pilot columns is monitored, and the lead column is backwashed when an excessive pressure drop is observed. Wastewater samples of the feed and effluent from each column are collected and analyzed. Based on this data, the design carbon-exhaustion rate, contact time, and filtration rate can be verified.

Finally, the spent carbon from the pilot plant should be subjected to a reactivation study to determine: (1) the spent-carbon characteristics, (2) the degree of difficulty in reactivating the carbon, (3) the existance of any air pollution problems, and (4) any potential for corrosion problems. Generally, batch laboratory reactivation tests are sufficient for generating this information, but in special cases, small-scale test furnaces can be used.

The author

Frederick E. Bernardin, Jr., is Technical Service Manager of the Adsorption Service group at Calgon Corp., Calgon Center (Box 1346), Pittsburgh, PA 15230, where he is responsible for providing technical service for the process performance of the firm's granular-activated-carbon adsorption service installations. Joining Calgon as a chemist in 1967, he was made group leader of the Activated Carbon R&D dept. before accepting his present position. He holds a B.S. in chemistry from Waynesburg College and an M.S. in sanitary engineering from the University of Pittsburgh.

Treating industrial wastewater with activated carbon

The authors cover the equipment and economics of systems used to adsorb organic contaminants from industrial wastewater.

Joseph L. Rizzo and *Austin R. Shepherd*, Calgon Corp.

☐ Treatment of industrial wastewater with granular activated carbon began in the United States, England and Germany during the early 1960s. The development of a high-surface-area carbon that could be reactivated made adsorption economically practical for removing dissolved organic contaminants from wastewater. (One pound of granular activated carbon made from select grades of bituminous coal can have a surface area equivalent to 125 acres.) Another improvement has been the increased hardness of the activated-carbon granules, which makes reactivation and repeated reuse feasible. Attrition and carbon losses during reactivation are kept to a minimum with carbons having high hardness.

Selectivity in adsorption

In solution, high-molecular-weight organic molecules of a given class adsorb preferentially to those in the same class having lower molecular weights. This rule applies especially to organic compounds having three or more carbon atoms, since they will in most cases respond to adsorption treatment.

If the molecule is nonpolar, adsorption will likewise be favored. Thus, aromatic, nonpolar contaminants of high molecular weight such as benzene can be easily removed from aqueous solutions; low-molecular-weight polar organics such as methanol are not effectively removed.

There is, however, a maximum molecular weight above which adsorption is adversely affected, particularly as the dissolved organic molecule approaches polymer size. The molecule may become so large that pore diffusion into the vast internal surface area of activated carbon is inhibited. If this phenomenon is suspected, laboratory tests should be run to evaluate the chances of success for adsorptive treatment.

Other major factors that affect adsorptive specificity are the pH of the water, and the contaminant's aqueous solubility. The specific adsorption characteristics of the contaminant will be altered if it is present in a mixture of organic contaminants.

Evaluating adsorption feasibility

The first step in assessing the feasibility of granular activated carbon for a specific application is to put together a liquid-phase adsorption isotherm in the laboratory.

Data for an isotherm are obtained by treating fixed volumes of the contaminated liquid with a series of known weights of carbon. The carbon-liquid mixture is agitated at a constant temperature. After the carbon and the liquid reach adsorptive equilibrium, the carbon is removed and the residual contamination in the liquid is measured by an appropriate analytical method. The data collected are then tabulated as shown in Table I.

The residual contaminant concentration, C, of the solution is measured directly. The amount of impurity adsorbed on the carbon, X, is the difference between the untreated contaminant concentration, C_0, and the residual contaminant concentration, C. Dividing X by M, the weight of carbon used in the test, gives the amount of contaminant adsorbed per unit weight of carbon.

The Freundlich isotherm equation empirically relates isotherm data and has been found to fit many adsorp-

Development of isotherm data			Table I
Carbon weight per solution volume, g/l (M)	Residual contaminant conc., mg/l (C)	Contaminant adsorbed per solution volume, mg/l (X)	Contaminant adsorbed per carbon weight mg/g (X/M)
0.0	75	—	—
0.1	63	12	120
0.2	53	22	110
0.4	36	39	97.5
1.0	8	67	67
2.0	1	74	37.0
4.0	0	—	—
10.0	0	—	—

Originally published January 3, 1977.

tions in dilute solutions. The equation is of the form:

$$X/M = KC^{1/n} \qquad (1)$$

Taking the log of both sides yields a linear equation whose slope is $1/n$ and whose intercept is K at $C = 1$.

$$\log X/M = \log K + \frac{1}{n} \log C \qquad (2)$$

When the data from the lab test are plotted on log-log paper (Fig. 1), the best straight line connecting the points is the isothermal relationship between contaminant concentration and carbon equilibrium capacity (Table II).

To estimate a contaminant capacity for calculating the weight of carbon needed for a column, one uses an X/M value that corresponds to the influent concentration, C_0. This value, X/M_{C_0}, represents the maximum amount of contaminant adsorbed per unit weight of carbon when the carbon is in equilibrium with the untreated contaminant concentration.

Once X/M_{C_0} is determined for a system, the theoretical carbon demand for a given volume of contaminated liquid can be calculated. For complete removal of the contaminant:

$$Y = C_0/(X/M_{C_0}) \qquad (3)$$

where Y = weight of carbon required per unit volume of contaminated liquid.

The isotherms of mixed-solute wastewaters, such as are commonly found in industry, are frequently nonlinear. Such plots may exhibit a series of straight lines with dissimilar slopes, each representing one of the components in the mixture. In such cases, each of the lines can be treated as a separate isotherm, and a theoretical carbon demand can be calculated by summing the weights required for removing each contaminant.

Dynamic column testing

The optimum operating capacity and contact time must next be determined in order to fix the column dimensions and the number of units needed for continuous treatment. Optimum contact time depends upon the rate at which the contaminant is adsorbed by the carbon, and can only be determined by dynamic column testing.

This test is usually conducted with a series of carbon columns whose contact times range from 15–60 min or greater. A typical pilot-column arrangement is shown in Fig. 2. Rate of flow per unit area through an adsorption column—termed surface loading—is not sensitive in most wastewater applications. However, it is advisable to make at least two dynamic column runs at different loadings before eliminating this variable.

Wastewater is pumped through the column system, and effluent samples are collected at appropriate intervals from each of the columns. The amount of impurity remaining in the samples is plotted as a function of volume throughput for each column. The result is a series of curves that vary in shape, depending on the dynamics of the system (Fig. 3).

The point at which the impurity in a column effluent exceeds the treatment objective is called the breakpoint. That part of the curve between the initial leakage and

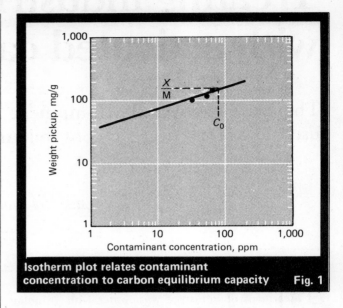

Isotherm plot relates contaminant concentration to carbon equilibrium capacity   **Fig. 1**

the point where the column effluent concentration is the same as the influent is called the breakthrough curve.

During adsorption, the upper section of a column saturates with impurities, whereas the lower section remains relatively virgin. Between these two extremes lies the adsorption zone, where removal of the impurity is actually taking place. As the column becomes saturated, the adsorption zone moves downward through the bed. The time required for appearance of the breakpoint in the effluent, as well as the breakthrough-curve slope, provides an index of the relative depth of the adsorption zone.

From the breakthrough curve, the volume of acceptable effluent collected for each contact time can be read directly. From isotherm curves, the carbon exhaustion rate (pounds of carbon used per unit volume of wastewater treated) can also be calculated. With these data, one can plot a curve of contact time vs. exhaustion rate (Fig. 4). Such a curve reveals at a glance whether there is anything to be gained from long contact times. At the point where the curve becomes asymptotic, there is no advantage in increasing adsorber volume to extend contact time.

Typical Freundlich-isotherm constants   **Table II**

	Intercept (K, mg/g)	Slope ($\frac{1}{N}$)
Aniline	25	0.322
Benzene sulfonic acid	7	0.169
Benzoic acid	7	0.237
Butanol	4.4	0.445
Butyraldehyde	3.3	0.570
Butyric acid	3.1	0.533
Chlorobenzene	40	0.406
Ethyl acetate	0.6	0.833
Methyl ethyl ketone	24	0.183
Nitrobenzene	82	0.237
Phenol	24	0.271
TNT	270	0.111
Toluene	30	0.729
Vinyl chloride	0.37	1.088

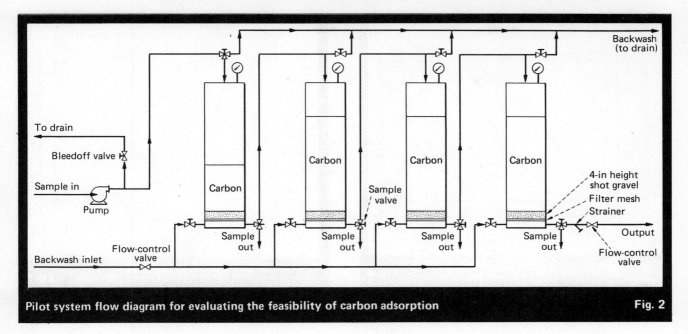

Pilot system flow diagram for evaluating the feasibility of carbon adsorption   Fig. 2

Adsorber systems

After the contact time has been established, one must choose between a single fixed-bed or countercurrent adsorber configuration. A tradeoff exists between the carbon exhausted and the capital spent for contacting equipment. Again, inspection of the breakthrough curves aids in the decision. Curves that are very steep favor fixed-bed, nonstaged systems, whereas those gradual in slope suggest countercurrent use of carbon to achieve the most economical, practically attainable exhaustion rates. The various adsorber configurations seek to exploit the carbon as much as possible, while avoiding hydraulic problems. Some of the most common configurations are shown in Fig. 5.

Fixed beds in series—In this contactor design, flow is downward through adsorbers connected in series. Each carbon bed is replaced as a complete unit, starting with the first unit—the recharged bed being then moved to the polishing position. The carbon thus moves countercurrent to the wastewater flow. This design is suited for systems that use small amounts of carbon, but require countercurrent efficiency. If backwashing is provided for, a limited amount of suspended material can be handled directly on the lead adsorbers.

Moving beds—This technique is a refinement of the countercurrent operating principle. Wastewater flows upward through the bed, with portions of the bed being periodically removed from the bottom of the unit. Fresh carbon enters at the top of the vessel. This design is most useful for systems requiring large amounts of carbon and countercurrent efficiency. Suspended material must generally be removed ahead of the moving bed to forestall hydraulic difficulties such as plugging and fouling.

Fixed beds in parallel—These units are operated downflow in parallel, and each carbon bed is replaced as a single batch. Startup of individual trains is staggered, so that exhaustion of the beds occurs sequentially. Blending effluent from columns containing fresh carbon with effluent from columns having partially exhausted carbon prolongs the useful life of the beds. This design favors large-volume plants where effluent blending will meet treatment standards. The adsorbers are generally equipped for backwashing, allowing a limited amount of suspended solids to be handled directly.

Upflow expanded beds—With this configuration, adsorbers operate upflow in series or parallel at surface loading rates of about 7–8 gpm/ft². Upflow operation expands the beds by about 10% of their packed volume. Exhausted units are reactivated batchwise, and are rotated according to a sequence similar to that used in downflow units in series or parallel operation. A water distribution scheme with backwash capabilities is required for each adsorber. The advantage of this type of system is its ability to pass suspended solids without developing excessive pressure drop.

Other design considerations

Adsorbers that hold granular activated carbon can be designed for both pressure and gravity flow. For fixed beds, flowrates usually run less than 5 gpm per square foot of carbon-bed surface area. Contact times for industrial wastewater mixtures average about 60 min., about twice the time employed for purification of domestic wastes. If suspended solids are present, they will be filtered out by the carbon bed. In such cases, the adsorbers can be designed to accommodate periodic backwashing and bed-cleansing procedures, such as air scouring or surface washing.

Pretreatment practices

Pretreatment ahead of activated-carbon contactors seeks to remove excessive contaminants such as suspended solids, oils, and greases. When present, these substances filter out on the carbon beds and boost the bed pressure drop. Chemical clarification, oil flotation, and filtration are pretreatments commonly used to eliminate this problem.

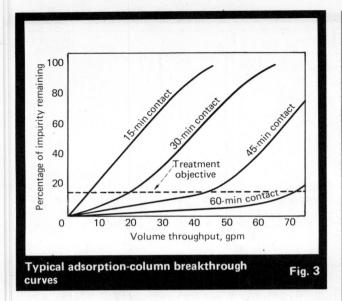

Typical adsorption-column breakthrough curves   **Fig. 3**

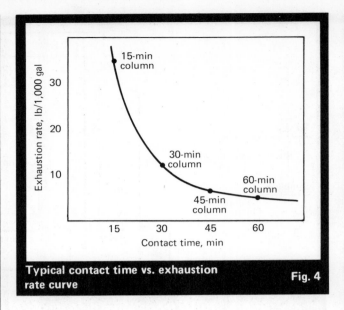

Typical contact time vs. exhaustion rate curve   **Fig. 4**

Flow equalization of the industrial wastewater stream is recommended for most adsorption processes. Although carbon systems can be designed to handle widely fluctuating inlet conditions, they can be installed more economically if hydraulic properties and organic concentrations will be stable.

Reactivation of granular carbon

Thermal oxidation in either multiple-hearth furnaces or rotary kilns reactivates exhausted carbon for reuse. The size of reactivation equipment depends on the carbon exhaustion rate (i.e., pounds of carbon exhausted per 1,000 gal of wastewater treated), and the weight of contaminant on the carbon. Excess capacity should be provided to handle variations in carbon usage that result from changes in wastewater flow and organic-loading, and from maintenance downtime.

Heating exhausted granular activated-carbon to 1,600–1,800°F drives off moisture, and volatilizes and oxidizes dissolved organic contaminants. Oxygen in the furnace is normally controlled at less than 1% (volume) to selectively oxidize contaminants over the activated carbon. About 5–10% weight-loss of activated carbon per reactivation cycle can be assumed in evaluating the economics of treatment systems. In addition, the reactivation package should include an afterburner and scrubber to eliminate the possibility of air pollution.

Centrifugal pumps and eductors, as well as hydraulic or pneumatic blowcases, can be used to transport granular carbon between the adsorbers and the reactivation equipment. The transport piping should include flush ports and wide-radius bends to prevent carbon attrition and plugging. Fig. 6 illustrates the process and functional features of a typical reactivation and carbon-handling system.

Adsorption-system options

Once the feasibility of carbon adsorption has been confirmed, there are four options regarding the handling of carbon after it has been exhausted.

Throwaway carbon—The simplest process uses carbon on a one-time basis. Virgin carbon is charged into the adsorbers and used until it becomes saturated with the adsorbed impurity. When the carbon becomes exhausted, it is removed from the adsorber, drained of liquid, and disposed of by landfilling. While this alternative has the advantage of not requiring thermal reactivation support, it is generally economically impractical when daily carbon usage exceeds several hundred pounds. Moreover, landfilling the carbon could be objectionable since the organic contaminants may leach from the carbon.

Offsite reactivation—The second option involves the reuse of carbon that has been reactivated offsite. Exhausted carbon is drained and shipped to an offsite reactivation facility for thermal reactivation and repackaging. Makeup carbon must also be provided, since 10–15% weight loss will be sustained during handling and reactivation. This high carbon loss is due primarily to extra handling steps that are not required with an onsite reactivation facility. Spent- and reactivated-carbon storage tanks are generally provided to ensure adequate onsite carbon inventory. Offsite reactivation improves project economics by permitting carbon reuse without the burden associated with the operation of an onsite reactivation furnace.

Onsite reactivation—A multi-hearth furnace or rotary kiln is used to thermally reactivate spent carbon, which is conveyed by eductors, slurry pumps, or a blowcase from the adsorbers to a furnace-feed storage tank. From the furnace-feed tank, the carbon passes to a dewatering screw that eliminates free water and reduces moisture content to about 50 wt%.

After adsorbed organics have been volatilized and oxidized in the furnace, hot carbon falls from the furnace into a quench tank and is transported to a reactivated-carbon storage tank to await reuse in the adsorbers. Any decision to use onsite reactivation must balance the benefits of carbon recovery against the relatively high cost of purchasing and installing reactivation equipment.

Adsorption service—The fourth option is to arrange for

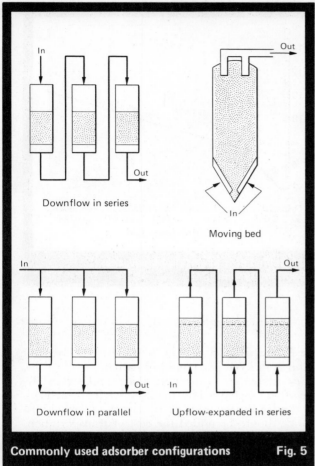

Commonly used adsorber configurations   **Fig. 5**

Downflow in series

Moving bed

Downflow in parallel

Upflow-expanded in series

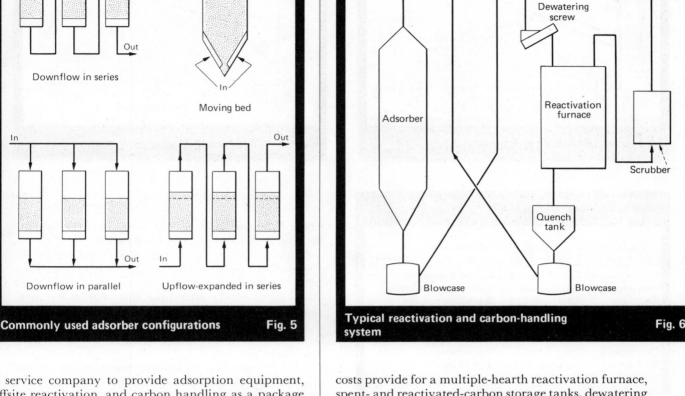

Typical reactivation and carbon-handling system   **Fig. 6**

Reactivation carbon storage

Spent carbon storage

Dewatering screw

Adsorber

Reactivation furnace

Scrubber

Quench tank

Blowcase

Blowcase

a service company to provide adsorption equipment, offsite reactivation, and carbon handling as a package that includes installation, startup and maintenance. The economics of operating a large central reactivation facility serving many clients enables this service to compete favorably with other carbon-handling options.

Adsorption economics

Costs associated with adsorption can be broken down into three categories:

- Capital cost of contacting system, including adsorbers and piping.
- Capital cost of reactivation facility, if provided.
- Ongoing costs associated with operation and maintenance.

Capital costs—Installed equipment costs have been estimated for two-stage, fixed-bed, downflow contacting systems of various capacities (Fig. 7). A contact time of 30 min per vessel has been assumed. The estimates include: epoxy-lined, carbon-steel pressure adsorbers; lined carbon-steel pipe; engineering and installation; and initial carbon inventory. Also shown are installed equipment costs for contacting systems with onsite, bulk carbon-storage. Bulk storage equipment must be specified if either throwaway or offsite reactivation is opted for. Winterization costs have not been included.

Fig. 8 presents estimates for the installed cost of an onsite reactivation and carbon-handling system. These

costs provide for a multiple-hearth reactivation furnace, spent- and reactivated-carbon storage tanks, dewatering screw, quench tank, carbon motive system, engineering and installation, and initial carbon inventory. Furnace sizing is based on a feedrate of 75 lb of carbon per square foot of hearth area per hour, and an operating factor of 0.75. The costs are based on January 1976 dollars and do not include winterization.

Annual operating costs—Table III summarizes the annual costs associated with the four carbon-handling options.

Capital annualization—Most industrial carbon-adsorption systems are depreciated over an 8–15 yr period. Over 11 years, at current interest rates (10%), the annualizing factor is about 6.5. (Capital annualization = installed capital costs/annualizing factor.)

Carbon makeup—Carbon-makeup requirements are derived from the amount of carbon exhausted by the system. For onsite reactivation facilities, this requirement runs about 5–10% (by weight) of the reactivation rate. For offsite reactivation 10–15% should be used for the estimate. With the throwaway option, the entire carbon-usage rate must be made up.

Fuel—Oil, natural gas or propane is burned in the reactivation furnace. Heat loads range from 6,000 to 8,000 Btu/lb of carbon reactivated, including afterburner demand.

Steam—The steam demand for onsite reactivation

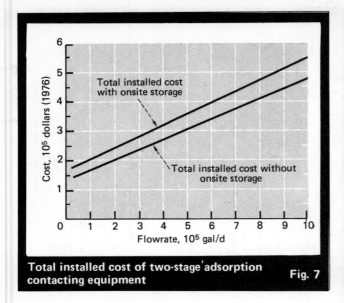

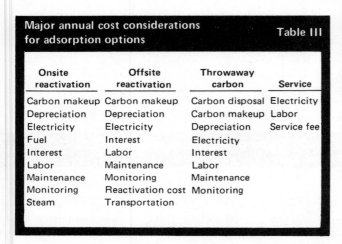

Total installed cost of two-stage adsorption contacting equipment   **Fig. 7**

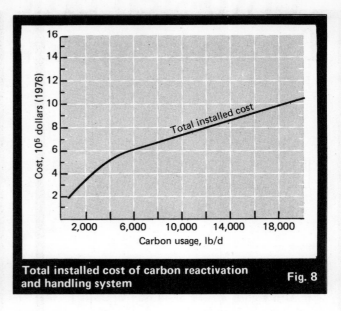

Total installed cost of carbon reactivation and handling system   **Fig. 8**

Major annual cost considerations for adsorption options   **Table III**

Onsite reactivation	Offsite reactivation	Throwaway carbon	Service
Carbon makeup	Carbon makeup	Carbon disposal	Electricity
Depreciation	Depreciation	Carbon makeup	Labor
Electricity	Electricity	Depreciation	Service fee
Fuel	Interest	Electricity	
Interest	Labor	Interest	
Labor	Maintenance	Labor	
Maintenance	Monitoring	Maintenance	
Monitoring	Reactivation cost	Monitoring	
Steam	Transportation		

$$\text{Electrical demand} = .015 \frac{C}{10,000} - 0.7 \qquad (4)$$

where C is activation rate in lb/d, and the electrical demand is in kWh per pound of carbon.

Offsite reactivation costs—This charge relates to the reactivation fee charged by an offsite service firm. Typically, the fee ranges from 25 to 35¢/lb of carbon. To this price should be added the cost of transportation, handling, and makeup carbon.

Maintenance, insurance and taxes—For typical adsorption facilities, these items average about 7–15% of the fixed investment per year.

Labor and monitoring—These requirements for adsorption systems vary widely, depending upon degree of automation and system size. Large onsite reactivation systems may require as much as four man-years of operating and monitoring labor per year. Small throwaway or offsite reactivation systems may need as little as one man-year per year.

Service fee—A monthly service fee is charged, eliminating the need for major capital expenditure on the part of the user. In some instances, the yearly fees can be less than the total annual costs of other adsorption alternatives. Moreover, for the client, many of the risks associated with designing, procuring, installing, operating and maintaining an adsorption system are eliminated.

systems averages one pound of steam per pound of carbon. The steam is introduced into the activating section of the furnace to help control volatilization and oxidation.

Electricity—Electrical needs arise from wastewater pumping, carbon reactivation, and control instrumentation. Pumping and instrumentation demands should be calculated for each case. That for reactivation can be estimated from the following correlation:

The authors

Joseph L. Rizzo is Marketing Director, Adsorption Systems, for Calgon Corp., Pittsburgh, PA 15230, directing the marketing and development of municipal and industrial wastewater-treatment systems. A member of the Water Pollution Control Federation and the American Assn. of Textile Colorists & Chemists, he received his B.A. in chemistry (with a minor in zoology) from Wabash College. He has done graduate work at University of Pittsburgh and the University of Indiana.

Austin R. Shepherd is Regional Manager of Engineering, Adsorption Systems, for Calgon Corp., concerned with the design and engineering of industrial wastewater-treatment systems. Previously, he was a project engineer for Calgon's Water Management Div. Mr. Shepherd is a registered P.E., and holds a B.T. in water resources engineering from Pennsylvania State University and an M.S. in environmental engineering from University of Pittsburgh.

A two-step process for toxic wastewaters

A combination of wet-air oxidation and biological treatment in the presence of powdered carbon has demonstrated an ability to handle chemical wastes.

A. R. Wilhelmi and *R. B. Ely*, *Zimpro Inc.*

☐ Although a variety of both biological and incineration waste-treatment processes have been highly refined and proven in many applications, there remains a large class of industrial and municipal wastewaters that do not lend themselves to treatment by such means. These wastes are biologically refractory or toxic to biological treatment organisms; and their large flow-volumes and low heat values bring on high costs when incinerated.

Wastewater from the manufacture of acrylonitrile typifies such wastes. This waste contains chemical oxygen demand (COD) in the range of 30,000–60,000 mg/l (4–6% ammonium sulfate) and 500–3,000 ppm cyanide in the form of hydrogen cyanide and organic nitriles. In Japan, these wastes are barged for ocean dumping, while in the U.S. all manufacturers but one dispose of them by deep-well injection. However, politi-

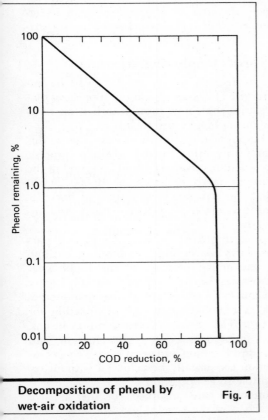

Decomposition of phenol by wet-air oxidation — **Fig. 1**

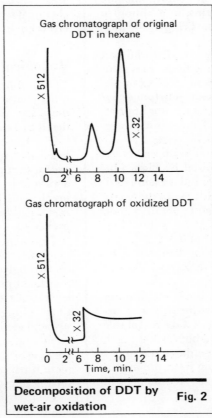

Decomposition of DDT by wet-air oxidation — **Fig. 2**

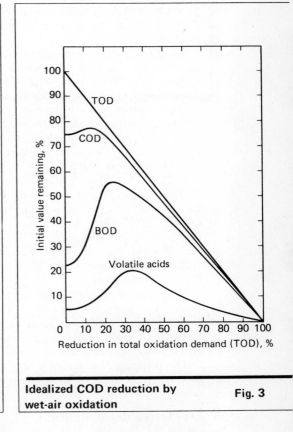

Idealized COD reduction by wet-air oxidation — **Fig. 3**

Originally published February 16, 1976.

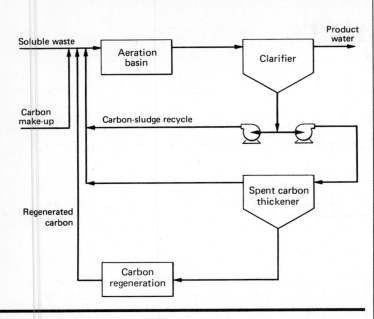

Biophysical treatment uses activated carbon   **Fig. 4**

cal pressures discourage such methods. Incineration has been employed, but the thermal decomposition of ammonium sulfate into ammonia and sulfur dioxide presents serious air pollution problems, in addition to contributing to a high fuel cost for the firing of the incinerator.

So-called night soil—chemically treated human wastes resulting from collection methods prevalent throughout the Far East—represents another example. This waste has been stored up to a month before delivery to a treatment plant. In terms of COD, it is approximately 100 times the strength of U.S. domestic sewage, more closely resembling many industrial wastes.

The ordinary treatment of night soil has been via extended digestion (30–60 days), followed by high dilution (20–40 times), and final treatment in an activated-sludge or trickling-filter system. However, the main problem with such treatment has been to obtain a good quality of digester supernatant liquor. While biochemical oxygen demands (BOD) of about 2,500 ppm were desired, these biochemical oxygen demands often ranged from as low as 2,000 ppm to BOD values as high as 10,000 ppm.

Furthermore, digested night soil does not dewater easily or respond well to aerobic treatment, partly because the raw waste has most of its COD in soluble form, has a high chloride content of about 5,000 mg/l, and contains a substantial nonbiodegradable fraction.

The broad group of wastes resulting from the manufacture of pesticides, herbicides, etc., represent still other examples.

Oxidative detoxification

Fortunately, the reaction kinetics of wet-air oxidation have provided a means of surmounting such waste treatment problems. During wet-air oxidation, higher-molecular-weight compounds are preferentially oxidized to lower-molecular-weight, biodegradable, intermediate products such as acetic acid, methanol, formaldehyde, and so forth. Some of the smaller but toxic radicals, such as cyanide, nitrile and thiocyanate, are readily oxidized; and odorous compounds of sulfur, such as the mercaptans, are quickly destroyed.

To illustrate, a solution of phenol was treated by wet-air oxidation at progressively higher temperatures. The percent phenol remaining was as shown in Fig. 1. It can be seen that 99.99 + % of the phenol is destroyed by the point where the chemical oxygen demand had been reduced 89%.

For another example, Fig. 2 shows a gas chromatograph trace of a solution of DDT before and after wet-air oxidation. Note that after oxidation there is no evidence of either the *ortho* or *para* DDT peaks, even after 16-fold amplification.

The fate of specific components is difficult to follow analytically in the complex mixture of components that one might find in an actual industrial wastewater. However, the reactions may be idealized by tracing the extent to which COD is converted to BOD. Fig. 3 presents such an idealized curve from the wet-air oxidation of acrylonitrile wastewaters. It can be seen that the BOD increases substantially as oxidation breaks up toxic and refractory components, in this case nitriles and polymeric substances. Also, the COD can increase initially, as organics resistant to the dichromate and acid-oxidizing conditions of the standard tests for chemical oxygen demand are attacked in the more rigorous environment that is characteristic of wet-air oxidation.

Biological versus biophysical treatment

This ability to detoxify wastes can be aided by another characteristic of wet-air oxidation: it oxidizes biomass and organics that have been adsorbed on powdered carbon during biophysical treatment.

The basic biophysical-treatment process (Fig. 4) differs little from a conventional activated-sludge process, except that powdered activated carbon is carried in the aeration basins at levels in excess of 10,000–20,000 mg/l. This has the advantages of improved flocculation, adsorbed toxic materials, adsorbed odors, and improved O_2 uptake.

Flocculation is improved because the carbon particles serve as heavy nuclei and thus promote settling. Accordingly, the active biomass of the biophysical treatment process can be carried at substantially higher levels than in conventional biological treatment—sometimes as high as 10,000–15,000 mg/l of mixed-liquor suspended solids. This means smaller aeration basins, greater sludge age and lower sludge production; and because of the rapid settling and great sludge age, complete nitrification can usually be achieved without resorting to multiple-stage operation.

Toxic components are adsorbed on the massive amounts of carbon in the system, and are thus inhibited from poisoning the bioorganisms. Then, as the toxic materials are biologically assimilated at the surface, they are desorbed. Or, they are carried out of the system on the carbon to subsequent destruction in a wet oxidation/regeneration step.

Odors, which can present problems as aeration strips out volatile organics, are adsorbed and thereby greatly reduced.

Oxygen transfer and uptake actually seem to improve in the presence of carbon. Though the mechanism is not well understood, it may involve an effect of extended biological surface or adsorption-desorption of either O_2 or inhibitory substances.

Even when it is possible to achieve the required effluent standards by conventional activated sludge treatment followed with adsorption by granular carbon, a simultaneous biophysical treatment with powdered carbon is likely to be more economical for two reasons: (1) Powdered carbon is much less expensive than granular, currently 15¢/lb versus 45¢/lb; and (2) if the powdered carbon is regenerated with wet-air oxidation, the biological solids are oxidized, and the secondary handling and disposal problem eliminated.

Wet-air oxidation

Wet-air oxidation should not be confused with submerged combustion. In the wet-air oxidation process, temperatures rarely rise above 400–600°F, and no flame is involved. The process is well established; more than 130 Zimpro wet-air-oxidation units are either in operation or under construction throughout the world. A number are dedicated specifically to the treatment of problem industrial wastes, such as those from the manufacture of acrylonitrile, monosodium glutamate, polysulfide-based rubbers, metallurgical coke, petrochemicals, and pharmaceuticals.

In the Zimpro process (Fig. 5), wastewater is pumped through an exchanger, where the temperature is increased to the point at which the reaction between oxidizable material and air will proceed autogenously. This oxidation, which takes place in the subsequent reaction vessel, raises the temperature due to the released heat of combustion. The degree of oxidation is primarily a function of reaction temperature and residence time. Operating pressure depends on temperature and, to some extent, on the organic concentration, ranging in actual installations from about 350 psig at 350°F to 3,000 psig at 610°F.

The released heat is recovered by exchanging the reactor effluent with the incoming waste stream. Then, spent air, carbon dioxide and steam are disengaged from the oxidized effluent in a separator.

Two-step process

Wet-air oxidation can be combined to economic advantage with either biological or biophysical treatment. Three basic systems are possible, depending on the character of the waste and the economics: wet-air oxidation followed by biological treatment; biophysical treatment, with wet-air oxidation used to regenerate the powdered carbon; and wet-air oxidation with biophysical treatment.

In the first system, wet-air oxidation typically reduces the COD 80–95%, with the remainder being removed biologically. Taking the ratio of five-day BOD to COD as an indication of suitability for biological

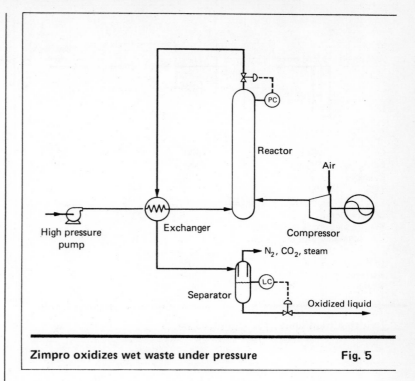

Zimpro oxidizes wet waste under pressure   Fig. 5

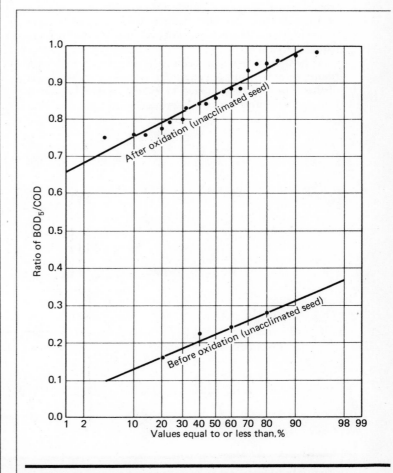

BOD/COD ratio before and after wet oxidation   Fig. 6

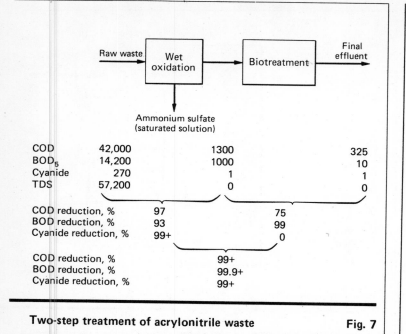

COD	42,000	1300	325
BOD$_5$	14,200	1000	10
Cyanide	270	1	1
TDS	57,200	0	0
COD reduction, %	97		75
BOD reduction, %	93		99
Cyanide reduction, %	99+		0
COD reduction, %		99+	
BOD reduction, %		99.9+	
Cyanide reduction, %		99+	

Two-step treatment of acrylonitrile waste　　　　　**Fig. 7**

treatment, the potential for this system is illustrated in Fig. 6, where the BOD$_5$/COD ratio of end-of-pipe waste from a pesticide-manufacturing complex is plotted on log-probability paper before and after wet-air oxidation. The average increase in BOD$_5$/COD is from 0.27 to 0.85.

Also, this system has been applied in five full-scale plants now treating acrylonitrile wastes. Future plants of this type will be designed to achieve a 95 + % reduction in COD, and a 99.9 + % reduction in total cyanide.

The effluent from wet-air oxidation of acrylonitrile wastes is low enough in COD to serve as a source of marketable recovered $(NH_4)_2SO_4$. The condensate from evaporations show BOD$_5$/COD and BOD$_{20}$/COD ratios of 0.76 and 0.96 respectively—quite biologically treatable. The overall performance of this two-step system is summarized in Fig. 7.

In the second system, the wet-air oxidation is used to regenerate carbon from biophysical treatment. During the regeneration process, adsorbed organics and biomass are largely oxidized (typically, 80% COD reduction). The remaining COD is solubilized and converted to an almost completely biodegradable form and returned to the aeration basins with the regenerated carbon. Residual ash is removed during a reactor blowdown.

In the third system, wet-air oxidation is combined with biophysical treatment in one of two flowschemes. Either a wet-air oxidation system dedicated to primary treatment of the raw wastes is slightly oversized for use in regenerating carbon in a blocked operation, or two oxidation units—one for primary treatment, one for regeneration—are installed (Fig. 8).

The improvements possible with either of these flowschemes are shown via experience treating night soil. Wet-air oxidation is now routinely employed for night-soil treatment, with 23 plants in Japan designed to accomplish a 50% reduction in COD by wet-air oxidation alone, and 85% in combination with activated sludge. However, an even better treatment has been obtained by converting the biological system to a biophysical system. With addition of powdered carbon, a stability and treatment capability has been achieved that was never found in the biological system.

A comparison of the COD and BOD values following biological and biophysical treatment of diluted,

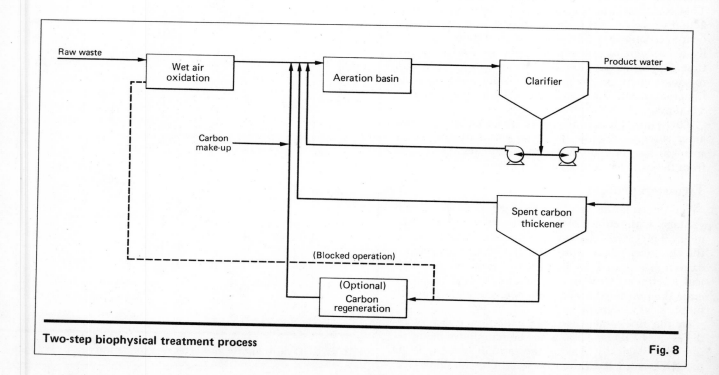

Two-step biophysical treatment process　　　　　**Fig. 8**

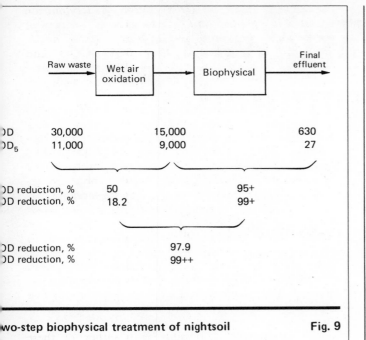

	Raw waste	Wet air oxidation		Biophysical		Final effluent
OD	30,000	15,000				630
OD$_5$	11,000	9,000				27
OD reduction, %	50		95+			
OD reduction, %	18.2		99+			
OD reduction, %		97.9				
OD reduction, %		99++				

wo-step biophysical treatment of nightsoil   Fig. 9

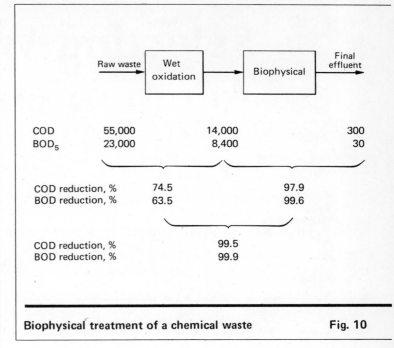

	Raw waste	Wet oxidation		Biophysical		Final effluent
COD	55,000	14,000				300
BOD$_5$	23,000	8,400				30
COD reduction, %	74.5		97.9			
BOD reduction, %	63.5		99.6			
COD reduction, %		99.5				
BOD reduction, %		99.9				

Biophysical treatment of a chemical waste   Fig. 10

wet-air-oxidized supernatant liquor, along with solids concentrations, shows:

	After dilution	Biological	Biophysical
COD, mg/l	1,350	412	130
BOD, mg/l	715	26	4
MLVSS*, mg/l		0	9,790
Biomass, mg/l		3,380	2,920

*Mixed liquor volume suspended solids of powdered activated carbon.

It has also been demonstrated that the wet-air-oxidized supernatant liquor can be treated in a biophysical system without dilution. This significantly reduces the size of the biophysical-aftertreatment equipment. The total system solids were carried at 18,000 mg/l (15,000 mg/l carbon, 3,000 mg/l biomass). The overall treatment that was obtained is illustrated in Fig. 9.

In another example of this system, the wastewater issuing from a series of processes for the manufacture of complex organic chemicals was found to be untreatable by conventional biological means. Biophysical treatment of the raw waste did work but gave only marginally acceptable results. Wet-air oxidation was applied to the raw waste, with a resulting 75% reduction in COD from 55,000 to 14,000 mg/l; and the BOD$_5$/COD ratio was increased from about 0.46 to about 0.70. Subsequent treatment by two-stage biophysical treatment further reduced the oxidized effluent to a COD level of 300 mg/l. Overall performance of this two-stage system is shown in Fig. 10.

In summary, wet-air oxidation combined with biological or biophysical treatment can render acceptable a large number of wastes otherwise resistant to treatment. The relative amount of treatment done by wet-air oxidation and that done biologically will be governed by economics and the character of the waste. Factors such as corrosiveness, ease of oxidation and toxicity come into play. Zimpro Inc. feels that one of the best applications of the wet oxidation process is in the treatment of wastes from the manufacture of herbicides, pesticides, etc. A considerable amount of development and demonstration work has been done in this.

The authors

Robert B. Ely is manager of industrial sales for Zimpro Inc., Rothschild, WI 54474, and is engaged in the application of wet-air oxidation and biophysical treatment to industrial wastewater disposal problems. Before joining Zimpro in 1972, he was involved in the evaluation of advanced wastewater technology with the Badger Co. He holds a B.S.Ch.E. from Purdue University and an M.S.C.E. from Northeastern University. He is a member of the AIChE.

Alan R. Welhelmi is an industrial applications engineer with Zimpro Inc., Rothschild, WI 54474. Formerly, he was with Amoco Chemicals Corp., where he served in research and development and manufacturing. He has a B.S.Ch.E. from the University of Illinois and an M.S.Ch.E. from Carnegie-Mellon University.

Selecting Wastewater Aeration Equipment

Each type of aerator for treating wastewater is suited for a given set of conditions. Here are guidelines for selecting the correct type, and equations for determining the number and size of aerators required, and the size and shape of treatment tanks.

RICHARD J. NOGAJ, Keene Corp.

The most common processes using air for wastewater treatment can be divided into the activated-sludge and aerated-lagoon categories. Depending on the circumstances, one of the several process variations within these categories is chosen as a treatment method. The aeration equipment is then selected, taking into account the characteristics inherent to each type. What follows discusses the aerators available for both the activated-sludge and aerated-lagoon processes, and suggests means for selecting and sizing such equipment.

WASTEWATER-AERATION PROCESSES

The most popular of the activated-sludge process variations are the conventional, extended, contact-stabilization, and complete-mixing types.

Conventional Activated-Sludge Process—In this process, the aeration detention period is approximately 6 hr., with a sludge-return rate from the final clarifier of approximately 25 to 50% of the wastewater flow. Reduction of BOD (biochemical oxygen demand), which is the most common measurement for organic matter,* is 90 to 95%. Both power consumption and amount of waste sludge are relatively high.

Extended Aeration—Here, a 24-hr. aeration detention period is provided. The process operates at high suspended-solid levels and low BOD loadings. Power consumption is high but the amount of sludge wasted is minimized.

Contact Stabilization—In this process, the return

* BOD is the quantity of dissolved oxygen in mg./l. required for the oxidation and final stabilization of organic matter decomposed by aerobic biochemical action.

Originally published April 17, 1972.

sludge is aerated separately before it enters the aeration tank. Aeration-tank sizes are thus reduced, which yields lower capital costs. Power consumption and sludge wasting are similar to the conventional process, but savings in space and in tank costs can be realized that are not possible with the conventional type.

Complete Mixing—Generally confined to systems with mechanical surface aerators, this activated-sludge treatment method involves the introduction of the waste material directly into the aeration tank to produce an instantaneous mixing effect. Uniform loadings are thus possible. This process allows for more variation in flow and loading, with a minimal shock effect on the aeration tank.

Aerated Lagoons—A relatively recent development for the treatment of wastewater, the aerated lagoon differs from the conventional activated-sludge process in that the detention time required for the lagoon is several days, as compared to hours in the conventional process. There is no final clarification step and no separate return of settled sludge.

Although air introduction devices are used in the lagoons to distribute oxygen throughout the basin, they are not sufficient to keep all solids in suspension. Nevertheless, 80% BOD removal from domestic sewage and some industrial wastes can be obtained in a six- to eight-day retention period. If sufficient land and space are available and state standards permit, the aerated lagoon is the least expensive of the wastewater-treatment processes. Biologically, the lagoon is similar to the activated-sludge process in that a sludge floc is established to reduce the BOD of the wastewater.

Characteristics of Available Aeration Equipment—Table I

Equipment Type	Equipment Characteristics	Processes Where Used	Waste-Type Application	Advantages	Disadvantages
Porous diffusers	Produce fine or small bubbles. Made of ceramic plates or tubes, plastic-wrapped or plastic-cloth tube or bag.	Large, conventional, activated-sludge process	Municipal sewage	High oxygen-transfer efficiency; good mixing; maintain high liquid temperature.	High initial and maintenance costs; tendency to clog; not suitable for complete mixing.
Nonporous diffusers	Made in nozzle, valve, orifice or shear types, they produce coarse or large bubbles. Some made of plastic with check-valve design.	All sizes of conventional activated-sludge process	Municipal sewage, industrial waste	Nonclogging; maintain high liquid temperature; low maintenance cost.	High initial cost; low oxygen-transfer efficiency; high power cost.
Mechanical, surface-turbine aerators	Low output speed. Large-dia. turbine, usually fixed-bridge or platform mounted. Used with gear reducer.	All sizes of conventional, activated-sludge & aerated-lagoon processes	Industrial waste, municipal sewage	Low initial and maintenance costs; high oxygen-transfer efficiency; tank-design flexibility.	Some icing in cold climates; more difficult to apply for total mixing.
Mechanical, surface-propeller aerators	High output speed. Small-dia. propeller. They are direct, motor-driven units mounted on floating structure.	Aerated lagoons, primarily	Municipal sewage, industrial waste	Low initial cost; simple to install and operate; moderate transfer efficiency; adjust to varying water level.	Some icing in cold climates; poor maintenance accessibility.
Submerged-turbine or combination units	Units contain a low-speed turbine and provide compressed air on sparge ring. Fixed-bridge application.	Conventional, activated-sludge process	Municipal sewage, industrial waste	Complete mixing; high-capacity input per unit volume; deep-tank application; moderate efficiency; wide oxygen-input range.	Require both gear reducer and compressor; tendency to foam; high total-power requirements.

TYPES OF AERATION EQUIPMENT

A summary of the characteristics and uses of the various kinds of aeration equipment is given in Table I. This equipment is classified as follows:

Air Diffusers	Mechanical Surface Units
Porous	Turbine plate-type
Nonporous	Turbine updraft-type
Perforated-pipe grid	Propeller Surface Units
Serrated flexible tubing	Combination Aerators

Air-Diffusion Devices

Generally installed below the liquid surface of the tank on either a fixed or retractable header, diffusers are used universally for aeration in wastewater plants of all types and sizes. Air is supplied to the diffuser by centrally located blowers, which provide sufficient pressure to overcome the static liquid head and all friction losses. Fig. 1(a) shows a typical flow pattern.

Diffusers are selected on the basis of a satisfactory balance between oxygen-transfer efficiency and maintenance characteristics. They can be classified as porous (small bubble), and nonporous (large bubble).

Porous Diffusers

Porous diffusers are available as ceramic plates or tubes, and plastic tubes (Fig. 2 shows ceramic tubes mounted on fixed headers). In some cases, a nonceramic type—consisting of plastic wrapping or plastic cloth and bags—is used.

The porous, fine-bubble diffusers have found application in large municipal waste-treatment plants that use the conventional activated-sludge process. To develop a spiral-flow pattern, the diffusers are usually mounted on one side of long narrow tanks. A standard diffuser delivers from 4 to 8 cfm. of air, with an oxygen-transfer efficiency (i.e., the ratio of oxygen transferred to oxygen in the delivered air) as high as 10 to 12%.

To maintain adequate circulation in the aeration tank, aerators are placed so that they will deliver a minimum of 3 cfm./lineal foot of tank, with diffuser spacing at a minimum of 6 in. and a maximum of 2 ft. Fine-bubble diffusers are subject to clogging either externally, by the tank materials, or internally, by particulate matter from the air supply. When the air supply is turned off, external clogging sometimes also results from ferric iron, high concentrations of calcium carbonate, fine sand, and deposits of soil and silt.

In addition to providing high oxygen-transfer efficiency, porous air diffusers also impart good mixing with high liquid-scouring velocities in the aeration tank. In general, air diffusers have the advantage over mechanical aerators in cold climates, because the mixed liquor is maintained at a higher temperature.

Since porous diffusers have much smaller openings than nonporous types, they tend to clog more easily. To avoid periodic maintenance cleaning, extensive air-filtering systems can be used. However, the overall efficiency

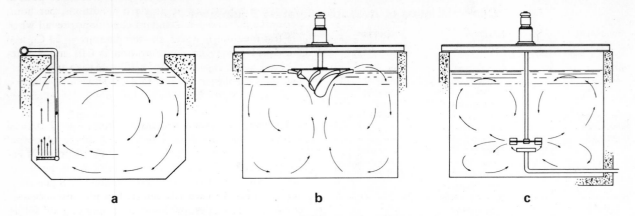

a   **b**   **c**

FLOW PATTERNS for (a) diffused-air, (b) mechanical-aeration and (c) combination turbine and diffuser-ring systems—Fig. 1

is somewhat reduced because of the power required for the air filters.

Nonporous Diffusers

Compared to the porous fine-bubble diffuser, which produces a relatively small bubble with corresponding high oxygen-transfer efficiency, the nonporous type produces large bubbles that have lower oxygen-transfer efficiency—between 4 and 8%. One big advantage is that the nonporous units do not clog internally from particulate matter, and also clog less frequently from external materials. Thus, from the maintenance point of view, they are more desirable than the porous types.

However, power usage with nonporous types can be twice the rate of the porous units, due to the lower oxygen-transfer efficiency. In smaller plants, where transfer efficiency is less critical, nonporous diffusers are economical, due to lower maintenance costs.

Nonporous diffusers may be of the nozzle, orifice, valve or shear type. Most of these diffusers are designed with a disk or check valve that closes when the air supply is shut off; this reduces clogging by preventing solids from entering the units. The diffusers are available in

metal or acrylonitrile-butadiene-styrene copolymer (ABS) construction, and can be applied over a wide range of air flowrates, as high as 25 cfm. per diffuser.

Nonporous diffusers are generally installed in similar fashion as the porous types, although in some cases fewer units may be required. Also, because of the large orifice openings, higher flowrates are attainable with smaller loss of head. For example, a typical nonporous, plastic diffuser may pass 10 to 15 cfm. at a loss of head of 6 in. of water, whereas the fine-bubble unit may only pass 4 to 8 cfm. Orifices range from $\frac{1}{8}$ to $\frac{3}{8}$ in. dia. and are sized for the application, depending on the air velocity required and the backpressure.

Other Diffuser Types

At least two other systems have found use in both the activated-sludge and aerated-lagoon processes. One of them consists of a baffled, perforated-pipe grid in a shallow-submersion aeration system. These grids—placed similarly to the standard aeration headers on one side of conventionally long, narrow aeration tanks—are installed at shallow depths by means of corrugated tank baffles that enhance liquid turnover. Preremoval of coarse material is necessary to avoid matting of the grids. Advantages of this system are the low pressure involved and the lightweight air ducts. Efficiencies are similar to those of nonporous diffusers.

The other system consists of serrated, polyethylene, flexible tubing that is usually installed at the bottom of aerated lagoons or oxygenation ponds. Compressed air passes through the tubing, which is laid in a grid pattern on the bottom of the lagoon or pond.

A variation of this system is another aerator, which is also primarily used in lagoons. Easy to install—it is lowered over the side of a boat—this static unit is claimed to have a low operating cost and to achieve high oxygen-transfer efficiency, while eliminating much maintenance and noise. Anchored to the bottom of the lagoon, the aerator floats vertically below the surface while compressed air rises through its tube. Mixing elements reduce the size of entrained air bubbles, thus providing more air contact for the water. The effective surface-

POROUS DIFFUSER TUBES made of ceramic material and mounted on fixed headers in aeration tank—Fig. 2

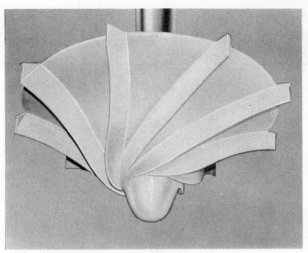

MECHANICAL SURFACE AERATOR of the updraft type, with mixed-flow, open-style impeller design—Fig. 3

radius of a unit is from 30 to 60 ft., depending on the size and depth of the aerator.

Mechanical Surface Types

Mechanical aerators have become more widely used in recent years, especially in industrial-waste applications. While air diffusers introduce air into the liquid, mechanical aerators—which are located near the liquid surface to provide high pumpage and agitation—expose the liquid to the air. Thus, oxygen is transferred through surface turnover, and by turbulently lifting large volumes of liquid above its surface, and exposing thin films to the atmosphere.

Turbine Surface Aerators

Turbine surface aerators can be classified as either plate or updraft type. The plate aerator causes a peripheral hydraulic jump that accomplishes oxygen transfer through entrainment. The updraft type pumps large quantities of water to the atmosphere at relatively low heads, similarly to a mixed-flow or centrifugal impeller. Fig. 3 shows an updraft surface unit and Fig. 1(b) shows the flow pattern it produces.

Used in both municipal and industrial activated-sludge and aerated-lagoon processes, mechanical aerators have wide application in the complete-mixing process because they are more capable of absorbing shock loading than conventional diffused-air systems. The additional advantage of high oxygen-transfer efficiency means lower operating and power costs over extended periods of time than for the diffused-air units. Mechanical aeration also requires simpler tank construction. This usually means lower initial tank costs, which may amount to savings of 20 to 30% over the diffused-air systems. One drawback of mechanical aerators is that to ensure necessary mixing and scouring, their application can be more difficult than that of the diffused-air units.

Individual aerators are supplied with their own motor gear-reducer; thus, each unit can be isolated from the remaining ones. Their efficiency is expressed in pounds of oxygen transferred at standard conditions, per hour, per brake horsepower. Standard conditions prevail when the test medium is water, the temperature is 20 C., and the dissolved-oxygen concentration is 0.01 mg./l. Mechanical aerators over 20 hp. exhibit oxygen-transfer efficiencies on the order of 3 to 4 lb./(hp.)(hr.), which is equivalent to an air-diffusion device with an efficiency of 10 to 12%.

Turbine rotors vary in diameter from 3 to 12 ft., and are available from 3 to 150 hp. The range of output speed varies from 30 to 60 rpm. A difficult problem encountered in the application of turbine units is the horsepower sizing relative to the tank geometry. Since these aerators must be carefully selected to provide necessary circulating velocities to maintain suspension of solids, the manufacturer's empirical knowledge of each aerator must be available to the user.

These aerators are generally capable of mixing the contents of large basins varying in depth from 3 to 18 ft. In some cases, deeper tank aeration can be used by means of lower mixing impellers and extended draft tubes. Turbine units are generally used in concrete tanks, being placed on fixed bridges or platforms. Care must be exercised in cold climates because some degree of icing may occur on accessory structures.

Propeller Surface Aerators

Another type of mechanical surface aerator that has increased in popularity is the floating-propeller kind, which is relatively simple in design and construction, and consists of a marine-type propeller directly driven by a high-speed motor (p. 144 shows aerator in motion).

The combination motor and propeller assembly is float-mounted (Fig. 4) and can be easily installed in a new or existing aeration tank. This aerator consists of a nonsubmersible motor, riser tube, glass-fiber or stainless-steel covered float, and a propeller-deflector assembly. Power is supplied from an onshore control station by means of a submersible power cable. The rotating element is supported in the liquid by an integral float, which au-

FLOATING SURFACE AERATOR (not in motion) of the mechanical, high-speed propeller type—Fig. 4

COMBINATION AERATOR with rotating impellers and lower stationary air-sparging ring—Fig. 5

Combination-Type Aerators

Another aeration device (Fig. 5) consists of a combination of a submerged-turbine aerator having a rotating impeller, and a stationary diffuser ring for the injection of atmospheric air. Fig. 1(c) illustrates the typical flow pattern. The turbine is powered by a directly connected motor and gear reducer.

Units having an optional upper-surface mixing impeller and a lower rotating impeller are available to more effectively distribute air bubbles released by the sparging ring. This is accomplished because the lower impeller serves as a shearing device for the compressed air, which is released by the ring at the periphery of the impeller. Oxygen is thus transferred by air bubbling at the sparge ring, as well as by the shearing action on the bubbles by the impeller located above the ring.

The combination-type unit has been successfully applied in activated-sludge processes treating municipal and/or industrial wastes. Since the aerator can transfer higher oxygen quantities per unit volume of aeration tank than other devices, it has become popular for applications at very high loadings. For deep-tank applications (over 18 ft.) the device transmits the necessary oxygen with efficient circulating patterns, thus ensuring complete mixing and suspension of all solids.

These units are limited to fixed-bridge mounting, usually in square or rectangular tanks made of concrete. Disadvantages are that shaft alignment depends on a steady bearing—which has to be maintained in a sludge environment—and that excessive foaming can result.

EQUIPMENT-SELECTION CRITERIA

The materials of construction and mechanical-drive components are of primary consideration in the selection of aerators. The basic requirements, however, are adequate mixing and oxygen transfer. Standard aerator-test procedures developed—as well as the data collected—by the industry for determining transfer and mixing efficiencies are essential in estimating the type and size of equipment needed.

Materials of Construction

Careful consideration of the materials of construction of aerators is especially critical in industrial-waste applications where pH levels and corrosive characteristics could cause premature equipment failure. Most devices today are designed with rotating elements fabricated from mild carbon steel, although in some instances units are available in stainless steel or with a plastic coating. A few manufacturers have developed rotating turbines in structural reinforced glass fiber (Fig. 3). This unit should be considered for longer life and less maintenance (no painting) due to its noncorrosive features.

Most air diffusers are available in various types of plastics, the most common being—especially for nonporous diffusers—the ABS plastic referred to as Cycolac. Nylons, polypropylenes and Dacron are also materials used, as well as stainless steel or PVC, as required.

Diffuser-header piping is generally made of hot-dip

tomatically adjusts to water-level changes. Materials of construction range from structural plastic or stainless steel for the rotating elements, to glass fiber or stainless steel for the flotation elements.

Although slow-speed turbine aerators have also been used on floating structures, the high-speed, direct-motor propeller is considerably more popular and lower in initial cost, being widely used in aerated lagoons. Applied in various depths ranging from 3 to 15 ft., the units usually have an oxygen-transfer efficiency of 3 lb./(hp.)(hr.) at standard conditions.

Several deflector designs can be used to meet various installation requirements. One design forms a fountain-like spray with controlled liquid-drop size, while another pumps large quantities of water with a lower trajectory. All propeller surface aerators operate on the updraft principle, pumping water from below the liquid surface up the riser tube and out through the discharge opening. Additional oxygen transfer takes place as the liquid returns to the surface of the tank, due to further entrainment and mixing. A modification of the propeller surface aerator has been successfully applied in cooling applications, using the liquid-spray principle.

Of all the aeration devices, the floating-propeller surface aerator is the lowest in initial cost, and the most simple to install and operate. However, as a floating unit, it is less accessible for periodic maintenance. Care must also be exercised in the use of watertight power cables and proper mooring lines. As with the fixed-type mechanical aerator, floating units are sometimes limited to warmer climates because of potential icing.

galvanized steel because of low cost and noncorrosive properties. The floating aerator is available is noncorrosive construction with material combinations of plastic, stainless steel, and structural glass fiber. Rotating elements come in both stainless steel and structural, high-density polyurethane.

Flotation pontoons are constructed of both stainless steel or reinforced glass fiber. In all cases, pontoons should be completely filled with polyurethane to ensure against leakage and possible submergence of the aerator.

Mechanical-Drive Components

Careful attention must be given to properly sizing mechanical-drive components. For diffused-air systems, blowers of the centrifugal or rotary two-impeller type can be used. Overall economy favors the centrifugal unit, especially in large applications of 10,000 cfm. or above; this type delivers air without pulsation. Being a positive-displacement device, the rotary type delivers a volume of air that varies directly with the speed. This unit, however, pulsates, which makes difficult the measurement of the air output.

To determine the necessary discharge blower pressure, the following pressure values must be summed up: (1) static water-pressure over the diffuser, (2) pressure loss through the pipe, and (3) pressure loss through the diffuser (which should amount to only 0.1 to 0.2 psi. for a clean unit). All diffuser manufacturers supply curves showing the loss of head through the diffuser at different air flowrates, with various orifice sizes.

For the slow-speed, mechanical, surface, turbine-type unit, careful consideration should be given to the type of gear speed-reducer used to rotate the surface turbine. In general, helical-gear reducers are favored over the worm type because of higher efficiency (less horsepower is required over extended periods).

Most units are equipped with constant-speed drive assemblies. The gear reducer is usually a double- or triple-reduction unit, with a gear ratio varying from 16:1 to 48:1. For helical gears, efficiency loss is ordinarily 2% per stage. Drive units can be of the vertical type, with integral or separate motor coupling. In some cases, two-speed motors are used to provide greater flexibility.

Another system used to vary the oxygen input works by controlling the water level in the aeration tank and thereby the submergence of the turbine. This can be done manually or automatically in conjunction with a dissolved-oxygen probe and/or a movable weir at the end of the aeration tank.

An important criterion in selecting the speed reducer is the service factor.* For gear reducers, a factor of 1.75 is usually considered to be a minimum but, in most cases, 2.00 is more universally used. Wherever possible, service factors above 2.00 should be considered as a sound investment for longer life.

For a high-speed, floating-type unit, it is necessary to consider the type and make of motor used. Experience has shown that a standard, totally enclosed, fan-cooled motor is not satisfactory for this kind of application.

*A service factor—for the purpose of selecting mechanical-drive components—is a multiplier applied to the actual average power requirement to account for peak loads.

Special aeration motors developed by several manufacturers include features such as water tightness and special insulation, bearings and seal designs. Water tightness must be an integral part of the motor construction because liquid ingestion can cause premature failure. All these special motor characteristics should be part of the aerator specification.

Oxygen-Transfer Ability

The two basic requirements for any aeration system are oxygen transfer and mixing. To make a proper aerator selection, actual test data should be available from manufacturers. It is important, though, to find out how the data were obtained and under what conditions.

In most cases, oxygen-transfer data are obtained under standard conditions in the manufacturer's test tank or facility. This gives a satisfactory indication of the aerator's capability under the conditions tested. However, when applying an aerator to field conditions, consideration must be given to the differences in tank geometry, location and liquid to be treated. A surface turbine may show different results if the horsepower per unit volume for a field application varies considerably from that of the test. For testing, 10,000 gal. of liquid per applied horsepower is most common.

If possible, field tests with clean water (to obtain accurate oxygen-transfer measurements) should be run prior to startup of a treatment plant. As an alternative, the operation of the aeration device may be witnessed at the manufacturer's test site to allow the user to judge the equipment's performance prior to its installation.

Oxygen-transfer efficiency for an air-diffusion device varies from 4% to 12% for the nonporous and porous diffusers, respectively. The efficiencies are approximately equivalent to 1.23 lb. O_2/(hp.)(hr.) for the 4% figure, and 3.70 lb. O_2/(hp.)(hr.) for 12%.

Surface turbine devices with high power ratings (over 30 hp.) commonly deliver 3.5 lb. O_2/(hp.)(hr.); smaller units have demonstrated efficiencies as high as 5.0. Floating, high speed, propeller units can deliver from 2.5 to 3.0 lb./(hp.)(hr.) for the larger units, and 3.0 to 3.5 for the smaller ones. Combination-type aerators—with speed reducer and compressor—generally have an overall transfer efficiency of 2.0 to 3.0 lb. O_2/(hp.)(hr.) at standard conditions.

Mixing Efficiency

In addition to satisfying oxygen-transfer requirements, thorough mixing must be provided to induce circulating velocities of sufficient magnitude to maintain the solids in suspension. In some instances—because of various tank sizes and different tank geometry—empirical information obtained by equipment manufacturers from field experience must be procured.

At times, large quantities of horsepower may have to be used beyond the oxygen requirements to maintain the necessary mixing. In activated-sludge applications, the mixing must maintain all solids in suspension; for this, a minimum bottom-scouring velocity of 0.5 ft./sec. is necessary. When primary treatment (sedimentation,

Nomenclature

a	Constant for a given waste, i.e., ratio of oxygen required (BOD) to actual oxygen applied
BOD	Biochemical oxygen demand—amount of dissolved oxygen required for the oxidation and final stabilization of organic matter decomposed by aerobic chemical action, mg./l.
C_s	Oxygen saturation in water at temperature T
C_l	Process-operation dissolved-oxygen level, mg./l.
E	BOD removal, %
K	Oxidation-rate factor expressed as the reciprocal of time, i.e., days^{-1}
L	Avg. daily BOD requirement, mg./l.
M	BOD requirement, lb./day
N	Amount of oxygen transferred at process conditions, lb./day
N_o	Amount of oxygen transferred at standard conditions, lb./day
q	Average flowrate, mgd.
T	Process operating temperature, °C.
t	Detention period, hr.
V	Volumetric load, lb. BOD/(day)(1,000 cu.ft.)
α	Ratio of oxygen transfer in waste to that in tap water at the same temperature
β	Ratio of oxygen saturation in waste to that in tap water at the same temperature

flotation, coagulation, etc.) is not resorted to and heavy industrial-waste sludge is being treated, the minimum scouring velocity may have to be 1.0 ft./sec.

Tank size is determined by the detention period required to achieve the desired degree of treatment. For activated sludge, whose detention time varies from 6 to 24 hr., a large horsepower per unit volume is required.

For best mixing with all mechanical surface aerators, there is an optimum spacing of units in multiple-cell applications. If the units are placed too close or too far apart, poor circulating patterns—with potential deposition of solids—are created. For instance, when the units are too close, an interference effect occurs that causes short-circuiting and lack of movement towards the bottom of the tank. When the units are too far apart, sufficient energy is not available to provide the necessary circulation. In aerated lagoons, deposition of light solids created by the conversion of biodegradable materials to suspended solids may be desirable, but aerators must be placed so that sufficient diffusion of oxygen takes place to prevent the creation of anaerobic zones.

AERATION-SYSTEM CALCULATIONS

Several useful equations are available for estimating oxygen needs and tank sizes of aeration systems. Based on such calculations, the size and number of aerators can be determined.

Oxygen Requirement

To estimate the amount of oxygen needed by a waste-treatment plant, the first step is to determine the daily pollution load in lb. BOD/day by the equation:

$$M = 8.34\, Lq \qquad (1)$$

where the number 8.34 is a conversion factor.

Relation of Oxygen-Transfer Efficiency to Overall Efficiency and Diffuser Air Requirements—Table II

Transfer Efficiency, %	Overall Efficiency Lb. O_2/(Hp.)(Hr.)	Diffuser Air Requirement,* Cfm./(Lb. BOD/Day)
4	1.23	1.665
6	1.85	1.113
8	2.46	0.833
10	3.08	0.667
12	3.70	0.555

* Based on the assumption $a = 1.0$ and $N_o = 1.6\, N$

It is then necessary to convert the BOD load to actual oxygen transfer by means of:

$$N = aM \qquad (2)$$

Most aeration devices are rated in terms of oxygen transferred at standard conditions per brake horsepower per hour; this is always so for mechanical surface aerators. However, most air diffusers use transfer efficiency as a measure of oxygen transfer; in other words, the quantity of oxygen transferred divided by the total quantity of oxygen in the delivered air. Transfer efficiency can be approximately converted to overall efficiency in lb. O_2/(hp.)(hr.) by means of Table II.

To determine the quantity of oxygen transfer required under actual process conditions, it is necessary to establish a value for the constant a. Most of the time, $a = 1.0$, especially for municipal applications. For industrial wastes, a may vary between 0.8 and 1.2; usually, a value of $a = 1.0$ is used for estimating purposes. If more-accurate a values must be determined, empirical data must be obtained from a pilot-plant or bench-scale study.

Since most devices are rated at standard conditions, it is necessary to convert from process conditions. As mentioned earlier, standard conditions are defined as oxygen transfer in clean tap water at 20 C., and at 0.01 mg./l. dissolved oxygen. To perform the conversion, this equation is used:

$$N_o = \frac{N}{\alpha(1.024)^{(T-20)}[(\beta C_s - C_l)/9.2]} \qquad (3)$$

For municipal activated-sludge processes, values for α and β of 0.9 and 0.95 are generally used; a process dissolved-oxygen level of 2.0 mg./l. can be used for design purposes. In aerated-lagoon processes, conservative values of α and β of 0.7 and 0.9 are more common.

For industrial wastes, values of α can vary considerably, from a minimum of 0.3 to a maximum of about 2.0. Again, laboratory tests on the specific wastes must be made to more accurately establish the alpha factor. For estimating purposes, usually a value of $\alpha = 0.7$ can be assumed. Substituting average values of α, β, C_s, T and C_l yields a constant for the conversion of oxygen requirements from process to standard conditions. This can be expressed as:

$$N_o = 1.6\, N \qquad (4)$$

For diffused-air applications, where air delivered in cfm. is a parameter for design, the above can be used to relate the transfer efficiency to the cfm. requirements. Based on the assumptions $a = 1.0$ and $N_o = 1.6\,N$, Table II provides a good design guide for diffused-air systems.

It is important to select aeration equipment with as great an efficiency as practical. In very large plants, the total connected horsepower becomes quite critical relative to overall power costs. Table III shows some power costs and how they relate to annual usage.

Tank-Detention Time

To determine tank size and geometry, the detention time of wastewater in the tank must be estimated. For activated-sludge processes use:

$$t = 1.5\,BOD/V \qquad (5)$$

For municipal wastes, a raw BOD concentration of 200 mg./l. can be used with a corresponding volumetric loading of 50 lb. BOD/(day)(1,000 cu.ft.). This load is considered average for acceptable levels of BOD removal (i.e., 90%) through a secondary, activated-sludge process. The detention period for the conditions just mentioned would be 6 hr.

For aerated lagoons, use the equation:

$$t = E/[K(100 - E)] \qquad (6)$$

The value of the factor K is usually determined from laboratory data. Some typical values are 0.5 for domestic waste; 0.75 for cannery waste; 0.80 for packing-house waste; and 1.0 for board-mill waste.

Calculation Examples

Example 1—It is desired to select nonporous diffusers for a conventional activated-sludge plant treating 1.0 mgd. of waste with a BOD of 240 mg./l. If $N_o = 1.6\,N$, and the oxygen transfer of the diffusers is 6%, how many units are needed, and what should their spacing be?

From Eq. (1), $M = 8.34 \times 1.0 \times 240 = 2,000$ lb. BOD/day.

From Table II (applicable only when $N_o = 1.6\,N$), the cfm. equivalent for 6%-efficiency units is 1.113 cfm./(lb. BOD)(day). Then, total air volume = $1.113 \times 2,000 = 2,226$ cfm. If a 12-cfm. diffuser is selected, the total number of diffusers required = $2,226/12 = 186$.

To determine the tank size, calculate the detention period by means of Eq. (5), assuming $V = 50$:

$$t = 1.5\,BOD/V = (1.5 \times 240)/50 = 7.25 \text{ hr.}$$

Tank volume = (7.25 hr.)(1,000,000 gal./day)(day/24 hr.) × (cu.ft./7.48 gal.) = 40,000 cu.ft.

If tank width = 30 ft. and side water-depth = 13 ft.:
Tank length = $40,000/(30 \times 13.3) = 100$ ft.

Then, diffuser spacing = $(100 \times 12)/186 = 6.5$ in., which is acceptable for coarse-bubble diffusers.

Example 2—It is desired to select mechanical aerators of the fixed, slow-speed, turbine type for the same plant as in Example 1. Determine the size and number of units needed, as well as the size of the tank.

Electricity Rates vs. Annual Cost/Hp.—Table III

Electric Rates, ¢/Kwh.	Annual Cost, $/Hp.
0.5	32.50
1.0	65.00
1.2	78.00
1.5	97.50
1.7	111.50
2.0	133.00

If $M = 2,000$ lb. BOD/day; $N = 1.0\,M$; and $N_o = 1.6\,N$; then $N_o = 3,200$ lb. O_2/day.

With 3.5 lb. O_2/(hp.)(hr.) as oxygen-transfer efficiency, the power required = $3,200/(24 \times 3.5) = 38.1$ hp. Therefore, a 40-hp., fixed, turbine aerator is selected.

For tank sizing, volume = 40,000 cu.ft., and the side water-depth = 13.3 ft. Therefore, select a square tank 55 ft. on the side.

Example 3—Floating, mechanical, surface aerators are to be selected for an aerated-lagoon system. The desired BOD removal = 85%. The material to be treated is a cannery waste ($K = 0.75$) with the following characteristics: $BOD = 1,000$ mg./l.; total flow = 1.2 mgd.; $\alpha = 0.70$; $\beta = 0.90$; $T = 25$ C.; $C_s = 8.4$ mg./l.; $C_l = 2.0$; and $a = 1.2$. Determine the tank size, and the number, size and spacing of the aerators required.

From Eq. (1), $M = (8.34)(1,000)(1.2) = 10,000$ lb. BOD/day, and from Eq. (2), $N = (1.2)(10,000) = 12,000$ lb. O_2/day.

Then, from Eq. (3):

$$N_o = \frac{12,000}{(0.7)(1.024)^5[(0.9 \times 8.4 - 2.0)/9.2]}$$
$$= 25,400 \text{ lb./day}$$

Using an efficiency of 2.8 lb./(hp.)(hr.):
Power required = $25,400/(24 \times 2.8) = 380$ hp.
To determine the detention period from Eq. (6):

$$t = 0.85/[(0.75)(1.00 - 0.85)] = 7.5 \text{ days}$$

Then, for 1.2 mgd.:
Volume = $(1,200,000 \times 7.5)/7.48 = 1,200,000$ cu.ft.
Side water-depth = 10 ft.
Area = $1,200,000/10 = 120,000$ sq.ft.

Therefore, select a width = 2 × (length); a tank 250 ft. × 500 ft. × 10 ft.; eight 50-hp. floating aerators; and place units equally spaced in approximately 125 ft. × 125 ft. square cells.

Meet the Author

Richard J. Nogaj is director of engineering of Keene Corp.'s Water Pollution Control Div., 1740 Molitor Rd., Aurora, IL 60507. Prior to joining Keene, he was manager of engineering at Clow-Yeomans Corp. Holder of a B.S. degree in civil engineering and an M.S. in sanitary engineering—both degrees from the Illinois Institute of Technology—he is a member of the Water Pollution Control Federation, where he has served on the Sub-Committee on Aeration in Wastewater Treatment. A registered professional engineer in the state of Illinois, he has authored a number of technical papers on wastewater treatment.

Controlling and monitoring activated-sludge units

Organic and toxic shocks are the major causes of upsets in industrial activated-sludge systems. Simple, field-proven methods are presented for monitoring these upsets and controlling sludge units.

George T. Thibault, Polybac Corp. *and* *Kenneth D. Tracy,* Environmental Dynamics Inc.

☐ Activated sludge is the most commonly specified biological treatment process for removing soluble and colloidal organic contaminants from process wastewaters. Wastes are stabilized by a culture of naturally occurring bacteria, although adapted, mutant bacterial strains can be used to augment the system. The culture is referred to as biomass or mixed-liquor suspended solids (MLSS). To operate these systems properly, it is necessary to consider upset conditions that result from organic and toxic shock loadings. Both of these conditions cause rapid changes in microbial activity and increase the amount of biomass required. Effluent quality deteriorates until either the upstream condition is alleviated or sufficient biomass is produced or introduced into the system.

Organic shocks are characterized by a rapid increase in microbial growth rate and dissolved oxygen uptake rate (DOUR). Fig. 1 shows the response of a refinery pilot unit to an organic shock of phenol.

Three secondary effects can delay the return to normal effluent quality: a dissolved-oxygen (DO) limitation in the aeration basin; poor solid-liquid separation in the clarifier, resulting from dispersed growth or filamentous bulking; and complete clarifier failure due to solids overloading.

Dissolved-oxygen problems occur when the DOUR of the culture exceeds the oxygen-transfer capabilities of the aeration system.

Dispersed growth results when the growth rate exceeds the rate of bioflocculation. In dispersed growth, bacteria do not flocculate adequately. Bacteria produce a polymer that normally glues them together—but at rapid growth rates, secretion of polymer cannot keep pace with the production of new bacteria, resulting in poor flocculation. Filamentous organisms prevent normal floc compression and severely hinder solid-liquid separation.

Finally, an increased MLSS level, which results from growth during organic shock, increases solids flux to the clarifier. This generally results in an increase in effluent suspended-solids level and may lead to complete clarifier failure if the design solids flux is exceeded—generally 15 to 25 lb/(ft²)(d).

Toxic shocks are characterized by a rapid decrease in microbial growth rate, and often in DOUR. Fig. 2 illustrates the response of a refinery pilot unit to a toxic shock of cyanide. Exceptions to this normal toxic response are shocks caused by toxins, such as sulfides, that exhibit a chemical oxygen demand in an oxygen environment. Sulfide shocks are characterized by a rapid decrease in microbial growth rate but are accompanied by an increase in DOUR. Fig. 3 illustrates the response of a pilot unit to a toxic shock of sulfide.

A material's toxicity to an activated-sludge culture depends on the material's background level in the wastewater. Also, certain toxins are consumed in their destructive reaction with the microorganisms (non-conservative) and others are not (conservative). Thus, the impact of a toxic shock is related not only to the duration of the upstream disorder but to the type of toxin involved.

Process-control objectives

All aerobic biological treatment processes use microorganisms to convert organic contaminants into carbon dioxide and water and to produce additional microor-

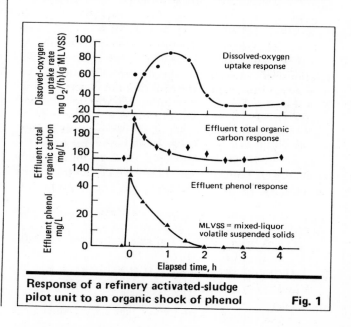

Response of a refinery activated-sludge pilot unit to an organic shock of phenol **Fig. 1**

Originally published September 11, 1978.

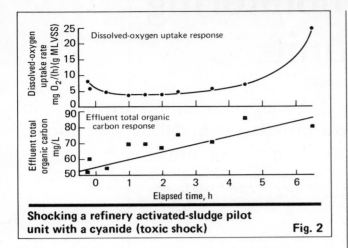

Shocking a refinery activated-sludge pilot unit with a cyanide (toxic shock)   Fig. 2

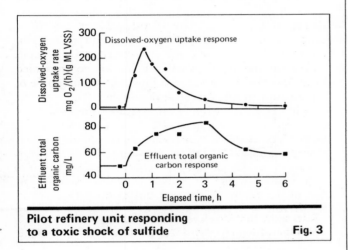

Pilot refinery unit responding to a toxic shock of sulfide   Fig. 3

sludge unit, this slow response must be accounted for, along with the potential impact of organic and toxic shocks.

Automatic process-control limitations

Recent emphasis has been directed toward automatic control of activated-sludge units. All control strategies—manual or automatic—match the mass of viable microorganisms to the mass rate of application of organic matter, in order to maintain a constant microbial growth rate. Online control strategies are limited by their reliance on an instantaneous source of additional viable organisms during shock conditions. Since the growth rate is on the order of hours, growth alone cannot be relied upon as a basis to mitigate toxic and organic shocks.

The clarifier contains a limited mass of viable organisms (about 15 to 25% of the total mass of microorganisms in the system), which may be tapped during a shock simply by increasing the recycle rate. However, this generally proves counterproductive, since increasing the rate increases the solids flux to the clarifier. Tracy [2] demonstrated that the response to a step increase in solids flux is a substantial increase in the sludge-blanket level.

New systems may be designed with a storage basin or an aerobic digester to hold sludge during shocks. However, if storage facilities are provided, additional clarifier capacity (surface area) must also be provided to prevent solids overloading, following the introduction of additional biomass.

Solids residence time—control strategy

Most industrial activated-sludge units are not designed for online addition of biological solids. Even if the treatment scheme includes an aerobic digester, it is often not possible to use the digester as a source of biomass. This is due to the solids-flux limitation on the existing clarifier. As a result of the slow growth rate, process-control decisions may practically be made only on a once-per-day basis. Therefore, emphasis must be on sound process monitoring for early detection of shock loadings.

Experience has shown that the simplest and most effective control strategy is a manual one in which the solids residence time (SRT) of the system is kept fixed, generally at 5 to 25 days.

SRT is defined as the average length of time the microorganisms spend in the treatment system. Use of the SRT concept as a basis for process control was first advanced by Lawrence and McCarty [3].

Holding SRT constant is equivalent to holding the microbial growth rate constant, as shown in Eq. 1:

$$\mu = \frac{1}{(\text{SRT})} + b \qquad (1)$$

where: μ = specific growth rate, d^{-1}; (SRT) = solids residence time, d; b = constant specific decay rate, d^{-1}.

This equation states that the microorganisms must grow at a rate at which they are lost from the system by wastage and decay. The solids residence time of a system is held constant by wasting a constant fraction of the biomass solids (1/SRT) from the system daily.

ganisms. Contaminant concentration is usually expressed in terms of one of the following parameters: five-day biochemical oxygen demand (BOD_5); chemical oxygen demand (COD), total oxygen demand (TOD), or total organic carbon (TOC). For wastewaters containing biodegradable organics, there is a simple and unique relationship between the specific growth rate of the microorganisms, μ, and the contaminant concentration, S, as shown in Fig. 4. This relationship was first demonstrated by Monod [1]. The objective of process control is to force the microbial population to grow at a rate, μ_e, that produces a desired effluent concentration, S_e.

In a completely mixed system, the contaminant concentration is homogenous. Thus, the concentration in the effluent equals that in the basin. This requires that the entire microbial population grow at the rate μ_e, which is normally very low. In a plug-flow system, the growth rate is highest at the inlet, gradually diminishing to μ_e at the outlet.

Most industrial activated-sludge units are completely mixed systems due to the shock-dilution capability inherent in such a scheme. Control of the growth rate is analogous to control of the reaction rate of any chemical process, but with one exception: The growth rate in the activated-sludge process is much slower than the reaction rate of most chemical processes. Thus, in considering a control strategy for an industrial activated-

For example, if the desired SRT is 10 days, one tenth of the solids in the system are wasted each day. Some unintentional wastage occurs continually, as suspended solids are always present in the effluent. This unintentional wastage is subtracted from the total mass to be wasted on a given day to determine the required intentional wastage.

SRT is defined at the mass of solids in the aeration basin, divided by the sum of the mass rates of unintentional and intentional wastage:

$$(\text{SRT}) = \frac{VX}{[(Q - Q_w)X_e + Q_w X_w] \times 1{,}440} \quad (2)$$

where: V = volume of the aeration basin, gal; Q = influent flowrate, gal/min; Q_w = sludge wastage rate, gal/min; X = total suspended solids in aeration basin, mg/L; X_e = total suspended solids in effluent stream, mg/L; X_w = total suspended solids in sludge wastage stream, mg/L.

It is not necessary to know the organic concentration in the feed stream in order to implement SRT control. Note that total suspended-solids concentrations are used rather than the concentrations of viable microorganisms. Eq. 2 is rearranged to Eq. 3 to calculate the sludge wastage rate:

$$Q_w = \frac{\dfrac{VX}{1{,}440 \times (\text{SRT})} - QX_e}{X_w - X_e} \quad (3)$$

This rate should be calculated and adjusted daily from current suspended-solids concentrations and the average flow rate, Q, for the preceding 24-h period.

Selecting an SRT

In selecting a solids residence time, it is necessary to consider the desired effluent quality, the nature of solid-liquid separation, the nature and duration of typical shocks, and economics. Since SRT is chosen to yield a growth rate at or below μ_e for a given average feed concentration, there is a lower limit on the SRT that will give the desired effluent concentration, S_e.

Solid-liquid separation can also impose a lower limit on SRT. At an SRT less than 5 days, growth rate of the bacteria is so rapid that dispersed growth results [4]. In setting a lower limit on SRT, the nature and duration of shock loadings are generally more important than normal loading conditions. Pilot and field experience in the petroleum and petrochemical industries has shown that the resistance of activated sludge to toxic and organic shocks increases with increasing SRT from 5 to 25 days. Fig. 5 illustrates the response of a refinery activated-sludge pilot unit to a 3-week organic shock of monoethanolamine. Beyond an SRT of 25 days, the improvement in shock resistance is minimal.

Improvement in shock resistance with SRT is not surprising, since a mathematical model of the process predicts an increase in the concentration of viable microorganisms with increasing SRT [5]. The model also predicts the decrease in incremental improvement that occurs beyond 25 days.

Fig. 6 shows the relationship between the concentration of viable microorganisms and predicted SRT. This

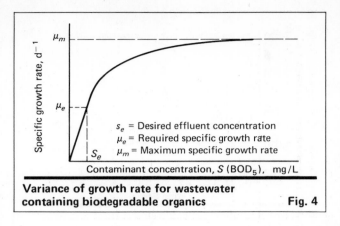

Variance of growth rate for wastewater containing biodegradable organics   **Fig. 4**

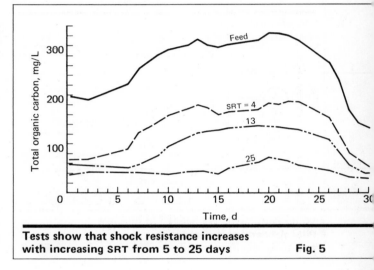

Tests show that shock resistance increases with increasing SRT from 5 to 25 days   Fig. 5

concentration is essentially constant beyond 25 days. The SRT where leveling occurs is a function of kinetic parameters and varies from wastewater to wastewater.

The solids-handling capacity of the clarifier may set an upper limit on SRT. As SRT increases, the solids purge-rate decreases. Since inert solids are present in the feed, lower purge rates result in higher levels of inert solids in the system (similar to a recirculating cooling-water system). This increases the MLSS concentration in the aeration basin, and yields an increase in the solids flux to the clarifier. Pflanz [6] showed a direct relationship between solids flux and effluent suspended-solids level. Most clarifiers are designed for a flux of 15 to 25 lb/(ft²)(d). Attempts to operate beyond this limiting flux may result in complete clarifier failure. The solids flux (SF) is given by:

$$(\text{SF}) = \frac{0.012 \times (Q + Q_r) \times X}{A} \quad (4)$$

where: (SF) = solids flux, lb/(ft²)(d); Q = influent flowrate, gal/min; Q_r = recycle rate, gal/min; X = total suspended solids concentration in aeration basin, mg/L; A = surface area of clarifier, ft².

Economics may also dictate the upper limit on SRT. As SRT increases, the rate of oxygen consumption also increases as the biological reaction approaches completeness. For an SRT greater than 15 days, a large part

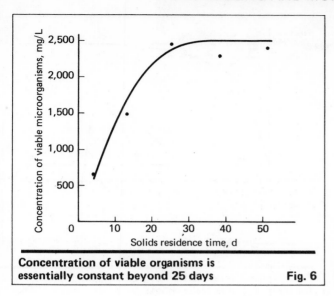

Concentration of viable organisms is essentially constant beyond 25 days   **Fig. 6**

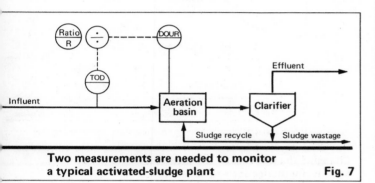

Two measurements are needed to monitor a typical activated-sludge plant   **Fig. 7**

of the oxygen requirement is due to bacteria that function in endogenous respiration, in which the bacteria consume other bacterial species as well as their own stored products. If the design of a given aeration basin is such that it is limited by the rate of oxygen transfer, rather than the rate at which energy is provided to achieve mixing, energy costs will increase with increasing SRT. However, most industrial activated-sludge units are designed to be mixing-limited to provide excess volume for dilution of shocks.

The cost of sludge disposal enters into the selection of an SRT, since as SRT increases, endogenous respiration increases and the mass of excess biological solids (waste sludge) produced in the reaction decreases. If oxygen transfer is not a problem, it is desirable to operate at the highest SRT the clarifier design will allow, to minimize sludge-disposal costs.

Process monitoring

We have shown that the slow response of activated-sludge systems limits the applicability of automatic control. Most existing units do not have a source of viable organisms to draw upon during shocks, and, even when a source does exist, there is a strong possibility of clarifier overloading.

The effectiveness of manual SRT control is a function of the degree of protection provided against severe shock loadings. Protection must consist of a process-

monitoring technique that provides a timely signal as to the state of the system. The approach suggested here is indirect monitoring of *changes* in the microbial growth rate. This should be done where the feed enters the aeration basin to provide the most sensitive signal.

There is no direct approach to online monitoring of absolute microbial growth rate. This is not a serious obstacle since process monitoring is more concerned with detection of rapid changes in growth rate. An effective way of detecting such changes is to measure the DOUR of the culture at the point of feed introduction. DOUR analyzers have been available for some time, and their reliability has been demonstrated [7,8]. However, because this method simultaneously measures oxygen consumed in metabolizing organic contaminants (exogenous respiration) and that consumed in oxidizing cell material (endogenous respiration), it therefore provides only an indirect indication of the growth rate.

Gaudy and Engelbrecht [9] have shown that the endogenous respiration rate is a function of the history of the system. During organic and toxic shock loadings, there is a shift in the relative rates of exogenous and endogenous respiration. However, the directional change in DOUR is as expected—it increases during organic shocks and decreases during toxic shocks (see Fig. 1 and 2). One exception is toxic shocks caused by a chemically oxidizable material (see Fig. 3).

Many researchers have proposed online monitoring of one of the measures of organic concentration, such as TOD or TOC. Like DOUR, these parameters are a rational first step in process monitoring; and reliable online analyzers are available.

Neither TOD nor TOC can be used to predict the actual microbial growth rate in a system because the rate is a function not only of the amount of carbonaceous material but also of the oxidation state and structure of the organic compounds, and of environmental factors such as pH and temperature. However, online organic analyzers are useful in detection of organic shocks.

TOC is the preferred parameter, since sulfides and other inorganics are included in the TOD measurement. However, TOD is used more often, because TOD analyzers are easier to maintain. TOC analyzers employ a more complex detector; also corrosion may be a problem since acidification of the sample is required. Owing to corrosion problems, TOC analyzers are not recommended for streams having a total dissolved solids content (TDS) greater than 15,000 mg/L.

The preceding discussion pointed out the advantages of DOUR and TOD (or TOC) as parameters for online detection of changes in growth rate. The following scheme is proposed:

1. Measure the TOD content of the feed stream at least every 15 min with a semicontinuous commercial analyzer.
2. Measure the DOUR in the aeration basin at least every 15 min with a semicontinuous commercial analyzer.
3. Use the DOUR and TOD signals in an electronic interface to calculate the ratio.
4. Print out each of the signals to develop trend charts.

Fig. 7 is a schematic of the instrumentation system.

Transient detection—typical toxic shocks

Typical toxic shocks are those caused by any toxin not chemically oxidizable in the aeration basin. Consider such a shock caused by a heavy metal, cyanide, or organic compound that is toxic at a low concentration and thus contributes little or no TOD to the feedstream. As the concentration of toxin increases in the aeration basin, the DOUR decreases accordingly. Since the TOD of the feedstream has not changed, the DOUR:TOD ratio also decreases. It is impossible to discuss the actual time response of the monitoring parameters here, because these factors are highly location-dependent.

Transient detection—atypical toxic shocks

Consider a shock caused by a toxin that is chemically oxidizable in the aeration basin, such as sulfide. If this type of shock is common at a particular location, it is essential to use TOC for monitoring the concentration of organics in the feedstream.

Chemically oxidizable toxins produce an increase in DOUR as their concentration increases in the system. Since the TOC of the feedstream has not changed, the DOUR:TOC ratio increases. The duration of the shock is affected by the same factors mentioned in the previous section, except that the chemical oxidation rate of the material also becomes important.

Transient detection—organic shocks

These are caused by a surplus of biologically oxidizable organics in the feed. For some systems, such a shock exists when the organic concentration of the feedstream increases twofold in a few hours, while in other systems, a fivefold increase is required.

Organic shocks are characterized by increases in both DOUR and TOD. The DOUR:TOD ratio may or may not change, depending on the biodegradability of the new TOD compared to that which exists normally. The threshold and duration of the shock are a function of the mass of organic material released upstream, the duration of the release, the degree of equalization, the types of organic compounds present, including their structures and oxidation states, the mode of operation of the system (plug-flow or completely mixed), and the hydraulic and solids residence-times of the system.

Decision analysis—toxic shocks

During any shock, operations personnel must determine its nature and source. This is most often accomplished by manual sampling at prearranged locations in the plant. During toxic shocks, sludge wasting should be ceased immediately to maintain as large a mass of viable microorganisms in the system as possible. All or part of the feedstream should be diverted to a holdup basin until the nature and source of the shock are determined.

The operator should try to maintain the DOUR:TOD ratio as close to its preshock value as possible. DOUR:TOD should be used to determine the amount of diversion required and the rate at which diverted material may be reintroduced. If the shock is caused by a chemically oxidizable toxin, more oxygen must be provided by turning on additional aerators or compressors.

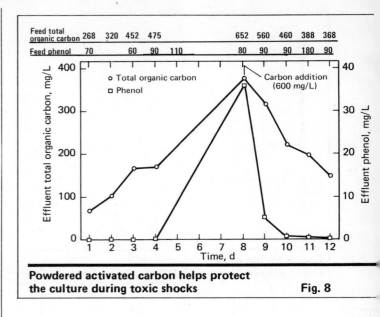

Powdered activated carbon helps protect the culture during toxic shocks   **Fig. 8**

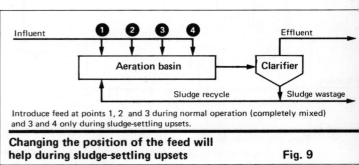

Introduce feed at points 1, 2 and 3 during normal operation (completely mixed) and 3 and 4 only during sludge-settling upsets.

Changing the position of the feed will help during sludge-settling upsets   **Fig. 9**

The addition of powdered activated carbon (PAC) to the aeration basin may protect the viability of the culture during shocks of toxic organic materials, such as cresylic acids.

In previous work, the authors [10] have demonstrated the shock-handling capabilities of PAC-treated units. Fig. 8 illustrates the response of an actual refinery activated-sludge unit during a shock following PAC treatment of 600 mg/L (aeration basin level). Recent pilot work by the authors [11] has shown the relationship between carbon dosage and the SRT of the system for units with continuous PAC addition.

If a large mass of viable organisms is destroyed in a toxic shock, the balance is upset between the microorganisms in the system and the rate of application of organics. Dispersed and/or filamentous growth may occur during recovery. The operator should monitor sludge-settling characteristics during recovery for possible dispersed and/or filamentous growth. An operations technique for combating sludge-settling problems is discussed later on.

Decision analysis—organic shocks

During an organic shock, the operator must maintain a residual DO level greater than 0.5 mg/L, and must ensure that viable microorganisms are not lost.

When there is no residual DO in the aeration basin, localized anaerobic growth can occur, resulting in

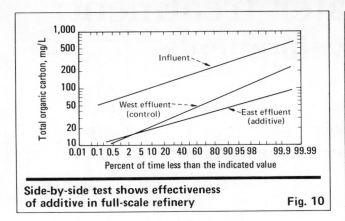

Side-by-side test shows effectiveness
of additive in full-scale refinery   **Fig. 10**

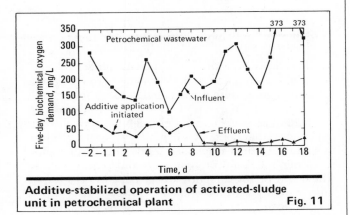

Additive-stabilized operation of activated-sludge
unit in petrochemical plant   **Fig. 11**

reduced treatment efficiency and possible sludge-settling problems. If viable organisms are lost during an
organic shock, the imbalance worsens between viable
biomass and food; dispersed growth is likely.

The simplest means of ensuring a residual DO is to
monitor DOUR at the point of feed introduction. With
experience, the operator will learn what is the maximum allowable DOUR for a given length of time. Thus,
DOUR can be used to determine the required feed-diver-
sion and the rate of diverted-feed reintroduction.

Sludge wastage should be stopped immediately during organic shocks to keep the growth rate in check. If
dispersed and/or filamentous growth occur and settling
problems result, reduce the solids flux to the clarifier to
lessen the chance of loss of the sludge blanket.

The most effective technique for reducing solids flux
is to change the point of feed addition from the head of
the basin to the exit, as demonstrated by Torpey [12]
and shown in Fig. 9. The short-term effect of this is the
rapid transfer of sludge from the clarifier to the aeration
basin, while reducing the MLSS at the exit of the system.
The long-term effect (days) is an improvement in
sludge-settling characteristics.

State of the art—biomass augmentation

Mutant bacterial additives are now available to augment the naturally developed bacterial population, and
thus improve process stability. Supplied in dry form, an
additive consists of several strains of bacteria, each
specific to a certain class of organic contaminants. Fig.
10 shows how such an additive reduced effluent TOC by

32% and improved stability in a 1-million gal/d refinery activated-sludge unit [13].

Inadequate oil removal in pretreatment operations
often causes problems with effluent BOD and suspended-solids levels. In such cases, bacterial additives
can enhance oil degradation and significantly reduce
effluent BOD. These additives also reduce foaming problems that can lead to clarifier upsets. Savings on antifoam chemicals can more than pay for the bacterial
additive program. Fig. 11 shows the improvement in
performance during acclimation in such a system.

References

1. Monod, J., The Growth of Bacterial Cultures, *Ann. Rev. of Microbiology*, Vol. 3, p. 371, 1949.
2. Tracy, K. D., Mathematical Modeling of Unsteady-State Thickening of Compressible Slurries, Ph.D. thesis, Clemson University, 1973.
3. Lawrence, A. W., and McCarty, P. L., Unified Basis for Biological Treatment Design and Operation, *J. Sanitary Eng. Div., Proc. of the Amer. Soc. of Civil Engineers*, Vol. 96, No. SA3, p. 757, 1970.
4. Tenny, M. W., and Stumm, W., Chemical Flocculation of Microorganisms in Biological Waste Treatment, *J. Water Pollution Control Fed.*, Vol. 37, p. 1,370, 1965.
5. Grady, C. P. L., and Roper, R. E., A Model for the Bio-Oxidation Process Which Incorporates the Viability Concept, *Water Res.*, Vol. 8, p. 471, 1974.
6. Pflanz, P., Performance of Secondary Sedimentation Basins, *Proc. of the Fourth Intl. Conf. on Water Pollution Res.*, p. 569, 1969.
7. U.S. Environmental Protection Agency, Advanced Automatic Control Strategies for the Activated Sludge Treatment Process, EPA-670/2-75-039, 1975.
8. Meagher, R. F., others, Instrumentation and Control at the Milwaukee South Shore Wastewater Treatment Plant, paper presented at 50th Ann. Meeting of the Central States Water Pollution Control Assn., 1977.
9. Gaudy, A. F., and Engelbrecht, R. S., Quantitative and Qualitative Shock Loading of Activated Sludge Systems, *J. Water Pollution Control Fed.*, Vol. 33, p. 800, 1961.
10. Thibault, G. T., Tracy, K. E., and Wilkinson, J. B., PACT Performance Evaluated, *Hydrocarbon Proc.*, Vol. 56, No. 5, p. 143, 1977.
11. Thibault, G. T., Steelman, B. L., and Tracy, K. D., Enhancement of the Refinery Activated Sludge Process With Powdered Activated Carbon, paper presented at 69th Annual Industrial Pollution Conf., Water and Wastewater Equip. Mfrs. Assn.
12. Torpey, W. N., Practical Results of Step Aeration, *Sewage Works J.*, Vol. 22, p. 781, 1948.
13. Tracy, K.D. and Shah, P.S., Application of Bacterial Additives to Refinery Activated Sludge, paper presented at 50th Ann. Meeting, Calif. Water Polln. Control Assn., 1978.

The Authors

George T. Thibault is vice-president at Polybac Corp., 505 Park Ave., New York, N.Y. 10022. He has worked as a project engineer for Exxon Research and Engineering Co., and is a graduate of Purdue U., where he earned a B.S.E. and an M.S.E. in environmental engineering. He is the author of several papers on the activated-sludge process, and is a member of the Intl. Assn. on Water Pollution Research, and Water Pollution Control Fed., among others.

Kenneth D. Tracy is a principal at Environmental Dynamics Inc., 1400 Cleveland St., Greenville, S.C. 29607. Previously, he was a senior project engineer at Exxon Research and Engineering Co. A registered engineer, he is the author of numerous papers on the activated-sludge process. He holds engineering degrees from Pennsylvania State U. (B.S., civil; M.S., sanitary) and Clemson U. (Ph.D., environmental). His affiliations include Chi Epsilon, Intl. Assn. on Water Pollution Research, and Water Pollution Control Fed.

Guide to wastewater treatment:
Biological-system developments

This article reviews such high-rate systems as activated sludge, trickling filter and rotating disks. It discusses preventive methods—e.g., equalization, auxiliary basins, hydrolysis—and lists design and operational variables that can enhance process performance.

Davis L. Ford and **Lial F. Tischler**, *Engineering-Science, Inc., Austin, Tex.*

☐ High-rate biological systems have been the mainstay in treating municipal and industrial wastewaters for over half a century. During this time, the technical literature has been describing new developments for optimizing the performance of such systems, particularly in the industrial sector.

This article will attempt a review of the most recent developments —i.e., of the biological methods that will be responsible for producing an effluent quality commensurate with the 1977 objectives set by the U.S. Environmental Protection Agency (EPA). These are the methods that will also provide the foundation for more-advanced treatment systems that will be required in the future.

Biological-system design

Suspended-growth (activated sludge) and fixed-growth (trickling filter, rotating disks) are two of the most popular high-rate systems used by industry to treat wastewaters with a high organic content. A brief discussion of the current state of the art for both processes will serve to explain the role they play in industrial waste treatment.

Suspended-growth systems. The completely-mixed activated sludge process is undoubtedly the most popular high rate biological treatment method for wastewaters with relatively high organic concentration. This is because of the route's flexibility, which allows adjustments in wasteload variations, and because of its ability to provide a measure of organic-load equalization.

The contact-stabilization version of activated sludge, in which the biological solids are contacted with the wastewater for a short period of time, then separated, and finally reaerated to degrade sorbed organics, has shown some success for industrial wastes with a high content of suspended and colloidal organics.

The aerated-lagoon system can approach or equal the organic removal capability of an activated-sludge process provided the unit is preceded by proper pretreatment, and the environmental factors in the plant's location are suitable. Aerated lagoons, however, require considerable land, and will probably only function well in a temperate climate, since they are much more sensitive to temperature than is activated sludge.

A problem common to the suspended-growth systems is how to concentrate the suspended solids in the effluent after gravity separation. Many industrial wastewaters tend to generate a relatively high portion of dispersed biomass that does not adequately separate, resulting in average effluent-suspended solids in the range of 40 to 75 mg/l. Careful pretreatment and process control, which will be subsequently discussed, can help to produce a highly clarified effluent. However, past experience indicates that some type of effluent polishing will be required for many biologically treated industrial wastewaters, if effluent limitations for total suspended solids of less than approximately 30 mg/l are imposed by regulatory agencies. Effective polishing systems include organic-polymer flocculants and granular-media filtration, the selection depending upon the degree of suspended-solids removal required.

Fixed-growth systems. The conventional trickling filter, using either rock or synthetic stationary media, has generally been used only for pretreatment of industrial wastewaters, because of its inability to provide the high effluent-quality usually required. It is still a viable and cost-effective biological roughing process for industrial wastewaters containing high quantities of relatively degradable organics.

In the past several years, there has been increasing interest in a variation of the fixed-growth system known as the rotating biological surface (RBS). Here, the biological film grows on the surface of thin, large-diameter plastic disks placed side by side on a rotating shaft. The bottom one-fourth to one-third of the rotating disks is immersed in a basin containing the wastewater. Aeration occurs when the surface of the disk is in the air, and there is some air entrainment in the basin.

This system overcomes many of the disadvantages of the stationary-media trickling filter, the most notable of which are: (1) the continuous shearing of the fixed biomass as the disks pass through the water, thus pre-

Originally published August 15, 1977.

venting accumulation of inactive material, and (2) the effective penetration of oxygen throughout the biomass under all operating conditions.

The RBS system has one advantage over suspended-growth systems—namely, lower energy requirements. It is also claimed that the suspended solids in the RBS effluents are lower than those in activated-sludge systems, although confirmation of this for a wide variety of industrial wastewaters has not yet been demonstrated.

The RBS system has been tested on industrial wastes of a wide range of organic concentrations and complexities. It appears at this time to be quite competitive for industrial wastewaters with relatively low concentrations of noncomplex organics. At high organic concentrations many disks are required, and the capital costs of the system may be quite high. If the organics tend to be refractory, and require long contact periods with the biomass for degradation, the effective biological mass-substrate contact time and the hydraulic detention time in the RBS system are possibly too short for optimal reduction.

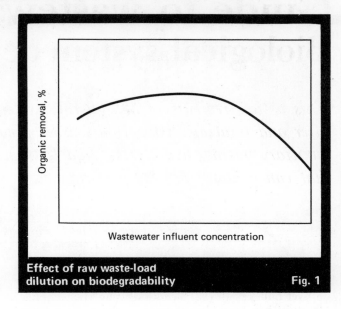

Effect of raw waste-load dilution on biodegradability   **Fig. 1**

Treating complex constituents

One of the primary limitations in applying the biological mode of treatment to industrial wastewaters is the failure to incorporate proper pretreatment and process flexibility into the basic design. Since the biochemical oxidative mechanisms are complex, and in many instances highly sensitive to external environmental factors, every effort must be made to accommodate the biological population to the maximum extent. Some of these preventive, rather than curative, steps are discussed in the following sections.

Equalization. The deleterious effect of transient loadings on biological systems is well documented [1,2]. Both hydraulic- and organic-load variation can be detrimental to performance. Past experience has underscored the necessity for equalizing the industrial raw-waste load prior to biological treatment. There are several rational methods which can be utilized in sizing equalization basins in order to sequester influent variations [3,4].

Utilization of auxiliary basins. It has recently become accepted practice to include auxiliary, off-specification basins to temporarily receive wastewaters of inordinately high organic concentrations, store this water, and then feed the inventory to the biological system at a controlled hydraulic rate. Diversion of the wastewater stream can be accomplished automatically using an on-line analyzer. Such basins, along with equalization, reduce the hydraulic and organic variations to the biological systems, and result in significantly higher overall efficiency.

Hydrolysis. Hydrolysis of selected organic wastewater streams, accomplished by adding caustic and exercising careful pH control, can enhance the biodegradability of the hydrolyzed stream [5]. This has been widely applied as a pretreatment step in the biological treatment of pesticide and herbicide waste streams, with positive results.

Dilutive reduction of influent organic concentration. When biochemical inhibition can be attributed strictly to influent organic concentration, rather than to constituent complexity or resistance to biodegradation, then predilution of high-organic streams with those of low organic concentration is a legitimate and appropriate pretreatment step to improve the overall performance of a biological system. The kinetics of removal are more concentration-sensitive than mass-sensitive, thereby justifying this approach. Typical diluent streams include utility blowdown, once-through cooling water, or fresh water from wells or surface supplies. An example of predilutive enhancement of treatment performance is illustrated in Fig. 1.

Stripping of select waste streams. Steam or solvent stripping of certain waste streams, with proper air emission control measures in an industrial plant can reduce high-organic loads, sequester organic-loading variations, and remove potentially toxic or inhibitory contaminants, greatly improving the amenability of the stream to biological treatment. Probably the most common example is sour-water stripping in petroleum refineries, but there are numerous other cases in the chemical process industries where this is an extremely effective pretreatment step. Examples of potentially toxic or refractory organics that have been successfully stripped from wastewaters include chlorinated hydrocarbon byproducts, and complex aldehydes.

Higher sludge inventory in aeration basins. A commonly used technique for providing additional resistance to biological upset is to increase the inventory of biological solids in the aeration basin of suspended growth systems. This is done in an activated-sludge system by increasing the sludge-recycle ratio and/or reducing sludge wastage. This increased inventory simply means that the quantity of biotoxic or biostatic constituents per bacteria is reduced. Design MLSS levels in activated-sludge systems are typically 2,500-3,000 mg/l, while many industrial systems, for the aforementioned reasons, are operating at 8,000-10,000 mg/l [8].

Removal of solid constituents. Any biological treatment system, with the possible exception of the contact-stabilization version of activated sludge, functions more effectively if all organic material to be removed is in soluble form. Since the biochemical mechanisms of

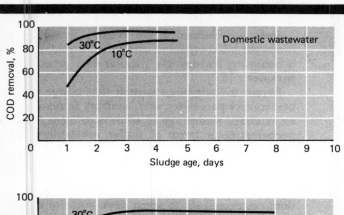

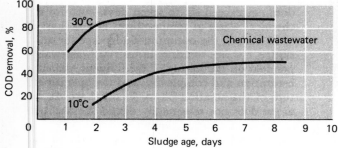

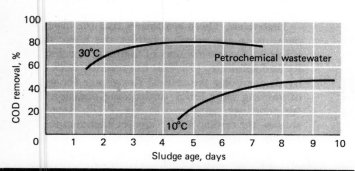

Effects of sludge age and temperature on biodegradability   Fig. 2

Sludge age. Defined as the average contact time between the microorganism and the substrate, sludge age is becoming increasingly popular as a process control parameter [6,7].

Recent investigations suggest that sludge age is the best control parameter, and, contrary to some theories, an extended sludge age of 40 days or more maximizes performance in terms of sludge settleability, process control, and organic removal efficiency [6]. Other studies show that the critical sludge age (defined as the minimum necessary to achieve maximum organic removal) is a function of substrate and temperature, but does not exceed six to seven days, even for a complex chemical waste at temperatures of less than 10°C [7].

Therefore, the optimum sludge age for an activated sludge system treating industrial wastewaters is dependent on the nature of the influent, namely, its concentration and complexity, and the operating temperature of the aeration basin. In other words, sludge age alone does not adequately define the ability of a suspended-growth biological system to provide maximum removal of organics from a specific wastewater. The food-to-microorganism ratio and hydraulic retention time are also important control criteria in some cases.

Temperature. There is a theory that the lower the temperature in the aeration basins, the lower the process efficiency for wastewaters of higher molecular complexity and solubility [8]. This has recently been confirmed by determining the critical sludge age for wastewaters of varying complexity undergoing aeration at several temperatures, as shown in Fig. 2 [7]. Based on these studies, the critical sludge age for each temperature and wastewater can be approximated as follows:

	Temperature, °C	Critical sludge age, days
Domestic wastewater	30°	2.
	10°	3.5
Chemical wastewater	30°	2.5
	10°	5.5
Petrochemical wastewater	30°	3.5
	10°	8

It is of paramount importance, therefore, that designers provide sludge ages which are adequate for maximum performance predicated on wastewater complexity and swings in operating temperatures. This is particularly important for systems with long hydraulic retention times, since aeration-basin temperatures will approach ambient air temperature even if the wastewater is quite warm before aeration.

Sludge bulking. The solids-liquid separation phase has always been one of the more important elements in biological treatment of wastewater. Sludge bulking is one of the main precursors to high-effluent TSS levels, and has consequently received much attention in the attempt to optimize biological treatment facilities.

A common occurrence in the food-processing and pulp and paper industries, bulking is very difficult to correct. A preventive approach based on adequate pH control, sufficient aeration, and an adequate supply of nitrogen and phosphorus is indicated, as filamentous

substrate utilization require soluble organics, it follows that the dissolution of solid and colloidal organics will require utilization of additional time and energy on the part of the microorganisms.

Although the impact of such organics (solid and colloidal) has not been quantified, there is an intuitive incentive to remove as much of them, as well as inorganic materials, as possible. This is done prior to biological treatment to enhance overall organic removal. A wide variety of secondary solids-removal processes, including chemically-assisted air flotation, granular-media filtration, and chemical coagulation with gravity sedimentation, can be used. Oily wastewaters can also be pretreated with these unit processes; fibrous-media coalescers are an additional option for oil removal.

Process optimization

Many design and operational variables can enhance process performance. In addition to the procedures described to minimize biological upsets, the following process and environmental factors influence attainable effluent quality.

microorganisms thrive at lower pH, oxygen tension and nutrient levels than do the flocculating microflora.

Preliminary studies indicate that the activated-sludge system is less prone to produce a filamentous population, and is therefore less susceptible to bulking. For this reason, conversion of contact stabilization to completely mixed activated sludge may enhance overall settleability and process performance.

Adding such chemicals as hydrogen peroxide to the aeration basin or recycle sludge to minimize bulking has had mixed success. But use of a controlled seeding of sludge microflora to a system that treats food and dairy wastewaters has recently been reported as an effective method. The practicality of this approach in terms of process and cost effectiveness should be verified before its use is considered.

Nitrification enhancement. Nitrification is necessary to keep effluent ammonia-concentration levels within permissible limits. But nitrifying microorganisms are very sensitive to pH and temperature, and since many process variables and trace chemicals affect their performance, it is difficult to base nitrifying strictly on sludge age or hydraulic retention time.

Two-stage activated sludge is one of the newer concepts being proposed for biological removal of ammonia nitrogen. Although it is considerably more expensive than a single-stage system, it provides more process control for enhancing nitrification.

A recent study indicates that aerobic submerged filters may be successful in nitrifying low-strength wastes. These systems [*11*] provide an upward flow of liquid through plastic or natural media, efficiently capturing solids, and controlling retention time.

The rotating biological surface has also been quite successful in biologically treating ammonia nitrogen. A properly designed RBS system inherently has the same advantages as a two-stage activated-sludge setup, allowing an enriched nitrifying population to develop in the latter stages of the RBS process.

It is recommended that bench- and/or pilot-scale treatability studies be performed on industrial wastewaters before incorporating a nitrification technique into a biological system. Such studies often identify certain waste streams and/or components that affect nitrifying bacteria. The wastewater can then be pretreated to remove these components.

Activated-carbon treatment. The addition of powdered activated carbon to activated-sludge aeration basins results in a substantial increase in process performance [*12*]. The most significant project/that uses activated carbon, and a biological mass and aeration system, is the 40-million-gal/d Du Pont Chambers Works (Deepwater, N.J.). Initial results there indicate good performance in terms of effluent quality. However, the dewatering, incineration and regeneration of the mixed biological-carbon sludge have not been tested, and these could be critical in determining how practical this approach is.

Required carbon dosages, and the ability to reuse the adsorbent material, will dictate the cost-effectiveness of the method, compared with other alternatives. Both bench/pilot treatability studies should be used to evaluate the application of powdered activated carbon.

References

1. Ford, D. L., and Eckenfelder, W. W., "The Effect of Process Variables on Sludge Floc Formation and Sludge Settling Characteristics," Water Pollution Control Federation Meeting, Kansas City, Missouri (Sept. 1966).
2. Ford, D. L., "Factors Affecting Variability from Wastewater Treatment Plants," Prog. Water Technology, Vol. 8, No. 1, pp. 91-111, Pergamon Press, London (1976).
3. LaGrega, M. D., and Keenan, John D., "Effects of Equalizing Wastewater Flows," Journal WPCF, Vol. 46, No. 1 (January 1974).
4. Speece, R. E., and LaGrega, M. D., "Flow Equalization by Use of Aeration Tank Volume," Journal WPCF, Vol. 48, No. 11 (November 1976).
5. Shell Chemical Company, Unpublished internal report (1975).
6. Grutsch, J. F., "A New Perspective on the Role of the Activated Sludge Process and Ancillary Facilities," *Proceedings of the Open Forum on Management of Petroleum Refinery Wastewaters,* sponsored by Environmental Protection Agency, American Petroleum Institute, the National Petroleum Refiners Association, and the University of Tulsa (January 1976).
7. Sayigh, B. A., "Temperature Effects on the Activated Sludge Process," Doctoral Dissertation, the University of Texas at Austin (May 1977).
8. Ford, D. L., "Water Pollution Control in the Petroleum Industry," *Industrial Wastewater Management Handbook,* p. 8-1, edited by H. Azad, McGraw-Hill, New York (1977).
9. Chambers, J. V., "Bioengineering an Activated Sludge Microflora to Improve Waste Removal Performance," Proceedings of the Fifth Annual Industrial Pollution Conference, WWEMA. Atlanta, Georgia (April, 1977).
10. Adams, C. E., and Eckenfelder, W. W., "Nitrification Design Approach For High Strength Ammonia Wastewater," Journal WPCF (March 1977).
11. McCarty, P. L., and Haug, Roger T., "Nitrogen Removal From Wastewaters by Biological Nitrification and Denitrification," Presented at *Society For Applied Bacteriology,* Liverpool, England (1971).
12. Rizzo, J. A., "Case History: Use of Powdered Activated Carbon in an Activated Sludge System," *Proceedings of the Open Forum on Management of Petroleum Refinery Wastewaters,* sponsored by Environmental Protection Agency, American Petroleum Institute, the National Petroleum Refiners Association, and the University of Tulsa (January 1976).
13. Davis, J. C., "Activated Carbon: Prime Choice to Boost Secondary Treatment," News Features, *Chemical Engineering* (April 11. 1977).

The authors

Davis L. Ford is Senior Vice-President, Engineering-Science, Inc., 3109 North Interregional, Austin, TX 78722. He has broad experience in the design and construction of wastewater treatment systems, and has been a consultant to industry, the United Nations, U.S. government agencies, and several foreign governments. He holds a Ph.D. in environmental health engineering from the University of Texas.

Lial F. Tischler is Project Manager and Program Director for major environmental engineering projects at Engineering-Science, Inc. He has been a consultant to the World Health Organization, and has taught courses at the University of Texas. He holds an M.S. in environmental health engineering, and a Ph.D. in civil engineering, both from the University of Texas.

How to control biological-waste-treatment processes

Active mass of solids is the preferred control parameter for biological treatment, because its relation to other variables is theoretically sound, and measuring it requires a minimum of laboratory work.

Harry S. Harbold, P.E.

☐ Passage of the Federal Water Pollution Control Act Amendments of 1972 (PL 92-500) calls for many chemical process companies to spend millions of dollars, either for new water-pollution-control equipment or for upgrading existing treatment plants. An engineer responsible for such treatment facilities must ensure that they are operated at maximum efficiency year-round to meet increasingly stringent discharge requirements.

Biological waste treatment is currently most widely used for removing organic pollutants from industrial wastes; and the most versatile and efficient of the available processes is the activated-sludge system, in which microorganisms present in the activated sludge remove organic matter from the liquid waste by synthesis into new protoplasm. The degradable organic matter is used for food by the microorganisms to synthesize new cell mass and obtain energy for metabolic functions by oxidation of the organic substrate to carbon dioxide and water (Fig. 1).

It is often difficult to meet regulatory-permit requirements during fluctuations in plant production or at increased treatment-plant loadings. As a result, there is need for a simple and reliable method of controlling the biological treatment process to ensure optimum treatment levels.

The early design technique for the activated-sludge process was primarily a hydraulic approach based on selection of an aeration detection time that did not relate to the fundamentals of the biological reaction. However, the biological solids-retention time, based on the Monod kinetic model, has been proposed for design and control of the activated-sludge process [1]; and this has been demonstrated both in the laboratory and in full-scale plant operations [2,3,4,5,6].

Basic theory

When bacteria's food supply is unlimited, bacterial growth is restricted only by the ability of the microorganisms to reproduce; and the organic removal and cell growth is directly proportional to the number of microorganisms present (Fig. 2). However, because the number of microorganisms increases rapidly during this

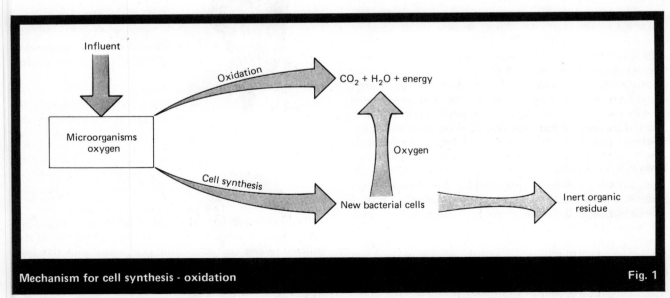

Mechanism for cell synthesis - oxidation

Fig. 1

Originally published December 6, 1976.

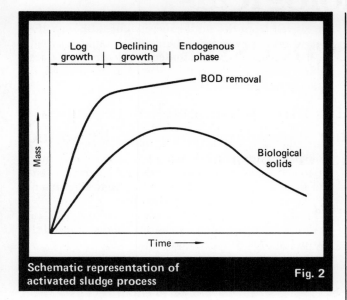

Schematic representation of activated sludge process   Fig. 2

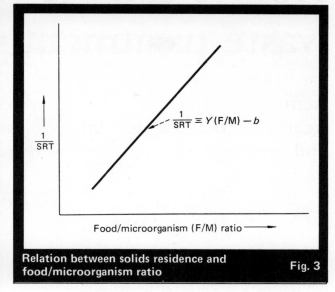

Relation between solids residence and food/microorganism ratio   Fig. 3

log-growth phase of the metabolism, the food supply becomes limited, and there is only enough food to keep the organisms alive. Next, the reaction enters the declining-growth phase, in which the food supply decreases; and this is followed by an endogenous phase, in which the cell mass goes through auto-oxidation by endogenous respiration to remain alive.

By thus using up its stored food supply, the cell mass is depleted of residual organic content, becomes inactive, and forms a heavy floc that may be separated by sedimentation. This settled floc of activated sludge is subsequently recirculated to be mixed with the treating-plant influent, so that cell metabolism and organic removal may continue.

The term solids retention time (SRT) represents the average residence time of active microorganisms in the system, and is defined by the equation:

$$SRT = \frac{\text{mass of active biological solids in the system}}{\text{mass of biological solids removed from the system per day}} \qquad (1)$$

One of the most important aspects of biological treatment is controlling the solids level in the process through the amount of sludge sent to waste daily from the system. Since SRT is a function of both the solids level in the system and the mass of solids wasted, it has a sound theoretical basis for use, as is revealed through the basic kinetics and control parameters of the system.

Process kinetics

In a biological treatment process, the relationship between the rate of growth of microorganisms and the rate of substrate utilization can be expressed by:

$$\frac{dX}{dt} = Y \frac{dF}{dt} - bX \qquad (2)$$

where: X = concentration of biological solids.

$\dfrac{dX}{dt}$ = net change in biological solids with time as a result of synthesis and endogenous respiration.

F = concentration of organic food substrate utilized by microorganisms.

Y = growth coefficient expressed as the mass of organisms produced per mass of waste consumed.

b = rate of endogenous respiration, expressed as time^{-1}.

t = time.

This equation indicates that the net production of biological solids is equal to the mass of new biological solids produced through food substrate utilization minus the biological solids lost through endogenous respiration.

Control parameters

Many methods are used to control the activated-sludge process, with each plant operator tending to develop his own technique. The three parameters in widest application for both design and operation are:

1. Food-to-microorganism ratio (F/M).
2. The quantity of either mixed-liquor suspended solids (MLSS) or mixed-liquor volatile suspended solids (MLVSS).
3. Solids retention time (SRT).

No matter which method is used, control is achieved by varying the rate at which sludge is wasted from the system. Also, all three methods can be described by the same basic equation for cell growth (Eq. 2).

Thus for the (F/M) method, Eq. 2 is divided by X:

$$\frac{dX/dt}{X} = Y \frac{dF/dt}{X} - b \qquad (3)$$

The term $(dF/dt)/X$ is the food-to-microorganism ratio, and is usually expressed as pounds of BOD$_5$ utilized over a period of time per pound of active biological solids. Control to a constant F/M ratio will result in a high-quality effluent and good sludge settleability, even with fluctuating wastewater strength, because it is a function of the solids level in the system. However, use of the F/M ratio requires considerable laboratory work since it is necessary to know both the amount of food removed and the mass of biological solids in the system. Since this requires tests for BOD, it is impossible to adjust to variable short-term changes in loading, since the BOD test takes five days for completion. The chemical oxygen demand (COD) test may be substituted for 5-day BOD, but a poor correlation is often obtained.

For the MLSS method, control actually attempts to maintain a constant mass of active biological solids in the aeration tank; and this can easily be determined by the suspended-solids test. The volatile fraction of the mixed-liquid suspended solids, for MLVSS control, is often taken as an approximation of the active mass of biological solids. This method has a severe disadvantage, however, if the strength of incoming waste does not remain constant.

Process control with SRT

The SRT method of control is theoretically sound and requires a minimum of laboratory work. From Eq. 3;

$$\text{SRT} = X/(dX/dt)$$

When plotted, Eq. 3 gives a straight line (Fig. 3). From this plot, we see that maintaining a constant SRT will result in a constant F/M ratio for solids control. SRT can easily be controlled by regulating the amount of sludge wasted from the system daily. Also, a minimum amount of laboratory work is required, since only the suspended-solids concentration in the aeration tank and return line needs to be measured.

In the activated-sludge system shown in Fig. 4, solids are wasted from the sludge recycle line, although they may also be wasted from the aeration tank. Also, biological solids are wasted unintentionally in the clarified effluent stream. Eq. 1 can be rewritten in terms of this system's parameters [1]:

$$\text{SRT} = \frac{VX}{X_e(Q - Q_W) + Q_W X_r} \tag{4}$$

This equation makes the assumption that the active biological solids in the system are measured in terms of the mixed-liquor suspended solids in the aeration tank, and that the system is well mixed, so that X represents an average concentration in the tank.

The relationship between solids recycle and SRT can be shown by performing a material balance around the clarifier at steady-state conditions:

$$X(Q + Q_r - Q_W) = X_e(Q - Q_W) + Q_r X_r$$
$$Q_r X_r = (Q_r - Q_W) X_r + Q_W X_r$$
$$X(Q + Q_r - Q_W) =$$
$$X_e(Q - Q_W) + Q_W X_r + (Q_r - Q_W)X_r \tag{5}$$

In a well-operated system, the mass of solids in the effluent, $X_e(Q - Q_W)$, is very low compared to the mass of solids in the aeration tank or waste sludge stream. If the solids lost in the clarifier overflow is considered negligible, Eq. 4 becomes:

$$\text{SRT} = VX/Q_W X_r \tag{6}$$

This can be combined with Eq. (5) as:

$$X(Q + Q_r - Q_W) = (Q_r - Q_W)X_r + VX/\text{SRT} \tag{7}$$

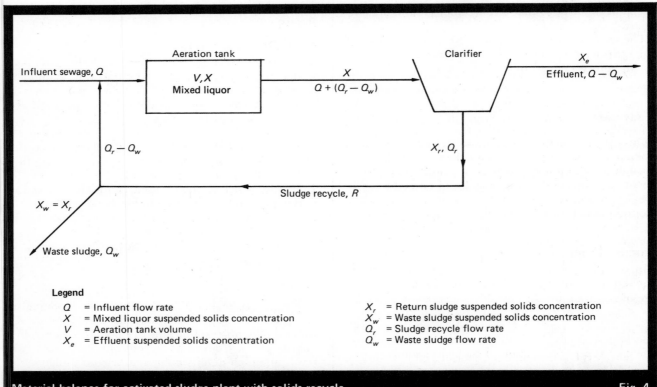

Legend

Q = Influent flow rate
X = Mixed liquor suspended solids concentration
V = Aeration tank volume
X_e = Effluent suspended solids concentration

X_r = Return sludge suspended solids concentration
X_w = Waste sludge suspended solids concentration
Q_r = Sludge recycle flow rate
Q_w = Waste sludge flow rate

Material balance for activated sludge plant with solids recycle   Fig. 4

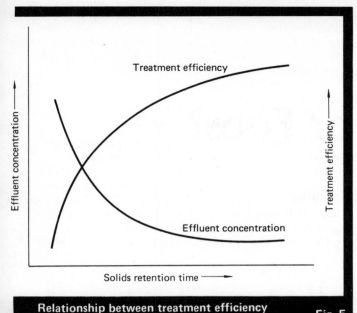

Relationship between treatment efficiency and solids retention time   **Fig. 5**

Dividing by X and rearranging terms yields:

$$1/\text{SRT} = (Q/V)(1 + R - RX_r/X) \qquad (8)$$

where R = recycle ratio = $(Q_r - Q_W)/Q$.

Thus SRT is shown to be a function of (a) the volumetric recycle ratio, R, (b) the ratio of MLSS, or X, and (c) the solids concentration in the thickened sludge, or X_r. The term X_r/X is a function of the efficiency of the final clarifier, and a function of the settling characteristics of the biological solids.

If Eq. 8 is rearranged to express R more clearly in terms of SRT, we get:

$$R(X_r/X - 1) = 1 - (V/Q)/\text{SRT} \qquad (9)$$

From this, it is apparent that as the biological SRT becomes large the right side of the equation approaches unity, and R becomes dependent on (X_r/X). The steady-state relationship between SRT and treatment efficiency (Fig. 5) suggests that the system should be operated at a high SRT to improve treatment efficiency and reduce waste sludge quantities and processing costs. However, a point is reached where further increases in SRT bring only minor improvement in effluent quality for a significant increase in the solids mass. Thus the SRT is limited by the solids loading on the clarifier.

Application of SRT

The normal procedure for using biological SRT to optimize a given plant would be to:

1. Select a value of SRT for the control point.
2. Determine X and X_r as total suspended solids, using 24-h composite samples if possible to account for diurnal flow patterns. (In any event, daily variations should not result in serious error, since SRT is normally in the range of 5–15 days.)
3. Using Eq. 9, determine the recycle ratio, R, for the system.
4. Using Eq. 6, determine the sludge wasting rate.

5. Maintain the predicted flowrates constant throughout the day, so as to waste sludge evenly from the system.

Example: Assume the process sequence in Fig. 4, with the aeration tank volume (V) = 0.33 million gal and the average daily flowrate (Q) = 0.5–1.0 million gal/d, depending on plant production. Step 1: Assume SRT equal to 5, 10, 15, and 20 days; and for each of these plus the given value of V, calculate the recycle rate, $(Q_r - Q_W)/Q$, as a function of X_r/X by means of Eq. 9, and plot the results as shown at the top of Fig. 6. Then use Eq. 6 to calculate the sludge wastage rate, Q_W, as a function of X_r/X for parameters of SRT, as shown at the bottom of Fig. 6. Choose an SRT of 10.

Step 2: Determine X and X_r. Assuming that the aeration tank MLSS was 2,500 mg/l and the sludge recycle concentration was 7,500 mg/l, the ratio X_r/X is 3.0.

Step 3: Using Fig. 6, which shows the solution of Eq. 9, read that at a plant effluent rate of 500,000 gal/d, the sludge recycle rate should be approximately 47% of the flowrate, or 235,000 gal/d.

Step 4: Similarly, the sludge wasting rate should be 11,000 gal/d.

Step 5: After operating under these conditions for several days, the mixed-liquor suspended-solids concentration has increased to 2,700 mg/l. The ratio of X_r/X is now calculated at 2.8/1.0. From Fig. 6, the sludge recycle ratio should be increased to approximately 53%, and the sludge wasting rate to 12,000 gal/d, in order to maintain SRT = 10.

In conclusion, SRT has been proven one of the best methods for controlling an activated-sludge system, based on both theoretical and practical considerations. As future treatment plants become more automated, this concept may be used as the basis for design.

References

1. Lawrence, A. W., and McCarty, P. L., Unified Basis for Biological Treatment Design and Operation, *Jour. Sanitary Engrg. Div., Proc. Amer, Soc. Civil Engrs.,* **96,** 757 (1970).
2. Keyes, T. W., and Asano, T., Application of Kinetic Models to the Control of Activated Sludge Processes, *Jour. Water Pol. Control Fedn.,* **47,** 2574 (1975).
3. Campbell, H. J., and Rocheleau, R. F., Waste Treatment at a Complex Plastics Manufacturing Plant, *Jour. Water Pol. Control Fedn.,* **48,** 256 (1976).
4. Walker, L. F., Hydraulically Controlling Solids Retention Time in the Activated Sludge Process, *Jour. Water Pol. Control Fedn.,* **43,** 30 (1971).
5. Bisogni, J. J., and Lawrence, A. W., Relationships Between Biological Solids Retention Time and Settling Characteristics of Activated Sludge, *Water Research,* **5,** 753 (1971).
6. Jenkins, D., and Garrison, W. E., Control of Activated Sludge by Mean Cell Residence Time, *Jour. Water Pol. Control Fedn.,* **40,** 1905 (1968).

The author

Harry S. Harbold has been employed by several government agencies, since 1970, as an environmental and chemical engineer in environmental and water-pollution control. A registered professional engineer in Pennsylvania, he holds a B.S.Ch.E. from Auburn University (1970) and is a member of the Water Pollution Control Federation and the Water Pollution Control Assn. of Pennsylvania.

Bacteria: Friends or Foes?

Microorganisms are essential for biological-wastewater treatment systems. But in some water systems, they cause corrosion, plug filters and perform other unfriendly acts. Here's what you need to know for responsible microbiological control.

GEORGE A. WOODS, BETZ Laboratories, Inc.

Industry is giving more attention to bacteria and other microorganisms as part of its increasing concern for the environment. Not only are these organisms instrumental in creating odors and buildup of heavy sludge, but they have the useful ability to reduce sludge volumes and deactivate toxic and other harmful chemicals. Greater knowledge of the functions of these organisms has resulted in greater control over them.

Until recently, microorganisms were actually contributing to pollution problems while the blame was being placed elsewhere. Various chemical techniques to resolve these problems were devised, while in reality responsible microbiological control would have been more effective. Some examples of faulty analysis:

■ A disposal well in Texas suddenly developed a high head. It had previously operated under vacuum. Calcium and magnesium compounds in the injection water were blamed for plugging the formation.

■ In California, a fresh-water well suddenly caused

IIIIIIIIIIIIIIIIIIIIIIIIIIIIIIIIIIIIIIIIIIIIIIIIIIIIIIIIIIIIIIIIIIIIIIIIIIIIIII

Steps in a Microbiological Control Program

1. Sample and identify specific organisms.
2. Know the life function of the organisms.
3. Understand the physiology and waste products.
4. Evaluate specific toxicants.
5. Monitor the program by plate counts.
6. Sample and analyze for the toxicants in discharge streams.
7. Detoxify or decharacterize the toxicant before discharge.
8. Reuse the water if possible.
9. Review control procedures.

IIIIIIIIIIIIIIIIIIIIIIIIIIIIIIIIIIIIIIIIIIIIIIIIIIIIIIIIIIIIIIIIIIIIIIIIIIIIIII

Originally published March 5, 1973.

plugged filters ahead of a demineralizer unit. The material was identified as elemental sulfur, and a nearby tank farm, loaded with high-sulfur crude, was blamed for leakage into the water table.

■ In Kansas, a water supply system suddenly became so fouled with iron deposits that the lines had to be replaced. A new well showing high iron content was blamed.

■ In Arizona, deionizing equipment became ineffective because of resin poisoning. Iron was detected, and leaching of iron from the formation was suggested.

Bacteria to Blame

In all of the above cases, the role of bacteria was completely ignored initially, and ineffective chemical additives were used to treat the symptom. More-careful analysis, however, revealed the real cause of the problem. For instance, slime-forming bacteria were causing the disposal-well plugging in Texas. Continuous addition of a chlorophenate-organic sulfur biocide not only eliminated the problem but also improved the well's porosity.

The sulfur-plugging problem in the California refinery was traced back to sulfur bacteria, *Thiobacillus thioxidans* (Fig. 1). These bacteria had contaminated the well reserve to such an extent that nearly one ton of chlorine was required in each well to eliminate the microorganism. Chlorine was added under controlled conditions to avoid plugging the well and avoid corroding piping and equipment, which could have resulted from the low pH and oxidizing effects of chlorine.

The problem in Kansas was found to be caused by iron bacteria of the genera *Sphaerotilus* and *Gallionella* (Fig. 2). The iron was coming from the new well, but iron bacteria had put it there. Sterilization of the well

SULFUR-PRODUCING bacteria (*Thiobacillu thioxidans*) can be controlled with chlorine — Fig. 1

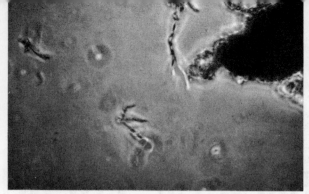

IRON BACTERIA (*Gallionella genes*) cause severe deposits in water supply systems — Fig. 2

with continuous chlorination was recommended to correct the problem.

In the case of the deionizer-resin poisoning in Arizona, sulfate-reducing bacteria created an acid environment in sections of the 5-mi. supply pipeline. This caused corrosion in the line, releasing iron to the system. The microorganisms were identified as *Desulfevibrio desulfricans* and *Clostridium nigrificans*. Not only did these organisms supply iron and corrode the piping, they also caused a hydrogen sulfide taste in the potable water.

In each situation, a specific toxicant controlled or eliminated the organisms. "Controlled" is more accurate because as long as nutrients and the proper minerals and physicial conditions exist, the bacteria will exist.

How Bacteria Cause Corrosion

In the metabolic process of any living thing, there is an electrodynamic balance, as energy is either required or released during the functioning of the organism. In some species of bacteria, these energy changes induce a measurable electric current. When corrosion-causing microorganisms reproduce on the surface of iron pipe, electrobiochemical removal of iron occurs, causing corrosion. Sulfate-reducing bacteria have been found to cause the electrode potential of stainless steel to shift 150 millivolts to the negative side.

Corrosion in a water system is electrochemical in nature. Anodic and cathodic areas form on the metal surface. At the anodes, the atoms of the metal release electrons that pass through the metal to the cathodes, where each electron displaces a positive ion, usually hydrogen. The positive metallic ions formed at the anodes go into the electrolyte solution.

Microorganisms can participate in iron and steel corrosion by creating conditions favorable to electrochemical reactions. For example, their metabolism can change the surface film resistance with such products as sulfuric and organic acids. In other instances, slime deposits may shield some areas so that they become anaerobic even with considerable oxygen present in the water. The anaerobic sites on the metal surfaces become anodic to the aerobic areas.

When two portions of the same metal receive oxygen at different rates, a corrosion cell is established and two types of oxidation may occur: dehydrogenation and loss of electrons. The hydrogen atom can disassociate into a proton and an electron. Electrons are transported by electron carrier systems, previously discussed, while the hydrogen ion is temporarily absorbed by the buffer medium of the cell.

Key Bacteria

The most important bacteria associated with corrosion are the sulfate reducers or sulfide formers. They are classified as either nonsporulating (*Desulfovibrio*) or spore-forming (*Desulfotomaculum*, which include the *Clostridium nigrificans*).

Sulfate-reducing bacteria use the sulfate ion as the terminal electron acceptor for their respiration. These organisms are also autotrophic in utilizing an inorganic source of carbon. Energy is obtained by the reduction of the sulfate ion by hydrogen, which can come from molecular hydrogen or organic compounds. In metabolism, the organisms can use molecular hydrogen for producing H_2S. Corrosion usually results by cathodic depolarization, and it is greatly accelerated by the organism's consumption of hydrogen.

Anodic
$$8H_2O \rightleftharpoons 8H^+ + 8(OH)^-$$
$$4Fe + 8H \rightleftharpoons 4Fe^{++} 8H$$

Cathodic Depolarization (bacteria)
$$SO_4 = + 8H \rightleftharpoons H_2S + 2H_2) + 2H_2O + 2(OH)^-$$

Corrosion Products
$$Fe^{++} + H_2S \rightleftharpoons FeS + 2H^+$$
$$3Fe^{+++} + 6(OH)^- \rightleftharpoons 3Fe(OH)_3$$

The net result is that iron is continuously removed from the anodic area, causing a pitting type of corrosion. FeS is cathodic to metallic iron, and its presence accelerates corrosion.

Other bacteria, such as *Thiobacillus thioxidans*, can oxidize H_2S to obtain energy for metabolism.

$$2H_2S + O_2 \longrightarrow 2S + 2H_2O$$
$$2S + 3O_2 + 2H_2O \longrightarrow 2H_2SO_4$$

The sulfuric acid produced is highly corrosive and forms an anodic concentration cell that further accelerates attack. The pH of this reaction has been recorded at less than 0.2.

The iron bacteria *Sphaerotilus*, *Gallionella* and *Crenothrix* (Fig. 3) use energy produced from converting ferrous oxide to ferric hydroxide. The ferric hydroxide forms a sheath on the organism and such sheaths deposit on the walls of pipes, creating small areas of iron hydroxide. Anaerobic conditions exist under such areas, developing a suitable environment for growth of sulfate-reducing bacteria.

With the continued production of iron, additional iron bacteria develop on the deposits, thus contributing to the

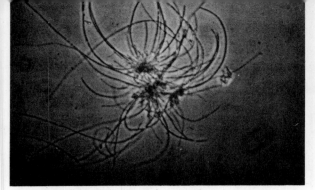

THESE ORGANISMS convert ferrous oxide to ferric hydroxide on walls of pipes and equipment — Fig. 3

CRATERS on heat-exchanger tube caused by iron bacteria (*Sphaerotilus, Gallionella and Crenothrix*) — Fig. 4

iron scale on the pipe. The cycle feeds upon itself and produces sharply defined pits. The craters on the steel heat-exchanger tube in Fig. 4 are an example of such action.

Establishing a Control Program

In a responsible control program, it is essential to know as much about causative factors as corrective action. Microbiological evaluation is achieved through sampling and identifying organisms. Preventive agents selected should control the organisms without any adverse effects of corrosion, fouling, or contamination in the system.

Many biocides, such as those listed in the table, are now available for the control of microorganisms.* They include oxidizing materials such as chlorine, hydantoin dichloride, hydrogen peroxide and sodium hypochlorite; the nonoxidizing materials such as dithiocarbamates; quaternary ammonium compounds; chlorinated phenols; bisthiocyanates; and heavy metals such as tin butyl oxides, organo silver and mercurials.

Biocides are effective in varying degrees, depending upon their application. The one selected should be most effective in controlling the organism at the lowest cost and concentration, with the least disturbance to the environment.

The residual effect of the toxicant system is very important. This is especially true where large volumes of water carrying the biocide are injected into disposal wells, open streams, or water floods for secondary oil recovery. In the last case, a key question is whether the biocide will show up in the recovered oil and affect refinery operations with respect to emulsions, catalyst life and ecological standards. Biocides considered ecologically safe where large volumes of water are being discharged include chlorine, hydrogen peroxide, formaldehyde, and acrolein.

Biocide Selection

In selection of biocides for toxicity, it is desirable to obtain samples of liquid containing the microorganisms. With iron bacteria, this is somewhat difficult as they usually attach themselves to piping and vessel walls. A small stream of the test liquid should be run continuously through filter paper for 24 hours. Usually enough bacteria will break away and be retained on the filter paper. The moist paper should be sealed in a sterile bottle for

* Also see, Controlling Biological Fouling in Cooling Systems, *Chem. Eng.*, Oct. 30, 1972, and Nov. 27, 1972.

later microscopic examination. Suppliers can then recommend toxicants most specific for the iron bacteria.

The liquid sample should also be analyzed for general bacteria contamination and for sulfate reducers. Then the sample is placed in nutrient solutions to which are added specific biocides in known concentrations. To measure the effect of the biocides, the samples are plated and counted after sufficient incubation.

After selecting a biocide, it should undergo field evaluation. Plate counts of the liquid after sufficient application time will indicate the field effectiveness of the biocide. Monitoring on a frequent basis is important to ensure that the organism is not building resistance to the toxicant system. The most effective biocides today are those that result in the least number of resistant strains, and at the same time exhibit a broad spectrum of biological kill.

Adverse Chlorine Reactions

In treating a well system loaded with iron contamination, indiscriminate use of chlorine can result in plugging. This happens when the powerful oxidizing effects of chlorine precipitate ferric iron after raising it from the ferrous (soluble) state to the ferric (insoluble). The acid effects of chlorine can also corrode the well casing and piping. Further, if chlorine is not sufficiently rinsed from a well, it can cause serious effects on animal life.

If sufficient sulfide-forming bacteria are in a well, chlorine can also oxidize the sulfides to elemental sulfur. The result can be serious, permanent plugging. (With iron fouling of the formation, acidizing can restore production. But if the formation is fouled with colloidal sulfur, acid will have no effect.)

Chlorination in petroleum-refinery water floods also presents a problem for the refinery handling the recovered petroleum. Because of chlorine's oxidizing properties, halogenation of unsaturated aliphatics is possible. This produces a crude with heavier viscosity and introduces the chloride ion, which is detrimental to platinum-catalyst life.

The reactions of hydrogen peroxide with ferrous ions are similar to those of chlorine. Also, hydrogen peroxide is unstable and relatively much more expensive than chlorine. It will not have an adverse effect on petroleum, however, nor will it create as corrosive an environment or affect pH as much as chlorine.

Formaldehyde has been used with limited success. It lacks the effectiveness of either chlorine or hydrogen peroxide at economical levels and does not have a broad spectrum of biological activity. Laboratory toxicity stud-

Typical Organic Biocides

Oxidizing Types (readily detoxified)	Comments
Dichlorodimethyl hydantoin	Slow chlorine donor. Very effective against bacterial slime.
Chlorinated isocyanurates	Fast chlorine donor. Very effective against bacterial slime.
Calcium hypochlorite	Fast chlorine donor. Very effective against bacterial slime.

Nonoxidizing Types	
Acrolein (can also be readily economically detoxified)	Competitive with chlorine. Excellent where chlorine and/or nonoxidizing biocides are ineffective.
Chlorinated phenols e.g., pentachlorophenol and 2, 4,	General-purpose biocides. Effective against fungi and sulfate reducers.
Quaternary ammonium compounds e.g., n-alkyl dimethylbenzyl ammonium chlorides	Relatively ineffective against fungi. Good against algae and sulfate reducers.
methylene bisthiocyanate	Very good against bacterial slime. Do not use under high-pH conditions.
Hexachlorodimethylsulfone	Excellent against bacterial slime.
Alkyl dithiocarbamates, Alkylene bisdithiocarbamates	Good general biocides. Effective in high-pH systems.

ies have shown formaldehyde to be only mildly effective in controlling sulfate reducers.

Advantages of Acrolein

The most recent biocide of great economic value is acrolein, which exhibits a wide range of effectiveness. It functions chiefly as a reducing agent and disrupts the enzyme system in living organisms, specifically the sulfurylase and hydrogenase enzymes.

The ecological advantage of acrolein is the ease of detoxifying it. In an oil-water system, acrolein is preferentially removed with the hydrocarbon. If the discharge water reaches fresh-water fish, oysters, or other marine life, detoxification will be necessary. The most effective method is with sodium bisulfite, using a ratio of three parts sodium bisulfite to each part of acrolein. The reaction is believed to be as follows:

$$NaHSO_3 + CH_2{=}CHCHO \longrightarrow CH_2{=}CH{-}CH(OH){-}SO_3Na$$

Acrolein has had broad application and success in water floods, disposal wells, recirculating cooling water, once-through cooling water, sewer systems, and other water systems. Microorganisms do not usually resist or develop tolerance to it. The effective concentration level has been from 0.1 ppm. to as high as 30 ppm. in a high-sulfate-bacteria water-flood application. Acrolein is effective against sulfate reducers, but it does require higher residuals to offset the chemical reaction of the sulfur compounds on the acrolein molecule.

The other biocides previously mentioned may be used if the discharge water is effectively controlled and the biocide is detoxified by chemical or biological oxidation. Biological oxidation can detoxify phenates, quaternary ammonium, carbamates, thiocyanates, and other organic sulfur biocides, provided there is sufficient dilution and adequate retention.

The practice of industry is to reuse cooling water by processing it through lime-soda clarifiers or similar systems. The water may still contain some residual biocide as well as corrosion inhibitors. Since the system is closed, however, no contaminants are released to the environment.

Bibliography

1. Ernest Beerstecher, Jr., "Petroleum Microbiology," Elsevier Press, Inc., Houston, Tex. 1954.
2. Robert A. Boehler, Role of Microorganisms in Industrial Corrosion, *Heating Piping and Air Conditioning*, Dec., 1969.
3. G.H. Booth and A.K. Tiller, "Polarization of Studies of Mild Steel in Cultures of Sulphate Reducing Bacteria." Transcript of the Faraday Society, 56 (11): 1689-1695, Nov., 1960.
4. L.L. Campbell and J.R. Postgate, Classification of the Sporeforming Sulfate Reducing Bacteria. *Bacteriological Reviews, Sept., 1965, 29: 359-363.*
5. C.K. Cloninger and Paul H. Krunperman, Microbiological Energy Transfers Related to Electro-Biochemical Corrosion, paper presented at Symposium on Microbiology of Petroleum. American Petroleum Inst., Long Beach, Calif., Dec. 4, 1962.
6. J.M. Donohue, A.J. Piluso and J.R. Schieber, New Type Biocide Produces Cost Savings in Slime Control, 22nd Annual NACE Conference, 1966.
7. Ted S. Felmann, Evaluation of Bactericides for Treatment Against Bacteria in Water Floods, paper presented at Fourth Symposium on Treatment and Control of Injection Waters. American Petroleum Inst., Anaheim, Calif., Dec. 1, 1970.
8. M.G. Fontana and N.D. Green, "Corrosion Engineering," McGraw-Hill Book Co., New York, 1967.
9. B.J. Mechalas and S.C. Rittenberg, Energy Coupling in Desulfovibrio desulfricans, Journal of Bacteriology, 80 (4): 501-507 (Oct., 1960.
10. Edith F. Schrot, The Effect of Growth Rate on the Energy Metabolism of Desulfovibrio Desulfricans, thesis, St. Bonaventure University, St. Bonaventure, New York, 1968.

Meet the Author

George A. Woods is Pittsburgh District Manager, BETZ Laboratories, Inc., Somerton Rd., Trevose, PA 19047. Before joining BETZ in 1954 as district manager for the Salt Lake City area, Mr. Woods was a high-school science teacher, and an instructor at Arizona State University. He received his B.S. and M.S. degrees in chemistry from Arizona State, and did gratuate work at the University of Minnesota. He currently conducts water-treatment seminars at Bakersfield College and Los Angeles College.

Section 5
TERTIARY TREATMENT METHODS

Treatment of industrial wastewaters

Although biological treatment is still the most widely used
procedure in handling wastewaters prior to discharge,
it cannot cope with many specific industrial pollutants.
Recent developments for treating such wastes include
solvent extraction, membrane processes
(reverse osmosis, ultrafiltration and electrodialysis)
and chemical oxidation.

Thomas J. Mulligan and *Robert D. Fox*, Hydroscience Associates, Inc.

☐ Better removal of chemicals from wastewaters was
mandated by Congress in the Water Pollution Control
Act of 1970 (P.L. 92-500). The first milestone in that
mandate is implementation of "best practicable technology" by 1977. That need, plus additional milestones
in 1983 and 1985, are catalyzing the development and
demonstration of improved methods for the removal of
chemicals from wastewaters.

The performance requirements for the various treatment technologies are being raised to new, higher
standards by the advent of legislation (and effluent
standards) aimed at regulating the amounts of toxic
and hazardous substances discharged to the nation's
waterways. Recent legal and enforcement efforts by
various government agencies to define toxic and hazardous substances give evidence of the demands that
will be placed on pollution-control technology. The
trend in effluent standards is definitely away from the
broad, nonspecific parameters (such as chemical oxygen
demand or biochemical oxygen demand), and toward
limits on specific chemical compounds.

This trend is exemplified in the listings of toxic pollutants issued or proposed in recent years. The one most
familiar occurs in the OSHA Standards [1] which identifies the 15 compounds shown in Table I as being
"Cancer Suspect Agents." Control of these chemicals is
required through every stage of their manufacture and
use. More chemicals can be expected to be added to this
list as testing for carcinogens continues.

Another list of toxic pollutants was the subject of a
recently signed consent decree, in U.S. District Court,
between the EPA and environmental groups [2]. Under
the agreement, the agency will write restrictions on the
discharge of 65 chemicals for 21 industrial-discharge

categories. Some of the more prominent organics listed
are included in Table II. The door was left open for
additions to or deletions from this list as new evidence
becomes available. In the same decree, standards were
also re-proposed for manufacturers and formulators of
aldrin/dieldrin, DDT (DDD, DDE), endrin, toxaphene,
and polychlorinated biphenyls.

A third list can be found in the proposed regulation
designating 306 substances thought to be hazardous in
natural waters [3].

Our purpose in this article is not to debate the relative merits of the regulations or the toxicity of the

---

Agents suspected of causing cancer [3]   Table I

4-Nitrobiphenyl
α-Napthylamine
4,4'-Methylenebis(2-chloroaniline)*
Methylchloromethyl ether
3,3'-Dichlorobenzidine (and its salts)
Bis(chloromethyl) ether
β-Naphthylamine
Benzidine
4-Aminodiphenyl
Ethyleneimine
β-Propiolactone
2-Acetylaminoflorene
4-Dimethylaminoazobenzene
N-Nitrosodimethylamine
Vinyl chloride

*Deleted by legal action.

Originally published October 18, 1976.

Glossary

BATEA	Best available technology economically achievable
B.E.S.T.	Basic extractive sludge treatment
BOD	Biochemical oxygen demand
BOD_5	Five-day biochemical oxygen demand
COD	Chemical oxygen demand
DDD	Dichlorodiphenyldichloroethane
DDE	Dichlorodiphenyldichloroethylene
DDT	Dichlorodiphenyltrichloroethane
DO	Dissolved oxygen
ED	Electrodialysis
EPA	U.S. Environmental Protection Agency
IX	Ion exchange
MBAS	Methylene-blue-active substance
MLSS	Mixed-liquor suspended solids
OSHA	Occupational Safety and Health Administration
PAC	Purified activated carbon
PCB	Polychlorinated biphenyl
RBC	Rotating biological contactor
RO	Reverse osmosis
TDS	Total dissolved solids
PA	Polyamide
TOPO	Tri-n-octylphosphine
UF	Ultrafiltration

Some organics in the EPA, NRDC consent order   Table II

Benzene	Isophorone
Chlorinated benzenes	Nitrophenols
Chlorinated ethanes	Nitrosamines
Chloroalkyl ethers	Phenol
Chlorinated phenols	Phenolate esters
Dichloroethylenes	Polychlorinated biphenyls
Ethylbenzene	Polynuclear aromatics
Haloethers	Tetrachloroethylene
Halomethanes	Toluene
Hexachlorobutadiene	Trichloroethylene
Hexachlorocyclohexane	Vinyl compounds
Hexachlorocyclopentadiene	

(Abridged, see Ref. 6 for details)

chemicals but to illustrate the direction that pollution control-technology must take to be responsive to the national needs. We want to highlight new, improved technology for water pollution control, and this review will focus on several aspects of this new, improved technology. Some will be truly "new" technology, new ways of dealing with old problems; some will be new applications of established technology or modifications of established technology that extend its usefulness.

Another important aspect we want to discuss is new, improved methods of evaluating water pollution control technology. These methods involve advances in the mathematical modeling of treatment systems and computer optimization and new methods of chemical analysis. Finally, the aspect of recovery versus destructive treatment has taken on new perspective in recent years due to the escalation in the value of chemicals, especially those derived from petroleum. Therefore, several new developments in recovery will be reviewed.

This increased emphasis on recovery and the trend toward specific-compound discharge regulations have created the need for new approaches to solving pollution problems. When the recovery of chemicals is considered and the need to reduce the discharge of chemicals is evaluated, it becomes expedient to consider capture and recovery of treatment of the chemical at the source in an undiluted form. Successful implementation of source control and recovery requires specific-compound analysis employing sophisticated analytical procedures such as gas chromatography, high-performance liquid chromatography, ultraviolet and infrared spectroscopy, and gas chromatography/mass spectrometry [4]. These same procedures will be required to identify and quantify specific compounds in wastewater discharges.

This article will focus on three major areas of water pollution control technology—biological processes, physical separation processes, and oxidation processes.

Biological treatment

Today, the more common wastewater treatment procedure is biological, whereby concentrated masses of microorganisms break down organic matter, resulting in the stabilization of organic wastes. These microorganisms are broadly classified as aerobic, facultative, or anaerobic:

■ Aerobic organisms require molecular oxygen for metabolism.

■ Anaerobes derive energy from organic compounds and function in the absence of oxygen.

■ Facultative organisms may function in either an aerobic or anaerobic environment.

The predominant species used in biological systems are known as heterotrophic microorganisms; these require an organic carbon source for both energy and cell synthesis. Autotrophic organisms, in contrast, use an inorganic carbon source, such as carbon dioxide or carbonate. The autotrophs may derive energy from the oxidation of inorganic compounds such as nitrogen or sulfur (chemosynthetic), or from solar energy (photosynthetic).

Basically, the biological processes involve either of two mechanisms to accumulate and store the biomass: as a flocculated suspension of biological growth known as activated sludge, which is mixed with the wastewaters, or as a biological film fixed to an inert medium, which the wastewaters pass over. The applied technology in the past has comprised essentially the activated sludge system, the trickling filter (a fixed film process), and the anaerobic digestion of waste organic solids. Recent advances in biological process-technology have primarily involved improvements in the basic technology, broadening of the applicability of the processes, and an advancement in the design, analysis, and operational procedures.

It should be noted that biological processes may have

limited application to specific industrial wastewater problems, especially those containing compounds presently classified as toxic substances. Cyclic organics, especially if halogenated, are resistant to biological degradation.

Microbial populations may be specifically adapted to certain compounds and successfully achieve oxidation. Examples of such compounds proven biodegradable under acclimated conditions are cyanide, phenol, formaldehyde, acrylonitrile and hydroquinone. The chlorinated organics, such as the pesticides, may prove toxic to a biosystem if discharged in high concentrations or under shock-load conditions. Complex cyanide compounds such as metallic cyanide complexes have been found highly resistant to biodegradation, even under strongly acclimated conditions. Slightly soluble components such as PCBs or the heavy metals can accumulate in a biological system through adsorption and bio-concentration and may reach levels inhibitory to the system.

Most organic industrial wastes are amenable to biological treatment. However, if specific toxic or inhibitory waste streams are present in the total stream, successful biological treatment will be hampered. Proper biological waste-treatment design procedures for complex industrial wastes (such as petrochemical, organic chemical, and pharmaceutical) should include characterization of the various waste streams contributing to the overall flow. Should treatability studies indicate the presence of an inhibitory or toxic component, procedures have been developed to determine the component or waste stream causing the problem, by using high-rate respirometry techniques [5]. The problem-causing stream can then be removed for separate treatment or pretreatment prior to the biological system.

Suspended growth processes

The activated-sludge process is a continuous system in which aerobic biological growths are mixed with wastewater and then physically separated by gravity clarification (or by air flotation). The concentrated sludge is recycled to the reactor to mix with the incoming waste. Oxygen is provided in a variety of ways; by diffused aeration, surface aeration, static mixers, etc., and may be introduced either as air, oxygen, or oxygen-enriched air. The waste product from the activated-sludge process is excess sludge, which must be ultimately disposed of.

As the concentration of organics in the remaining wastewaters decreases, the rate of biological removal also decreases, since the remaining organics are progressively more difficult to remove. In the treatment of industrial wastes, a significant fraction of the organics may prove refractory, i.e., nonbiodegradable. Thus, although the BOD removal may be excellent, the removal of COD may be low. Process evaluations prior to system design should center on the establishment of removal rates, oxygen requirements, sludge production, nutrient requirements (nitrogen and phosphorus), and solids settleability.

The activated sludge process is broadly applicable. It is attractive in the treatment of soluble organic wastes from many industries including food processing, meat packing, pulp and paper, refining, leather tanning, textiles, organic chemicals, and petrochemicals.

Pure-oxygen activated-sludge process

A major modification to the conventional air-aerated activated-sludge process has been the use of high-purity oxygen. The development of a pure oxygen waste-treatment system has advanced to the present wide availability of commercial systems, notably the Union Carbide Unox process, which incorporates a series of concurrent gas-liquid contacting stages in a sealed aeration tank. Some advantages of this system compared to conventional air-aerated systems are: lower land requirements, the ability to supply oxygen at high rates (making it amenable to high-strength wastes) and increased self-neutralization of highly alkaline wastes. It follows, therefore, for low-alkalinity wastewaters that increased neutralization costs may be incurred. A heat buildup will also occur in the enclosed basin. Ordinarily, biological treatment will operate well over the approximate temperature range of 15 to 25°C. Temperature buildup above this range will retard the biological process and at approximately 48°C cause failure.

The use of pure oxygen in activated-sludge systems requires somewhat different analytical and design procedures from those accepted for the conventional activated-sludge processes. With the covered basin, the gas phase is not vented continuously. Thus, carbon dioxide buildup occurs in the aeration tank, causing a drop in the pH of the mixed liquor and a reduction of the gas-phase oxygen partial pressure. Improved insight and evaluations of the covered-basin pure-oxygen treatment system has been obtained through the mathematical modeling [6] of the biological process. This model incorporates the interrelationship of the gas-transfer kinetics, the biological reactions, and the chemical reactions. Use of this analytical technique is extremely valuable and provides the necessary insight and information to estimate system performance. Hence use of these techniques provides for better system design. Application of the model to specific industrial applications [7] highlighted the relative importance both of wastewater alkalinity and pH, and of carbon dioxide production—when analyzing and evaluating the pure-oxygen system.

Nitrification-denitrification processes

Increasingly stringent effluent criteria now require the removal of the nutrients phosphorus and nitrogen prior to discharge to a natural receiving water. Activated-sludge systems have been successfully operated in a variety of configurations to obtain carbonaceous BOD removal, nitrification, and denitrification.

Nitrification is achieved by the autotrophic microorganisms: *nitrosomonas,* which converts reduced nitrogen (as NH_3) to nitrite (NO_2), and *nitrobacter,* which oxidizes the nitrite to nitrate (NO_3). The organisms require inorganic carbon for synthesis. Oxygen consumption approximates the stoichiometric oxygen requirement for the oxidation of NH_3 to NO_3 (4.57 lb of oxygen per lb nitrogen). Sufficient alkalinity must be available to offset the production of the nitrous and nitric acids.

The growth rate of the nitrifiers is significantly lower

than that of the heterotrophic bacteria used in the breakdown of carbonaceous organic matter. Thus in the presence of significant amounts of carbonaceous organic material, the nitrifiers are unable to compete successfully with the heterotrophs and cannot accumulate as significant populations. The process configuration may be a single- or a two-stage system. In the latter carbonaceous removal is achieved in the first stage and nitrification is done in the second. Interstage clarification allows segregation of the two types of bacteria. Single-stage BOD removal and nitrification is practicable when the organic loading is kept suficiently low (less than 0.2 lb BOD/(lb MLSS) (day). In effect, this implies maintaining a sludge age (mass of sludge in reactor/daily solids wasting rate) high enough for development and maintenance of the nitrifying bacteria.

A summary [8] of several industrial applications of biological nitrification shows successful treatment of a variety of wastes in single-stage systems. Organic loadings were all less than 0.2 lb BOD_5/(lb MLSS) (day). Sludge ages were greater than 20 days. Table III summarizes these results.

Denitrification is an anaerobic biological process in which the nitrite and nitrate forms of nitrogen are reduced to nitrogen gas. Typical facultative, heterotrophic bacteria found in activated sludges can cause denitrification. As these organisms require an organic-carbon source for cell growth, and because this process generally follows secondary treatment, some form of organic compound must be present or added. Approximately 4.5 lb of chemical oxygen demand (COD) per pound of nitrate are required to reduce the nitrates to nitrogen. Typically this is methanol. Suspended growth (mixed tanks) and fixed film (biological disk contactors or columns) are used.

Fixed-film processes

The predominant fixed-film biological process has, in the past, been the conventional trickling filter. It is a packed bed, of some medium covered with a biological slime, over which wastewater is passed. Oxygen and organic matter diffuse into the slime film, where oxidation and synthesis occur. End products such as CO_2, NO_3, etc., counter-diffuse back out of the film and appear in the filter effluent.

Trickling filters have found limited application to industrial wastes. Removal rates for soluble industrial wastes are typically low, making filters unattractive for high BOD-removal efficiency. The process is capable of accepting highly variable loadings and thus may be used as a roughing filter to provide partial treatment.

Rotating biological contactors (RBC)

The rotating disk process operates as a fixed-film biological reactor. Plastic disks are mounted on a shaft and placed in a tank conforming to the general shape of the disks. The disks are slowly rotated while approximately half immersed in the wastewater to be treated. Rotation brings the attached culture into contact with the wastewater for removal of organic matter. Rotation also provides a means of aeration by exposing a thin film of wastewater on the disk surface to the air.

The RBC units are typically operated in staged configurations. Already in wide use in Europe, they are gaining acceptance in the United States. Advantages demonstrated by the RBC system are low maintenance, low power-consumption, and stable operation. Advances have been made in nitrification and denitrification applications. The latter stages of a staged configuration receive low-strength-organic-content wastes, thus allowing a viable environment for the growth of nitrifiers. These stages are followed by fully submerged media where denitrification can take place.

Analyses of the fixed-film mechanism, with specific application to the RBC process, has been developed [9]. The studies associated with the model development indicate the strong influence of mass-transfer mechanisms on the efficiency of the biofilm processes. This is in contrast to the suspended-growth processes, where kinetic removal rates are the dominant factor.

Biological fluidized bed

In biological fluid-bed process, a proprietary system by Ecolotrol, Inc., wastewater passes upward through a

---

Industrial examples of biological nitrification   **Table III**

Waste source and strength	Influent BOD, mg/l	Influent Organic N, mg/l	Influent NH_3–N, mg/l	Loading rate, lb/day BOD / lb MLSS	Sludge age, days	Nitrate and nitrite production, lb N/day / lb MLSS	Average effluent NH_3–N, mg/l	Average effluent NO_3–N + NO_2–N, mg/l
I. Domestic wastes	150-220	12-18	15-20	0.1-0.3	6-18	0.02-0.03	0-1	12-20
II. Industrial wastes								
Organic acids	300	3-8	25	0.12	11-28	0.01	low	30
Pharmaceutical and organic chemicals*	250	80	25	0.05-0.1	>20	0.01	2	34
Synthetic fibers and polymers	1,000	80	low	0.15	>20	0.003	5	18
Synthetic fibers	5,000	700	50	0.17	30+	0.017	2	500+
Nylon	2,850	200	50	0.15-0.2	>10	0.012	2	230
Pharmaceutical and organic chemicals†	200	75	625	0.02	>20	0.05	50-100	300-400‡
Inorganic wastes	10	0	500-1,000	0	>30	0.2	<1	500-1,000

*Nitrification has also been demonstrated for refinery and meatpacking wastewaters.
†Second stage of two-stage system.
‡Predominantly NO_2–N

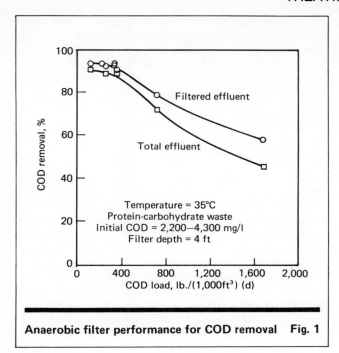

Anaerobic filter performance for COD removal   Fig. 1

A recent research study [12] demonstrated COD removals of 90% and 50% at loadings of 200 to 1,700 lb COD/1,000 (ft³) (d) respectively, at an operating temperature of 35°C. This performance for a protein and carbohydrate waste mixture is shown on Fig. 1. In order to properly design this system, it is important to be aware of the chemistry interactions [12] occurring in the filter among nitrogen, volatile acids, alkalinity, carbon dioxide, and pH.

In the treatment of industrial wastes, successful operation was reported for the pretreatment of high-strength waste streams from the organic acids, brewery, food processing, pharmaceutical, detergents, and paper and pulp industries.

Combined biological-carbon systems

Addition of powdered activated carbon (PAC) to aeration tanks of suspended-growth systems has been used to improve activated-sludge operation and performance [13]. The Du Pont PACT process involves a continuous addition of PAC to the aeration tank. A buildup in the aeration tank occurs (a function of the sludge retention-time), providing a substantial reservoir of carbon in the system. The fresh carbon continually added to the aeration tank gives the process the capability of adsorbing nonbiodegradable organic matter present in the wastewaters, thereby providing a degree of tertiary treatment. A recent study [14] indicated that PAC addition to activated-sludge systems provides exceptional resistance to shock-loading by trichlorophenol, presumably due to the large reservoir of carbon carried in the MLSS.

A study [15] completed on the organic-chemical wastewater from the Du Pont Chambers Works indicated that the PACT system with carbon regeneration by pyrolysis was economically preferable in achieving the desired effluent quality to columnar, granular and activated-carbon systems, both preceding and following activated sludge.

The Zimpro Wet-Air-Oxidation-Biophysical treatment process [16], schematically shown on Fig. 2, has

reaction vessel partially filled with fine-grained media. The velocity of the liquid is such that it fluidizes the bed [10]. On the medium's surface, a biological mass grows as a firmly attached active body that effectively consumes the waste as it passes. To date, activated carbon and sand have been used as support media. Fluidizing the medium allows solids to pass through and overcomes the problems associated with packed-bed systems, such as high head-losses, plugging, and frequent need for backwashing. The fluid bed offers a high available-surface area, since there is no contact between particles—intimate contact of the entire surface with the waste stream is assured. Detention times are short because of the very high biomass carrying-capacity of the system.

The process is essentially in the development stage and has not as yet been applied to an industrial waste. Extensive pilot studies, conducted on a municipal wastewater, have demonstrated the feasibility of using fluidized-bed technology for carbonaceous BOD removal, nitrification, and denitrification. The significant advantage to the system lies in the very low area-requirements. In the treatment of municipal wastes, 90% BOD, 99% NH_3—N, and 99% NO_3—N removals were reported for a reaction time of less than one hour. Land area requirements are reduced by approximately 80% [11].

Anaerobic filter

The anaerobic filter is basically an upflow column or tower filled with rock or a synthetic medium. Recent research on this process indicates it is usable for pretreatment of high-strength industrial wastes, including high-temperature wastes. Compared to aerobic systems, the anaerobic filter can accomplish significant organics removal at a low operating cost and with very little sludge production. It has a high capacity for highly variable or shock loadings.

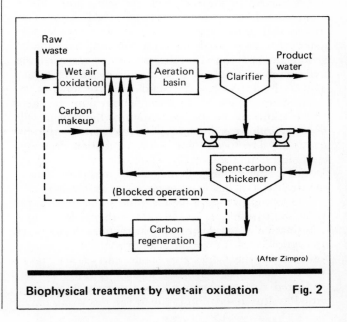

Biophysical treatment by wet-air oxidation   Fig. 2

been successful in the treatment of cyanide, acrylonitrile, and pesticide wastes. The incoming wastes are subjected to wet-air-oxidation, where the large molecules and cyclic compounds are broken down to intermediates such as aldehydes and organic acids. These wastes are more amenable to the biological treatment system that follows. The process is modified to include the addition of powdered activated carbon. The presence of the carbon is reported to serve as a "toxic sink" for substances that may have passed wet-air oxidation. The components are adsorbed and pass out on the surface of the carbon. The combined biological sludge and carbon are treated by the wet-air-oxidation process, where the sludge is reduced to a small fraction of its original weight, after which the carbon is reactivated and returned to the aeration tank.

Adsorption processes

Adsorption processes, and in particular those using activated carbon, are finding increased use in wastewater treatment for removal of refractory organics, toxic substances, and color. The use of activated carbon is expected to increase further in response to the 1983 EPA effluent guidelines, which call for the application of "Best Available Technology Economically Achievable" (BATEA) in wastewater treatment. The effluent quality specified in many 1983 BATEA regulations is not achievable with biological treatment. Some degree of advanced treatment will be required and, at present, carbon adsorption is the most economical and technically attractive method available.

Adsorption theory

In adsorption, dissolved substances in the wastewater are attracted to and adhere to a solid adsorbent surface. The primary driving forces for adsorption are a combination of the (1) hydrophobic (water disliking) nature of the solutes (dissolved organics) in the solvent (wastewater) and (2) the affinity of the solute for the solid.

The solubility of the dissolved organics is the property most directly related to hydrophobic behavior—organics of low solubility are most readily adsorbed. Since the water solubility of organics generally decreases with increase in chain length within a homologous series, adsorption is greater for the higher the molecular weight within the series. For example, for the organic acids: formic, acetic, propionic, and butyric acid; adsorption increases in the order listed.

The degree of ionization of the solute also affects adsorption. Ions are very poorly adsorbed as are polar solutes, such as the sugars. Surface-active agents, such as detergents, have both hydrophobic and hydrophilic groups and are termed amphoteric. When adsorbed, the hydrophobic part of the molecule adheres to the surface and the hydrophilic part extends into the solution.

Since adsorption is a surface phenomenon, the effect of the solute on the surface tension of the solvent will have an influence. It can be shown, from thermodynamic considerations, that if surface tension decreases upon addition of solute, the solute will tend to migrate toward a solid surface where it can be adsorbed. This is characteristic of dissolved organics in wastewater.

The second driving force for adsorption, the affinity of the solute (sorbate) for the solid (sorbent), is due to a combination of (a) electrostatic attraction, (b) physical adsorption, and (c) chemical adsorption. Electrostatic forces occur when sorbate molecules carry a net charge and are attracted to surface sites of opposite charge. Activated carbon commonly has negatively charged sites. Physical adsorption is due to van der Waals forces of attraction and occurs at low temperatures with low energies of adsorption and weak surface bonds. The sorbate is not fixed to any specific surface site but is free to move over the surface. Chemisorption occurs from chemical interaction at specific sites, is normally characterized by strong bonds and high adsorption energies, and is favored by high temperatures. Most adsorption phenomena are combinations of the above three types.

An effective adsorbent must possess high surface-area per unit weight. Activated carbon, the most effective adsorbent used in wastewater treatment, has this property, possessing on the order of 1,000 m^2 of surface area/g. It can be manufactured from a variety of carbonaceous materials, but the present tendency in wastewater treatment is to use carbons from coal and coke because of lower cost and better availability.

Adsorption isotherms

The quantity of sorbate that can be adsorbed by a given weight of activated carbon is related to the concentration of this sorbate in the external liquid phase. Adsorptive capacity increases with liquid-phase concentration. An adsorption isotherm is a relationship at constant temperature between the equilibrium concentration of sorbate in solution, c, and the quantity adsorbed per unit weight of carbon, q. Isotherms are determined experimentally by contacting weighed amounts of carbon with wastewaters over a period of time sufficient to achieve equilibrium. The resultant data of q versus c is commonly correlated by use of either the Langmuir equation or Freundlich equation. An example of an isotherm is shown in Fig. 3 for a pure sorbate, phenol.

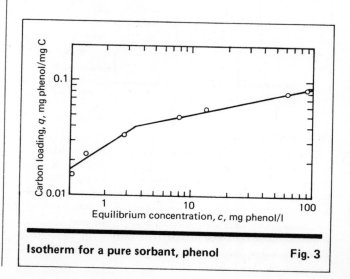

Isotherm for a pure sorbant, phenol   **Fig. 3**

The Langmuir isotherm equation is based on a model that assumes sorbate can attach to the sorbent surface in, at most, a single layer of molecules, regardless of how high the sorbate concentration is in the liquid phase. A uniform energy of adsorption is also assumed. The Freundlich equation corresponds to a variation of adsorption energy with quantity of sorbate adsorbed.

Types of materials adsorbed

As described above, the adsorption of organic chemicals is markedly influenced by a variety of factors relating to the character of the organic molecule, the adsorbent, and the conditions of contact. For organic molecules, factors such as functional groups, molecular weight, and side-chain branching affect adsorption. Surface chemistry and treatment, pore size, pore distribution, particle size (and, for activated carbon, the source of carbon) are important adsorbent properties. Temperature, pH, and contact time are important conditions affecting adsorption.

Much has been published on the effect of these parameters on adsorption of organics; Guisti and coworkers [16] have compiled data evaluating several of these parameters for activated carbons, at equilibrium conditions, on 93 petrochemicals. Recently published data [17] on treating dye wastes give further evidence on the importance of careful evaluation of adsorption performance. For example, adsorption at higher temperature had no significant effect on COD removal, but

Process configuration used for carbon columns   Fig. 4

color removal was increased almost 100%; a lignite-source activated carbon was reported to perform better than a bituminous source for high-molecular-weight (above 700) dye-waste compounds.

In general, activated carbon has been found to adsorb a wide variety of organic chemicals. Only the smaller organics, C_1—C_2, and large complex molecules, adsorb poorly, or not at all. In between are a broad range of chemicals that can be removed from water by activated carbon. Many of the "toxic" organic pesticides have been tested and found to be adsorbable [18], but at the low concentrations desired in a waste effluent (<0.1 ppm) the loading is only around 1%.

Synthetic adsorbents are reported [19] to produce better effluent quality than activated carbon for certain organic compounds. In actual waste treatment applications, so many factors can affect the small differences reported that relative performance advantages need to be better defined.

In assessing the applicability of adsorption as a tertiary treatment for the removal of organic chemicals from wastewater, recent data are available [20] from continuous activated-carbon-column studies that show the importance of adsorption kinetics and competitive adsorption effects. These tests on effluents from a variety of organic-chemical plants show the possible problems associated with treating complex mixtures by adsorption. The need for specific analytical characterization of wastes to be treated is evident, as is the need for source control of wastes in more-concentrated, less-complex mixtures.

Design of carbon adsorption systems

Carbon adsorption systems consist of columns loaded with granular activated carbon. Flow can either be down (fixed-bed operation) or up (expanded-bed operation). Loading rates are typically within the range of 2–5 gal/min per ft² of bed cross-section. The columns can be arranged in several different process configurations (as shown schematically in Fig. 4):

In parallel operation, each column receives the same feed. However, the columns are placed onstream at staggered time-intervals to ensure that each is at a different level of carbon exhaustion. The effluents from each column are combined to form a composite product. When a column becomes saturated, as evidenced by breakthrough of sorbate, the carbon in the column is replaced by virgin or regenerated carbon. With a sufficient number of columns in parallel, a combined high-quality effluent can be obtained even though that from an individual column may contain high concentrations of sorbate.

When columns are arranged in series, the lead column is removed from service when its carbon is exhausted. In a properly designed system, the effluent from the last column will still be of high quality at the time the lead column becomes saturated. After removing the lead column from service, a fresh column is placed at the end of the series and the second column becomes the lead column. (In practice this is accomplished through complex piping and valving.)

The pulsed-bed column is similar in concept to columns in series. However, instead of multiple columns, a

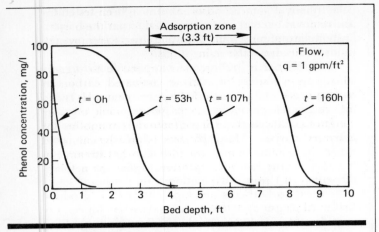

Phenol adsorption zones in a 2-in. column   **Fig. 5**

through data is illustrated in Fig. 6. The data were obtained in a 1.5-in. laboratory column charged to a height of 28 in. with 12×40 mesh carbon. The column was operated in the downflow mode at a loading rate of 10 gpm/ft^2.

Thermal regeneration techniques

Granular activated carbon—The conventional method for regeneration of spent activated carbon is to thermally oxidize the sorbed organics in a controlled atmosphere that minimizes the oxidation of the carbon itself. The process for thermal regeneration of granular activated carbon begins with mechanical conveyance of the spent carbon to a furnace (usually a multiple-hearth type, although rotary kilns have been used). In the furnace the carbon passes through zones where it is first dried and then where the organics undergo pyrolysis and oxidation in an oxygen-starved atmosphere. The hot furnace gases usually require a secondary combustion to complete the oxidation. If acid gases or salt particles are generated, scrubbing is required. The regenerated carbon is quenched in water and returned to the adsorber. Carbon losses occur by oxidation in the furnace and as fines caused by attrition; they generally range from 5 to 10% per cycle.

The major operating cost elements for thermal regeneration, shown in Table IV, are those for fueling the

single column is employed with provision for bottom removal of spent carbon and top addition of fresh carbon. By removing small amounts of carbon in periodic pulses, it is possible to achieve a high degree of saturation of the carbon with a minimum carbon inventory.

The design of activated-carbon adsorption systems requires a knowledge of the equilibrium capacity (isotherm) of the carbon and the overall rate of adsorption. The adsorption rate is dependent on transport of sorbate molecules, first from the liquid phase to the solid interface and then into the particle to internal-surface-sites of adsorption. These rate processes give rise to an adsorption zone within the column over which sorbate concentration is reduced from influent level to zero. When the adsorption rate is very low, a deep bed of carbon is required to provide sufficient detention time. Adsorption zones for phenol adsorption at a loading rate of 2 gpm/ft^2 are shown in Fig. 5.

Carbon adsorption systems can be effectively designed through the use of mathematical simulation models coupled with laboratory testing to obtain isotherm data and rate constants. We have developed a model that simulates the operation of adsorption columns and yields the necessary information for design.

This model incorporates the following system parameters. A mass transfer coefficient, k_f, is employed for transport of sorbate from the liquid flowing through the column to the outer surface of carbon particles. This coefficient is calculated within the model from equations on mass-transfer in packed and fluidized beds. These equations employ the sorbate diffusivity in water, the column loading rate, the average particle size, and the bed porosity. When sorbate diffusivity is unknown, it is determined experimentally in a small laboratory column.

Equilibrium is assumed between solid-phase concentration and liquid-phase concentration at the outer surface of the particle. An experimentally determined isotherm is employed. Transport of sorbate into the carbon particles is incorporated into the model. Transport is assumed to be proportional to the radial gradient in the solid-phase concentration.

The ability of the model to represent column break-

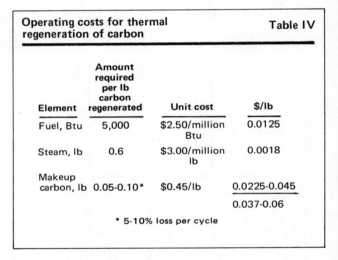

Operating costs for thermal regeneration of carbon   **Table IV**

Element	Amount required per lb carbon regenerated	Unit cost	$/lb
Fuel, Btu	5,000	$2.50/million Btu	0.0125
Steam, lb	0.6	$3.00/million lb	0.0018
Makeup carbon, lb	0.05-0.10*	$0.45/lb	0.0225-0.045
			0.037-0.06

* 5-10% loss per cycle

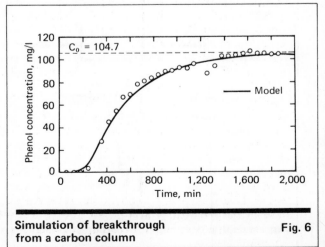

Simulation of breakthrough from a carbon column   **Fig. 6**

furnace and for supplying makeup carbon. Both fuel needs and carbon makeup costs are dependent on the ease of pryolysis and oxidation of the organics sorbed on the carbon. Difficult-to-oxidize organics will require longer furnace residence times, leading to higher-than-normal fuel consumption and larger carbon losses. Incomplete removal of the organics will lead to reduced adsorptive capacity of the regenerated carbon and more frequent regeneration. These parameters are difficult to evaluate in small-scale regeneration tests and feedback from full-scale systems must be relied upon.

In addition to these variable operating costs, thermal regeneration can require sizable capital investment and operating labor costs. Despite this, numerous plants have installed activated-carbon-adsorption/thermal-regeneration-systems in recent years. A recent addition to the options available to plants considering activated carbon treatment of their wastes is the Calgon Adsorption Service. For a monthly service fee, Calgon will install, maintain, monitor, operate, and own an adsorption system on the client's site. The spent carbon is removed from the system and replaced with a fresh load of regenerated carbon. The spent carbon is trucked to a central regeneration furnace, regenerated, and then returned to the client, along with enough makeup carbon to refill the system. Such a service will appeal to a plant short on capital or with a low rate of carbon usage.

Powdered activated-carbon—Successful and economical regeneration of powdered activated-carbon continues to be the goal of development activities in waste treatment operations where it is used. Recent articles [21,22] have described the advances in powdered-carbon regeneration by a thermal transport reactor (cocurrent fluidized bed) and by wet-air oxidation.

In-place, nondestructive regeneration

As the name suggests, in this technique for regeneration of adsorbents the adsorbent is not removed from the adsorption vessel, and the adsorbed chemicals are removed by procedures that do not destroy them. By avoiding physical transport of the adsorbent, losses due to attrition are avoided. Nondestructive regeneration also offers the opportunity to recover valuable organics that have been adsorbed.

There are three general types of in-place, nondestructive regeneration: pH change, solvent, and steam. The latter is most familiar as it is the regeneration method used in the recovery of solvent vapors from air streams, by adsorption of activated carbon followed by steam regeneration.

Regeneration by use of pH change

The pH change method involves adsorption of an ionic organic chemical as the free acid or base, and elution of it from the adsorbent as the ionized salt by using an alkaline or acidic regeneration stream. This procedure has found full-scale application in the recovery of phenols and cresol from process waste streams [23–26]. Doubtless, other applications exist. An application to the removal of acetic acid from a waste sodium chloride brine had high costs and was sensitive to feed conditions [24].

Synthetic polymeric adsorbents have been tested for the removal of color bodies and BOD from the bleaching effluents of pulp mills. Caustic is used to remove the color bodies from the resin [27].

The pH-change technique is applicable to organic chemicals whose adsorption on activated carbon or synthetic adsorbents is affected by change in pH. It can be used to selectively adsorb and desorb ionic organics and to separate one type of compound from another for recovery purposes. Its usefulness is greatly enhanced when the resultant concentrated regenerant stream can be directly recycled in the process, so that the net consumption of chemicals is zero.

Solvent regeneration

Solvent regeneration of adsorbents involves the use of a solvent phase to desorb the organic chemicals from the adsorbent. The solvent is then removed (as in conventional solvent-recovery systems) by steam. This solvent regeneration technology has received increased attention recently. The process involves carrying out a sequence of conventional unit operations common to the chemical process industries.

The only requirement is that the organics to be removed be soluble in a common solvent. Solvents such as methanol, acetone, and benzene have been used. A noteworthy aspect of this technology is that water-miscible solvents can be used to, in essence, separate organics from water. Evaluation of the effectiveness of solvent regeneration can be evaluated initially by means of isotherm tests.

The generalized process is shown schematically in Fig. 7. The saturated carbon bed is contacted with a

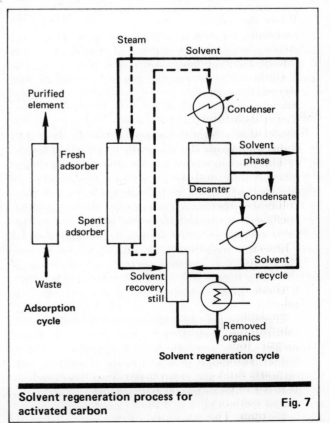

Solvent regeneration process for activated carbon   **Fig. 7**

quantity of whichever solvent regenerant has been found to be most effective. The solvent regenerant containing desorbed organics is distilled to recover solvent, leaving an organic concentrate suitable for recovery or incineration. When a solvent already employed in the process is used to regenerate, added equipment for distilling the solvent may not be needed. The carbon, now saturated with solvent, is restored to adsorptive capacity by steaming off the solvent. The steam-plus-solvent vapors are condensed and separated by conventional techniques. With solvent regeneration, the cycle of adsorption-desorption can be repeated many times before the carbon has to be replaced. Typical costs factors for this process are as shown in Table V. Similar steam and solvent costs have been reported for synthetic polymeric adsorbents [28]. Over a wide range, the costs are nearly independent of organic concentration in the waste.

Applications for solvent regeneration

This solvent regeneration technology is currently being evaluated for recovery of drug intermediates [29] with activated carbon, and is being designed into two systems for the removal of problem organics from wastewaters [30,31]. These latter two applications involve the use of synthetic polymeric adsorbents. In one case [30], a system for removal of toxic chemicals consists of filtration, followed by activated-carbon adsorption beds followed by polymeric adsorbent beds. The spent carbon is burned, whereas the polymeric adsorbents are regenerated with methanol, which is then burned.

Typical costs for solvent regeneration of carbon		Table V
	Use rate, lb/lb adsorbent	Typical cost, $/lb adsorbent
Fuel	0	0
Steam @$3/1,000 lb	2	0.006
Solvent makeup @10¢/lb @1.0% loss	0.07	0.007
Carbon replacement @45¢/lb after 100 cycles	0.01	0.0045
		0.0058

A variation on the solvent regeneration process is being installed to recover the organic byproducts in the foul condensate from a Wisconsin pulp mill [32,33]. In this application the adsorbed organics (chiefly furfural and acetic acid) are removed from activated carbon by hot ethanol vapors. Acetic acid reacts, on the carbon, with ethanol vapor to form ethyl acetate, which is then swept out of the bed. The resultant regenerant, containing furfural, ethyl acetate, ethanol, and some acetic acid is fractionated to recover ethanol (for reuse) and the byproducts (to be sold).

These emerging techniques of in-place, nondestructive regeneration offer new opportunities to use the organic removal properties of adsorbents to recover chamicals from process wastewaters and help meet water quality criteria.

Solvent extraction

Recent developments in the application of solvent extraction processes to the removal of organic chemicals from waste effluents have occurred in two areas: new applications for conventional extraction systems and new extraction systems.

New applications

Relatively new from the standpoint of the availability of the technology is a more efficient process for the recovery of phenols from aqueous wastes. Extraction of phenols has long been practiced in the treatment of refinery wastes [34], where phenols are extracted by a light-oil refinery stream or by a solvent such as benzene. The phenol is recovered from the extract as sodium phenate by back-extraction with caustic soda. However, these processes require a high solvent-to-waste ratio, usually around 1:1, and achieve removal efficiencies of only 90–99%, insufficient for waste treatment applications where direct discharge of the treated waste is desired.

By employing a proprietary solvent having a very high distribution-coefficient for the phenols in coke-plant waste, Jones and Laughlin Corp. [35] was able to develop an extraction process having a very low solvent-to-waste ratio of 1:18. Phenol removal efficiencies are reported to be in the 99.7–99.9% range, producing an effluent containing 1 to 4 ppm from a feed containing 1,500 ppm. The schematic of this process, Fig. 8,

resembles any conventional solvent-extraction process. Its major unit operations consist of a continuous countercurrent extractor, two distillation columns to separate the solvent from the recovered organics, and a steam-stripping column to remove soluble solvent from the raffinate.

In conventional extraction processes, the principal operating costs are for energy and makeup solvent. The principal energy use is steam for the distillation operations, and the amount is highly dependent on the concentration of the organics in the waste and the ease with which the solvent can be separated from the organic extract and the aqueous raffinate. Similarly, solvent makeup to replace losses is a function of the amount of solvent required in the solvent circuit. Solvent losses occur due to residual solvent in the raffinate stripper bottoms, residual solvent in the recovered organics, escape from vents, decomposition, and leakage.

A cost analysis for the recovery of a compound such as phenol from aqueous wastes has been developed based on use of a solvent with a high distribution coefficient so that the solvent/waste ratio was low (0.12–0.06) and also one that was relatively easy to strip from the aqueous raffinate. The steam and solvent makeup costs for this hypothetical system are shown in Table VI. These cost data show that conventional extraction systems are very sensitive to feed concentrations. In addition, it is not difficult to discern the cost impact of solvent price and the amount that must be used. For the example of phenol, where solvents with high distribution coefficients are available, extraction can be an economical process for recovery at higher concentrations.

Not included in this cost analysis was the impact of capital-related costs, labor, and maintenance on the system. Obviously, these costs will significantly affect recovery costs, especially for small systems. Also, not included was the cost of additional treatment of the aqueous raffinate. It is unlikely that solvent extraction alone will reduce the concentration of organics, especially toxic substances, to levels where direct discharge will be permissible. Some form of polishing treatment will be required.

New extraction systems

Additional applications for recovery of chemicals from waste streams are emerging as a result of research and development on chemical extractants with unique properties. These new chemical extractants have the ability to separate (from water) both organic and inorganic chemicals that are poorly, if at all, separated by conventional solvents.

Several examples of these new extraction systems have been described recently, such as the Hydroscience process for recovery of acetic acid and other organic acids by extraction with tri-n-octylphosphine oxide (TOPO) [36]. Acetic acid is recovered by fractionation of the extract with the more volatile acetic acid taken overhead as distillate. The TOPO extractant is dissolved in a carrier solvent. Because TOPO and the carrier solvent are both virtually insoluble in water, an aqueous raffinate steam stripper is not needed. This process makes recovery of acetic acid more economical than extraction with conventional solvents, because of its lower energy requirements.

Amines are another class of extractant chemicals finding increased use in separation and recovery applications. Use of organic amines to shift chemical equilibria reactions by chemically tying up acid halides and to chemisorb hydrogen fluoride from dilute aqueous solution is described by Bailes and coworkers [37]. The same article describes pilot-plant work on the use of tributyl phosphate in kerosene to recover nitric acid and HF from stainless-steel pickling solutions.

An amine extraction process has also been developed [38] for the recovery of cyanide and metal cyanides from plating waste streams. The process uses high-molecular-weight quaternary amines, dissolved in a carrier solvent, to extract zinc and cyanide. The chemicals are then recovered for recycle by chemical stripping with caustic.

An interesting aspect of these new extraction processes is that developmental success hinged on the use of a

Effect of phenol concentration on costs of recovery by means of solvent extraction		Table VI
A. Energy consumption		
Phenol concentration, %	Steam consumption, lb/lb phenol	$/lb phenol
1.0	6	0.018
0.1	50	0.15
0.01	500	1.50
B. Solvent Losses @ 0.15%		
	Solvent loss, lb/lb phenol	$/lb phenol
1.0	0.02	0.004
0.1	0.09	0.018
0.01	0.9	0.18

carrier solvent for the chemical extractant. Carrier solvents can reduce the solubility of the extractant in the aqueous raffinate, improve phase-density difference, and reduce viscosity and freezing point.

Another new process that takes advantage of the unique extractant properties of amines is called Basic Extractive Sludge Treatment (B.E.S.T.) [39]. The process takes advantage of the ability of aliphatic amines to extract water at one temperature and release it at a higher temperature. The process removes essentially all of the water and oils from inorganic and organic sludges. It requires an extractor, separation of the solids, heating of the extract to cause separation, stripping of the water layer to recover amine solvent, and distillation of the solvent to separate solvent and oils. The process converts sludges to nearly dry solids and water (plus recovered oils, if any are present).

The increased emphasis on solvent extraction in wastewater treatment is exemplified by the variety of solvent extraction evaluations that have been and are

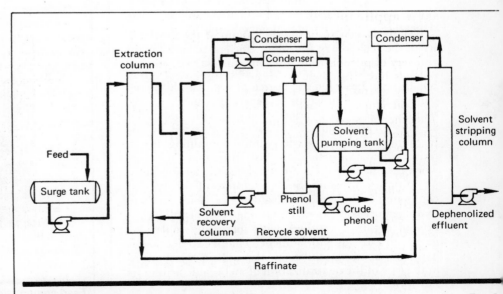

Solvent extraction of phenols from coke waste   Fig. 8

being funded by the Robert S. Kerr Laboratory of the EPA. Applications include solvent extraction using volatile solvents under pressure for the purification of wastewaters from the manufacture of acetic acid, caprolactam, acrylonitrile/acetonitrile, and various chlorinated hydrocarbons.

With the increased value of chemicals and the more stringent discharge regulations, it is certain that chemical processes will use new and improved solvent-extraction operations to achieve greater removal of products and byproducts from effluents than has been heretofore practiced. The use of solvent extraction will be principally as a recovery step applied to individual (or segre-

gated) process waste streams rather than as an overall removal method. As a source-control technology, it will generally be a roughing treatment, capable of high degrees of chemical removal from wastewaters, but requiring a polishing treatment to assure compliance with discharge criteria. As a physical separation method to remove chemicals from wastewater, its successful application in environmental control will depend heavily on knowing the fate throughout the process of all the chemical constituents in the waste. This is another example of how new technology in environmental-control engineering depends on sophisticated analytical procedures.

Membrane processes

The three membrane processes [40] that are finding increased interest in pollution-control applications as end-of-pipe treatment and for inplant recovery systems are ultrafiltration (UF), reverse osmosis (RO) and electrodialysis (ED).

Although various processes are available for producing a concentrated solution from a relatively dilute feed, each process tends to dominate in a particular region of application due to economic as well as technological considerations. Fig. 9 indicates these approximate regions. As seen, ion exchange (IX) and reverse osmosis overlap somewhat in the lower concentration ranges while electrodialysis has greater application in the intermediate range and distillation is most effective in the upper range.

Reverse osmosis and ultrafiltration

There is no sharp distinction between UF and RO. Both processes remove solutes from solution much as a conventional filter separates suspended matter from a liquid suspension. Unlike a filter, however, the separated phase does not accumulate on the surface of the medium to any great extent but rather emerges in the form of a more concentrated solution while the permeate emerges as a diluted stream. In UF the separation is based primarily on the size of the solute which, depending on the particular membrane porosity, can range from about 2 to 10,000 millimicrons (nanometers).

In the RO process the size of the solute molecule is not the sole basis for the degree of removal, although 0.04 to 600 millimicrons is generally considered the useful range. Thus, nonelectrolytes with molecular weights greater than 200 tend to be rejected by the membranes. Aliphatic compounds containing up to four carbon atoms tend to pass through membranes when they have hydrogen-bonding characteristics similar in nature to water. Some examples of this type of compound include alcohols, aldehydes, acids, amines, hydrogen peroxide, urea, acetamide and phenol. Rejection of organics improves with increasing molecular length and size (as with branched chains). Reduction of hydrogen-bonding tendencies also aids in rejection, as in the case of the neutralization of an organic acid to its salt form.

The ability of RO membranes to reject electrolytes increases with an increase of valency. Thus, among

cations, calcium and magnesium are rejected to a greater extent than sodium, while for the anions, sulfate is rejected more than chloride. Free gases that tend to diffuse through the membranes fairly easily include carbon dioxide (pH $\leq$ 4), oxygen, sulfur dioxide (pH $\leq$ 3), chlorine (pH $\leq$ 7), and hydrogen sulfide. It thus appears that the basis for rejection is more complicated than that of a simple molecular sieve, but the exact mechanism is still in dispute.

To understand why the process has been called reverse osmosis, it is necessary to first review the process of ordinary or "forward" osmosis. If a vessel is divided into two compartments by means of a partition in the form of a semipermeable membrane (one that permits passage of the solvent but not the solute) and then one compartment is partially filled with a solution while the other compartment contains only the pure solvent, it will be observed that the level of the solution will start to rise, while the level of the pure solvent will fall by a corresponding amount. The change in levels has obviously occurred due to the diffusion of solvent into the solution. This natural tendency toward the equalization of concentrations is in accordance with the laws of

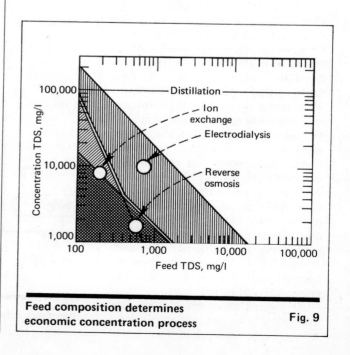

**Feed composition determines
economic concentration process**

Fig. 9

thermodynamics regarding the passage of solvent from an area of higher partial-free-energy to an area of lower partial-free-energy. This diffusion, or osmotic flow, will continue until the difference of levels on both sides of the membrane creates a sufficient pressure to counteract the net solvent flow. The system is then in dynamic equilibrium with the difference of levels representing the osmotic pressure of the solution. By applying a hydraulic pressure to the solution side, the natural tendency of osmosis can be stopped and then reversed so that solvent can be forced to flow from the solution. In order to obtain reasonable flowrates through the membranes, RO systems are operated at from 300 to 1,500 psi while UF systems require only about 10 to 100 psig pressure.

The first material found to offer any promise for the desalination of salt solutions was cellulose acetate. This was discovered by C.E. Reid and J.E. Breton [41] at the University of Florida in 1958 after screening all commercially available films. Since the original material had exceedingly low water-transport rates, S. Loeb and S. Sourirajan [42] sought and developed methods to modify the structure in order to improve the water transport or flux without sacrificing the inherent salt-rejecting abilities. The structure developed consisted of a very thin, dense "skin" supported by a more porous under-layer.

More recently, other polymers, notably an aromatic polyamide, have been developed for reverse osmosis applications; these have more chemical resistance but lower flux. Dynamic membranes that are formed in place by reacting a mixture of hydrous zirconium oxide and polyacrylic acid have shown potential for higher productivity, and inorganic membranes are available that can withstand very severe operating conditions.

The earliest RO and UF units were based on flat membrane sheets in arrangements similar to that of a plate-and-frame filter press. More efficient assemblies in the form of scrolls, straight tubular, helical tubular, and hollow fibers have since emerged. Except for the latter, they all require some form of membrane support to withstand the operating pressures.

The water flux is dependent upon the applied pressure, whereas the solute flux is not directly pressure-dependent. The degree of salt rejection, as well as flux, tend to improve, therefore, as pressure is increased.

Proceeding along the membrane from the feed inlet toward the outlet end, as more water is removed, the salt concentration increases until the final brine concentrate is discharged. This increase of concentration causes a corresponding increase of osmotic pressure, which in turn reduces the flux and permeate quality. Another factor that should be noted is that the concentration at the membrane interface is always greater than that in the bulk of the solution, since a gradient is established as water permeates the membrane. It is this polarization concentration that most affects flux, salt rejection and also scaling or fouling tendencies. In order to minimize the boundary-layer thickness, high flowrates and turbulence are required.

In addition to structural considerations, another constraint placed on the operating pressure is that of membrane compaction. It has been found that excessive pressures accelerate the rate at which the substructure compacts, thus reducing the long-term water flux.

Other factors that can affect performance include temperature, chlorine, fouling or scaling tendencies, and pH. the effect of increasing the operating temperature will be to increase the water flux about 1.4% per °F within the safe range of 32 to 86°F. This agrees with the reduction in the viscosity of water. Low concentrations of chlorine (up to 1.0 ppm) can be tolerated and are beneficial in reducing slime growths on cellulose acetate membranes but cannot be used with the polyamide type above 0.05 ppm. Other methods to control fouling and scale formation are various periodic flushings. Enzyme detergents are effective against many organic slimes, warm solutions of 1–2% citric acid can remove some types of scale deposits, and sodium hydrosulfite at 1–5% is effective for iron scales at ambient conditions. Some UF systems are designed to permit a backflush cleaning of the membranes with water.

The ability to tolerate suspended solids can vary widely. Large-bore tubular systems are relatively insensitive to suspended matter (and some are designed to permit mechanical cleaning with sponge balls), while hollow-fiber systems (even with feed on the outside of the fibers) require more stringent pretreatment for removal. For optimum performance, pH should be kept in the range of 4.0 to 7.5 to minimize hydrolysis of the

---

Applications of reverse osmosis and ultrafiltration   **Table VII**

Metal industry

Recovery of gold, silver, platinum, nickel, chrome, zinc, aluminum, and cadmium from plating wastes

Reuse of rinse waters

Recovery of waste nickel, nickel sulfamate, copper sulfate, copper pyrophosphate, nickel fluorinamate, zinc chloride

Recovery of detergent for reuse in cleaning operations in metal phosphating; oil concentration

Removal of sulfates from acid mine drainage

Textile industry

Recovery of polyvinyl alcohols, size and mineral oils

Removal of dyes

Paper and pulp

Color removal

Solids concentration from white liquor

Clean water from black liquor

Food industry

Concentrate whey from dairy wastes

Sugar recovery in candy manufacture

Miscellaneous

Desalting

Chromate removal from cooling-tower blowdowns

Concentration of radioactive wastes

Recovery of water-based paints from rinse waters of the electrophoretic deposition process

Removal of emulsified oils

cellulose acetate membranes (4 to 10 with polyamide types). There are many cases, however, in which it is economical to operate with a much wider pH range at a somewhat reduced membrane life.

Inorganic UF membranes, more-recently introduced, are claimed to operate up to 200°F or more, and over a pH range of 1 to 14.

Although developed primarily for the desalting of brackish waters, reverse osmosis is finding an increasing number of applications for pollution abatement, often with product recovery helping to defray expenses. In some instances the substances recovered would otherwise pose a problem in conventional wastewater-treatment processes.

Table VII lists some of the applications where reverse osmosis and ultrafiltration process have been used.

Although RO membranes reject most ionized salts, some complexes, such as those of cyanide, may pass through quite easily [43]. Thus, although gold chloride is more than 95% rejected, a potassium gold cyanide complex is not rejected at all at a pH of 6.0. As pH is increased above 7.0, the degree of dissociation increases; it becomes nearly complete above pH 11.0. The ability to reject the cyanide closely follows the dissociation curve, so that for rejection to exceed 90% the pH must be adjusted to a value of 10.5. This pH exceeds the limitations for cellulose acetate membranes (except for very brief duration), although it may fall within the upper limits of the polyamide type. The ability of RO membranes to preferentially pass cyanide complexes probably may find use in the separation of the more-noble metals.

Electrodialysis

In contrast to the membrane processes of reverse osmosis and ultrafiltration, electrodialysis (ED) employs the removal of the solute (with some small amount of accompanying water) from solution rather than the removal of the solvent. The other major distinction is that only ionic species are removed.

The advantages of ED are primarily due to these distinctions. Operating costs tend to be lower since less energy is required to transport the salt (as compared with the water) because this component is present to a much smaller extent. The fact that only ionized salts will be removed allows for the simultaneous separation of these substances from any neutral or un-ionized organic matter that may be present. Although these features are similar to ion exchange, the advantage is that in ED the process is continuous, with nothing to be regenerated—hence no chemical additions are required.

In ED, unlike the other membrane processes discussed, two different membranes must be used together: the anion-permeable membrane permits passage of anions only, while the cation-permeable membrane allows only cations to pass. Fig. 10 shows how the two types of membranes are arranged alternately to form many compartments between the electrodes at each end. As shown, the feed is introduced into alternate membrane compartments. Under the influence of an applied dc voltage across the membrane assembly or stack, the anions (negatively charged ions) are attracted toward the anode (positive electrode) while the cations are attracted to the oppositely charged cathode. The anions are prevented from reaching their respective electrode, however, because they cannot pass through the cation membrane. Similarly, the cations cannot reach their electrode since they cannot pass through the anion membrane. This situation causes the formation of alternate concentrate compartments and dilution compartments to be established. The compartments into which the feed was introduced become the dilution side and the stream exiting from these compartments is referred to as the dilution stream (any organic matter present will remain in this stream). The stream issuing from the concentrate compartments is the brine.

As anions, for example, pass through an anion membrane, they will form concentration gradients on both sides of that membrane. In leaving the dilution side they form a layer depleted of ions, while on the opposite surface there will be a concentration higher than that of the bulk brine. These gradients or polarization effects will become more pronounced under the influence of higher applied voltages or current densities (which cause the rate of ionic transport to increase). The effect on the brine side will not be detrimental unless the solubility limit of a component is exceeded and thereby causes a scale to form. More often, however, a problem is created on the dilution side, since an absence of ions to conduct the current through the solution causes an increase in resistance. This may start to cause hydroxyl ions (from the dissociation of water) to be transported through the membrane (which reduces the efficiency of the process), and may also increase the pH, thereby adversely affecting the membrane structure. This effect

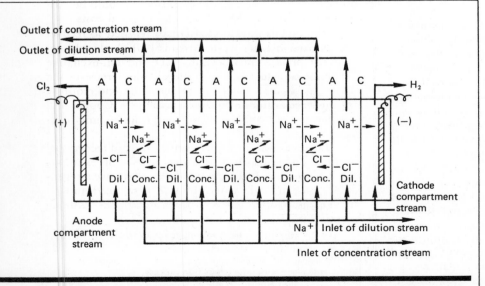

Schematic of electrodialysis apparatus　　　　　　　　　　**Fig. 10**

places limits on the current density that can be used.

To minimize these polarization effects, the spacing between the membranes is kept narrow and flowrates are maintained at relatively high levels. Some designs incorporate tortuous-path arrangements to increase turbulence while others have relatively unobstructed flow channels with only screen spacers to serve as membrane separators. In order to provide high flowrates, the dilution stream is partially recycled through the stack, whereas recycle of the brine is not always required.

Since unwanted side-reactions of hydrogen-gas and alkali formation take place at the cathode while chlorine and acid form at the anode, these two compartments must be provided with separate flush streams. In large commercial units the end effects do not constitute a serious loss, and overall current efficiencies of over 80 to 90% are generally obtained. Current efficiencies are determined by comparison with that theoretically predicted by Faraday's law.

Although ED is not a new process and has, in fact, been used in laboratories since the early 1900s, it was not until after World War II that practical synthetic ion-exchange membranes were developed. Most of the stable membranes today are based on copolymers of divinyl-benzene-stryene with ion-exchange groups. The cation membranes are mostly of the sulfonic type, while the anion membranes are of the quarternary ammonium type and are usually reinforced with plastic fibers. Newer membranes of perfluorosulfonic acid polymers that have improved stability and chemical resistance are also available.

Most membranes permit operation up to 50°C, although membrane development activity is attempting to raise that limit. The ability of the different types of membranes to withstand pH extremes varies considerably. For example, some are stable in acid environments ranging from 5 to 35% H_2SO_4, and from 4% up to concentrated HCl. On the alkaline side some membranes cannot be used at all, while others are claimed to be stable in 50% NaOH. Almost all of the ED membranes are quite sensitive to oxidizing solutions. Some permit short-duration exposure to 1 ppm chlorine while needing less than 0.1 ppm for long-term operations.

In addition to scaling caused by exceeding solubility limits, fouling can result from other factors. Suspended matter should be removed down to 50–100 microns (μm). The physical configuration of the ED stack affects its susceptibility to solids fouling. Iron and manganese should be limited to 0.2 ppm combined. Trivalent ions, such as aluminum and phosphate, may cause increased electrical resistance. Some organic matter, including ionic surfactants (MBAS), and tannic, fulvic, and humic acids, as well as butyrate and larger esters, can cause membrane blockage. Although microorganisms do not attack the membranes, they can form an interfering slime layer that will require periodic cleaning.

The order of performance-deterioration indicators are generally as follows: changes in selectivity, voltage increase, and then pressure-drop increase. An increase of pressure drop on the dilution side is caused by an accumulation of suspended matter, whereas an increase on the brine side is caused by organic fouling or scaling.

Applications of electrodialysis

While the emphasis in the U.S. has been on the production of fresh water from saline sources, in Japan the major application of ED is for concentrating seawater to recover the salt. In fact, ED is now used to produce that nation's entire domestic salt supply (1 million tons/yr), with large systems having been employed for more than fifteen years. Although, ordinarily, divalent ions will be concentrated to an even greater extent than monovalent ions, special membranes have been developed that can reverse this preference, as is required for this application. Another commercial application for ED is the desalting of cheese whey, where the salt content of lactose solutions is reduced sufficiently to permit their use in food applications.

New applications of this established technology to chemical recovery and wastewater treatment is receiving increased attention. The principal action of ED—to concentrate ionic materials—enables consideration of systems to accomplish the following separations: to reduce the volume of brine waste streams, to recover inorganic salts, to recover organic salts, to remove inorganic salts from waste streams to facilitate further treatment, to separate and recover ionic materials on the basis of valence, to separate and recover ionic materials from complex aqueous solutions containing neutral organics, and to concentrate acids. ED's ability to remove inorganic salts from liquid organics with no chemical consumption enables it to be considered as a purification process for organic chemical synthesis processes. Examples of some potential applications follow.

ED might be used for reduction of the blowdown from those cooling towers where solar ponds are used to dispose of the blowdown. The use of ED to concentrate the discharge into a much smaller volume is a very attractive alternative. The net capital savings from reducing the required pond size can be substantial. Also, less makeup water would be needed. Although it is difficult to qualify all of the costs (due to variables that differ from case to case), the total savings could be 50% of the conventional blowdown and disposal costs [44].

In the metals industries, ED can be used to recover metals from plating waste streams [45]. A closed-loop recovery process incorporating ED and ion-exchange permits complete recovery of nickel, allows reuse of the dilute stream for rinsing, and recovers the acid required for the regeneration of the ion exchangers. Copper and chromium have also been recovered from etching processes.

ED can be used for the concentration and recovery of hydrochloric, sulfuric and other acids. For example, spent pickle liquor can be processed to produce a regenerated pickling acid while the iron is reduced and removed by plating out on the cathode. The cathode reaction is:

$$Fe^{++} + 2e^- \longrightarrow Fe^0$$

while at the anode:

$$H_2O \longrightarrow \tfrac{1}{2}O_2 + 2H^+ + 2e^-$$

As discussed previously, cyanide complexes require a pH of about 11.0 for complete dissociation into ions.

Since ED membranes are more resistant to this environment than are the RO type, this process appears to offer promise for the economic control of these toxic substances [46].

A process has been developed for the recovery of the components of sulfite pulp liquor. Byproducts such as lignosulfonic acid, lower-molecular-weight organics, pulping chemicals, and purified and reusable water have been recovered. In other paper-manufacturing processes, zinc chloride has been recovered from treating solutions.

With water-recovery rates of 80 to 90%, greatly reduced disposal volumes (or closed-loop operation), the independence of chemical additives and low energy requirements, ED systems can be an economically feasible means of meeting the parallel challenges of wastewater treatment and recovery through brine concentration.

Chemical oxidation

The chemical oxidants in wide use today are chlorine, ozone and hydrogen peroxide. Their historical use, particularly for chlorine and ozone, have been for disinfection of water and wastewaters. They are, however, receiving increased consideration for removing from wastewaters organic materials that are resistant to biological or other treatment processes.

Ozone

Ozone is a powerful oxidizing agent that reacts rapidly with many compounds present in wastewaters. It has a relatively low solubility in water (at standard conditions of temperature and pressure) and is unstable, having a half-life of several minutes. Ozone generation, therefore, must be done at the plant site.

Ozone is found to be effective in many applications for color removal, disinfection, taste and odor removal, iron and manganese removal and in the oxidation of many complex organics, including lindane, aldrin, surfactants, cyanides, phenols and organo-metal complexes. With the latter the metal ion is released and can be removed by precipitation.

In ozone oxidation of organic materials, it is generally expected that only one atom of oxygen from the O_3 molecule is highly reactive. On this basis, the oxidation of 1 lb of COD should consume 3 lb of ozone and yield 2 lb of molecular oxygen as a byproduct. Actual usage of ozone varies with the organic being oxidized.

Chlorine oxidation

Chlorine (and its more readily storable form, hypochlorite ion, or bleach) is a well-known chemical oxidant and has long been used to purify water, destroy organisms in wastewater and swimming pools, and oxidize chemicals in aqueous solutions. The destruction of cyanide and phenols by chlorine oxidation is well known in waste-treatment technology. It has recently come under intense scrutiny and debate as to its true merit for these applications because of the uncertainty in knowing and predicting the products of the chlorine oxidation reactions and their relative toxicity.

Recent developments in the use of chlorine oxidation for wastewater treatment are characterized by a greater understanding and control of chlorine oxidation chemistry. Chlorine oxidation reactions and kinetics are highly dependent on a variety of process variables associated with the aqueous chemistry of chlorine itself. One of the most important is pH. Fig. 11 shows the various forms of chlorine in a 1.0M sodium chloride solution at room temperature [47]. In acidic solutions, hydrated chloride molecules and trichloride ions exist as shown.

Weakly acidic solutions contain hypochlorous acid. In neutral or alkaline solutions, HOCl dissociates into hypochlorite ions. As the brine concentration decreases, the hydrolysis of chlorine to HOCl on the acid end of the pH scale becomes dependent not only on pH but also on the absolute concentration of chlorine. As the brine concentration increases at low pH, the trichloride ion, Cl_3^-, increases so that in 5.32M NaCl brines the relative fractions of Cl_3^- and Cl_2 shown in Fig. 11 are reversed. Consequently, the concentration of chloride ion can affect reactions that involve Cl_3^-.

Other important variables are temperature, chlorine (or hypochlorite) concentration, the presence of inhibitors of the various reactions, and rate-improving or mechanism-directing catalysts.

The basic chemistry of aqueous chlorine as a chemical oxidant involves the consumption of one mole of chlorine to generate one atom of oxidizing power, according to the general equation $Cl_2 + H_2O \longrightarrow 2 HCl + [O]$. Thus, 71 lb of chlorine will generate 16 lb of oxygen. In addition, control of pH during the reaction can require alkalinity, adding to the chemical cost of chlorine oxidation. The chemical cost of oxidation and pH control at today's prices is approximately 80¢/lb of chemical oxygen demand.

Other reactions can increase or decrease the consumption of chlorine in aqueous oxidation reactions. One that increases consumption is the disproportionation to chlorate, which is fastest near pH 7.5. Direct reaction of chlorine with organics in solution to make chlorinated hydrocarbons is another example of the possible mechanisms. Chlorine consumption will also be affected by the efficiency of mass transfer and the kinetics of the various reactions.

Despite the costs and complexities of chlorine oxidation systems, chlorine does have the advantage of feasibility because it is a powerful chemical oxidant and it usually works in destroying organics.

Recent developments illustrate various ways in which the oxidation of organics in aqueous waste streams by chlorine has been unravelled and the state of technology has been advanced. Much of the recent work has been done by researchers at the Dow Chemical Co. on purification of waste brines (as evidenced by patents and publications). U.S. Patent No. 3,910,999 teaches the stepwise chlorine oxidation of mixtures of compounds present in industrial wastes, such as glycols, chloroalcohols, organic acids, and ketones, by controlling the pH in the alkaline range initially to oxidize those most favorably destroyed by hypochlorite ion, then allowing the pH to drop to the acid range to destroy those oxidized more rapidly at low pH. Defining a process such as this requires the ability to identify and follow the fate of the various organic species in the waste.

The purification of the waste sodium chloride brine from 2,4-dichlorophenoxyacetic acid by chlorination has been reported [48]. The chief organic contaminant in the brine being destroyed by chlorination is glycolic acid, a byproduct generated by the hydrolysis of chloroacetic acid.

U.S. Patent No. 3,839, 169 teaches the oxidation of carboxylic acids in brines by chlorination in the presence of ultraviolet light catalysis. By careful control of reaction conditions, the destruction proceeds by a mechanism that consumes only one-fourth of the chlorine that would be required for total oxidation. The more efficient reaction is:

$$CH_3COOH + Cl_2 + NaOH \xrightarrow[\text{UV}]{\text{pH 4}}$$
$$CO_2 + CH_3Cl + NaCl$$

It is favored by precise control of pH and chlorine concentration, and by high temperature. The quantum yield increases dramatically in the temperature range near the boiling point. This process is an example of the use of UV catalysis as a method of increasing the reactivity of chlorine with organics in aqueous wastes.

The demonstration of this technology on an industrial scale was supported by a grant from the EPA—the report has not yet been issued. A two-stage reaction system purified 200 gpm of the waste brine resulting from the manufacture of phenol by the hydrolysis of chlorobenzene.

The reaction of chlorine with amino groups is complicated because of the number of reaction pathways that exist. Chlorination of a methylamine solution has been described [49], emphasizing the need for pH control to optimize reaction rate and to obtain the desired reaction pathway. Similar problems arise in the chlorination of wastewaters containing ammonia, especially in disinfection applications. Inadequate control of chlorine dosage leads to the formation of chloramines. Chloramines are much less reactive to microorganisms than the free chlorine, and they are also toxic to fish.

These problems with disinfection by chlorine have spawned a search for alternative disinfection treatment methods. A recently completed EPA-funded study at the Wyoming, Mich., wastewater treatment plant showed that sulfur dioxide, ozone, and a new halogen disinfectant, bromine chloride, were acceptable substitutes for chlorine. That is, all three gave adequate disinfection and the effluent from each treatment was not toxic to fish. Because of the overall reactivity of ozone, performance of that system required suspended-solids removal to reduce the consumption of ozone.

Bromine chloride reacts in water to generate hypobromous acid and hydrochloric acid. The advantage of bromine chloride over chlorine is that bromamines, rather than chloramines, are formed. Bromamines are as reactive as the free halogen in disinfection reactions, and they also have a short half-life, thereby minimizing their presence in the treated effluent. Bromine chloride is a stoichiometric mixture of bromine and chlorine, and is easier to handle than either halogen alone.

Another source of chemical oxidation power for removal of organics from water is chlorine dioxide. It has been reported [50] to be a selective oxidant for industrial wastewaters containing cyanide, phenol, sulfides, and mercaptans. It does not react with many other organics, such as alcohols, glycols, ketones, aldehydes, and organic acids. Typical values for the consumption of chlorine dioxide were found to be: 1.5 lb to convert 1 lb of phenol to quinone, 2.5 lb to convert 1 lb of cyanide to cyanate, and 4–5 lb to oxidize 1 lb of sulfide or mercaptan. As with chlorine reactions, pH control is very important.

Chlorine dioxide is a gas that must be generated onsite. Because of its cost, approximately $4/lb, its use will be as a selective, polishing treatment for destruction of trace amounts of specific compounds in industrial effluents.

There is one other new use of chlorine in industrial pollution control that should be mentioned, even though it deals with disposal of organic residues rather than wastewaters. The process, called chlorinolysis, is being considered for the destruction of highly chlorinated residues. The technology has been developed by Hoechst Uhde in Germany, and the EPA has funded an engineering evaluation of the design considerations for a regional chlorinolysis facility, with an emphasis on materials of construction. The process involves the exhaustive chlorination of highly chlorinated residues at high temperature and pressure in carbon tetrachloride solvent. The reaction products are chiefly carbon tetrachloride and hydrogen chloride. Carbonyl compounds react to give phosgene. The reactor effluent is quenched with carbon tetrachloride and then taken to a separation train for recovery of the various reaction products for reuse and sale. It is anticipated that this process can convert residues such as "hex" tars, chlorinated still bottoms, pesticide and herbicide residues, and CB agents to innocuous, usable byproducts.

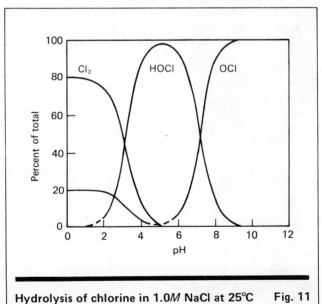

Hydrolysis of chlorine in 1.0M NaCl at 25°C   **Fig. 11**

Acknowledgements

The authors acknowledge the assistance of those who contributed to this paper. In particular, to Jack Famularo for his input to the carbon-adsorption discussion, Robert Kalish for his input to the discussion of the membrane process, and Karl Scheible for his input to the discussion of biological treatment processes.

References

1. 29 CFR [Code of Federal Regulations] 1910.93; 39 FR [Federal Register] 23540, June 27, 1976.

2. *Natural Resources Defense Council v. Train,* U.S. District Court, District of Columbia No. 73-2153. Consent Agreement, *Environmental Reporter Decisions,* 8 ERC, p. 2120, June 26, 1976.

3. 40 FR [Federal Register] 59960, Dec. 30, 1975.

4. Hall, J. R., others, Role of New Techniques in Wastewater Analysis, presented at 31st Purdue Industrial Waste Conference, May 4, 1976.

5. Scheible, O. K., and Campbell, G. R., Biodegradability of Selected Photoprocessing Chemicals, presented at Soc. of Photographic Scientists & Engineers 14th Annual Fall Symposium, Washington, D.C., Oct. 23–26, 1974.

6. Mueller, J. A., others, Gas Transfer Kinetics of Pure Oxygen Systems, American Soc. of Civil Engineers

7. Watkins, J. P., others, Pure Oxygen and Conventional Air Activated Sludge Treatment—Pilot Plant Evaluation for a Pulp and Paper Mill Waste, presented at the 31st Purdue Industrial Waste Conference, May 6, 1976.

8. Mulligan, T. J., Removal of Nitrogen from Municipal and Industrial Wastewaters, presented at the Texas Water Pollution Control Federation Conference, College Station, Tex., May 23, 1976.

9. Famularo, J., Totating Biological Contactors, presented at the 21st Summer Institute on Biological Waste Treatment, Manhattan College, New York, N.Y., May 1976.

10. Jeris, J. S., Biological Fluid Bed Technology, presented at the 21st Summer Institute on Biological Waste Treatment, Manhattan College, New York, N.Y., May 1976.

11. Ecolotrol, Inc., Summary of Operation—Biological Treatment Using Fluidized Bed Technology at Nassau County's Bay Park Facility, sponsored by Nassau County Dept. of Public Works, presented at New York American Soc. of Civil Engineers Conference, Apr. 23, 1976.

12. Mueller, J. A., and Mancini, J. L., Anaerobic Filter Kinetics and Application, presented at the 30th Purdue Industrial Waste Conference, May 1975.

13. Mueller, J. A., Activated Carbon Adsorption, presented at the 21st Summer Institute on Biological Waste Treatment, Manhattan College, New York, N.Y., May 1976.

14. Ferguson, J. F., Combined Powdered Activated Carbon–Biological Contact Stabilization Treatment of Municipal Wastewater, 31st Purdue Industrial Waste Conference, May 1976.

15. Barry, L. T., and Flynn, B. P., Finding a Home for the Carbon: Aerator (Powdered) or Columnar (Granular), presented at the 31st Purdue Industrial Waste Conference, May 1976.

16. Giusti, D. M., others, Activated Carbon Adsorption of Petrochemicals, *J. Water Poll. Control Fed.,* Vol. 46, No. 5, p. 947.

17. De John, P. B., Factors to Consider When Treating Dye Wastes with Granular Activated Carbon, presented at the 31st Purdue Industrial Waste Conference, May 6, 1976.

18. Hager, D. G., and Rizzo, J. L., Removal of Toxic Organics From Wastewater by Adsorption with Granular Activated Carbon, presented at Technology Transfer Session on Treatment of Toxic Chemicals, Atlanta, Ga., Apr. 19, 1975.

19. Treatment of Effluent from Manufacture of Chlorinated Pesticides with a Synthetic Polymeric Adsorbent, *Env. Sci. Tech.,* Vol. 7, No. 2, p. 138.

20. Scherm, M., and Lawson, C. T., Activated Carbon Adsorption of Organic Chemical Manufacturing Wastewaters After Extensive Biological Treatment, presented at Third National Conference on Complete Water Reuse, American Isntitute of Chemical Engineers, June 1976.

21. Smith, S. B., Activated Carbon Regeneration: The Thermal Transport Process, *Chem. Eng. Progr.,* Vol. 71, No. 5, p. 87 (May 1975).

22. Gitchel, W. B., others, Activated Carbon Regeneration: Carbon Regeneration by Wet Air Oxidation, *Chem. Eng. Progr.,* Vol. 71, No. 5, p. 90 (May 1975).

23. Halting Phenol Pollution, *Chem. Eng.,* Oct. 20, 1969, p. 60.

24. Recondition and Reuse of Organically Contaminated Waste Sodium Chloride Brine, U.S. Environmental Protection Agency, EPA-R2-73-200, May 1973.

25. Baker, C. D., others, Recovery of *p*-Cresol from Process Effluent, presented at 74th National American Institute of Chemical Engineers Meeting, Mar. 1973.

26. Gaska, R. A., Process Changes in an Organic Chemicals Production Plant to Improve Yields and Decrease By-Products, presented at American Institute of Chemical Engineers Workshop, Industrial Process Design for Pollution Control, No. 4, Oct. 1971.

27. Chamberlin, T. A., others, Color Removal from Bleached Kraft Effluents, presented at TAPPI (Technical Assn. of the Pulp and Paper Industry) 1975 Environmental Conference, May 1975.

28. Fox, C. R., Remove and Recover Phenol, *Hydrocarbon Proc.,* July 1975, p. 109.

29. Fox, R. D., Chemical Recovery—Key to Water Purification, presented at 4th Annual Industrial Pollution Conference of Water and Wastewater Equipment Manufacturers Assn., Mar. 31, 1976.

30. Kennedy, D. C., others, Functional Design of a Zero Discharge Wastewater Treatment System for the National Center of Toxicological Research, presented at 31st Purdue Industrial Waste Conference, May 6, 1976.

31. *Chem. Proc.,* Vol. 35, No. 9, p. 13 (Sept. 1972).

32. Treatment of Sulfite Evaporator Condensate for Recovery of Volatile Components, U.S. Environmental Protection Agency, EPA-660/2-73-030, Dec. 1973.

33. *Business Week,* June 28, 1976, p. 113.

34. "Manual on Disposal of Refinery Wastes," Chap. 10, American Petroleum Institute, Washington, D.C., 1969.

35. Lauer, F. C., others, Solvent Extraction Process for Phenols Recovery from Coke Plant Aqueous Waste, presented at Eastern States Blast Furnace and Coke Oven Assn. Meeting, Pittsburgh, Pa., Feb. 14, 1969.

36. Kohn, Philip M., New Extraction Process Wins Acetic Acid From Waste Streams, *Chem. Eng.,* Mar. 15, 1976, p. 58.

37. Bailes, P. J., others, Liquid-Liquid Extraction: Nonmetallic Materials, *Chem. Eng.,* May 10, 1976, p. 115.

38. Cyanide Tamer Looks Promising, *Chem. Week,* Apr. 14, 1976, p. 44.

39. Ames, R. K., others, Sludge Dewatering/Dehydration Results with Mini-B.E.S.T., presented at 30th Purdue Industrial Waste Conference, May 6, 1975.

40. Lacey, Robert E., Membrane Separation Processes, *Chem. Eng.,* Sept. 4, 1972, p. 56.

41. Reid, C. E., and Breton, J. E., *J. Appl. Sci.,* 1959, pp. 1–133.

42. Loeb, S., and Sourinajan, S., UCLA Dept. of Engineering Report 60-60, 1960.

43. Spatz, D., Electroplating Wastewater Processing With Reverse Osmosis, *Products Finishing Mag.*

44. Jordan, D. R., others, Blowdown Concentration by Electrodialysis, *Chem. Eng. Progr.,* Vol. 71, No. 7, p. 77 (1975).

45. Membrane Processes for Treating Metal Finishing Wastes, U.S. Environmental Protection Agency, Project 12010 HJQ to the American Electroplaters Soc. (Progress Reports).

46. Towiner, S. B., Investigation of Treating Electroplaters Cyanide Waste by Electrodialysis, U.S. Environmental Protection Agency, Grant Project 12010 DFS, Environmental Protection Technology Series, EPA-R2-73-287, Dec. 1973, MTIS PB 23i 263/AS.

47. Moyer, J. R., Purifying Industrial Waste Brine for Recycling: Chlorination Processes for the Destruction of Organic Impurities, presented at International Businss Contact Club, International Technico-Economical Symposium on Environmental Chemistry, Oct. 24, 1973.

48. Theis, J. M., Waste Problems in 2,4-D Manufacture Solved by Use of High Purity Intermediates and Total Recycle, presented at 75th National Meeting, American Institute of Chemical Engineers, June 4, 1973.

49. Exner, J. H., Chemistry of Pollution Control at the Source, presented at 172nd National Meeting of American Chemical Soc., Sept. 1, 1976.

50. Wheeler, G. L., Chlorine Dioxide: A Selective Oxidant for Industrial Wastewater Treatment, presented at 4th Annual Industrial Pollution Conference of Water and Wastewater Equipment Manufacturers Assn., Apr. 1, 1976.

The authors

Thomas J. Mulligan is Vice-President and Secretary of Hydroscience, Inc., 2630 Old Hook Rd., Westwood, NJ 07675, where he is principal engineer in charge of process engineering and laboratory operations, and where he supervises a wide variety of wastewater management planning studies. He holds both a B.Ch.E. and M.S.E. in environmental engineering from Manhattan College, New York City, and is a member of the New York Water Pollution Control Federation, the American Soc. of Civil Engineers, and the American Water Works Assn. He is a licensed Professional Engineer in a number of states.

Robert D. Fox is Manager of Source Control/Resource Recovery for Hydroscience Environmental Systems, 9041 Executive Park Drive, Suite 226, P. O. Box 11685, Knoxville, TN 37919, with the responsibility for the development and application of source control methodology and technology for the recovery of chemicals from industrial process wastes. He holds a B.S. in chemical engineering from Purdue University.

Physical and Chemical Methods of Wastewater Treatment

Here is an evaluation of a group of processes developed to deal with wastewater contaminants—e.g., heavy metals, complex organic compounds—that usually pass untreated through the most sophisticated biological systems.

Peter B. Lederman, Research-Cottrell, Bound Brook, N.J.

☐ Chemical and physical treatment processes have emerged for handling increasingly complex wastewaters that contain ever-higher quantities of non-biodegradable dissolved wastes. Ideally, these methods do not suffer the same limitations—e.g., sensitivity to temperature, feed-concentration changes—that burden the conventional biological processes.

Many of the physical-chemical techniques—in a list that includes adsorption, chemical reaction, filtration, ion exchange, reverse osmosis, electrodialysis—have had wide chemical process industries (CPI) use, and are now finding application in waste treatment. These processes have not yet gained total acceptance in either the municipal or industrial fields, but are nevertheless being used more widely for specific applications.

This article will discuss some of the above-mentioned techniques from an applications viewpoint, stressing new developments in each field, and giving examples of industrial use whenever possible. Also included is an analysis indicating several directions in which physical-chemical treatment may develop in the future.

Adsorption

Adsorption is probably one of the most common of the so-called advanced waste-treatment processes. It is primarily used to remove organic pollutants that are not amenable to bacterial attack. Activated carbon is the most popular commercial adsorption-medium. Several full-scale processes based on activated carbon, as well as an integrated service system with offsite contract regeneration, are onstream [2].

The advanced stage of this technology is typified by

Originally published August 15, 1977.

Coke-plant wastewater is treated with granular activated carbon in this new unit for Republic Steel Corp.

the carbon adsorption unit at the Hercules Naval Stores facility in Hattiesburg, Mississippi [3]. Wastewater high in aromatics is treated in a primary clarifier and air flotation clarifier prior to passing through the carbon adsorption towers. The chemical oxygen demand (COD) is typically reduced from 750 to 160 mg/l in the adsorbers. Carbon usage is 21,000 lb/d for removal of 12,800 lb of COD from 2.6 million gal of wastewater. Unfortunately, no specific compound-removal figures are available.

Hercules has reported two major problems: bacterial growth on the carbon under anaerobic conditions, and severe maintenance problems with the multihearth carbon-regeneration furnace. The first is correctable by control of dissolved oxygen in the adsorber feed. The second would probably not be a problem if the furnace were used on a continuous basis; Hercules is switching to a fluid-bed regeneration system that should improve the situation.

Calgon Corp. utilizes a multihearth furnace to regenerate the spent carbon used in its contract carbon-adsorption wastewater-treatment service. The company has fixed-bed adsorbers, and exchanges fresh for spent carbon for offsite regeneration.

American Cyanamid Co. is constructing a 20-million-gal/d moving-bed carbon-adsorption treatment facility in Bound Brook, N. J., to reduce refractory organics in the effluent; this unit is preceded by pH adjustment, precipitation, activated-sludge treatment, and filtration.

Activated carbon can be added directly to a waste stream, which eliminates the need for carbon-adsorption beds or columns. Du Pont has patented, and is installing, its PACT (Powdered Activated Carbon Treatment) system at its Chambers Works [4]. This 40-million-gal/d unit will treat waste from a plant producing a wide variety of organic and inorganic products. The system was built because Du Pont found that activated

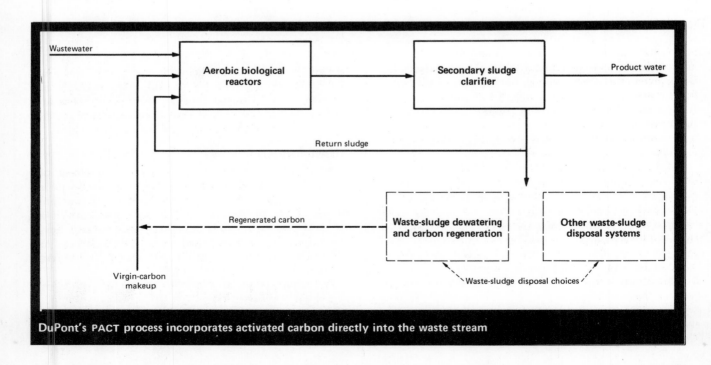

DuPont's PACT process incorporates activated carbon directly into the waste stream

carbon would not remove the simpler organic molecules such as methanol and acetic acid, which can easily be removed in a biological system.

In this process, powdered activated carbon is added directly to the wastewater feed to the aeration tanks of the activated sludge system. Typically, the COD in the effluent from the combined biological/physical-chemical system is reduced by 50% by the addition of carbon, in comparison with that obtained by conventional activated-sludge treatment. Typical carbon-dosage ranges between 50 and 220 ppm. Combined carbon regeneration and sludge incineration use a multihearth furnace. The lack of an effective regeneration system had previously prevented the large-scale use of the less-expensive powdered activated carbon.

Carbon adsorption can also be used to remove heavy metals from plating wastes, although economic regeneration is still a problem [5]. Carbon can also serve as a site for promoting chemical and bacterial reactions. Much work remains to be done to optimize these techniques. The effects of bacteria are still not well understood; a preponderance of carbon or bacteria appears to be synergistic, but heavy bacterial growth on fixed beds is undesirable, as flow characteristics are disturbed, pressures soar, and odors are often generated.

Chemical reaction

A multitude of treatment processes fall under this label. Among the more traditional ones are chemical coagulation, neutralization and chlorination. These are in wide use today, and will not be discussed further in this paper. Instead, some of the newer techniques will be mentioned briefly.

Ozonation is becoming more attractive, particularly as a substitute for chlorine. The reaction products are not organic chlorides, and as a general rule have a lower toxic potential. Ozonation in the presence of activated carbon, ultraviolet light, or other promoters shows promise of reducing reaction times and improving ozone utilization. The latter is very critical because of the expense of producing ozone. To date, the major resistance to ozonation has been the need to produce it onsite, its expense, energy consumption, and unfamiliarity with its use as a chemical agent. These factors are being overcome. Ozonation is being used successfully for the destruction of cyanide in the plating wastes of Boeing Corp. Ozonation for reduction of organic material must, of course, compete with oxidation by other routes, but in some instances it can prove to be economical because of its much greater activity.

Oxidation, in the form of aeration, has been practiced in municipal and industrial waste treatment plants for years. It is used successfully to reduce both biological and chemical oxygen demand (BOD and COD). Other chemical oxidizing agents continue to be widely used. Chlorine, a perennial favorite for disinfection, oxidizes many organic molecules. But it has fallen out of favor because of possible side reactions and the resulting production, in trace quantities, of chlorinated hydrocarbons. Hydrogen peroxide has not been widely used because, like O_3, it was relatively expensive and unavailable in concentrated form. Du Pont is now marketing the Kaston process, which utilizes a hydro-gen peroxide and formaldehyde mixture in the treatment of plating wastes; cyanide is reduced to below 0.3 mg/l, which is equivalent to 99.9+% reduction. Such heavy metals as zinc and cadmium precipitate, while the cyanide is oxidized [6].

The oxidation of organics appears to be enhanced by the presence of both activated carbon, which provides reaction sites, and bacteria, as demonstrated in a number of applications. Other oxidation promoters are also beneficial; for example, the oxidation of cyanide in the presence of activated carbon is enhanced by the addition of $CuCl_2$ [7]. Oxidation by incineration is viable in many cases, but the energy costs for incineration of dilute organic wastewater streams makes this route less and less attractive.

There are a number of other chemical reaction systems worth mentioning. These include the Sulfex process, in which FeS is added to wastewater streams containing heavy metals; the metal sulfide precipitate is relatively easy to filter, in comparison with the hydroxides resulting from pH adjustment by lime addition. Sulfide precipitation is also widely used for mercury removal in the chlor-alkali industry. Starch xanthate may be a viable substitute for metal removal from plating wastewater; its use is now being developed in pilot scale by the U.S. Department of Agriculture at its Peoria laboratory [8, 9]. Cementation has been successfully used for years to recover copper from wastewater. More recently, the same reaction has been used successfully to reduce hexavalent chromium [10].

Probably the most pressing waste-treatment problem at this time is the destruction of organic chlorides. Until recently, high-temperature incineration—above 1,600°C. at a residence time of 2 to 5 s—was the only viable treatment. Now three chemical processes appear to show promise. The first involves the iron-catalyzed reduction of chlorinated pesticides to produce HCl, aromatics and aliphatics. This process has been developed by Envirogenics under an EPA contract [11]. Destruction of a wide spectrum of materials, from chloroform to Kepone and polychlorinated biphenyls has been demonstrated in the laboratory. The second technique is the Hoechst Chlorolysis process, which degrades chlorinated hydrocarbons in the presence of chlorine at high pressure (3,400 psig) and temperature (up to 620°C) to HCl and CCl_4. A 50,000-ton/yr plant has been operational for the past year [12].

Membrane processes

The use of membranes is still somewhat limited, but significant strides have been made in recent years in membrane development. These make reverse osmosis, ultrafiltration and electrodialysis attractive in certain applications.

New significant work has been done in the cleanup of both chromium and copper cyanide plating and rinse solutions [13]. This application has led to closed-loop operation with recycle of plating chemicals at New England Plating Co. (Worcester, Mass.) [14]. Other plating solutions, plus pickle liquor and acid mine drainage (iron oxides plus sulfurous acid) have also been successfully cleaned up via reverse osmosis.

The use of electrodialysis has been tested at full scale

at the Risdon Manufacturing Co. (Waterbury, Conn.) to recover nickel successfully, but at Keystone Lamp Co. (Slatington, Pa.), recovery of zinc cyanides from rinse water has been unsuccessful; both projects are supported by EPA. Ultrafiltration using porous membranes appears to have potential in the separation and recycle of large, dissolved, organic molecules from wastewater.

Membrane process-utilization will depend largely on the development of new membranes that are stronger and less subject to chemical attack, particularly from strong acids, alkalis and solvents.

Ion exchange

Ion exchange is usually more expensive than the preceding processes. However, it may be the treatment of choice in closed-loop recycle systems for dilute solutions. It has been successfully applied to the recovery of chromate from wastewater in pigment manufacturing. Chromic acid from process wastewater has been successfully recovered by ion-exchange resins, which are then regenerated with a sodium hydroxide solution; the resulting sodium chromate is exchanged over a cation resin to yield chromic acid.

New areas for exploration

Further developments in physical-chemical treatment will probably take the following routes:
- Pretreatment for small operations discharging to municipal systems.
- Selective waste treatment for such particularly toxic materials as pesticides and heavy metals.
- In plant waste separation, recovery and recycle.
- Optimization of adsorption processes and regeneration.

First in line is to develop inexpensive, self-contained, easy-to-operate systems that selectively remove those contaminants that are unacceptable and untreatable in the typical municipal treatment facility. Emphasis will be on toxic, chemical and heavy-metal removal.

The second goal will place emphasis on processes to remove toxic and hazardous materials—particularly nitrogen-containing organics, halogenated organics and organometals. The work on chlorinated organics has already been discussed; to date, treatment processes to handle brominated and fluorinated compounds have not been developed. There is some indication that ozonation of the activated-sludge effluent, followed by carbon adsorption or ozonation prior to activated-sludge treatment, may be successful in breaking down complex nitrogen-containing organic molecules. This sequence is being tested by two major producers of these types of compounds, but a lot of work lies ahead.

There is every indication that in-process waste treatment prior to mixing of the waste streams, followed by recycle and recovery, may be technically and economically more efficient. Except for metal platers, few have worked on the development of integrated treatment. The impetus for this approach in other applications exists because of raw material and energy constraints. It should, therefore, receive increasing attention.

Lastly, adsorption will continue to play an important role. New mixed processes, such as adsorption coupled with oxidation, should be developed. However, adsorp-

tion applications need to be broadened and optimized. Efficient regeneration schemes and handling of the stripped adsorbate require particular attention.

To date, activated carbon might be termed the universal adsorbent. Others must be tried or developed. These include alumina, which appears promising for fluoride removal, and the XAD resins.

Other process options will also be explored farther for specific problems. Hyperfiltration, use of ultrasonics, development of energy- and cost-effective evaporation processes, use of UV treatment, and development of new membrane configurations, both conventional and liquid, are but a few options that will be applied in special situations.

Physical-chemical treatment of wastewaters is still in the early stages of application both in the industrial and municipal treatment fields. It will never displace biological treatment, nor should it. Both will continue to play significant roles.

Acknowledgments

The author appreciates the assistance received from former colleagues at EPA: Dr. Herbert Skovronek, Ms. Mary Stinson, Mr. Paul DesRoisiers and Mr. Marshall Dick. Without their help, this article would not be as complete.

References

1. Rudolfs, W. & Turbrick, E. M., Sewage Works Journal, 7, 5, 852 1935.
2. Ford, D. L., Current State of the Art of Activated Carbon; Proc. Manag. of Petroleum Refinery Wastewater, Univ. of Tulsa, 1976.
3. Gardner, F. H., "Activated Carbon Treatment of Wastewater From a Wood Naval Stores Plant," AIChE, 69th annual meeting, Chicago (1976).
4. Heath, N. W., Jr., Combined Powdered Activated Carbon-Biological Treatment of 40 MGP of Industrial Waste.
5. Landrigen, R. B. & Hallowell, J. B., Removal of Chromium from Plating Rinse Water Using Activated Carbon, EPA-670/2-75-055 Cincinnati (June 1975).
6. Patterson, J. W., Wastewater Treatment Technology, Ann Arbor Science, Ann Arbor, Mich. (1975).
7. Huff, J. E. and Fochtman, E. G. Cyanide Removal from Refinery Waste Using Powdered Activated Carbon, ACS mtg., New Orleans March 20-25, (1977).
8. Wing, R., Water-insoluable Starch Xanthate, ... CA-NRRL-41 (RGY.) USDA, Peoria, Ill. (1974).
9. Wing, R., IAG-PS-0714-USDA, Removal of Heavy Metals from Industrial Wastewater with Insoluble Starch Xanthate, Report in Prep.
10. Joster, T. L. and Duff, D. H., Indus. Waste Treat. at Scoville Manuf. 28th Ind. Waste Treatment Conf., Purdue Univ. (1973).
11. Sweeny, K. H. & Des Resiers, P. E., Reductive Degradation for the Treatment of PCB Containing Wastewater, ACS, Montreal (May 29, 1977).
12. Weber, H., DesRosiers P., & Rebhan D., Resources Recovery Through Chlorolysis of Chlorohydrocarbon Waste Residue, ACS, Montreal (May 29, 1977).
13. Donnely, R. G., et. al., A Treatment of Electroplating Wastes by Reverse Osmosis, USEPA-600/2-76-261, Cincinnati (Sept. 1976).
14. Gollan, Z. et. al., Reverse Osmosis Field Test, EPA report to be published.

The author

Peter B. Lederman is Manager of Technical Development for the Utility Gas Cleaning Div., Research-Cottrell, P.O. Box 750, Bound Brook, NJ 08805. He has had prior experience, as Director of the Industrial and Extractive Processes Div. of the Environmental Protection Agency, in the development of industrial wastewater treatment processes. He has been active in teaching, and is an AIChE member.

Recovering Organic Materials From Wastewater

In these days of fuel and material shortages, polymeric adsorbents permit the recovery of valuable organic materials from wastewater, with savings in energy.

BRUCE W. STEVENS, Rohm and Haas Co., and JEFFREY W. KERNER, Alberts & Associates

Over the last decade, the use of synthetic polymeric adsorbents in separating and recovering organic compounds from wastewater in the chemical and pharmaceutical industries has increased rapidly. These adsorbents solve industrial-waste-treatment problems, while meeting two needs not usually attainable with activated carbon: nondestructive adsorption, which permits recovery of costly or short-supply products; and nonthermal regeneration, which reduces fuel bills.

Capital costs of synthetic-adsorbent systems and those of activated carbon are comparable. However, operating costs indicate that polymeric adsorbent methods are more economical than carbon systems when the level of dissolved organic adsorbates is high.

Polymeric Adsorbents Vs. Activated Carbon

Synthetic adsorbents are porous, spherical beads (average dia. = 0.5 mm), available in a range of surface polarities, surface areas, and pore sizes. Contrary to activated carbons (derived from coal, petroleum residues and coconut shells), polymerizable monomers are selected to yield desired adsorptive properties. While the surface area and pore size of activated carbons are generally set,* with synthetic adsorbents such parameters are controlled by suspension-polymerization means.

Each polymeric bead is a collection of microspheres whose diameter is as small as 10^{-4} mm. The size and packing factor of the microspheres determine the surface area and porosity of the adsorbent. Average pore diameters range from 50-235 Å, and surface areas can be as large as 780 m^2/g (Fig. 1).

Activated carbons are usually good adsorbents only for nonpolar substances because of their hydrophobic origin, but polymeric adsorbents are good for both polar and nonpolar materials. The synthetics are made from either unsaturated aromatics, mainly styrene and divinylbenzene—which impart a very hydrophobic nature to the ad-

*Bituminous-coal granules have a high surface area and small pores (ideal for adsorbing lower molecular-weight nonpolar compounds); lignite granules have a much lower surface area but larger pores, which pick up larger particles.

Originally published February 3, 1975.

sorbent—or from acrylic esters that, due to the presence of carboxyl groups throughout the resin matrix, introduce a fair degree of hydrophilicity to the surface.

Aromatic-based adsorbents are particularly effective for adsorbing nonpolar organics from aqueous solutions, whereas acrylic-based adsorbents show good attraction for more-polar solutes, and are even effective in adsorbing polar substances from nonpolar solvents.

If each organic molecule is thought of as having both a hydrophobic and a hydrophilic end, then the hydrophobic end will be attracted to the more-hydrophobic, aromatic-based adsorbents. Materials of intermediate polarity, based on acrylic esters, have an attraction for both the hydrophobic and the hydrophilic end of a molecule. Sometimes these adsorbents function well both in adsorbing hydrophobic materials from aqueous media and hydrophilic substances from nonaqueous media. For example, certain adsorbents are able to adsorb phenol—a fairly polar molecule—from either water or hexane, with adsorption taking place by attraction to the hydrophobic benzene ring during the aqueous cycle, and to the hydrophilic hydroxyl group during the hexane cycle.

Thus, synthetics adsorbents exist that are either nonpolar or have intermediate polarity. They may have a high

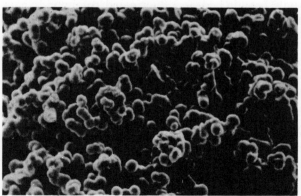

PHOTO shows collection of microspheres—Fig. 1

II

Typical Properties of Adsorbent Materials

Material	Porosity, Cm³/G	Surface Area, M²/G	Skeletal Density, G/Cm³	Nominal Mesh Sizes
Polystyrene*	0.69	100	1.06	20 to 50
Polystyrene*	0.69	330	1.08	20 to 50
Polystyrene*	0.99	750	1.09	20 to 50
Acrylic ester†	1.08	450	1.25	20 to 50
Acrylic ester†	0.82	140	1.26	25 to 50
Carbon, high-density	0.85	950-1,050	1.3-1.4	8 to 30
Carbon, high-density	0.94	1,000-1,200	1.3-1.4	12 to 40

*Nonpolar
†Intermediate polarity

IIIIIIIIIIIIIIIIIIIIIIIIIIIIIIIIIIIIIIIIIIIIIIIIIIIIIIIIIIIIIIIIIIIIIIIIIIIIIIIIII

or a low average pore size, which allows the user to choose the type most suitable for his needs. It is important to consider the size of the organic compound to be adsorbed with that of the average pore size of the adsorbent. Poor adsorption occurs if the ratio of the average-pore dia. to sorbate dia. is too small [1]. The table compares properties of some synthetics to activated carbons.

Binding energies of the synthetics are lower than those of activated carbons for the same organic molecules [2]. Enthalpies of adsorption for the larger-pore, nonpolar adsorbent on the table are only –2 to –6 kcal/mole for such compounds as butyric acid, sodium naphthalenesulfonate and sodium anthraquinone sulfonate. Thus, small energy inputs are needed for desorption, which is the key to the value of these resins in waste-materials recovery.

Usually, thermal reactivation of activated carbons is the only successful way of regeneration [3,4]—because it overcomes high-binding energies—but it also destroys the adsorbed species. With their inherently lower binding energies, synthetic adsorbents can be regenerated through nonthermal means, such as water-miscible organic solvents, which set the stage for a total materials-recovery process. The most efficient regenerants are low-molecular-weight alcohols and ketones.

In general, the lower the solubility parameter (δ) of a solvent, the better it is as a regenerant. Some solvents follow in order of regenerating efficiency [1]:

Solvent	Solubility, δ Cal$^{0.5}$/Cm$^{1.5}$
2-butanone	9.3
2-propanone	10.0
1-butanol	11.4
1-propanol	11.9
Ethanol	12.7
Methanol	14.5
Water	23.2

Almost always, methanol is a good regenerant. It washes off the adsorbents easily, and is about the most available and least-expensive solvent of those listed. Solubility of the sorbate in the regenerant is also important. Not only must van der Waal's attractive forces binding the sorbate to the adsorbent be overcome, but the solubility must be high enough to permit rapid dissolution after the solvent diffuses to the adsorption site.

Careful selection of the solvent often allows recycling to the process a regenerant stream laden with an adsorbed organic compound. Thus, what would have been a nonproductive pollution-control step is transformed into a closed-loop, materials-recovery process. Even when recycling the regenerant stream is not feasible, solvent and sorbate can be separated by common liquid-liquid separation techniques. These adsorptive methods make good sense today, when shortages and high prices of raw materials plague the chemical industry.

Adsorbent Regeneration and Losses

Solvent regeneration of polymeric adsorbents is carried out most effectively by passing the regenerant through the in-situ bed. In treating an exhausted bed, it matters little whether the regenerant passes through cocurrent or countercurrent to the direction followed by the exhausting waste stream. In fixed-bed, ion-exchange systems, ions adsorbed in one cycle tend to "leak" off a bed during the next exhaustion cycle, unless countercurrent regeneration or excesses of regenerant are used. In contrast, organics that are left on a polymeric-adsorbent bed after regeneration do not desorb in the subsequent cycle.

Other advantages of in-situ over thermal regeneration are: (1) no investment in slurry furnace or quenching and conveyance equipment; (2) no adsorbent losses due to handling or thermal oxidation; (3) no corrosion problems due to thermal oxidation or chlorinated hydrocarbons; (4) no adsorbent losses due to irreversible fouling by formation of inorganic ash during thermal regeneration; (5) lower cost of operation as the adsorbate concentration in the waste stream increases.

There is also very little chance of long-range losses due to resin degradation, fouling or attrition. Unlike ion-exchange resins, which swell considerably from one ionic form to another and are subject to osmotic shock, polymeric adsorbents do not depend on swelling, which is low, to expose more surface area.

Barring exposure to strong oxidants for long periods—which would attack chains in the polymer and decross-link the beads—their life should be as good as that of strongly-acidic cation resins for water treatment. By comparison, thermal regeneration of activated carbon involves losses of from 3-10% [5,6] per regeneration cycle.

Organically complexed metals, or chelates, also play an important part in adsorbent loss. For example, premetallized dyestuffs are often found in waste streams from dye and textile industries. Not only do these materials provide color, but they also contain transition metals whose discharge is regulated. Among these are chromium, cobalt and copper. The presence of these organically complexed metals often explains why residual-metal concentrations (after liming is carried out) are not as low as solubility constants would predict.

While these dyestuffs can be readily adsorbed by activated carbon from aqueous wastes, thermal regeneration of the carbon at 1,200-1,800°F is likely to result in the formation of inorganic metal-oxide residues that build up as ash in the carbon. Since fouled carbon has to be replaced unless a suitable acid wash exists, carbon adsorptive capacity decreases, as well as its loss from cycle to cycle. Solvent regeneration of such compounds from

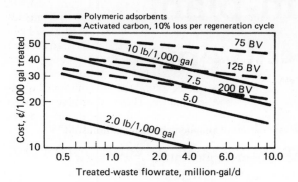

COSTS of synthetics and activated carbon—Fig. 2

polymeric adsorbents is one answer to the problem. A solvent-regenerated resin, waste-treatment system based upon an acrylic ester has been running for over a year at a major dyestuff manufacturer, where premetallized dyes contribute highly to effluent metal discharges [8].

Operation and Capital Costs

Cost calculations for solvent-regeneration of polymeric adsorbents show that solvent distillation and recovery facilities, distillate-bottoms evaporation and concentration, and waste-residue incineration were the analogues of the thermal regeneration furnace, along with quenching and carbon-makeup facilities.

In Fig. 2, the cost of thermal regeneration/lb of carbon is expressed as lb/1,000 gal to yield the cost of thermal regeneration as a function of waste volume. These regeneration costs per volume treated are shown in Fig. 2 at 10% carbon losses per regeneration cycle. Operating costs for polymeric adsorbent systems were superimposed on such 10% carbon-loss data. Bed volume, or BV (total volume of resin bed, including voids), of waste liquor treated serves as the dosage parameter for the superimposed polymeric adsorbent data. Differences in bulk densities of activated carbon and polymeric adsorbents do not permit ready comparison on a weight basis.

A recent survey of industrial wastes by a major carbon manufacturer [10] shows carbon-dosage rates as high as 83 lb/1,000 gal to remove residual color from certain textile waste streams, and 185 lb/1,000 gal to remove phenol from some inorganic waste streams.

It is in this area of higher dosages, though not necessarily as high as the above, that more frequent, high-cost carbon regeneration—along with high losses of carbon from cycle to cycle—results in higher operating costs than for solvent-regenerated polymeric systems. The reason for this is the more economical operation of distillation columns, compared to higher maintenance costs of startup and shutdown of thermal furnaces.

It is clear that, as the carbon dosage increases much beyond the low rates (0.5 to 5 lb/1,000 gal) normally found in municipal sewage or very dilute industrial wastes, adsorption with solvent-regenerated polymeric adsorbents will be more economically attractive.

Concerning capital-cost comparisons for the two systems, there is not much difference between them because adsorption trains are sized primarily on volumetric waste flowrate. While polymeric adsorbents are priced higher than granular activated carbons (about $1.60/lb vs. 35¢/lb), the volume of adsorbent installed is frequently much smaller than that of carbon, due to the polymerics' ability to adsorb much more rapidly.

Regeneration facilities can differ greatly in capital cost, primarily due to the large cost of multiple-hearth or rotary-kiln furnaces in which the carbon must be reactivated. Polymeric adsorbers need only economically constructed distillation columns that treat only the spent solvent. Distillation-column bottoms dictate whether additional concentration and/or incineration facilities are required. When such additions are needed because of nonrecyclability of recovered residues, capital costs for the two regeneration systems may not differ greatly.

References

1. Gustafson, Albright, et al., Adsorption of Organic Species by High Surface Area Styrene-Divinylbenzene Copolymers, *I&EC Research and Development*, June 1968, pp. 107-115.
2. Ibid.
3. Loven, A. W., Perspectives on Carbon Regeneration, *Chem. Eng. Progr.*, **69**, No. 11, pp. 56-62.
4. Hager, D. G., Industrial Wastewater Treatment by Granular Activated Carbon, *American Dyestuff Reporter*, Nov. 1973, pp. 69-74.
5. Hutchins, R. A., Economic Factors in Granular Carbon Thermal Regeneration, *Chem. Eng. Progr.*, **69**, No. 11, pp. 48-55.
6. Cohen, J. M., and Kugelman, T. J., "Physical-Chemical Treatment for Wastewater," **6**, pp. 487-492, Pergamon Press, Great Britain (1972).
7. Sebastian, F. P., The Future Role of Activated Carbon, *Water & Waste Treatment*, May 1972, pp. 9-12.
8. Montaro, R. A., and Noble, A. H., "Removal of Color and Heavy Metals From a Dyestuff Waste Stream by Adsorption," paper presented at American Assn. of Textile Chemists and Colorists Symposium, Washington, D.C., May 22-24, 1973.
9. Himmelstein, Fox and Winter, In-Place Regeneration of Activated Carbon, *Chem. Eng. Progr.*, **69**, No. 11, pp. 65-69.
10. Hager, D. G., "A Survey of Industrial Wastewater Treatment by Granular Activated Carbon," paper presented at the 47th Joint Chemical Engineering Conference (AIChE and Canadian Soc. of Chem. Engs.), Vancouver, B.C., Sept. 10, 1973.

Meet the Authors

◄ **Bruce W. Stevens** is product manager for waste control in Rohm and Haas Co.'s Fluid Process Chemicals Dept., Independence Mall West, Philadelphia, PA 19105, where he is responsible for marketing resins and fluid processes developed by Rohm and Haas for waste-treatment applications. He has B.S. and M.S. degrees in chemical engineering from Massachusetts Institute of Technology. After joining the Engineering Div. of Rohm and Haas as a process engineer, he has participated in plant-design projects, dealing with aspects of pollution-abatement technology for all new Rohm and Haas installations. He is a member of AIChE and Sigma Xi Soc.

Jeffrey W. Kerner is a partner in the chemical and process-equipment firm of Alberts & Associates, 9551 Bustleton Ave., Philadelphia, PA 19115. Prior to this, he held positions in technical development, production supervision, sales and market management with Rohm and Haas Co. Before, he worked for E.I. du Pont de Nemours & Co. as research engineer. He holds a B.S. degree in chemical engineering from Cooper Union and an M.S.Ch.E. from the University of Michigan. Chairman of the Philadelphia AIChE Section, he is also a registered professional engineer in Pennsylvania. He has written articles primarily on new adsorption processes. ►

Fast payout from in-plant recovery of spent solvents

Environmental control and rising prices for many solvents prove major incentives for recovering and reusing waste solvents formerly discarded or incinerated by many processing plants.

James M. Teale, J. M. Teale Associates

☐ Solvent-laden spent streams now face stiffer regulations from agencies of federal, state and local governments. These streams were formerly disposed of via trucking, dumping and burning, practices of disposal that have encouraged wastefulness and environmental pollution.

Worldwide demands on a depleting oil reserve and general inflation mean solvent shortages and increased prices. As a result, we must take a hard look at raw-materials cost-reduction devices to combat rising prices and short supply.

Many small- to medium-sized industries are feeling the "crunch" even more severely, because they use relatively small amounts of such materials and often do not have the engineering talent to design chemical recovery systems suitable for their needs.

We will review the problem encountered by a small printing company in the Northeast and offer a practical solution to its solvent shortage and the accompanying disposal of solvent wastes.

Analysis of the problem

Let us consider a fresh mixture of n-propyl acetate (10%) and ethyl alcohol (90%) being purchased at the rate of 150 gal/d at $1.75/gal. Fifty gallons of this mixture is used to clean press rollers of paint inks, and the balance is used to dilute color concentrates. Washings contain about 95% solvent, and are stored in 5-gal cans for trucking to incinerators.

Our objective is to design a small distillation unit to recover solvents, establish the capital-equipment and operating costs, and determine whether the payback

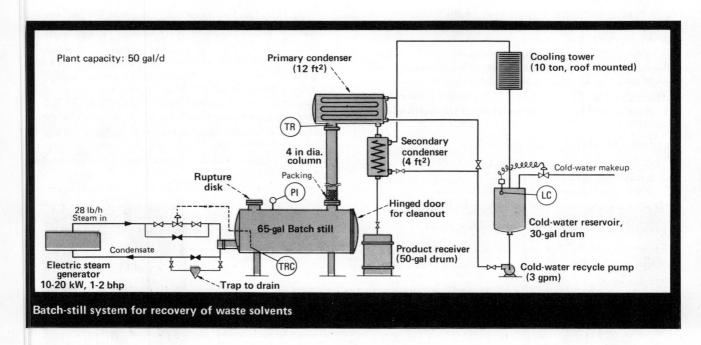

Batch-still system for recovery of waste solvents

Plant capacity: 50 gal/d

Primary condenser (12 ft²)

Cooling tower (10 ton, roof mounted)

TR

4 in dia. column

Secondary condenser (4 ft²)

Cold-water makeup

Rupture disk

Packing

PI

LC

28 lb/h Steam in

Hinged door for cleanout

65-gal Batch still

Cold-water reservoir, 30-gal drum

Condensate

Product receiver (50-gal drum)

Electric steam generator 10-20 kW, 1-2 bhp

TRC

Cold-water recycle pump (3 gpm)

Trap to drain

Originally published January 31, 1977.

period (based on annual savings) warrants such a venture. The design has the following objectives:

- Distillation package must recover the major portion of solvents from the washings.
- System cost, including labor and utilities, must be more economical than private-vendor recovery.
- System must be simple enough to run with existing personnel, using a minimum amount of their work day.
- System must meet approval by insurance companies and local enforcement agencies.

An alternative solution to the problem suggests trucking the solvents to an outside recovery firm. While this would alleviate the resupply situation somewhat, it would not satisfy the price situation—especially when charges of 40 to 50¢/gal for this service were added to the first cost. The economics of outside recovery, including freight costs, are generally unjustified.

A more feasible solution is in-house recovery, with a packaged unit modified to meet particular requirements. Since the facility is already handling solvent, it is equipped electrically and structurally from a safety standpoint, and personnel are familiar with the hazards of solvent handling.

For this problem, the data are available from which we will develop design criteria, and establish parameters for the still, the steam source and the cooling-water tower. Selection of equipment, factory layout, specifications and quotations follow the design criteria:

1. System will be sized to operate on a single shift, 5 d/wk.
2. Composition of the washings will approximate 90% ethyl alcohol and 10% n-propyl acetate.
3. Washings to be fed to the still amount to 50 gal/d.
4. Steam generator must be provided, since there is no steam source available.
5. There is no way to sewer the condenser cooling water. So a tower must be provided to chill and recycle cooling water. Raw water is available at 50 to 70°F.
6. Overly concentrated pigments do not pour easily and may even dry out to a hard consistency. Therefore, we must establish a maximum amount of solvent intended for recovery to avoid this situation. Consistency is determined at 75 s on a #2 Zahn-cup viscosimeter.
7. System will be located apart from the main plant.

Equipment sizing and selection

The following calculations determine utility usage and equipment size.

1. *Feed composition to the still*—We calculate the amount of spent washings on a weekly basis from:

$$(50 \text{ gal/d})(5\frac{1}{2} \text{ d/wk}) = 275 \text{ gal/wk}$$

The recovered solvent portion of spent washings will contain:

n-Propyl acetate:

$$275(0.1)(8.34 \times 0.886) = \quad 200 \text{ lb/wk}$$

Ethyl alcohol:

$$275(0.9)(8.34 \times 0.789) = \underline{1,630} \text{ lb/wk}$$
$$1,830 \text{ lb/wk}$$

Plus 5% pigments added $\underline{\quad 90} \text{ lb/wk}$

Total washings 1,920 lb/wk

2. *Outflow from the still*—We will assume that only 80% of the washings will be recovered as solvents, in order to achieve the viscosity requirement in the concentrates for easy flow from the still as bottoms. Therefore, the net solvent to be recovered will be $(0.8)(1,920) = 1,536$ lb/wk. Recovery operations at 5 h/d for 5 d/wk gives a recovery rate of $1,536/(5)(5)$, or 61.5 lb/h, for the total solvent-mixture outflow.

3. *Distillation conditions*—The composition of the azeotrope from the still will be 85% ethyl alcohol and 15% n-propyl acetate, with a boiling point of 173°F. Rate of recovery for each component of azeotrope will be:

Ethyl alcohol: $0.85(61.5) = 52.5$ lb/h

n-Propyl acetate: $0.15(61.5) = \underline{\quad 9.0}$ lb/h

Total   61.5 lb/h

We calculate the sensible heat required, H_s, for each component from:

$$H_s = mC_p\Delta t + mC_p\Delta t$$

where m = lb/h of solvent, C_p = specific heat of solvent, and Δt = temperature difference between boiling point and room temperature.

$$H_s = 52.5(0.75)(173 - 70) + 9(0.46)(173 - 70)$$
$$H_s = 4,480 \text{ Btu/h}$$

We will determine the heat of vaporization required for the mixture, H_v, by assuming that the spent solvent is entirely ethyl alcohol, for which $H_v = 366$ Btu/lb. Therefore:

$$H_v = 366 \times 61.5 = 22,510 \text{ Btu/h}$$

Total heat input to the still becomes:

$$q = H_v + H_s$$
$$q = 22,510 + 4,480 = 26,990 \text{ Btu/h}$$

We recommend the purchase of a 65-gal still, 10 ft² removable heating bundle, 6 ft² heating jacket, with all required vents, sight holes, gages, etc. (see diagram).

4. *Steam requirement*—Let us assume that we will use low-pressure steam at 15 psig, temperature = 250°F, and an enthalpy, h, of 945 Btu/lb. Hence, the quantity of steam required becomes $26,990/945 = 28$ lb/h.

We recommend the purchase of an electrically powered steam generator (packaged unit), rated at 10 kW to deliver 34.5 lb/h at 1 bhp, with a condensate return system, and automatic flush and drain system (see diagram).

5. *Cooling requirements for condenser*—Let us assume an overall heat-transfer coefficient, U, of 100 Btu/(h)(ft²)(°F), and a cooling-water temperature rise of 20°F. The area required for the condenser, A_1, is:

$$A_1 = H_v/(U)(\Delta t)$$
$$A_1 = 22,510/(100)(20) = 11.2 \text{ ft}^2$$

The secondary cooler will chill product distillate from 173° to 100°F. Assuming a 10°F cooling-water rise and an overall heat-transfer coefficient of 80 Btu/(hr)(ft²)(°F), we calculate the cooler area, A_2, from:

$$A_2 = \frac{52.5(0.75)(173 - 70) + 9(0.46)(173 - 70)}{(80)(10)}$$
$$A_2 = 4 \text{ ft}^2$$

Cooling-water rates for the primary condenser, R_1, and the secondary cooler, R_2, are calculated as:

$$R_1 = \frac{22,510}{(20)(8.34)(60)} = 2.25 \text{ gpm}$$

$$R_2 = \frac{3,176}{(10)(60)(98.34)} = 0.65 \text{ gpm}$$

Total cooling-water flow is $2.25 + 0.65 = 2.9$, or say 3 gpm.

We recommend a 10-ft^2 primary condenser and 4-ft^2 secondary cooler with controls, and the purchase of a 10-ton water-cooling tower complete with 1/3-hp motor, makeup valve and liquid-level float for either roof or local mounting (see diagram).

Summary of capital estimate

65-gal still, vapor column, condensers $3,300
Cooling tower, cold-water makeup, recycle
pump 1,440
Electric steam generator, 10kW, with
condensate return 1,360
Valves and piping   840
Total f.o.b. (without tax) 6,940

Installation costs are not included in the above estimate because packaged units can usually be accommodated by existing maintenance staff.

Operating costs for solvent recovery

Recovery of spent solvents by distillation from the raw waste would ordinarily constitute an added operation in a department that is able to absorb this operation without increasing the labor force and without committing much capital.

An estimate of incremental costs provides a measure of how much extra money must be spent in order to realize the benefits of a certain course of action. The following estimate for incremental costs should be applied to project evaluation with the understanding that it is conditional.

Incremental costs undoubtedly present a sharper appraisal for arriving at this decision. However, we will also include a direct-labor charge of 2 h/d at $5.50/h in our estimate, to show that such an in-house operation is favorable even with this factor considered.

Utility costs depend upon the locality and size of the facility. We will use an estimate based upon a medium-sized factory having rates of 75¢/1,000 gal for cooling water, and 6¢/kWh for electricity. Then:

Direct labor . . . .$0.275/gal solvent
Utilities . . . . 0.092/gal solvent
Total incremental cost . . . .$0.367/gal solvent

Payback period

Savings resulting from recovery cost vs. fresh-solvent-replacement purchase are $1.75/gal minus $0.367/gal, or $1.38/gal of solvent recovered. The amount of spent washings that must be processed to the breakeven point for the capital expenditure is:

$$\frac{\$6,940}{\$1.38/0.8} = 4,000 \text{ gal washings}$$

We will calculate the payback period (PBP) as the time needed to process 4,000 gal washings.

$$(PBP) = \frac{4,000}{(50)(5)} = 16 \text{ wk} = 3.7 \text{ mo}$$

Hence, for a minimal capital outlay of $6,940, we can solve many of the price and availability problems in solvent purchasing.

Other applications

Depending upon the application, distillation units can recover more than one solvent by using additives to develop azeotropic mixtures having boiling points spread enough for separation, or by changing the type of packing, or by employing plates in the rectification column.

Choice of column internals depends upon feed composition, solvents to be recovered, and the size of the column required. Sieve trays are not generally used because of the low reflux ratios. Columns smaller than 2 ft dia. are usually packed to give the best ratios. In columns over 2 ft dia., ballast trays become feasible.

These columns can be effective for recovering such common solvents as ethanol, acetone, methanol, isopropyl alcohol, ketones and toluene. Industrial applications are found in the manufacture of paints, adhesives, photographic chemicals, printing inks, vegetable oils, and for processing animal byproducts.

Advantages of packaged systems

Batch-distillation systems are easily adaptable to modular design, so that preassembly of the basic units and piping can be done by the supplier.

Because control of batch operations is quite simple, controls are usually minimal, and often the units are manually operated.

The columns are versatile in that plates or packing can be used to provide sufficient rectification.

Batch distillation is useful for feed compositions containing residuals that foul the column internals. These residuals remain in the still. Injecting steam directly into the still can also be used should the residues be very heavy and the solvent to be recovered compatible with steam distillation.

Preassembly of the units reduces installation costs by up to 25% in many instances, and decreases installation time. Simplicity of component parts often allows in-house maintenance crews to do the job.

The author

James M. Teale is the principal engineering and management consultant for J. M. Teale Associates, 48 Pinecrest Drive, Woodcliff Lake, NJ 07675. Previously, he spent a 20-year period with several major firms in engineering and management positions. He has a B.S. in chemical engineering from Clarkson College of Technology, an M.S. in industrial engineering from Stevens Institute of Technology, and is a Professional Engineer in New Jersey. He is a member of AIChE, American Inst. of Industrial Engineers and Soc. of Cosmetic Chemists.

Using wastewater as cooling-system makeup water

No magical skill is required to adapt wastewater streams for cooling-tower makeup. What is normally required is an in-depth investigation, and the application of good water-treatment technology.

Edwin W. James, William F. Maguire and *William L. Harpel*, Betz Laboratories, Inc.

☐ Water is called wastewater when it has received some sort of contaminant that makes it undesirable for conventional uses. However, the water may still be useful, and some applications may actually serve as an effective treatment for the contaminant. Proper definition of the contaminants is thus the key factor in determining the water's suitability for cooling-tower makeup.

The microorganisms in a cooling tower, for example, may remove 90% or more of the phenols and reduce effluent organic carbon and oxygen demand, but this ability decreases with rising amounts of sulfide because sulfides apparently kill the better phenol-oxidizer microorganisms [2]. Also, hydrogen sulfide is corrosive to cooling-water equipment, reduces chromate inhibitors to ineffective chromic oxide sludge, and causes an odor. Therefore, when sulfide levels exceed 2 ppm in the wastewater, the water should be diverted from the cooling system.

The first step in evaluating a waste stream for cooling-water makeup is a thorough analysis of that stream. If the stream passes the analysis test, the next

Secondary sewage effluent*	Table I
Total hardness as $CaCO_3$, ppm	192
Calcium as $CaCO_3$, ppm	128
M-Alkalinity as $CaCO_3$, ppm	210
Silica as SiO_2, ppm	22
Orthophosphate as PO_4, ppm	8
pH	7.3
Chloride as Cl, ppm	172
Sulfate as SO_4, ppm	180
Specific conductance, micromhos at 18°C	900
Ammonium as N, ppm	10
Iron as Fe, ppm	0.5
Zinc as Zn, ppm	0.2
Copper as Cu, ppm	0.04
Suspended solids, ppm	8
BOD, mg/l	15
COD, mg/l	40
Nickel as Ni, ppm	0.18
Turbidity	5
Nitrate as N, ppm	0.3
Nitrite as N, ppm	3.1
Chloroform extractable, ppm	2

*Representative of water used in laboratory studies

Recirculator study (7 days)	Table II
Analysis of cycled water	
Calcium ($CaCO_3$)	615 ppm
Magnesium ($CaCO_3$)	240 ppm
pH	7.2
Chromate (CrO_4)	13.5 ppm
Chloride (Cl)	920 ppm
Orthophosphate (PO_4)	20 ppm
Treatment	
Chromate-zinc-phosphonate-polymer	
Corrosion data	mpy
Unpretreated and pretreated high carbon steel	1
Unpretreated and pretreated low carbon steel	1
304 Stainless steel	0
Admiralty	2
Welded high carbon steel	1
Comments	
1. Admiralty corrosion not localized.	
2. Some evidence of biological growth.	
3. Heat transfer coefficient decreased by only 10%.	
4. Minor foaming occurred and was eliminated by use of antifoam.	

Originally published August 30, 1976.

Foam potential studies Table III
(Secondary sewage effluent at 5 cycles)

Chromate-zinc-phosphonate-polymer-biocide program

Antifoam	Antifoam level (ppm)	Foam height (mm)
None	0	175
A	1	0
B	1	0
C	1	0
D	1	55
D	5	15
D	7	0

Phosphate-phosphonate-polymer-organic-biocide program

Antifoam	Antifoam level (ppm)	Foam height (mm)
None	0	175
A	1	0
B	1	0
C	1	0
D	1	0

Comments: foaming tendency was determined using an inverted cone into which the solution under study was added. Foam was generated by bubbling air at a constant rate through the solution.

Ion exchange unit effluent Table IV

Total hardness as $CaCO_3$, ppm	2
M-Alkalinity as $CaCO_3$, ppm	60
Sulfate as SO_4, ppm	10
Chloride as Cl, ppm	5
Silica as SiO_2, ppm	18
Total phosphate as PO_4, ppm	0.5
Orthophosphate as PO_4, ppm	0.5
pH	7.0
Specific conductance, micromhos at 18°C	200
Nitrate as NO_3, ppm	60
Ammonia as N, ppm	20

Kellogg ammonia plant boiler blowdown Table V

M-Alkalinity as $CaCO_3$, ppm	32
Silica as SiO_2, ppm	0.1
pH	9.7
Iron as Fe, ppm	0.05
Sodium as Na, ppm	17.5
Specific conductance, micromhos at 18°C	80
Phosphate as PO_4, ppm	15

step is usually a feasibility study that would, via various tests, determine the stream's potential for corrosion, general fouling, scale formation, biological fouling and foaming, and what treatment would be needed to control such problems. Obviously, the nature of the contamination is indicated by the plant process generating the wastewater, but sophisticated organic analyses may be required to pinpoint unusual contaminants and concentrations.

Also, the variability of the composition of the wastewater is important in determining the treatment methods, and analyses should be extended over as long a period as possible. This will define the consistency of the effluent composition, and at the same time provide information helpful toward planning additional treatment. If water reuse is being considered part of a new plant, the cost of improved water quality may be compared to the savings it allows in the cost of heat exchangers or other equipment.

The problems tend to be magnified as cycles of concentration are increased in the cooling tower. In some cases, increased cycles mean the addition of an external treatment system, such as cold lime softening for phosphate removal and hardness reduction. In all cases, increased cycles call for an application of proper internal water-treatment methods such as application of biocides, deposit-control agents, antifoams, as well as blowdown and pH control.

A good understanding of the problems and practical answers may be achieved by actual in-plant runs or laboratory evaluations. Any in-plant runs should follow a gradual approach, with slowly increasing proportions of wastewater in the makeup water and with close monitoring, so as to avoid disruption of plant operations. If laboratory evaluation is the method, a test cooling-tower might be involved. The size of the test tower varies, but is normally about 15 gpm. Cooling-water conditions can be altered quite readily in these towers, and detrimental effects monitored in test heat exchangers and other monitoring devices.

While the foregoing principles apply to almost any instance of wastewater application as cooling-tower makeup, the variables inherent in water, plant operating conditions and metallurgy, and the cooling-tower environment, combine to make extensive generalization impossible. Instead, it will be more instructive to give some examples of waste streams used as cooling-tower makeup: domestic-sewage effluent, nitrogen-fertilizer-plant effluent, boiler blowdown and process condensate, chemical-plant effluent, bottoms from refinery sour-water strippers, and effluent from dissolved-air flotation systems.

Domestic-sewage effluent

Among wastewater streams, sewage-plant effluent typically provides large volumes of water that may serve as the sole source of makeup for a cooling tower. Such reuse offers two benefits: ecological, in that fresh water is conserved; and economic, in that the sewage effluent may be available at no charge, or at a greatly reduced cost, compared with fresh water.

Sewage-plant effluent has been used by a number of plants over the years, and first-hand experiences are documented in the literature [10, 11, 12, 13, 14, 17]. A literature review will give a general idea of what to expect with this type of water; however, a more specific

Ammonia plant process condensate		Table VI
	Gulf coast plant	Alaskan plant
Ammonia as N, ppm	350	824
Carbon dioxide as CO_2, ppm	360	2,200
pH	6.7	7.1
Specific conductance, micromhos at $18°C$	1,600	5,000
Iron as Fe, ppm	0.0	0.7

Southern chemical plant waste stream	Table VII
Total hardness as $CaCO_3$, ppm	32
Calcium as $CaCO_3$, ppm	24
Magnesium as $CaCO_3$, ppm	8
M-Alkalinity as $CaCO_3$, ppm	24
Sulfate as SO_4, ppm	3
Chloride as Cl, ppm	8
Silica as SiO_2, ppm	3
Total phosphate as PO_4, ppm	0.4
Orthophosphate as PO_4, ppm	0.2
pH	7.2
Specific conductance, micromhos at $18°C$	80
Copper as Cu, ppm	0.05
Iron as Fe, ppm	0.05
COD, mg/l	450
BOD, mg/l	150
TOC, ppm	135
Aromatics, ppm	125
Total hydrocarbons, ppm	200

Southern chemical plant waste stream corrosion study results				Table VIII
Cycles	Mild steel (mpy)	Copper (mpy)	Admiralty (mpy)	304 SS (mpy)
3.0	0.5	0.4	0.4	0.0
6.0	0.8	0.4	0.5	0.0

Experimental conditions: $120°F$, 2 ft/sec., pH 6.5, spinner studies with CrO_4-Zn-Phosphonate-Polymer treatment. Chromate level was 30 ppm.

Spinner study results*		Table IX
	Corrosion rate, mpy	
Treatment	Steel	Admiralty
Chromate-phosphate-Zn (pH 6.5)	1	5
Phosphate-phosphonate (pH 7.5)	17	10
Cycled water analysis		
Calcium as $CaCO_3$, ppm	480	
Magnesium as $CaCO_3$, ppm	430	
Silica as SiO_2, ppm	130	
Chloride as Cl, ppm	135	
Sulfate as SO_4, ppm	20	
Organic carbon as C, ppm	150	

*All studies at $120°F$ with process condensate and simulated cycled water mixed at 1:6 ratio.

characterization is usually necessary. Knowledge of the type of sewage-plant treatment—primary, secondary, tertiary—is required, along with analyses.

Water quality from a sewage treatment plant may vary, depending on the percentage of industrial waste, the percentage of domestic waste (which may vary according to traditional washdays), and the runoff that is handled by the given plant. Thus the analyses should cover as long a period as possible.

Corrosion is usually less severe with sewage effluent than with fresh water. Although such effluent typically contains high orthophosphate concentrations, its tendency for calcium phosphate scaling is inhibited by the stabilizing effect of organic material that is also typically present in high concentrations. So, proper control of cycles of concentration and pH, along with the addition of deposit-control agents, usually permits higher calcium phosphate loadings than may be carried in fresh-water systems.

The nature of sewage effluents presents a severe microbiological fouling problem when such water is used as cooling-tower makeup. This can be controlled, but at increased biocide cost. A combined program of chlorine and nonoxidizing biocide addition is typically used to maintain good control.

A representative analysis of secondary sewage effluent is presented in Table I. Based on prior experience with water of this type, a review of the analysis indicates a tendency for: biological fouling (indicated by the BOD level), calcium phosphate deposition (indicated by the calcium and orthophosphate concentrations), and foaming (indicated by the suspended solids and MBAS concentrations).

A laboratory evaluation of sewage effluent is presented in Table II. This indicates a need for corrosion control with a chromate/zinc/phosphonate/polymer program. A well-chosen nonchromate approach may also be used in this instance. Although there are reports of some very severe, hard-to-contain foaming problems, most systems can be controlled with a small amount of antifoam. The data in Table III support this.

Fertilizer-plant effluents

Many plants have waste streams that are treated for quality improvement and subsequent discharge—but why throw this treated water away?

In fertilizer plants, process waters typically contaminated with high levels of ammonia and nitrate are treated by ion exchange to reduce this type of contamination. Some ammonia and nitrate may remain along with some urea, but, overall, the quality of this treated water is good. A sample analysis is presented in Table IV. There is no reason to discard this water; it may be

Waste treatment system effluent	Table X
Ammonia as N, ppm	10
Total hardness as $CaCO_3$, ppm	212
Calcium as CaCO3, ppm	160
Magnesium as $CaCO_3$, ppm	52
M-Alkalinity as $CaCO_3$, ppm	1540
Sulfate as SO_4, ppm	112
Chloride as Cl, ppm	152
Silica as SiO_2, ppm	17
Total phosphate as PO_4, ppm	29
Ortho phosphate as PO_4, ppm	28
pH	8.2
Copper as Cu, ppm	0.03
Iron as Fe, ppm	0.05
Suspended solids, ppm	18
COD, mg/l	306
Chlorine demand, ppm	108
BOD, mg/l	17
Triethylamine, ppm	1
Sodium as Na, ppm	790
Chromium as Cr, ppm	0.0

Analyses of sour water stripper bottoms					Table XI	
	1	2	3	4	5	6
Chlorine demand, ppm	—	—	—	400	—	—
Ammonia as N, ppm	28	30	30	12	50	50
BOD, ppm	—	—	288	121	—	—
COD, ppm	—	—	387	203	—	—
Total hardness as $CaCO_3$, ppm	14	1.2	92	5	6	2
Calcium as $CaCO_3$, ppm	12	0.8	32	5	2	2
Magnesium as $CaCO_3$, ppm	2	0.4	60	0	0	0
Phenolphthalein alkalinity as $CaCO_3$, ppm	0	0	16	51	0	42
Methyl orange, alkalinity as $CaCO_3$, ppm	88	140	68	100	46	94
Sulfate as SO_4, ppm	16	37	125		32	34
Chloride as Cl, ppm	39	1.5	260	28	142	93
Silica as SiO_2, ppm	<2.0	1.3	4.4	—	—	—
Total phosphate as PO_4, ppm	3.0	0.4	2.1	—	—	—
Total inorganic phosphate as PO_4, ppm	—	0.2	0.5	—	—	—
Ortho phosphate as PO_4, ppm	—	0.1	0.5	0.06	—	—
pH	6.8	7.7	8.7	9.6	6.8	9.
Specific conductance, micromhos at 18°C	250	500	1,200	280	—	425
Total carbon, ppm	—	—	130	—	—	—
Inorganic carbon, ppm	—	—	11	—	—	—
Organic carbon, ppm	—	—	119	32	—	—
Chloroform extractable, ppm	13	—	71	—	—	—
Phenol, ppb	110,000	33,000	79,800	50,000	—	—
Total copper as Cu, ppm	.05	0.02	0.12	—	0.1	—
Total iron as Fe, ppm	1.6	0.6	0.5	—	9.4	32.
Soluble zinc as Zn, ppm	0.0	0.0	0.0	—	0.0	—
Nitrite as NO_2, ppm	0	253	276	—	—	—
Nitrite as NO_3, ppm	<1	<1	<1	—	—	—
Hydrogen sulfide as H_2S, ppm	.01	0.1	0.1	0.1	0	0.6
Suspended solids, ppm	28	—	—	6	—	—

used as part of the cooling-tower makeup. Depending on the plant, this water may amount to only about 30% of the makeup stream; otherwise, the percentage may depend on the quality of the treated water. Nutrients in the form of ammonia, nitrate and urea present a substantial potential for microbiological growth, requiring closer monitoring of biological fouling, and a more intensive biocide program to minimize slime outbreaks.

Boiler blowdown and process condensates

Table V presents the characteristics of a boiler blowdown stream that was added to an ammonia plant's cooling-tower water circulating at 45,000 gpm. In this plant, deionized water had been used for boiler makeup, along with a coordinated phosphate/pH treatment program. The major concern was the phosphate concentration, because of its potential for phosphate deposition. The reconciling factor was the small size of the boiler blowdown—only 1% of the required cooling-tower makeup. Thus the blowdown is diluted, and the use of this water is appropriate.

Condensate streams are prime candidates for water reuse. The typical mineral salts are not present to create problems, although condensate streams usually do contain contaminants of some sort and must be evaluated. An ammonia plant condensate is shown in Table VI. The substantial concentrations of ammonia and carbon dioxide in this condensate provide nutrients for biological growth and create an increased biocide demand. In addition, the ammonia concentration would present a corrosion potential for copper alloys.

The increased biocide requirement will raise costs approximately 25–75% over the cost of fresh water, but microbiological fouling can be controlled. Although a substantial portion of the ammonia will be volatilized over the tower, the remaining concentration will still be enough to attack copper and admiralty-copper metals. While the exact nature and degree of this attack remain to be defined, system metallurgy should be thoroughly reviewed before proceeding with water of this quality.

Chemical-plant effluents

The possibility of reusing a particular water stream may not be easy to establish, even with extensive laboratory work. Table VII presents an analysis of a plant waste stream that has good mineral quality but is high in aromatic content. It was theorized that much of the aromatic contamination would be volatilized over the cooling tower. A laboratory study using wastewater cycled both three and six times confirmed volatilization of the organics and further indicated that a chromate-based treatment would provide satisfactory corrosion protection. These results are illustrated in Table VIII.

However, those laboratory tests also indicated problems from a gummy type of deposit, and this required further studies on a test tower to establish that the deposits would not be a serious problem.

Table IX presents the results of another study conducted for a plant that contained significant amounts of admiralty tubing. The corrosion rate indicated by this study was prohibitively high, due (it was thought) to the organic content of the water, principally formic acid. However, additional studies indicated that a substantial portion of the organics would be volatilized over the cooling tower. The preliminary study actually presented the worst case possible and not the average condition that might be experienced in the cooling tower.

The next step was to experiment reusing the water in the plant cooling tower, starting at a low percentage of the makeup, and building toward the final desired percentage, with constant monitoring of corrosion and fouling. In this actual operation, the predicted corrosiveness of the water did not materialize, and the water was successfully reused in its entirety, which amounted to 10–20% of the cooling-tower makeup.

Table X presents the analysis of the effluent from a bio-oxidation pond. Significant features of this analysis include high alkalinity, high chemical oxygen demand and high phosphate levels. The latter occurred when phosphoric acid was added as a nutrient to the activated-sludge basins. A brief review of this analysis did not indicate any insuperable problem.

Chromate inhibitors provide the best protection against corrosion. However, preliminary testing indicated that in this case the water had a high chromate demand, which precluded chromate treatment. Additional studies would be required to define the corrosion, fouling and biological growth potentials for the system. These studies would include laboratory evaluations and a gradual approach to the use of this water in the actual system by increasing the proportion of reused water to makeup water, with close monitoring.

Sour-water stripper bottoms

Refineries commonly use the sour-water stripper to remove dissolved gases from process waters. Typically, these strippers will remove 99% of the sulfides, 80–95% of the ammonia, and 30–60% of the phenol. Except for the phenol and ammonia contaminants, with their resultant contributions to COD, BPD and TOC, the bottoms from these towers are essentially distilled water. Analyses for six such bottoms are shown in Table XI.

Dissolved air flotation effluent			Table XII
	DAF inlet	DAF effluent without polymer	DAF effluent with 5 ppm polymer
Oil	80	14	4
Suspended solids	95	28	8
Dissolved solids	700	690	695
Total hardness, as $CaCO_3$	210	205	210
Calcium, as $CaCO_3$	140	138	142
Magnesium, as $CaCO_3$	70	67	68
"P" Alkalinity	0	0	0
"M" Alkalinity	60	62	60
Chloride	220	214	215
Sulfate	230	216	222
Sulfide	trace	None	None
pH	6.3	6.3	6.3
Iron	6.5	1.4	0.3
Ammonia	15	13	12
Phenol	12	10	10
COD	400	265	240
BOD	120	77	73
TOC	110	68	60

This wastewater is commonly reused to conserve fresh water and to remove up to 98% of the phenol from the refinery effluent. Alternate applications for reuse are: crude-oil desalting, process wash-water, and cooling-tower makeup. Since the cooling tower provides very effective bio-oxidation of the phenol, but is hampered in this by the presence of sulfides, the efficiency to which sulfides, ammonia and phenol are removed in the stripper determines if the stripper bottoms should be used as cooling-tower makeup.

As mentioned earlier, this water should be diverted from the cooling system whenever sulfide levels exceed 2 ppm. Also, ammonia has three important effects: Its

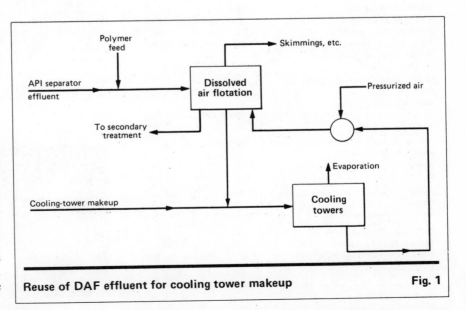

Reuse of DAF effluent for cooling tower makeup   **Fig. 1**

fluctuation causes pH changes; it is quite corrosive to copper and copper alloys; it is a nutrient for fouling organisms. During upset conditions, therefore, the sour-water stripper bottoms should be diverted away from the cooling tower and to the API separator or other waste-treatment facilities.

Microbiological fouling will occur as a result of the bio-oxidation of phenols, so that biocides are necessary for preventing massive growth of slime. Any tower biocide program must be carefully controlled, however, to achieve the proper phenol bio-oxidation. Chlorine can be injected intermittently to prevent slime growth and still permit efficient oxidation of the phenols, but this can result in chlorinated phenols and chloramines. Consequently, non-oxidizing biocides may be preferred over chlorine to prevent the formation of toxic compounds.

Dissolved-air-flotation effluent

The dissolved-air-flotation (DAF) unit, which normally follows the API separator in a refinery waste-treatment plant, serves primarily to remove oils and suspended solids from refinery wastes. Typically, it removes 50–90% of these contaminants, but that removal can be improved with low concentrations of polyelectrolytes.

Since the cooling tower will remove phenol and re-duce oxygen demand of the DAF effluent, and since this effluent is usually a large volume of water that can significantly reduce demand for fresh water, some refineries are now using it as cooling-tower makeup. The major disadvantage of this use is fouling. Oil remaining in the water tends to agglomerate other solids into masses that restrict flow and retard heat transfer. This can be minimized by oil detectors to divert this water during upset conditions. Deposit control agents and wetting agents should be used continuously in the cooling water.

In the case of the system shown in Fig. 1, cooling-tower blowdown also goes to the DAF unit, so that a corresponding amount of DAF effluent must be diverted away from the tower. Towers should be controlled, limiting the levels for hardness, dissolved solids, suspended solids, and oil, in order to minimize fouling and corrosion.

In summary, industry is finding that many waste streams provide very satisfactory cooling-tower makeup water, if good water-treatment policies are used to compensate for the potentially deleterious features of such waters. An ideal example is the increased use of secondary sewage effluent as makeup water. With the proper approach, more and more plants will find themselves using wastewater streams for tower makeup, and obtaining very satisfactory results.

References

1. Hart, James A., Waste Water Recycle for Use in Refinery Cooling Towers, *Oil & Gas Jour.*, June 11, 1973, pp. 92-96.
2. Mohler, E. F., and Clere, L. T., Development of Extensive Water Reuse and Bio-oxidation in a Large Oil Refinery, National Conference on Complete Water Reuse (AIChE), Washington, D.C., April, 1973.
3. Willenbrink, Ron, Waste Water Reuse and Inplant Treatment, Annual Petrochemical and Refining Conference (AIChE), New Orleans, Mar., 1974.
4. Petrey, E. Q., Waste Water and Pollution Control, 12th Annual Liberty Bell Corrosion Course, Philadelphia, 1974.
5. Maguire, W. F., Harpel, W. L., and Carter, D. A., Aspects of Wastewater Reuse as Makeup for Boiler and Cooling Systems, National Conference on Complete Water Reuse (AIChE), Chicago, May, 1975.
6. Maguire, W. F., Minimize Plant Effluent Through Proper Water Engineering, Annual Petrochemical and Refining Conference (AIChE), Houston, March, 1975.
7. Maguire, W. F., Reuse Sour Water Stripper Bottoms, *Hyd. Proc.*, Sept., 1975.
8. Harpel, W. L., and James, E. W., Waste Water Reuse as Cooling Tower Makeup, International Water Conference, Pittsburgh, Oct. 1973.
9. Harpel, W. L., Waste Water Treatment, Regulation and Reuse, Industrial Fuel Conference, Purdue University, Oct., 1973.
10. Gray, H. J., McGuigan, C. V., and Rowland, H. W., Treated Sewage Serves as Tower Makeup, *Power*, May, 1973.
11. Hofstein, H., and Kim, K. B., Treated Municipal Waste Water as a Major Water Source for Industry, National Conference on Complete Water Reuse (AIChE), Washington, DC, April, 1973.
12. Weddle, C. L., and Masri, H. N., Reuse of Municipal Waste Water by Industry, *Ind. Water Engrg.*, June/July, 1972.
13. Humphreys, F. C., Sewage Effluent in Use as Power Plant Circulating Water, Proceedings of the 14th Industrial Waste Conference, Purdue (1959), reprinted in *Betz Indicator*, 29, No. 3.
14. Cecil, L. K., Sewage Treatment Plant Effluent for Water Reuse, *Water & Swg. Wks.*, 111, 421 (1964).
15. Carnes, B. A., Eller, J. M., and Martin, J. C., Reuse of Refinery and Petrochemical Waste Waters, *Ind. Water Engrg.*, June/July, 1972.
16. Osborn, D. W., Nitrified Sewage Effluents: Their Corrosiveness and Suitability for Use as Power Station Cooling Water, *Jour. Inst. Swge. Purif.*, 243 (1964).
17. Terry, S. L., and Ladd, K., City Waste Water Reused for Power Plant Cooling and Boiler Makeup, National Conference on Complete Water Reuse (AIChE), Washington, DC, April, 1973.
18. Petrey, E. Q., The Role of Cooling Water Systems and Water Treatment in Achieving Zero Discharge, Cooling Tower Institute Meeting, Houston, Jan., 1973.
19. Weisberg, E., and Stockton, D. L., Water Reuse in a Petroleum Refinery, National Conference on Complete Water Reuse (AIChE), Washington, DC, Apr., 1973.

The authors

William L. Harpel is Assistant Director, Product Development, for Betz Laboratories, Inc., Somerton Road, Trevose, PA 19047. A member of the AIChE, the National Association of Manufacturing Engineers, the Air Pollution Control Assoc., and the National Catalysis Soc., he has presented technical papers at a number of national conferences. He holds a B.A. in chemistry and a B.S. in chemical engineering from Lehigh University and did additional graduate work at Stanford University and the University of Pennsylvania.

Edwin W. James is Product Manager, Cooling Water Services for Betz Laboratories, where he has worked since 1968, first as engineer and Assistant Product Manager, Cooling Water Services. An author of several papers dealing with cooling water systems, he holds a B.S. in chemical engineering from Drexel University.

William F. Maguire is Assistant Market Manager, Hydrocarbon Processing Industry for Betz Laboratories. A member of the AIChE and the API, he has prepared and presented papers at several conferences. He holds a B.S. in chemical engineering and a masters degree in business administration.

Design Factors in Reverse Osmosis

Performance of a reverse-osmosis system for water purification depends
on several engineering factors—osmotic pressure, membrane properties,
temperature, degree of fouling, and concentration polarization

EDGAR C. KAUP, Burns and Roe, Inc.

After 200 years as a laboratory curiosity, reverse osmosis (RO) has come into its own as a commercial unit operation. RO can be used to recover valuable byproducts, recycle wasted reactants, help reduce polluting effluents, concentrate beverages and drugs, and most important at this time produce potable water from brackish and saline sources.

The osmotic phenomenon was first noted in 1748 by Abbe' Nollet, who found that water would diffuse through a pig bladder into alcohol spontaneously. Further experiments during the next 100 years were hampered by the unreliability of animal membranes until 1864, when Traube prepared a "selective" membrane of copper ferrocyanide. This was used by Pfeffer in his precise experiments with sucrose and other solutions that linked osmotic pressure with temperature and solute concentration. His data are still referenced in physical chemistry texts as an illustration of osmotic phenomena.[1] Later, J. H. van't Hoff developed a mathematical relationship for osmotic pressure.

Interest in the osmotic process waned after the turn of the twentieth century because reliable membranes were not available. In the 1950's, amid predictions of water scarcity, the U.S. Dept. of Interior established its Office of Saline Water (OSW) to evaluate water purification methods. Reverse osmosis was attractive because of its simplicity and, theoretically, low energy requirements—no energy-wasting phase change takes place. All that was needed was a strong, corrosion-resistant, reliable, cheap, selective membrane.

Cellulose Acetate Studies

In 1958, Professor C. E. Reid[2] at the University of Florida suggested seeking a suitable membrane from commercially available films. In the ensuing program, he and J. E. Breton demonstrated that secondary cellulose acetate (CA), one of the first plastic films manufactured in this country, had the desired selectivity. However, CA had disappointingly low water transport and a very short productive life.

Sidney Loeb and associates,[3] working on a similar project at the University of California, enhanced the water permeability or "flux" of CA membranes and substantially extended their usefulness by heat treating the film and adding swelling agents to the casting formulation. They found that when hydrophilic compounds such as potassium perchlorate and formamide were added to the casting "dope," the resulting film was swollen with water bound to the polymer chains and had 10 to 20 times the water transport rate.

This work demonstrated that the technology necessary to purify water by reverse osmosis was in hand. From this point on, the major advances have been development, engineering and marketing of RO systems, although much investigation and argument still goes on concerning mode of solvent transport, membrane formulations and operating conditions.

Today's Applications

Reverse-osmosis systems for water treatment are now commercially available for any moderate need, ranging from a 2.5-gal./day unit for home drinking supply, to a 150,000-gal./day auxiliary source for municipal water, a 350,000-gal./day plant for vacation resorts and 800,000-gal./day plants for industrial water. Such units have been sold, with quality and quantity of water guaranteed by the manufacturers.

While most applications today are for treating brackish water, i.e., 1,000 to 10,000-ppm. total dissolved solids (TDS), the scope of RO systems is rapidly being extended. Seawater desalting is OSW's next target. This involves problems with high operating pressures, calcium sulfate scaling, and organic fouling. OSW is testing a two-stage plant. The first stage reduces the 35,000-ppm. seawater feed to brackish water with 3,500-ppm. salt, and the second stage yields 350-ppm. TDS product.

Originally published April 2, 1973.

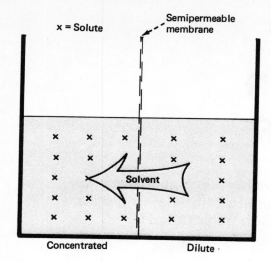

DIRECT OSMOSIS: solvent flows spontaneously through semipermeable membrane—Fig. 1

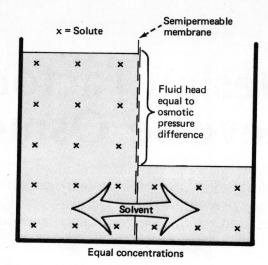

SOLVENT FLOW stops when the osmotic pressures of two solutions are equal—Fig. 2

Semiconductor fabricating plants, which use great quantities of ultrahigh-purity water for washing and rinsing, have reduced the cost of operating demineralizers 90% by pretreating feedwater with a RO unit. Demineralizing costs are determined by regenerant cost; regenerant consumption, in turn, is directly related to ionic composition of feedwater. Therefore, lowering the feed TDS to one tenth by RO pretreatment reduces operating cost proportionally. Some other applications are:

• Food processing—recovery of protein from cheese whey; concentration of maple sap, fruit juices, coffee and tea; concentration of drugs and biological products.

• Pollution control—removing chromate from cooling tower blowdowns; removal of sulfates from acid mine drainage; retrieval of gold, silver, platinum and other precious metals from electroplating solutions and rinses.

• Water reclamation—treating of secondary sewage effluent. An RO plant is operating at the San Diego municipal disposal plant, successfully reducing phosphate in the discharge.

Calculating Osmotic Pressure

Direct osmosis, the spontaneous passage of a solvent from a dilute solution through a semipermeable membrane into a more concentrated one, is depicted in Fig. 1; the opposite process, the moving of a pure solvent through the membrane by an external force on the concentrated solution, is represented in Fig. 3; and the intermediate step, osmotic equilibrium, is shown in Fig 2.

At equilibrium, the external force plus the osmotic pressure of the solution equals the osmotic pressure of the solvent on the right side of the membrane. Therefore, the total force necessary to stop the spontaneous flow of pure solvent into a solution is equivalent to the osmotic pressure of that solution. Osmotic pressure is stated relative to the pure solvent, but in actual practice it is the effective pressure between two solutions. Its magnitude is proportional to the amount of dissolved substances in

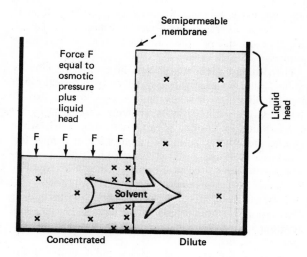

REVERSE OSMOSIS requires applied force equal to osmotic pressure plus liquid head—Fig. 3

the solution and to the temperature of the solution, and is completely independent of the membrane. In 1886, van't Hoff formulated an equation to calculate osmotic pressure, based on data for sugar solutions and the similarity of dilute solutions to ideal gases:

$$\pi = \frac{n}{v} RT$$

where: π = Osmotic pressure, atm.

$\frac{n}{v}$ = Ionic concentration, $\frac{\text{Moles}}{\text{liter}}$

T = Temperature, °K.

R = Proportionality factor, $0.083 \, \frac{\text{atm. liter}}{\text{moles °K.}}$

This equation follows the ideal gas law so closely that the value of the gas law constant is used with reasonable

How Osmosis Works

Osmosis (which comes from the Greek word osmos, meaning "to push") is the tendency of a solvent to flow through a semipermeable membrane from a dilute solution to a concentrated one. It is a simple dilution operation that continually occurs in nature, but has always been a bit mystical because of the strong hydraulic forces that can be developed.

To clearly understand osmosis, a review of physical laws relating to liquids, solutes and solvents is helpful. All matter tends to distribute itself uniformly throughout the universe. Applied to solvents and solutes, this means that an unrestrained solute will disperse itself (diffuse) uniformly throughout a solvent and the two will in time form a homogeneous solution. The distribution rate depends on the solute (or solvent) concentration difference.

A homey example is a cube of sugar dropped into a cup of coffee. Sugar molecules diffuse rapidly from the saturated solution near the solid surface of the cube, throughout the sugar-less, "pure" coffee. Neglecting thermal effects that agitate the solution, the diffusion begins spontaneously, and the highest rate occurs at the beginning, when there is "pure" coffee in one place and pure sugar in another. The rate decreases as the sugar and coffee mix to form a homogeneous solution. Without stirring, uniform sweetening takes a long time because diffusion is a slow process.

One way of describing this effect is by saying the difference in sugar concentration causes the sugar molecules to diffuse. A more-fundamental explanation involves the free energies of the individual species. The partial free energy of the solvent molecule is greatest in the pure state, and is reduced in solution in proportion to the amount of solute added. As more solute is added to a solution, the partial free energy of the solvent becomes smaller and the partial free energy of the solute increases. Thus, if a pure solvent (or a weaker solution) is separated from a solution of that solvent by an impermeable membrane, there is a finite energy difference between the two liquids.

When the barrier is removed, the two species spontaneously begin to diffuse, seeking lower energy levels. Mixing is rapid at first because the energy difference is greatest, but it slows logarithmically as the concentration differences, and therefore the energy differences, throughout the solution decrease. Mixing stops when partial free energy throughout the "new" solution is uniform.

Role of the Membrane

To use the capability of solutions to move spontaneously, there must be present a natural "trick" that will harness movement of the molecules. The "trick" that makes osmosis work is a film with the ability to discriminate between molecules. This film is called semipermeable because it allows free passage of one species, such as water, while holding back dissolved substances such as salts and sugars. Such films are common in nature, e.g., in plant and tree cells and plasma membranes, but not until recently were any permanent man-made membranes available.

Back to our homey example. If a sack made from a semipermeable membrane is partially filled with a saturated sugar solution and then put into a cup of sugarless coffee, we will find that the water from the coffee will spontaneously pass into the sugar solution. Since the membrane is permeable only to water, it will allow water into the sack but not allow the sugar out. Furthermore, if the coffee supply is unlimited and the membrane is without leaks, the water level will continue to build up until the sugar solution is infinitely dilute, i.e., no osmotic pressure difference, or until some restraining force is placed on the sugar side.

Movement of solvent through a semipermeable membrane under the influence of a difference in solution concentrations is called osmotic flow. The process can be stopped at any time by applying a force equal, in our example, to the difference in osmotic pressures of the coffee and the sugar solutions.

When excessive pressure is exerted on the sugar solution, pure water moves into the coffee side of the system and, since the membrane is a barrier to the high-molecular-weight sugar molecules, the sugar solution is concentrated. This situation in which solvent (water) is transferred by an outside force from a solution with high solute concentration (a lower free energy level for the solvent) to a solution with low solute concentration (a higher free energy level for the solvent) is called reverse osmosis.

As water is "osmoted," the sugar solution becomes more concentrated, and more force is required to push out each drop. Carried to the extreme in an ideal system using an infinite amount of force, all the water could be squeezed out, leaving crystals of sugar on one side of the membrane and weak coffee on the other.

---

success. The similarity of the van't Hoff equation to the ideal gas law is so obvious that to lessen confusion the Greek Pi is used, instead of P, for pressure. Over a broader range and especially at higher solute concentrations, the following logarithmic expression gives more-accurate pressure values:

$$V = RT \ln \frac{P_0}{P_1}$$

where: P_0 = vapor pressure of the pure solvent
P_1 = vapor pressure of the solvent in solution

This relationship is not as convenient to use, since vapor-pressure measurements of "practical" solutions are not usually available. Also, since most RO applications are limited to weak, aqueous solutions (brackish water contains 0.1% to 1.0% dissolved solids, and seawater, 3.5%), it is usually adequate to roughly estimate the osmotic pressure from simple chemical analyses of the feed. The exact-pressure value is of little concern to the operation of a RO system. Therefore, refinements to the van't Hoff equation will not be discussed any further because for practical applications these improvements are inconsequential.

The osmotic pressure is only the barest minimum force required to obtain pure water from a solution. In current work, the operating pressures of RO systems are many times the osmotic pressure. For example, the osmotic pressure of the 2,000-ppm. brackish-water feed used for testing RO units at OSW's Roswell Test Facility in New Mexico is 32 psi., but the operating pressures of the units range from 250 to 800 psi. Table I lists osmotic pressures for several pure solutions and Roswell brackish waters calculated by van't Hoff's equation.

Types of Semipermeable Membranes

Many natural materials have semipermeable characteristics. Animal and plant membranes mentioned earlier are, of course, the best known examples. Collodion, cellophanes, porous glass frits, finely cracked glass, and inorganic precipitates such as copper ferrocyanide and zinc and uranyl phosphates have been used historically. All, however, have the common faults of leaks, short-lived selectivity and poor reproducibility.

Since Reid and Breton's discovery[4] that cellulose acetate possessed good ion selectivity, CA has become the most universally studied membrane material and the most successful. It is now the standard against which other developments are measured.

Numerous polymers have been found selective. Their suitability as a membrane is being investigated along with all sorts of modifications to cellulose acetate. Most of the research and development is sponsored by the U.S. Dept. of Interior's OSW, with the expectation that private industry will exploit the uses once feasibility is demonstrated.

One development that has received very little federal support until recently is Du Pont's Permasep membrane. Originally, Du Pont developed a system based on a nylon derivative that had moderate semipermeability. The membrane was spun in a unique form of a hollow, hairlike fiber to realize enough surface area to make it reasonably productive. This membrane was not successful because, even with its large surface area, water transport was very low (0.1 gal./day/sq.ft.) and rejection of the chloride ion was poor. Du Pont developed a second generation of Permasep membranes in 1969 with a new aromatic polyamide fiber that largely overcame the past failings. The system worked well enough so that in 1971 the company received a contract to work on a single-pass membrane system for producing potable water from seawater. To date, the published results have been quite optimistic.

One serious restriction of today's semipermeable membranes is their loss of ion selectivity (ability to reject specific ions) when operating above room temperature. The practical lower temperature limit is 40 to 45 F., a range in which diffusion is very sluggish. Increasing the temperature from 50 to 77 F. raises productivity 50%. Above 85 F., most membranes become unstable, show poor discrimination and lose strength.

General Electric is investigating polyphenylene oxide and sulfonated phenylene oxide polymers that show moderate selectivity at higher temperatures. The maximum operating limit has been increased to about 130 F. This is still not very hot, but it yields a three- or four-fold productivity increase.

Higher productivity is promised by dynamic membranes, which are loosely-structured polymeric mats with semipermeable characteristics. They are held in place on a filter support by the force of the liquid flowing through them. Dynamic membranes, however, have little cohesion and dissipate into the feed when flow stops.

Dynamic-membrane systems are operated continuously until cleaning is necessary. Then flow is stopped, the membrane dissipates and is flushed away. A new membrane is formed by reacting a mixture of hydrous zirconium oxide and polyacrylic acid in place. Production rates are very high, with fluxes on the order of 100 to 200 gal./day/sq.ft.

Properties of CA

Cellulose acetate, with an established technology gained through 60 years of experience in casting massive quantities of film, has good selectivity, dope formulations amenable to variation, good availability of raw materials and relatively low cost.

Chemically, cellulose acetate is a hydroxylic polymer made up of long chains of β-glucoside units (30,000–60,000) that have been acetylated with acetic anhydride and then hydrolyzed to reduce acetylation to about 40%. Most often, the CA powder supplied for RO membrane casting is in this partially acetylated form, which is known as 2.5 cellulose acetate. RO membranes are also cast from the fully acetylated form (triacetate). These perform equally well.

When cast as RO membranes, CA is a film about 4-mils thick. It is asymmetric—that is, the film has a thin, dense layer of about 0.25 microns above a thick, porous layer. Water passes easily from the dense layer through to the porous one, but with difficulty the other way. Unlike the thick, amorphous underlayer, the dense layer on top of the membrane is made up of tightly packed and organized chains of CA polymer that attract and hold water. Thus, water and solute are separated because the water molecules can form hydrogen bonds with the acetyl groups on the polymer, while many other species can not.

One simplified explanation of the separation process is: The CA polymer chains within the dense layer are highly organized because the membrane receives an annealing treatment that shrinks the film and crystallizes the polymer. The long molecules are somewhat separated and relatively immobile because of the interplay of van der Waals' forces. In this state, the polymer chains are close enough to crosslink with water molecules in the casting formulations and the annealing media.

These water molecules bridge across adjacent chains by forming strong hydrogen bonds with the acetyl groups. In this way, the voids between chains are filled with bound-water molecules and no foreign substance can pass. If the polymer chains were not densely packed, water molecules would bond to acetyls on the same chain and leave voids through which foreign ions could pass.

Water molecules move through the membrane by an applied pressure that pushes the water from a bond with one acetyl group to the next. Only a moderate force is necessary because the bonds are transferred, not broken. Dissolved ions or molecules that do not hydrogen bond cannot enter into attachments with bonding sites (acetyl groups) and are left to concentrate at the membrane surface.

Water Flux

Two parameters for characterizing an RO system are (1) production per unit area of membrane and (2) product quality. Production is measured by water flux, defined as the amount of product recovered per day from a unit area of membrane. English units (gal./day/sq.ft. are used in field work, while metric units (g./sec./sq.cm.) are used in the laboratory.

The flux through a particular membrane is determined by its physical characteristics (e.g., thickness, chemical composition, porosity) and by the conditions of the system (e.g., temperature, differential pressure across the membrane, salt concentration of solutions touching the membrane, and velocity of the feed moving across the membrane). In practice, the properties of the membrane and the solutions are relatively constant, and water flux becomes a simple function of pressure. Quantitatively, it is described by:

$$F_{H_2O} = A(\Delta P - \Delta \pi)$$

where: A = a permeation coefficient for a unit area of a membrane. This term includes the physical variables of the membrane and is relatively constant.

$\Delta P = (P_F P_p)$ = pressure exerted on the feed solution, P_F, less the pressure on product, P_p.

$\Delta \pi = (\pi_F - \pi_p)$ = osmotic pressure of the feed solution, π_F, less the osmotic pressure of the product, π_p.

The equation can be simplified for high pressures (above 600 psi.) by lumping all the constants into the coefficient A'. Flux for brackish water at constant temperature is very approximately:

$$F_{H_2O} \sim (A')(P_{Feed})$$

Temperature Factor

Flux is also affected by feed temperature. Water permeability of the membrane increases about 1½% per °F. The flux of a membrane is usually given by the manufacturer for 75–77 F., and a correction factor is applied at other temperatures. The correction can be derived on theoretical grounds from diffusivity and viscosity values, but experimentally determined corrections are held in more confidence.

The correction factor is applied to the area computed for feed at 25 C. (77 F.). This multiplier is obtained for the lowest expected operating temperature from a curve such as the flux/temperature correction curve in Fig. 4. This curve was developed by the Gulf Environmental Systems Co. for modified cellulose-acetate membranes. Application of the data is illustrated in the following membrane-area calculation:

It is desired to specify the membrane area for a 100,000-gal./day RO system to treat brackish water. Records show the lowest water temperature expected for any prolonged period is 68 F. (20 C.). The cellulose acetate membrane chosen for this system is expected to have an average flux of 15 gal./day/sq.ft. over its two-

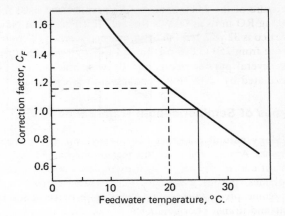

FLUX/TEMPERATURE correction curve—Fig. 4

year-life, operating at 600 psig. The average flux was determined at a base temperature of 77 F. The required membrane area is:

$$\text{Membrane area} = \frac{\text{Water production, gal./day}}{\text{Water flux, gal./day/sq.ft.}}$$

at 77 F.

$$\text{Area} = \frac{100,000}{15} = 6,667 \text{ sq.ft.}$$

The correction factor, C_F, for feed at 20 C., determined from Fig. 6, is 1.15.

$$\text{Area at 68 F.} = 6,667 \times 1.15 = 7,667 \text{ sq.ft.}$$

If the additional product is not desired or necessary when temperature is above 68 F., the operating pressure of the system can be reduced.

Flux Decline

Flux falls off during the life of the membrane for several causes. First, there is an initial decrease because capillary water is squeezed out as the operating pressure collapses the porous layer of membrane. CA has some "memory" and if the pressure is released before the membrane is permanently set, it will swell and regain some of the capillary water.

Following this initial compaction, there is a long-term compaction that slowly reduces the water flux. This results from densification of the thin, air-dried, membrane layer and probably corresponds to narrowing of pores through which water must pass. As the channels narrow, flow decreases.

Another cause of flux decline is the hydrolysis of acetyl groups that continually occurs during the life of the membrane.* This reaction results in a loss of hydrogen bonding sites, which further reduces the water transport. The reaction is also a source of salt leakage because there

*RO membranes are limited to a pH operating range of 3 to 7, outside of which rapid hydrolysis and membrane degradation occur. Presently, the optimum range is believed to be pH 5–6.

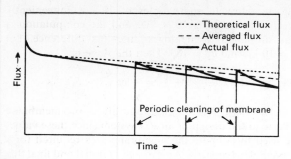

EFFECT of membrane cleaning on flux—Fig. 5

are fewer water bridges blocking the passage of foreign materials through the pores.

Loss in productivity happens to every membrane; it occurs slowly and is permanent. Chemical rejuvenation and low-pressure operation to relax and swell the matrix have been tried without success; the membrane simply ages and flux decreases, until economics dictate replacement.

Fouling: Causes and Cures

Fouling or temporary flux reduction is caused when foreign materials coat the membrane surface and interfere with inward movement of water. Materials accumulate because only hydrogen-bonding substances (water, ammonia) pass through the membrane's discriminating pores. Nonbonding materials are left in the quiescent film known as the liquid boundary layer.

The composition of deposits in the boundary layer reflects the composition of the feedwater. As expected, the most common constituents are calcium carbonate and sulfate scales, hydrates of iron oxides and aluminum, silicates, miscellaneous particulates, and biological growths. Most can be minimized by pretreating the feed to remove iron and control pH; by limiting the process to nonscaling concentrations of waste; by filtration; and by injection of small amounts of biocide.

Low concentrations of chlorine have been used to prevent organic growths on CA membranes, without discernable adverse effects. But chlorine definitely is not recommended for the polyamide type. Its use with Permasep membranes voids all guarantees.

Unfortunately, no matter how thorough the protection, fouling always occurs. The usual cleaning procedure is to first flush the membrane with feedwater at reduced pressure and two or three times normal velocity; the turbulent action of the fluid loosens the foulants and carries them away. Water flushing is a preliminary step in every cleaning operation.

Hardness scales (carbonate and sulfate salts of calcium and magnesium) can sometimes be removed by simply soaking the unit in distilled water for long periods. But this is time-consuming, and most often a warm solution of 1–2% citric acid is vigorously circulated through the unit. The acid dissolves large amounts of metallic ions and keeps them in solution by chelation.

Microbiological growths occur in most natural waters

and are a particular problem when treating secondary sewage. They are often removed by recirculating washes of enzyme detergents. Organic fouling almost always reoccurs, and periodic cleanings are necessary to keep flux at a high level.

Flux degradation by fouling is an additional loss superimposed on the permanent losses. As seen in Fig. 5, the actual flux curve follows a declining, saw-tooth pattern when the membrane is cleaned periodically. Without cleaning, flux would follow the lowest curve, a projection of the initial smooth decline. The flux from a membrane that is never fouled is shown by the upper line, the theoretical flux curve that touches only the peaks of the saw-toothed curve.

Prediction of Flux

Unlike other unit operations in which production improves after an initial break-in period, water output from an RO system begins to decline as soon as the membrane is pressurized, and continues to degrade slowly throughout the membrane's life. The loss is irreversible, and if more flux is required the feed pressure must be increased. This alternative is somewhat self-defeating since additional pressure, while producing more water, also compresses the membrane further and hastens flux decline. Normal practice is to overspecify membrane area slightly and to keep operating pressure constant as long as possible, resorting to additional pressure late in the life of the membrane.

For a desalting system to be useful, its water production must, of course, be reliable. The output of a membrane is predictable because, as suggested above, the decline in productivity from a unit area of membrane is quite uniform and can be projected.

A linear plot of flux vs. operating time at a particular pressure yields a curve with an initial steep descent, followed by a prolonged, moderate decline. On exponential paper, this curve becomes a straight line and is reasonably constant for one and probably two years. Because of this constancy, flux can be predicted for the life of a membrane once the initial flux and curve slope are known.

Manufacturers provide initial flux values and estimates of curve slopes at various operating pressures. These slopes are experimentally determined for various simple salt solutions or local tap water. They can be applied only as guides for other waters.

Initial flux is the production for the first 24 hr., divided by the membrane area in the test unit. The decline slope is computed or graphically determined from flux values taken at time intervals such as 10, 100 and 1,000 hr. The decline rate is given as the slope of a log-log plot of flux in gal./day/sq.ft. vs. operating time in hours. The equation for the slope of the flux curve is:

$$m = \frac{\log F_i - F_x}{\theta_i - \theta_x}$$

where: F_i = initial flux in gal./day/sq.ft.
F_x = flux at time x hr.
θ_i = operating time, hr., for initial flux
θ_x = operating time, in x hr.

Some Reasons for Solute Rejection

- **Large nonelectrolytes are not permeable.** Water-soluble molecules that are too large to fit through the pores in the membrane are rejected whether they hydrogen bond or not. The historical example is, of course, Pfeffer's osmosis experiments with sucrose and dextrose. Blunk estimates the radius of an average pore to be about 3.6 A. and suggests sucrose is rejected because its radius is 4.4 A. Sourirajan showed that, besides large sugar molecules, sorbital glycerol, pentaerythritol are also well rejected. Molecular weight is sometimes used as a *rough* measure of permeability; nonelectrolytes with molecular weights over 200 are well rejected, those below are not.

- **Small, hydrogen-bonding, nonelectrolytes are permeable.** Simple straight-chain organics (four carbons or less) pass through membranes if they possess hydrogen-bonding abilities. Examples of molecules that permeate by hydrogen bonding are alcohols, aldehydes, acids, amines. Others are hydrogen peroxide, urea, acetamide and nitrates. Rejection of organic materials improves as the molecules become large, sterically complex and/or polyfunctional. Although organic acids and amines permeate easily in their free state, they are well rejected when neutralized to the salts. For example, acetic acid readily permeates CA membranes as the free acid but is 98–99% rejected when combined into the ionic sodium salt.

- **Dissolved gases such as carbon dioxide, sulfur dioxide, oxygen, chlorine and hydrogen sulfide are permeable in their free state.** Gases that are dissolved in water under conditions amenable to combinations with dissolved molecules are rejected. Carbon dioxide is poorly rejected from solutions with pH 4 or below but is 80–98% rejected when the acidic gas is neutralized to sodium bicarbonate. A similar permeability is found for sulfur dioxide at pH values below 3. Chlorine is poorly rejected below pH 7, but its permeability can be used to advantage when producing potable water with cellulose-acetate-membrane RO systems. Small amounts of chlorine added to the feed of a RO unit not only protect the membrane from biological growths but also purify the product.

- **Rejection of electrolytes increases with valences of their ionic species.** Di- and trivalent ions are more thoroughly rejected, no matter how strong their hydrogen-bonding tendencies, than monovalent ions. For instance, the divalent cations calcium and magnesium are more strongly rejected than the monovalent sodium and potassium. Similarly, the sulfate anion is much more strongly rejected than the monovalent chloride ion. A few small univalent species readily permeate membranes under most conditions because of their strong hydrogen-bonding tendencies. Hydronium, ammonium and fluoride ions and boric acid, phenol and acetic acid are examples of species that permeate easily. Acetic acid and phenol permeate so readily that they have shown some indications of concentrating in the product.

Time intervals in multiples of 10 are conveniently selected because log 10 = 1.00 and the computation is thus simplified to the log of the flux difference. Let $\theta_i = 10$ hr. and $\theta_x = 100$. Then the slope is:

$$m = \frac{\log F_i - F_x}{(1.0 - 2.0)} = -\log (F_i - F_x)$$

Determination of flux over the life of the membrane is basic to the design of an RO system since these values are used to estimate the membrane area required for a desired plant capacity. An average of initial and final flux can be chosen as a reasonable compromise to obtain an average area. But when such a choice is made, it is expected that the flux will be augmented during the later life of the membrane by increasing operating pressure. A second design approach is to specify membrane area from the final or smallest flux value. In this case, although fixed costs are highest, lower system pressures will decrease operating costs.

Another alternative is to select the initial flux as the design basis. This minimizes membrane area and capital costs but raises operating costs because pressures will have to be increased to maintain production. This is a logical choice for intermittent and short-term projects in which initial costs must be minimized.

Salt Permeability

Product quality from a semipermeable membrane is measured by the amount of solute or salt in the product. This depends on selectivity of the membrane and its imperfections. Theoretically, the salt passing through the membrane, or simply salt flux, is a function of the membrane's permeability coefficient, and the difference between the salt concentration in the feed and the product. (The coefficient is based on the physical characteristics of the membrane, such as thickness, salt diffusion, and the distribution of solute between the membrane and the solution.) The amount of salt passing through a unit area of membrane is:

$$F_{salt} = (C_H - C_L) = \beta \Delta C \approx \beta C_H$$

where: F_{salt} = salt flux, g./sq.cm./sec.
β = salt permeability coefficient, cm./sec.
C_H = concentration of solute on high-pressure side of membrane, e.g., concentrate side, g./cc.
C_L = concentration of solute on low-pressure side of membrane, e.g., product side, g./cc.

The equation says that salt passage is the result of the difference in salt concentrations across the membrane. It is believed that most salt moves by diffusion through the same membrane channels as water.

From the equation, normal salt flux (unlike water flux) is independent of pressure. Theoretically, if the pressure of the RO system is increased, the salt will diffuse at a constant rate, while the water flow will increase. The result will be greater production of purer water. Careful experimental work shows that these principles are correct, but often improvements are not seen because imperfections in the membrane leak substantial amounts of salt into the product. The problem of leakage is discussed later.

Rejection Factor

The measure of membrane selectivity is salt rejection, the ratio of solute rejected by a membrane to the solute in the feed. It is the most common method of evaluating a membrane's ability to separate salt, because the determination is simple and can be done as accurately in the field as in the laboratory.

Salt rejection is computed by dividing the difference in salt concentrations between the high- and low-pressure sides of the membrane by the high side. The result is expressed in percent. The solute concentrations can be obtained by chemical analyses or total dissolved solids (evaporation to dryness) but usually are determined by conductivity. This is a rapid, reliable method that requires a minimum of equipment and skill.

There are two possible bases for computation. One is the concentration of the feed solution as it is fed to the unit, which is an accurate representation if the membrane area is small. If, however, the membrane area is large, and the path between the inlet and outlet is long, there can be a considerable difference in feed concentration by the time the feed emerges from the discharge point. Since the latter portions of the membrane system are "seeing" a more concentrated solution, it follows that salt flux near the discharge of an RO unit will be greater than it was at the inlet.

To compensate for this increase in solute concentration, and give a more accurate description of the membrane's selectivity, most manufacturers use the feed-brine average between inlet and outlet as the basis for calculating salt rejection. This method always "improves" the salt rejection by a few percent over the simpler inlet-feed method. Field operations usually use the inlet basis because the calculation is simpler and the evaluation is of an RO system, not the membrane alone. Both methods of computing salt rejection are valid. Choice should be influenced by whether the system or membrane is being evaluated.

Influences on Salt Rejection

The ability of a membrane to reject salt is a complicated problem that seems to depend on combinations of physicochemical characteristics of the solute, the membrane and the water. The properties of the solute that have the most influence on the rejection of individual species are:

• Valence charge: Rejection increases with size of charge on ion.

• Molecular size: Rejection increases with physical size.

• Hydrogen bonding tendency: Permeation increases with strength of hydrogen bonding.

The membrane itself restricts solute, by limiting the physical size of the species that will pass through the pores, and by showing preference for solutes that hydrogen bond. The water molecules present in the membrane can also affect rejection. It appears that when water molecules become strongly bonded, as in a direct "bridge" between two acetyl groups, the water loses its

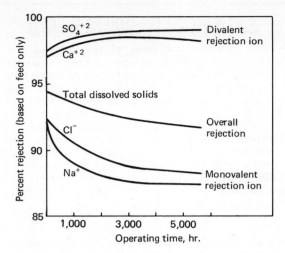

SALT REJECTION varies with time—Fig. 6

solvating ability and rejects salts that it formerly dissolved as a free molecule.

Also, water in some instances may be able to discriminate. It is believed that water molecules can form weak, secondary bonds to those that are strongly attached elsewhere. The weakly-bonded molecule will retain part of its solvency and be able to dissolve small, univalent ions. Thus, weakly-bound water molecules could act as a vehicle for some solutes and "carry" them through the membrane as part of the water flux.

The ability of a membrane to reject salts decreases with time. The decrease is noted first for small, univalent ions such as sodium and chloride. These are normally among the most permeable, show the lowest initial rejection, and have the highest rate of decline. Divalent cations such as calcium and magnesium, and anions such as sulfate, are initially rejected very well and show a very slow rate of decline.

Membrane units tested in Webster, S.D., show a constant overall rejection of 98% because the water is very hard and contains low salinity (sodium chloride). In another location, where both salinity and hardness are high, the rejection of divalent ions was at the same level but the overall rejection was only 91%. Because of the high saline content of the feed, rejection was initially lower and declined at a rate approaching that of sodium chloride.

The variation of rejection with time is illustrated in Fig. 6. Note the decline in rejection of monovalent ions as compared with the constant (and even improving) rejection for divalent ions. The overall salt rejection tends to follow the monovalent curve. The water in this test contained 950 mg./l. sodium chloride, 800 mg./l. hardness as $CaCO_3$, and total dissolved solids of 2,000 mg./l.

The progressive decrease in salt rejection may be caused by hydrolysis of the membrane, with subsequent loss in bonding sites. Another cause may be an increase in pore size due to membrane swelling. Most probably, it is the result of both effects.

At the present time, there are no quantitative laws governing the overall rejection of mixed species. Qualitative statements are made based on extensive experiments.

Costs of Reverse Osmosis Systems

| | Operating Costs, ¢/1,000 gal. | | | |
| | 50,000 gal./day | | 1 million gal./day | |
Power, 400 psi.,	Spiral	Hollow Fiber	Spiral	Hollow Fiber
Power, 400 psi., 1¢/kwh.	8	8	8	8
Chemicals	2	2	1	1
Membrane replacement, including labor	18	18	10	10
Labor, $8/hr. Operations	18	18	1	1
Maintenance	3	12	2	8
Total	49	58	22	28

Data from Gulf Environmental Systems. Based on surface water with a residual feed turbidity of 1–5 Jackson units, and 75% recovery. Chemical costs based on conditioning with sulfuric acid at 1 lb./1,000 gal. Maintenance costs assume cleaning twice a week for hollow fiber units and once every two weeks for spiral systems.

Data have been reported by workers such as Pfeffer,[1] Reid and Breton,[2] Loeb and Sourirajan,[3] Lonsdale,[4] Keilin,[5] Glueckauf,[6] Hodges,[7] and Blunk.[8] Their results often agree, but often do not; sometimes for explainable reasons but sometimes not. The qualitative statements in the box concerning rejection of specific solutes by cellulose acetate membranes were distilled from the experiences of this author and many other investigators.

Membrane Leakage

Calculated salt rejection is always higher than the experimental values, even those that are determined under the most careful conditions. For example, the theoretical rejection of sodium chloride for a modified CA membrane is calculated as 99.7%, but experimental results show no better than 97–99% exclusion. The difference is believed caused by imperfections in the membrane (minute holidays) through which the pressurized brine can flow and contaminate the product water.

All membranes have imperfections; these are probably not manufacturing faults but a "property" of the membrane that must be adjusted or optimized to suit a particular service. Hence, very porous CA membranes are used to screen out large molecules (20–500 A. dia.) and very small particles for ultrafiltration applications; less-porous membranes are selected for high water flow and moderate salt-rejecting service in "saltless softeners"; and membranes heat-treated to low porosity are used for applications requiring high rejection.

Fortunately, most imperfections are small and easily plugged. Distribution runs from numerous holes with diameters of 100 A. and smaller, to a few with 1-micron diameters. The major source of product contamination results from salt passage through the larger holes because, by Poiseuille's law for viscose flow, salt leakage increases as the fourth power of the pore diameter. In addition to the heavy salt flow through large pores, considerable salt must also pass through the numerous small holes because the small-sized sodium and chloride ions always appear in the product, no matter how minor their concentration in the feed.

A number of ways have been tried to stop membrane leakage. Heat treating the membrane and modifying the dope formula seem to be the only permanent ways to improve semipermeability. Membranes for seawater must be highly heat treated to yield a film that can reject 99.5% of the salt in the feed.

Rejection can be temporarily improved by adding to the feed certain chemicals whose molecular size is large enough to plug up the leaking pores. Loeb[9] discovered that trace amounts of aluminum salts that occur naturally in Los Angeles tap water plugged the leaking pores of test membranes and improved salt rejection. Zephiran, a tetralkyl ammonium chloride, was used by Loeb and Manjikian[10] to gain a similar and more reproducible effect. Other materials that will improve rejection include polyvinyl methyl ether and Dowfax.

Unfortunately, leak-stopping additives have serious drawbacks that limit their use: They are more effective on low-flux, standard CA membranes than on the modified ones now used almost exclusively; they dissipate quickly and must be regularly replenished; most of them reduce water flux as they reduce salt leakage; and worst of all, they are expensive. These additives are useful only in special situations in which cost is not a factor.

Concentration Polarization

In all desalting processes, there are dynamic, localized concentrations of solute at those points where the solvent leaves the solution. In distillation processes, there are small increases in solute concentration at the solution surface when the water vapor leaves, but this is soon dispersed by bubble and thermal agitation. In membrane processes, however, the solute accumulates at a fixed boundary (the membrane) in a relatively stable layer (the boundary layer) and brings about an effect called concentration polarization. This is defined as the ratio of the solute concentration at the membrane wall relative to that in the bulk of solution.

Initially, the boundary layer is at the same concentration as the bulk solution, but since the wall (membrane) is permeable to solvent (water), the boundary layer becomes heavily populated with solute as solvent leaves through the channel wall. The layer grows thicker and more concentrated because the rate of solute diffusion away from the membrane cannot keep up with solvent flow through the membrane.

The layer at the wall can be reduced by higher feed velocities and turbulence. The extra flow through the unit at higher feed velocity results in overall lower product recovery, i.e. smaller ratio of product to feed. This increases power consumption and the amount of waste produced.

Turbulence promoters are a more attractive way to reduce the boundary layer. Tubular membrane units have small plastic balls in the annular space to break up the smooth flow of the feed solution. RO units with sheet membranes often place waffle-like polyethylene screens in the flowpath. With turbulance promoters, good filtering of the feed is necessary so that particulates or precipitates do not get lodged in narrow flow spaces and plug up the membrane element.

Effects of Polarization

When feed salinity increases, the product shows a corresponding increase in salt content. This effect is noted from the equation for salt flux, which shows that the salt or solute passed is almost entirely dependent upon feed concentration. Thus, the concentration of the solution nearest the membrane governs the amount of salt in the product. Unfortunately, the highest concentration is at the wall. For this reason, feed flow is adjusted to give the least concentrated solution in the boundary layer. The usual concentration polarization ratio is 1.2 to 2.0, which means that the concentration at the wall is 1.2 to 2.0 times that in the bulk of the feed.

Water flux will fall off when the salt content of the feed-brine solution increases, because the amount of salt affects the driving force of the flux. (Net driving force is the difference between the net pressure and the net osmotic pressure, and the salt content of the solutions determines the osmotic pressure.) The solution adjacent to the membrane is the most influential in the system because the osmotic pressure of this solution sets the driving force of the system.

As an example, let us examine what can occur in an RO unit treating seawater at 1,000 psi., when the product recovery is increased. Assume the seawater (35,000 mg./l. TDS) has an osmotic pressure of 350 psi. and that the water recovery of the unit is normally 17%. If the concentration polarization factor is 1.2, the salt concentration at the wall of the membrane is 42,000 mg./l. TDS, and the osmotic pressure at the membrane becomes 420 psi. If the feed is slowed so that half the water in it is recovered, the solution at the wall is concentrated 100%, and the local osmotic pressure is 700 psi. As a result of the increased concentration at the wall, water flux (production) will fall off almost 50% because the driving force has dropped from 580 psi., (1,000 − 420 = 580), to 300 psi., (1,000 − 700 = 300).

Thus, the power requirements of an RO system vary with concentration polarization. When the salt concentration at the membrane increases, either the operating pressure must be raised to compensate for higher osmotic pressure, or the pumping rate must be increased to gain more turbulence and thereby dilute the concentrated salt solution.

When dealing with natural brackish waters containing 2,000 to 10,000-ppm. dissolved solids, the concentration polarization effect is small because the net pressure is large and the osmotic pressure is low. The operating pressures of brackish-water units usually range from 250 and 800 psi., and the osmotic pressures are from 30 to 100 psi.

A very common problem caused by high concentration polarization is the plugging of membranes by sparsely soluble salts. If the solution at the wall becomes supersaturated, insoluble salts precipitate. These deposits can stop water flux and cause permanent membrane deterioration. The most common scales are calcium carbonate and calcium sulfate, since they occur in almost all natural waters. They have low solubilities, 14 mg./l. and 1,500 mg./l. respectively, which are easily exceeded if natural waters are overconcentrated.

Some of the precautions taken to prevent scaling are: (1) pH adjustment by adding acid to the feed, which breaks down the bicarbonate ion and prevents formation of carbonates in the boundary layer; (2) keeping the product recovery, i.e. product-to-feed ratio, low (less than 50% for most natural waters) to assure that a saturated calcium sulfate solution is not formed in the boundary layer; and (3) adding antiscale substances such as sodium hexametaphosphate to prevent deposits. Deposits can be dissolved by chelating agents described earlier; but after cleaning, the membrane always shows permanent damage.

References

1. Glasstone, Samuel, "Textbook of Physical Chemistry," D. Van Nostrand Co., New York, pp. 651–683.
2. Reid, C. E., and Breton, J. E., *Journal of Applied Science,* 1959, pp. 1–133.
3. Leob, S., and Sourirajan, S., UCLA Department of Engineering Report 60-60, 1960.
4. Lonsdale, H., et al, Reverse Osmosis for Water Desalination, Office of Saline Water Research and Development Report No. 111, Pb. 18-1696, 1964.
5. Keilin, B., Mechanism of Desalination by Reverse Osmosis, Office of Saline Water Progress Report No. 117, Pb. 166395, 1964.
6. Glueckauf, E., Proceedings of First International Desalination Symposium, Paper SWD/1, Wash., D.C., Oct. 3–9, 1965.
7. Hodges, R. M. Jr., Polymer Films As Reverse Osmosis Membranes For Sea Water Desalination, Ph.D. Thesis, M.I.T. Dept. of Chemical Engineering, Cambridge, Mass., 1964.
8. Blunk, R. W., Study of Criteria for the Semipermeability of Cellulose Acetate Membranes to Aqueous Solutions, UCLA Dept. of Engineering Report 64-28, 1964.
9. Leob, S., Seawater Desalination by Means of a Semipermeable Membrane, UCLA Dept. of Engineering Report.
10. Leob, S., and Manjikian, S., Brackish Water Desalination by an Osmotic Membrane, UCLA Dept. of Engineering Report 63-37, 1963.

Meet the Author

Edgar G. Kaup is a senior chemical engineer at Burns and Roe, Inc., Oradell, NJ 07649. He is presently field manager for an acid-mine drainage-control project in Hawk Run, Pa., and previously was resident engineering manager at Roswell, N.M., where he evaluated membrane systems for the Office of Saline Water. Mr. Kaup has a B.S. degree in chemistry from Lehigh University, and B.S. and M.S. degrees in chemical engineering from Newark College of Engineering. He is a registered engineer in Pennsylvania and New Jersey.

Choosing a process for chloride removal

The presence of traces of organic chlorides in wastewater now looms as a major problem in many segments of the chemical process industries. Here is a summary of techniques for removing these compounds.

M. F. Nathan, Crawford & Russell, Inc.

☐ Many desirable properties—such as inflammability, biological inertness, useful vapor pressures, and solvent power—have for years made organic chlorides attractive both to chemical processors and domestic consumers for a wide range of uses varying from chemical intermediates, industrial solvents and dry-cleaning fluids to aerosol sprays and an anesthetic. In these uses, the organic chlorides have been generally considered inert and relatively harmless.

Examples of such chlorides: chlorobenzene, an early precursor for production of phenol; vinylidene chloride, a monomer for the production of plastic film; trichloroethylene, a vaporous degreasing agent for fabricated metal products; perchloroethylene, a dry-cleaning solvent for clothing; trichlorofluoromethane, an aerosol; chloroform, an anesthetic; and polychlorinated biphenyls (PCBs) as transformer fluid and condenser dielectric.

However, organic chlorides are no longer considered harmless. Various governmental regulations now require that they be kept out of the environment. Vinyl chloride monomer in air is now under stringent regulation by OSHA and EPA. Chlorofluorocarbons have recently been banned for use as propellants. OSHA is reported to be reducing exposure to chloroform, carbon tetrachloride and trichloroethylene. PCBs are banned. Further, the chlorination of water as a hygienic practice has come under scrutiny and probably will be curtailed since chlorination converts hydrocarbons in the water to carcinogenic chlorinated organics.

Consequently, removal of organic chlorides from water has become an important problem. This problem is compounded by existence of some of the same properties that in the past have made organic chlorides so desirable. Such chlorides are considered to be resistant to biodegradation and in some cases are completely nonbiodegradable.

Also, although the distribution of organic chlorides between water and hydrocarbons is favorable for extraction, such extraction results in a wastewater that contains solvent, and thus requires further treatment. Although organic chlorides can be adsorbed on activated carbon and polymeric resins, these adsorbents are too expensive to be used once and discarded, so that

regeneration is necessary. In the case of carbon, the cost of first dewatering and then regenerating it in a multi-tiered reducing-atmosphere furnace, followed by an oxidizer to destroy the material, becomes expensive. In the case of polymeric resin, regeneration involves an expensive distillation system for recovering a regeneration solvent and rejecting the water that ends up in the regeneration streams.

Finally, although the high activity coefficients of many organic chlorides make steam stripping possible, these same high activities mean that tray efficiencies will be low or that the heights of packed towers will be excessive.

Match the process to the chloride

On the other hand, mutant bacteria have been developed to destroy many organic chlorides; and when

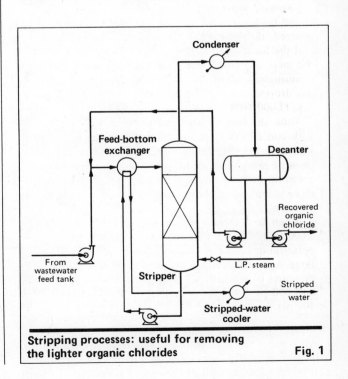

Stripping processes: useful for removing the lighter organic chlorides

Fig. 1

Originally published January 30, 1978.

217

These properties help identify the process for removing organic chlorides from wastewater   Table I

Organic chloride	Formula	Mol. weight	Vapor pressure		Organic solubility in water, weight %	Water solubility in organics, weight %
			mm Hg	@ °C		
Methyl chloride (Chloromethane)	CH_3Cl	50.49	4028	25	0.46 @ 20°C	
Methylene chloride (Dichloromethane)	CH_2Cl_2	84.94	420	25	1.96 20°C	0.167 25°C
Chloroform (Trichloromethane)	$CH Cl_3$	119.38	197.4	25	0.8 @ 20°C	0.932 @ 25°C
Carbon tetrachloride (Tetrachloromethane)	CCl_4	153.84	115.25	25	0.16 @ 25°C	0.0116 @ 25°C
Ethyl chloride (Chloroethane)	CH_3CH_2Cl	64.92	1199	25	.574 @ 20°C	
Vinyl chloride (Chloroethylene)	$CH_2=CHCl$	62.5	2660	25	0.68 @ 20°C	0.9 @ 20°C
Ethylidene chloride (1,1, Dichloroethane)	CH_3CHCl_2	98.97	225	25	0.537 @ 30°C	0.115 @ 30°C
Ethylene dichloride (1,2 Dichloroethane)	CH_2ClCH_2Cl	98.97	82	25		0.187 @ 25°C
1,1,1 Trichloroethane (Methyl chloroform)	CCl_3-CH_3	133.42	125	25		0.0339 @ 20°C
Trichloroethylene (Trichloroethene)	$CCl_2=CHCl$	131.4	74.3	25	0.11 @ 25°C	
Tetrachloroethylene (Perchloroethylene, Ethylene tetrachloride)	$Cl_2C=CCl_2$	165.83	19	25	0.015 @ 25°C	0.0105 @ 25°C
Ethylene chlorohydrin (2-Chloroethanol)	$ClCH_2CH_2OH$	80.52	8	25	miscible with water	
Propyl chloride (1-Chloropropane)	$CH_3CH_2CH_2Cl$	78.54	335	25	0.27 @ 20°C	
Isopropyl chloride (2-Chloropropane)	$CH_3CHClCH_3$	78.54	505	25	0.31 H_2O @ 20°C	0.33
Allyl chloride (2-Chloro-1 propene)	$CH_2=CHCH_2Cl$	76.53	360	25	0.1	0.1
Propylene dichloride (1,2-Dichloropropane)	$CH_3CHClCH_2Cl$	112.99	50	25	0.27 @ 20°C	0.06 @ 20°C
Chloroacetone (1-Chloro-2-propanone)	$ClCH_2COCH_3$	92.53	20	30	9	
1,3-Dichloro-2-propanol (Glycerol dichlorohydrin)	$CH_2ClCHOHCH_2Cl$	128.99	1	28	9.9 @ 19°C	
n-Butyl chloride (1-Chlorobutane)	$CH_3CH_2CH_2CH_2Cl$	92.57	105	25	0.07	
sec-Butyl chloride (2-Chlorobutane)	$CH_3CH_2CHClCH_3$	92.57	155	25	0.1 @ 25°C	
tert-Butyl chloride (2-Chloro-2-methyl propane)	$(CH_3)_3CCl$	92.57	300	25	sparingly soluble in water	
Isobutyl chloride (1-Chloro-2-methyl propane)	$(CH_3)_2CHCH_2Cl$	92.57	142	25	insoluble in water	
Butylidene chloride (1,2-Dichlorobutane)	$CH_3CH_2CHCl_2$	127.02	760	113.5	practically insoluble in water	
Chlorobenzene (Benzene chloride)		112.56	760	131.7	0.0512 @ 25°C	0.049 @ 30°C
o-Dichlorobenzene		147.01	1.282 / 760	25 / 180.5	0.0145 @ 25°C	
m-Dichlorobenzene		147.01	760	173	0.0123 @ 25°C	
p-Dichlorobenzene		147.01	0.4 / 760	25 / 174	practically insoluble in water	
1,2,3 Trichlorobenzene		181.46	1	40	insoluble	

the chlorides represent a small fraction of the total organic content of the wastewater, such bacteria can be added daily to a biochemical oxidation system to destroy the unwanted compounds.

Also, some organic chlorides, such as chlorophenols, exhibit a solubility great enough to justify distillation and solvent loss, as well as the cost of preventing solvent from entering the environment. In the case of low concentrations, a nonregenerable carbon bed can be used at reasonable operating cost; or the regeneration of

carbon or a polymeric resin can be accomplished with caustic (for chlorinated phenols, etc.). Finally, the one theoretical tray that can be achieved in a flash drum is often sufficient to attain adequate stripping of organic chlorides.

The key, therefore, to solving the problem of chloride removal lies in identifying the chlorides in a wastewater with their properties, like vapor pressures and solubilities. Once the compounds and their properties are identified, they can usually be matched to a suitable re-

These properties help identify the process for removing organic chlorides from wastewater — Table I cont'd

Organic chloride	Formula	Mol. weight	Vapor pressure		Organic solubility in water, weight %	Water solubility in organics, weight %
			mm Hg	@ °C		
o-Chlorophenol	(structure)	128.56	2.25	25	2.77	
m-Chlorophenol	(structure)	128.56	1	40	2.53 @ 20°C	17.7 @ 23.1°C
p-Chlorophenol	(structure)	128.56	1	50	2.63 @ 20°C	15.98
2,4-Dichlorophenol	(structure)	163.01	1	60	0.45 @ 20°C	
2,4,6 Trichlorophenol (Dowicide 25)	(structure)	197.46	1	77	<0.1	
Phosgene (Carbonic dichloride)	$COCl_2$	98.92	1418	25	slightly sol. in water; hydrolizes	
Aldrin	(structure)	364.93	760	104	insoluble in water	
Chlordane	(structure)	409.8		155 (m.p.)	insoluble in water	
3,3' Dichlorobenzidine	(structure)	253.13		132-33	almost insoluble in water	
Dieldrin	(structure)	380.93	1.8×10^{-7}	25	<0.1 ppm	
Endrin	(structure)	380.93	2×10^{-7}	25	<0.1 ppm	
Heptachlor	(structure)	373.35	4×10^{-4}	25		
Lindane (γ isomer)	(structure)	290.85	9.4×10^{-6}	20	insoluble	
p,p'-DDD (TDE)	(structure)	320.05		109-10 (m.p.)		
DDE	(structure)					
DDT	(structure)	354.5	1.5×10^{-7}	20	practically insoluble	
Toxaphene (Chlorinated camphene—	Mixture of at least 175 poly-chloro-deriv-atives: $C_{10}H_{10}Cl_8$			65–90 (m.p.)	practically insoluble	

moval process. A list of potential organic chlorides and their properties is shown in Table I. Some processes for their removal are stripping, adsorption, biological oxidation, and extraction.

Stripping

A typical stripper system operating at atmospheric pressure is shown in Fig. 1. Wastewater containing organic chlorides is preheated against bottoms and fed to the top of a packed stripper tower. Low-pressure steam is introduced at the bottom of the tower. The mixture of steam and vaporized organic chlorides leaving the top is condensed and separated in a decanter, from which the water is returned to the feedstream while the organic layer is removed for disposal. After passing in heat exchange with the feedstream, the tower bottoms can be further cooled.

This system can also be operated under vacuum. Although vacuum operation eliminates the need for a feed-bottoms exchanger, it incurs the added costs of the

These data are typical of processes to strip ethylene dichloride (EDC) from wastewater   Table II

Flowrate of wastewater stream	10 gpm
Temperature of wastewater stream	90° F
EDC Content of wastewater stream	0.87 weight %
Recovery of EDC	99.9%
Average stripping-tower temperature	221°F
Solubility of EDC in Water at 221° F	1.95 lbs/100 lbs water
Solubility of water in EDC at 221 °F	0.99 lbs/100 lbs EDC
Vapor pressure of wastewater at 221°F	25 psia
Tower design data	
Diameter	28 in.
Packing	20 ft of 3/4 in. Raschig rings
Low-pressure steam flow	1,565 lb/h
Bottoms pressure	2 psig
Bottoms temperature	225° F
Overhead pressure	1.5 psig
Overhead temperature	216° F
$K = \frac{y}{x}$ for hydrocarbon	408
Activity coefficient for hydrocarbon in water	269
Overhead vapor	865 lb/h of water vapor plus 43.4 lb/h EDC
Theoretical trays	less than 2

equipment and utilities for maintaining the vacuum. Also, the quantity of stripped water is increased with vacuum, since the vacuum steam-jet gets contaminated.

Finally, low tray-efficiencies often make it desirable to substitute a flash drum for a stripping tower.

In calculating a stripper, the tower temperature can be assumed equal to the boiling point of water at the tower pressure. Normally, the pressure differential over the tower is small, and an average equilibrium constant at an average tower temperature can be assumed for the organics. The presence of the hydrocarbon can be ignored in performing the heat balance to obtain the rate of vapor traffic in the tower. Vapor loads can be assumed constant, except for the top tray. The combined feed is generally at a temperature lower than the tower temperature, with consequent condensation of some of the stripping steam on the top tray.

As an alternative to ignoring the hydrocarbon in calculating the vapor load, a correction in top-tray vapor can be made on the basis that most of the hydrocarbon entering the tower is vaporized on the top tray. Thus the top-tray vapor can be condensed to supply the heat required to vaporize all the hydrocarbon, and the hydrocarbon vapor can then be added to the water vapor to obtain the total top-tray load.

Hydraulics for the stripping tower are the same as for other distillation systems; i.e., conventional packed-tower correlations or tray calculations are used to set pressure drop as a function of tower traffic.

In addition to the liquid and vapor loads for sizing tower diameter, the equilibrium constant for the organic chloride is needed to calculate the number of theoretical trays.

Equilibrium constant: When a water phase and a hydrocarbon phase coexist at the same temperature, the total pressure is equal to the sum of the vapor pressures of the two pure liquids. There is, however, a small amount of hydrocarbon in solution in the water phase, as well as water in the hydrocarbon phase; and the fugacity of the hydrocarbon in the water phase must equal its fugacity in the hydrocarbon phase.

In terms of activity coefficients, this fugacity relationship is shown as

$$(\gamma_{hc} P_{ohc} x_{hc})_{w\phi} = (\gamma_{hc} P_{ohc} x_{hc})_{hc\phi}$$

where γ_{hc} is the activity coefficient of the hydrocarbon, x_{hc} is the mol fraction of the organic chloride, and P_{oho} is the vapor pressure of the pure hydrocarbon. The subscripts $w\phi$ and $hc\phi$ refer to water phase and hydrocarbon phase, respectively.

As the mol fraction of organic chloride in the hydrocarbon phase approaches 1.0, the activity coefficient approaches 1.0. Thus the activity coefficient of the organic chloride in water can be determined from the saturation solubility data for the two phases:

$$(\gamma_{hc})_{w\phi} = (x_{hc})_{hc\phi}/(x_{hc})_{w\phi}$$

This activity coefficient can be assumed to apply from the saturation solubility down to zero solubility of the organic chloride in the water phase at a given temperature. The equilibrium constant for the organic chloride, $K_{hc} = y_{hc}/x_{hc}$, can be obtained from the vapor-phase/water-phase relationship:

$$(\gamma_{hc} P_{ohc} x_{hc})_{w\phi} = y_{hc} \pi$$

where π is the total pressure, and y_{hc} is the mol fraction in the vapor phase. Then:

$$K_{hc} = y_{hc}/x_{hc} = \gamma_{hc} P_{ohc}/\pi$$

The pressure at any point in the tower determines the temperature at that point. (It should be noted that the solubility of other hydrocarbons in the wastewater has the effect of reducing an activity coefficient calculated from pure hydrocarbon-water data.)

Theoretical trays: Assuming that the amount of organic chlorides in the stripped water has been set to meet some standard, there are infinite combinations of theoretical trays versus stripping steam that will remove the hydrocarbon from the water. The choice should be based on both economics and the variations expected in day-to-day operation. Also, there are practical limitations on the number of theoretical trays, since tray efficiencies for stripping organic chlorides are generally low, with correspondingly high packing height for a theoretical tray.

The stripped-water composition plus the amount of stripping steam can be used to determine the tower's bottom-tray composition through heat and material balances. This composition plus the equilibrium constant can then be used to calculate the concentration of organic chlorides in the vapor leaving the bottom tray; and a material balance on the organic chlorides entering and leaving the bottom tray then determines the amount of organic chlorides entering that tray in the liquid from the tray above. Since the volume of that liquid is determined by heat and material balances, the quantity of contained organic chlorides sets the composition, which with the equilibrium constant can be used

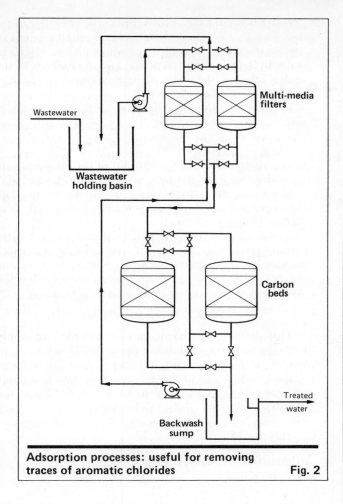

Adsorption processes: useful for removing traces of aromatic chlorides   **Fig. 2**

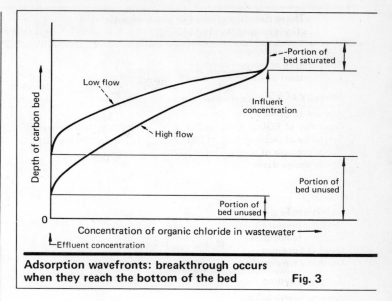

Adsorption wavefronts: breakthrough occurs when they reach the bottom of the bed   **Fig. 3**

to calculate the organic chlorides leaving the second tray from the bottom.

This tray-to-tray calculation, starting from the bottom, is continued up the tower, until the calculated concentration of organic chlorides in the liquid equals or exceeds the concentration in the incoming feed. If such calculations show that the concentration of organic chlorides in the liquid from a tray above is the same as that leaving the tray just calculated, a pinch-point has been reached, and the calculations must be repeated using more stripping steam.

Once a tower has been calculated in this manner, day-to-day variations should be assumed and checked against the design. It may be desirable to design for more steam and fewer theoretical trays, rather than have an unstable situation, even though optimum economics may not be achieved.

Table II summarizes the design data for a system that will remove 99.9% of 0.87 wt % ethylene dichloride in a 10-gpm stream of wastewater occurring at 90°F. A feed/bottoms exchanger was not used in this example; its use would cut steam requirements substantially.

Adsorption on activated carbon

Activated-carbon adsorption of organics dissolved in water proceeds slowly, so that the plot of organic concentration versus length of passage through an adsorbent bed shows a gradual curve, or long wave-front

(Fig. 3). At any given time, the first layer of carbon is saturated, while the last layer remains unused until the wave reaches that layer and "breakthrough" occurs.

In a typical carbon-adsorption flowsheet (Fig. 2), wastewater is first pumped downward over one of two parallel multi-media filter beds to remove suspended particles and undissolved hydrocarbons. From the filters, the wastewater passes through two carbon beds functioning in series. Each of these beds should be sized so that it alone can remove the organic chlorides to the desired level. Then when breakthrough occurs from the first bed, that bed can be bypassed and removed from service while its carbon is replaced, and subsequently returned to service downstream of the other bed. Maintaining two beds in series permits better carbon usage, since each bed can be saturated before it is removed from service.

A sump and pump, which are provided for backwashing the filter, should also be sized to permit periodic backwashing of the carbon beds, since they tend to accumulate suspended solids over a period of time, despite presence of the filters. The backwash stream is preferably returned to the wastewater holding basin, so that its solids may settle out. Otherwise, a separate filter must be provided to filter particles out of the backwash water.

Adsorption of chlorinated organics from water occurs in surface pores of the granulated activated carbon. These pores must thus be large enough to permit the organic chlorides to enter. Commercially available activated carbons, made from a variety of base materials, typically have surface-pore areas from 500 to 1,400 m^2/g [3]. In addition to its effect on pore size and surface area, the choice of base material also affects the hardness of the carbon granules and thus their resistance to abrasion. Abrasion resistance is particularly important when the carbon is to be transported, as a slurry for example, in a regeneration process.

For any given carbon, the organic compounds differ in their tendency to be adsorbed, according to polarity, molecular structure and molecular weight [4]. Highly

| | | Wastewater concentration | | | Carbon's |
Compound	pH	Initial microgram/1	Effluent microgram/1	% reduction	adsorptive capacity*
Aldrin	7.0	48	<1.0	99+	30
Dieldrin	7.0	19	0.05	99+	15
Endrin	7.0	62	0.05	99+	100
DDT	7.0	41	0.1	99+	11
DDD	7.0	56	0.1	99+	130
DDE	7.0	38	<1.0	99+	9.4
Toxaphene	7.0	155	<1.0	99+	42
Arochlor 1242 (PCB)	7.0	45	<0.5	99+	25
Arochlor 1254 (PCB)	7.0	49	0.5	99+	7.2

Adsorption isotherms for toxic organic chlorides on activated carbon   Table III

*Mg of toxic chemicals adsorbed/g carbon

polar substances are generally more soluble and do not adsorb easily. Aromatic rings, on the other hand, adsorb well, and the addition of chlorine to such molecules increases this tendency. Higher molecular weights are also conducive to adsorption, so long as the molecules are smaller than the pores. Large molecules have a limited surface-area capacity [4].

Suppliers of carbon will usually recommend that laboratory isotherms be run, and column tests carried out, for a given wastewater. In such tests [5], it is important that the carbon be ground, in order to allow the test bed to reach equilibrium within a reasonable time. The column tests will permit estimates of contact time, carbon usage rates, and requirements for pretreatment. These tests thus permit sizing the bed in terms of time onstream before being bypassed for carbon renewal.

Contact times can vary considerably, from as little as 20 minutes to several hours and more. Flowrates over carbon generally range from 2 to 10 gpm/ft². Beds as long as 40 ft have been reported [5] and even longer beds can be used. A flow of 2 gpm/ft² in a 40-ft bed results in a contact time of 150 min.

The data in Table III [6] are based on isotherms for various pesticides and polychlorinated biphenyls, showing 99 + % removal and 7 to 130 milligrams of toxic chemicals adsorbed per gram of carbon. Based on these data for DDE, and assuming a contact time of 60 min., calculations indicate the carbon would have to be changed once in 15 years. Actually, it would probably have to be changed more frequently. But even so, this is essentially a nonregenerative system.

Pressure drop through the bed—a function of hydraulic loading, bed depth and particle size—is typically 0.5 to 2.0 in. of water per foot at a flow of 2 gpm/ft², and 3.5 to 10.0 in. of water per foot at a flow of 10 gpm/ft². Carbon suppliers generally have data on pressure drop per foot of bed, and will make such data available.

The EPA has published a process design manual for carbon adsorption [2], indicating that carbon-contacting systems can be upflow or downflow, series or paral-

lel, single- or multi-stage, pressure or gravity downflow, packed-bed or expanded-bed upflow. The advantages and disadvantages of each type, and how they should be related to a specific problem, are outlined.

Depending on how the carbon beds are managed, they can be nonregenerated or regenerated onsite, or regenerated under contract. If they are regenerated onsite, multiple-hearth reducing-atmosphere furnaces or rotating kilns are required [29]. While carbon can adsorb most organic chlorides from water, its major use will probably be for wastewaters or relatively small volumes of waste containing large molecules of low vapor pressure. The carbon for such systems is replaced as often as several times a year; i.e. the systems are not regenerable onsite.

Since carbon is abrasive, long-radius bonds are recommended for slurry service; and the piping should be designed for multiple cleanouts and flushing connections. The EPA design manual recommends ball valves and plug valves for slurry service, and suggests controlling flow by throttling the water supply rather than throttling the slurry.

Finally, industrial users have an alternative to owning a carbon adsorption unit. At least one of the carbon suppliers will design and build such a unit tailored to the need, using a modular system. The unit is leased and operated by the carbon supplier, who also makes guarantees, does analytical work, carries out maintenance, and removes and regenerates carbon offsite for a monthly fee [8].

Adsorption on polymeric resins

The flowsheet for polymeric adsorption will be similar to that for carbon-bed adsorption, with a multi-media filter ahead of the adsorption beds. However, regeneration with a solvent requires a distillation system to separate the solvent from the adsorbed organic chloride as well as from water brought into the regeneration system during rinse cycles. Thus one tower is needed to distill out the water, and a second tower to separate solvent from the organic chloride.

Pressure drops through polymeric adsorption beds range from about 8 in. of water per foot of bed at 2 gpm/ft² to about 75 in. of water per foot of bed at 10 gpm/ft². Data for the adsorption of chlorinated phenols [11] are shown in Table IV, along with information on the effects of sodium chloride in the wastewater [12].

When a bed is ready for regeneration, excess water is drained to a level that will keep the bed wet. Then, solvent is passed at a low flowrate (e.g., 1 to 4 bed-volumes/h) through the bed. A rinse with washwater is recommended before the bed is put back onstream. The regeneration liquids are then separated in the distillation system.

Although the polymeric resins are more expensive per pound than activated carbon and generally have a lower adsorptive capacity, they also have a higher density—about 44 lb/ft³ compared with 30 lb/ft³—so that the volumetric capacity is more nearly equivalent.

A new carbonaceous adsorption material has been developed that can be regenerated with steam after removing low levels of halogenated hydrocarbons from drinking water. Such steam regeneration, however, would apply only to light materials, such as chloroform.

Suppliers of resin will aid in engineering the system and in performing test work required. The patent situation and the need to license must be investigated by the user.

Biological oxidation

For organic chlorides to be successfully removed in a biological oxidation system, the microbial population of that system must contain components capable of withstanding and of degrading the toxic organic chlorides. The rate at which these organisms develop, in relation to other organisms and to the total amount of food available, will determine their relative concentration in the biomass. On the other hand, mutant bacteria for degrading specific organic chlorides can be added daily to a conventional activated-sludge system in order to enable the system to handle the organic chlorides.

A sufficient concentration of mutant bacteria must be maintained to destroy the organic chlorides. The amount of such bacteria required for any given waste system must be determined via laboratory studies.

Mutant bacteria capable of degrading many of the aromatic organic chlorides have been developed by commercial laboratories (Tables V, VI) [13]. As more chlorine is attached to the benzene carbon rings, the resultant compounds become more difficult to degrade [14]. However, 100% ring disruption has been achieved for mono-, di-, and even tri- chlorinated benzene; and once the ring has been broken, the chlorinated material is biodegradable. Table VI shows that chlorinated phenols can be destroyed.

Extraction

Wastewater is fed to the top of an extraction tower (Fig. 4), while solvent enters the bottom and flows countercurrent to the downward-flowing water. Rich solvent from the top of the extraction tower passes to a distillation tower, where the organic chloride is separated, and the regenerated solvent then returns to the extraction tower.

Raffinate water from the bottom of the extraction tower meanwhile passes to a steam stripper, where most dissolved solvent and unextracted chlorides are vaporized overhead along with water, and both are condensed. Solvent and water are then separated in a decanter, with the solvent from the decanter returning to the main solvent system, while the decanted water is recycled to the feed to the stripper.

Either a polar or an aromatic solvent can extract organic chlorides from water. Polar solvents, such as n-butyl acetate, have much higher distribution coefficients than aromatic solvents, such as benzene and toluene; and higher distribution-coefficients result in a lower solvent-to-water ratio for a given extraction, with consequent lower energy costs for separating the extracted organic chloride in the distillation column.

Whether aromatic or polar, however, some of the solvent is dissolved in the raffinate water, which is why the stripper is provided (Fig. 4). This stripper also removes part of the organic chloride not extracted from the water, and thereby serves to counterbalance the

---

Adsorption of chlorinated phenols on a polymeric resin   Table IV*

Flowrate: 0.5 gpm per cubic foot (15 min. contact time)

Material	Solubility in water, ppm	Solute in influent, ppm	Solute adsorbed, % at 0 leakage
m-Chlorophenol	26,000	350	5.45
2,4-Dichlorophenol	4,500	430	11.6
2,4,6-Trichlorophenol	900	510	27.2
m-Chlorophenol (13% NaCl)		350	7

*Source: Ref. 8

Times required to biodegrade organic chlorides   Table V*

Organic chloride	Concentration mg/l	Ring disruption, %		Time in hours	
		Parent	Mutant	Parent	Mutant
Monochlorobenzene	200	100	100	58	14
o-Dichlorobenzene	200	100	100	72	20
m-Dichlorobenzene	200	100	100	96	28
p-Dichlorobenzene	200	100	100	92	25
1,2,3-Trichloro-benzene	200	87	100	120	43
1,2,4-Trichloro-benzene	200	92	100	120	46
1,3,5-Trichloro-benzene	200	78	100	120	50
1,2,3,4-Tetrachloro-benzene	200	33	74	120	120
1,2,3,5-Tetrachloro-benzene	200	30	80	120	120
Hexachlorobenzene	200	0	0	120	120

* Biodegradation by mutant Pseudomonas species at 30°C
Source: Ref. 9

Times required to biodegrade phenol and chlorophenols   Table VI*

Compound	Concentration mg/l	Ring disruption,%		Time in hours	
		Parent	Mutant	Parent	Mutant
Phenol	200	100	100	25	8
o-Chlorophenol	200	100	100	52	26
m-Chlorophenol	200	100	100	72	28
p-Chlorophenol	200	100	100	96	33
2,4-Dichlorophenol	200	100	100	96	34
2,5-Dichlorophenol	200	60	100	120	38
2,3,5-Trichlorophenol	200	100	100	100	52
2,4,6-Trichlorophenol	200	100	100	120	50
Pentachlorophenol	200	7	26	120	120

* Biodegradation by mutant Pseudomonas species at 30°C
Source: Ref. 9

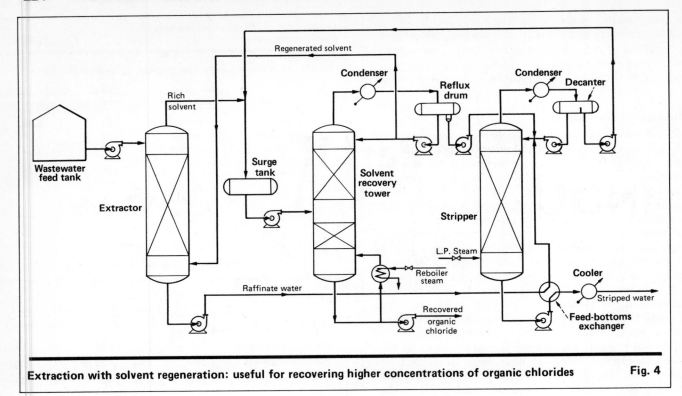

Extraction with solvent regeneration: useful for recovering higher concentrations of organic chlorides   **Fig. 4**

difference in distribution coefficients between polar and aromatic solvents. Although the polar solvents leave less organic chloride in the raffinate water, they are more water-soluble than the aromatic solvents, and the result is about the same total of organic-chloride-plus-hydrocarbon solvent removed by the stripper. Thus there is not much difference between polar and aromatic solvents for the flowsheet shown in Fig. 4.

A dual-extraction process has also been suggested to minimize energy requirements [15]. However, that approach requires two extraction towers and two distillation towers.

J. P. Earhart, others [11] have presented data on distribution coefficients, based on experimental data from the literature, together with solvent boiling points and solubility of solvent in water. The distribution coefficient is the weight-fraction of solute in solvent per weight-fraction of solute in the water.

Corrosion

Organic chlorides are not considered corrosive at normal temperatures, so carbon steel is permitted. However, dilute HCl is formed to the extent that organic chlorides are hydrolyzed in water solutions, and Monel is recommended up to temperatures of about 250°F and for cases where no O_2 is present.

If the HCl is neutralized, the resulting sodium chloride presents a potential for stress corrosion. Thus, 316L stainless steel has been used under process conditions where salts are not naturally concentrated, or up to 140 ppm sodium chloride at temperatures not more than 140°F. Above 140°F, and in situations where the sodium chloride is more concentrated, Carpenter 20 or titanium are often specified. In any given situation,

however, a metallurgist should be consulted to make sure that a seemingly innocuous component doesn't alter the corrosiveness of the system.

References

1. U.S. Environmental Protection Agency, "Process Design Manual for Carbon Adsorption," Oct., 1973.
2. Calgon Corp., Pittsburgh, Pa., "Basic Concepts of Adsorption on Activated Carbon," Brochure 27-18.
3. Bernardin, F. E., Jr., Selecting and specifying activated carbon adsorption systems, *Chem. Eng.,* Oct. 18, 1976, p. 77.
4. Calgon Corp., Pittsburgh, Pa., Bulletin 20-52.
5. Hager, D. G., Waste Water Treatment via Activated Carbon, *Chem. Eng. Prog.,* Oct., 1976, p. 57.
6. Calgon Corp., Pittsburgh, Pa., "Calgon Adsorption Service," Brochure 2702B.
7. Kennedy, D. C., Treatment of Effluent From Manufacture of Chlorinated Pesticides with a Synthetic Polymeric Adsorbent, Amberlite XAD-4, *Enviro. Sci. Tech.,* Feb., 1973, p. 138.
8. Rohm & Haas, Philadelphia, Pa., "Amberlite XAD-4," Technical Bulletin, 1976.
9. Polybac Corp., New York, N.Y., "Polybac," Technical Bulletin, 1973.
10. Zanitsch, Roger H., and Lynch, Richard T., "Selecting a thermal regeneration system for activated carbon," *Chem. Eng.,* Jan. 2, 1977, p. 95.
11. Earhart, J. P., others, Recovery of Organic Pollutants Via Solvent Extraction, *Chem. Eng. Prog.,* May, 1977, p. 67.

The author

M. F. Nathan is manager of environmental engineering for Crawford & Russell, Stamford, CN 06904, where he has worked in the capacity of a process design engineer and manager since 1970. Prior to joining Crawford & Russell, he was manager of the Process Development Div. of M. W. Kellogg, with broad experience in refining and petro-chemical process design. He holds a B.S. in chemical engineering from The Pennsylvania State University, as well as an M.S. and Ph.D. in chemical engineering from the University of Illinois.

Part II
INDUSTRIAL SOLID WASTE CONTROL

Section 1
INTRODUCTION

A guide to the resource conservation and recovery act
How industry can prepare for RCRA

A guide to the Resource Conservation and Recovery Act

RCRA "closes the circle" of environmental control that began with the Clean Air and Clean Water Acts. Here is an outline of its key provisions.

Robert S. Glaubinger, Assistant Editor

☐ The Resource Conservation and Recovery Act (RCRA) is the first comprehensive federal regulation of solid waste. It amends the Solid Waste Disposal Act of 1965, which was essentially limited to R & D programs for waste disposal and resource recovery. RCRA was enacted to achieve two basic objectives: (1) protection of public health and the environment, and (2) the conservation of national resources.

RCRA provides three major programs to help reach these goals: (1) the establishment of a hazardous-waste-control program, to be administered by the States or, where States choose not to do so, by EPA; (2) the establishment of a land-disposal regulatory program in each State; and (3) the initiation and support of resource-conservation programs by State and local governments to conserve resources and reduce the amount of solid waste requiring land disposal.

The impact of the programs and regulations will be significant. The universe to be covered includes: 30–45 million tons of hazardous waste, 429,000 generators of such waste, and over 20,000 transporters and storers. Onsite as well as offsite disposal facilities will be covered. Of major interest are the regulations for hazardous-waste management (Subtitle C) and land disposal of waste that is not hazardous (Subtitle D).

Subtitle C sets forth a management scheme that provides for "cradle to grave" regulation of hazardous wastes. Basically, this subtitle provides for:

1. An identification and listing of "hazardous wastes" according to specified criteria (Sect. 3001).

2. Standards of performance (Sect. 3004) for those who store, treat or dispose of such wastes, including bonding by the disposal site operator to provide for closure costs of the site and long-term care thereof, and a liability-insurance coverage ($5-million limits) for sudden or nonsudden occurrences.

3. Permits (Sect. 3005) for storage, treatment and disposal facilities—not for transporters, and only for generators having onsite waste-handling facilities.

4. A manifest system (Sect. 3002), which must be complied with by all who handle hazardous waste, to ensure that the wastes are transported from the waste generator to only a "permitted" disposal facility. The manifest is a special form of shipping paper that is filled out by the waste generator. It directs and tracks the movement of the waste from the point of generation to ultimate disposal.

5. Notification requirements (Sect. 3010) for generators, transporters and treatment, storage and/or disposal facilities.

In addition, the hazardous-waste program will establish standards on the basis of disposal methods (incineration, landfilling, etc.) and not by national industry-specific standards.

The states will be authorized to run the hazardous-waste program (as with the Clean Water Act approach); EPA will implement it until the states are authorized.

Wastes that are not classified as hazardous are covered by Subtitle D. This is basically a State regulatory program with certain federally imposed constraints, the most significant of which is the phasing-out of open dumps. Sect. 4004 will set criteria for sanitary landfills, and any disposal facility not meeting those criteria will be considered an open dump. Open dumps have to be upgraded or phased out.

Thus, all land-disposal sites will have to meet the Subtitle D sanitary landfill criteria and, if a site handles hazardous waste, it will have to meet the additional, more-stringent requirements of Subtitle C.

Of relevance to both subtitles are the definitions set forth in Sect. 1004. The term "solid waste" is broadly defined to include "solid, liquid, semisolid and contained gaseous materials," and "other discarded material." Specifically excluded are "materials in industrial discharges . . . subject to NPDES [National Pollutant Discharge Elimination System] permits." The interpretation of "discarded" and the NPDES-related exclusions are likely to be key issues.

By definition, "hazardous waste" is a subclass of "solid waste." Hazardous wastes cause or significantly contribute to death or serious illness, or pose a

Originally published January 29, 1979.

Subtitle A—General Provisions

Sect. 1004—Definitions

Sect. 1004(3) Disposal—the discharge, deposit, injection, dumping, spilling, leaking or placing of any solid waste or hazardous waste into or on any land or water so that such solid waste or hazardous waste or any constituent thereof may enter the environment or be emitted into the air or discharged into any waters, including ground waters.

Sect. 1004(5) Hazardous waste—a solid waste, or combination of solid wastes, which because of its quantity, concentration or physical, chemical or infectious characteristics may:
(a) cause, or significantly contribute to, an increase in mortality or an increase in serious irreversible, or incapacitating reversible, illness; or
(b) pose a substantial present or potential hazard to human health or the environment when improperly treated, stored, transported or disposed of, or otherwise managed.

Sect. 1004(12) Manifest—the form used for identifying the quantity, composition, origin, routing and destination of hazardous waste during its transportation from the point of generation to the point of disposal, treatment or storage.

Sect. 1004(14) Open dump—a site for the disposal of solid waste that is not a sanitary landfill within the meaning of Sect. 4004.

Sect. 1004(26) Sanitary landfill—a facility for the disposal of solid waste that meets the criteria published under Sect. 4004.

Sect. 1004(26A) Sludge—any solid, semisolid or liquid waste generated from a municipal, commercial or industrial wastewater treatment plant, water supply treatment plant or air pollution control facility, or any other such waste having similar characteristics and effects.

Sect. 1004(27) Solid waste—any garbage, refuse, sludge from a waste treatment plant, water supply treatment plant or air pollution control facility, and other discarded material, including: solid, liquid, semisolid or contained gaseous material resulting from industrial, commercial, mining and agricultural operations, and from community activities, but does not include solid or dissolved material in domestic sewage, or solid or dissolved materials in irrigation return flows or industrial discharges that are point sources subject to permits under Sect. 402 of the Federal Water Pollution Control Act, as amended (86 Stat. 880), or source, special nuclear or byproduct material as defined by the Atomic Energy Act of 1954, as amended.

Sect. 1004(33) Storage—the containment of hazardous waste, either on a temporary basis or for a period of years, in such a manner as not to constitute disposal of such hazardous waste.

Sect. 1004(34) Treatment—any method, technique or process, including neutralization, designed to change the physical, chemical or biological character or composition of any hazardous waste so as to neutralize or render it nonhazardous, safer for transport, amenable for recovery, amenable for storage or reduced in volume. Also included is any activity or processing designed to change the physical form or chemical composition of hazardous waste so as to render it nonhazardous.

Sect. 1006—Application and integration with other acts

Sect. 1006(a) Application of act
RCRA shall not be construed to apply to any "activity or substance" that is subject to:
1. Federal Water Pollution Control Act.
2. Safe Drinking Water Act.
3. Marine Protection, Research and Sanctuaries Act.
4. Atomic Energy Act.
Except to the extent that application "is not inconsistent" with these acts.

Sect. 1006(b) Integration with other acts
EPA shall "integrate" and "avoid duplication to maximum extent possible" of RCRA with all acts which it administers.

Subtitle C—Hazardous-waste management

Sect. 3001—Identification and listing of hazardous wastes

Sect. 3001(a) Criteria for identification or listing
Takes into account:
1. Toxicity.
2. Persistence.
3. Degradability in nature.
4. Potential for accumulation in tissue.
5. "Other related factors such as flammability, corrosiveness and other hazardous characteristics."

Sect. 3001(b) Identification and listing
EPA shall identify the characteristics of hazardous waste and list particular wastes that shall be subject to the provisions of Subtitle C.

Sect. 3001—Proposed regulations: 43 *Fed. Reg.* 58954 (Dec. 18, 1978)
Includes within the definition of hazardous, wastes that are ignitable, corrosive, reactive or toxic when tested by the criteria set forth therein. Additionally, lists of hazardous wastes (e.g., water-based paint wastes), hazardous-waste-generating processes (e.g., still bottoms from final purification of acrylonitrile) and sources are set forth. [If one of your materials is listed you would have to handle it as a "hazardous waste" unless you could satisfy the delisting provisions under a section entitled "Demonstration of non-inclusion in the hazardous-waste system."]

Sect. 3002—Standards applicable to hazardous-waste generators

Requires regulations providing for:
1. Record keeping that accurately identifies quantities, constituents and disposition of hazardous wastes.
2. Labeling of containers used for storage, transport or disposal.
3. Use of appropriate containers.
4. Furnishing of information on the general chemical composition of the hazardous waste to persons transporting, treating or disposing of such wastes.
5. Use of the manifest system, unless disposed of onsite.
6. Reporting of quantities and disposition of wastes generated during a particular time period.

Sect. 3002—Proposed regulations: 43 *Fed. Reg.* 58975 (Dec. 18, 1978)

Sect. 3003—Standards applicable to hazardous-waste transporters

Sect. 3003(a) Requires regulations providing for:
1. Record keeping.
2. Transporting of properly labeled wastes only.
3. Compliance with the manifest system.
4. Transporting to a permitted disposal site only.

Sect. 3003(b) Requires that the regulations be consistent with the hazardous-materials transportation act and DOT regulations.

Sect. 3003—Proposed regulations: 43 *Fed. Reg.* 18506 (Apr. 28, 1978)

Sect. 3004—Standards applicable to hazardous-waste treatment, storage and disposal facilities

Requires regulations setting performance standards for:
1. Keeping records of hazardous wastes handled.
2. Reporting, monitoring and inspection, and compliance with manifest system.
3. Treatment, storage and disposal pursuant to operating methods, techniques and practices "as may be satisfactory to the administrator" of EPA.
4. The "location, design and construction" of the facilities.
5. Contingency plans to minimize unanticipated damage.
6. Maintenance of operation of the facilities, and qualifications as to: ownership, continuity of operation, training of personnel and financial responsibility.
7. Compliance with the Sect. 3005 requirements.

Sect. 3004—Proposed regulations: 43 *Fed. Reg.* 58994 (Dec. 18, 1978)

Sect. 3005—Permits for treatment, storage or disposal facilities

Sect. 3005(a) Permits required
No treatment, storage or disposal of hazardous waste without a permit (effective date 6 months after regulations are promulgated).

Sect. 3005(b) Permit application requirements:
1. Estimates of composition, quantities and concentrations of waste that will be handled.
2. Information regarding the site.

Sect. 3005(c) Permit issuance
1. Issued if in compliance.
2. Compliance schedule if modifications are needed.

Sect. 3005(d) Permit revocation for noncompliance

Sect. 3005(e) Interim status for facilities that:
1. Were in existence as of Oct. 21, 1976 (the date of promulgation of RCRA).
2. Complied with preliminary notification requirements of Sect. 3010(a).
3. Filed a permit application before effective date of regulations.

Sect. 3006—Authorization of State programs

Sect. 3006(a) Federal guidelines for State programs to be published

Sect. 3006(b) State program to be authorized unless:
1. It is not equivalent to Federal program.
2. It is not consistent with the Federal or State programs in other states.
3. It does not provide adequate enforcement.

Sect. 3006(c) Interim authorization for State programs
1. "Substantially equivalent" to Federal program.
2. In existence as of July 20, 1978 (90 days after the date required for promulgation of regulations under Sect. 3002-3005).

Sect. 3006—Proposed regulations: 43 *Fed. Reg.* 4366 (Feb. 1, 1978)

Sect. 3007—Inspections

Sect. 3007(a) EPA is given access entry to facilities where hazardous waste is generated, stored, treated, transported or disposed of or otherwise handled. Includes right to copy records, inspect and take samples.

Sect. 3007(b) Information is available to the public except if subject to confidentiality under 18 USC 1905 [1948 law that protects, by criminal prohibition, trade secrets, confidential business and technical information, etc.].

Sect. 3007—Proposed regulations: 43 *Fed. Reg.* 2637 (Jan. 18, 1978)

Sect. 3008—Federal enforcement

Sect. 3008(a) Compliance orders
1. Compliance orders may be issued for any violation of Subtitle C.
2. Failure to provide corrective action subjects violator to up to $25,000/d in civil penalties.
3. Injunctive relief can also be sought.

Sect. 3008(d) Criminal penalties
Criminal penalties of up to $25,000/d and/or up to 1 year in jail (doubles after 1st conviction) for "knowingly":
1. Transporting hazardous waste to a nonpermitted facility.
2. Disposing of hazardous waste without a permit.
3. Making false statements in a permit application, label, manifest, record, etc.

Sect. 3009—Retention of State authority

No State or political subdivision may impose any requirements less stringent than those authorized under Subtitle C.

Sect. 3010—Preliminary notification/ effective date

Sect. 3010(a) Preliminary notification
1. By generators and transporters, and treatment, storage and/or disposal facilities.
2. Within 90 days after Sect. 3001 regulations identifying or listing hazardous wastes are promulgated.
3. Notification "stating location and general description" of activity and the hazardous wastes handled.
4. If notice is not given, the transporting, treating, storing or disposing of the waste is prohibited [generating is not listed].

Sect. 3010(b) Effective date
1. Regulations of generating, transporting, treating, storing and disposing of hazardous wastes (Sec. 3002-3005) are effective 6 months after promulgation.
2. Disposal-facility permit applications must be filed by effective date.

Sect. 3010—Proposed Regulations: 43 *Fed. Reg.* 29908 (July 11, 1978)
(i) States that owners of inactive hazardous-waste treatment, storage and disposal facilities are not required to notify.
(ii) Includes sample of Notification Form.

Subtitle D—State or regional solid waste plans

Sect. 4001—Objectives
The objectives of this subtitle are to assist in developing and encouraging methods for the disposal of solid waste that are environmentally sound, and to maximize the utilization of valuable resources and encourage resource conservation.

Sect. 4002—Guidelines for plans
Proposed regulations: 43 *Fed. Reg.* 38534 (Aug. 28, 1978)

Sect. 4004(a) Criteria for sanitary landfill
At a minimum, such criteria shall provide that a facility may be classified as a sanitary landfill and not an open dump only if there is no reasonable probability of adverse effects on health or the environment from disposal of solid waste at such facility. Such regulations may provide for the classification of the types of sanitary landfills.

Sect. 4004(b) Disposal required to be in sanitary landfill

Sect. 4004—Proposed regulations: 43 *Fed. Reg.* 4942 (Feb. 6, 1978)
Criteria under Sect. 4004(a)
A sanitary landfill shall not be located in wetlands, floodplains, permafrost areas, critical habitats or sole-source aquifers, unless within specified exceptions.

Sect. 4005—Upgrading of open dumps
Sect. 4005(a) Open dumps
Open dump means any facility or site where solid waste is disposed of that is not a sanitary landfill under the criteria promulgated in Sect. 4004 and which is not a facility for disposal of hazardous waste.

Sect. 4005(b) Inventory
EPA shall publish an inventory of all disposal facilities or sites in the United States that are open dumps.

Sect. 4005(c) Closing or upgrading of existing open dumps
Any facility listed in the inventory as an open dump must be closed or upgraded within five years according to a State-established compliance schedule containing an enforceable sequence of actions leading to compliance.

Subtitle G—Miscellaneous Provisions

Sect. 7001—Employee protection
No person shall fire, or discriminate against, any employee who has caused to be filed or instituted any proceeding under RCRA or who has or will testify in any proceeding.

Sect. 7002—Citizen suits
Sect. 7003—Imminent hazard
EPA can bring suit to take necessary action if the handling, storage, treatment, transportation or disposal of any solid or hazardous waste is presenting an imminent and substantial endangerment.

substantial hazard to human health or the environment when improperly managed. Sect. 3001 will translate this definition into practice.

Also of importance is the relationship between RCRA and other laws. Sect. 1006(a) provides that nothing in RCRA "shall be construed to apply to any activity or substance which is subject to the Federal Water Pollution Control Act, the Safe Drinking Water Act, the Marine Protection and Sanctuaries Act or the Atomic Energy Act, except to the extent that such application is not inconsistent with the requirements of such acts."

Furthermore, Sect. 1006(b) requires EPA to "integrate" the administration and enforcement of RCRA and "avoid duplication to the maximum extent possible" with all of the other statutes administered by EPA, specifically including the four acts mentioned in Sect. 1006(a) and the Clean Air Act and the Federal Insecticide, Fungicide and Rodenticide Act.

In practice, these two sections mean, for example, that an industrial point-source-discharge covered by an NPDES permit would not also require an RCRA permit for that discharge, but disposal of sludge from that water-treatment facility would be covered by RCRA.

The handling of inactive or abandoned dumpsites is not expressly dealt with by RCRA. However, the proposed regulations for Sect. 3010 indicate that owners of inactive hazardous-waste facilities are not required to give notice. Proposals for Sect. 4002, dealing with State solid-waste-management plans, require that states must include measures to ensure that adverse health or environmental effects from abandoned facilities are minimized or eliminated. From these statements and other comments made by EPA personnel, it is probable that EPA's current plan is to utilize the imminent-hazards provisions of Sect. 7003 to deal with serious inactive-site problems, but not to stretch the coverage to all inactive sites. This issue bears careful attention.

There are significant penalties for failure to comply with RCRA. Compliance by a generator of wastes would not necessarily insulate the generator from liability to third parties in the event of improper disposal by a disposal contractor. The liability issue is a complex one, focusing on negligence, strict liability, nuisance, and independent-contractor relationships.

How industry can prepare for RCRA

Subtitle C of the Resource Conservation and Recovery Act will forever change industrial waste-disposal practices, while increasing expenses and workloads. Get a headstart by taking positive action now.

Richard Sobel, *Allied Chemical Corp.*

☐ While awaiting the promulgation of RCRA regulations, you can take many practical actions, both short- and long-term, to minimize their impact on your company. The intent and direction of the rules are apparent and they should not be treated casually. They will force us to reconsider an aspect of plant operations that has not received close scrutiny in the past.

Certainly, the cost of disposal will increase, as will administrative and technical workloads. The parts of Subtitle C that will create the largest workload and expense are Sect. 3001—Criteria and Listing; Sect. 3004—Standards for Owners or Operators of Treatment, Storage and Disposal Facilities; and Sect. 3005—Permits. The key to Subtitle C is: (1) waste identification and (2) notification to EPA.

Short-term actions

Here are steps you should begin taking immediately.

Inventory your wastes

What kind are you generating, storing, treating or disposing of? List them all, even the ones thought to be innocuous. Starting with a complete list, you can make a decision later about individual items. Do not forget the trash picked up by the local collector, because empty containers that previously contained hazardous wastes or any of the 300 or so DOT poisons, pesticides or priority pollutants are considered hazardous unless cleaned properly at a permitted facility. Most plants probably dispose of their empty bags and fiberdrums with the regular trash.

You may want to include on your list the information (waste type and description) called for in the notification forms being considered under Sect. 3010. This will help you meet the 90-day notification lead time, which could be in effect before 1979 is over.

Where does the waste come from? What process, unit operation or piece of equipment? What is the volume or weight/y or the cubic yards or number of trash-bin loads? Go through your processes unit-by-unit.

Where is the waste disposed of, onsite or offsite? By whom? Who hauls it? Where is it stored or transferred?

Do not forget to include surface impoundments, even though they are part of NPDES treatment facilities. It is possible, for instance, that contaminated earth or accumulations of settled sludges in impoundments will be covered by RCRA rules. The Dec. 18, 1978, proposed Sect. 3001 regulations defined "other discarded material" to include waste oils, so you should list these in your inventory. Also, hazardous waste stored by generators onsite for more than 90 days prior to offsite disposal will be subject to all requirements, and the storage site will need a permit.

Inventory carefully; it will be well worth the effort.

Analyze your wastes

In addition to the general description in the inventory, summarize your existing analytical information describing the waste and get additional data on composition ranges for major waste-components and others of concern. Get data on viscosity, flash point, heating value, screen size, etc. For large-volume wastes, detailed analytical data will provide a basis for decisions on storage, treatment, disposal and other regulatory requirements. You may want to start the delisting test procedures under Sect. 3001 if you believe a waste is non-hazardous. For small-volume wastes, testing may not be economical, but do sufficient analysis to satisfy the notification and manifest requirements, and those of your offsite-disposal contractor.

Originally published January 29, 1979.

Review your inactive sites

RCRA is not clear on retroactive situations, such as abandoned waste deposits or inactive facilities. According to EPA, such situations would be excluded from the hazardous-waste-management system. The proposed Sect. 3010 notification rules spell this out—only persons conducting a hazardous waste activity at the time of promulgation of the Sect. 3001 criteria would be required to notify.

If there is an obvious environmental problem from an inactive site, you should be working on it. This is specified in the proposed rules under Sect. 4002(b), State Solid Waste Management Plans, published Aug. 28, 1978. These require that abandoned facilities that continue producing significant adverse health and environmental effects be subject both to classification as open dumps and to action by the state. The proposed rules do not call for making an inventory of abandoned sites. However, some states are doing this.

It might be prudent to inventory your inactive facilities and old waste deposits after consulting with your management and legal people. There are probably few plants that do not have wastes stored in drums or buried on the "back 40." Some plants have been in operation since the 1890's or early 1900's, and waste practices were quite different then. If you decide to conduct an inventory, interview your older plant personnel and retirees, and review files, maps and photographs. List the nature of the material, volume, where buried or stored, about when, and who told you. Include old covered lagoons, buried tanks, deposits of contaminated rubble, collections of wastes stored in containers, flyash basins, and maybe even abandoned sewers and buried pipes. You may find your new unit is built on an old waste bed.

Do not forget about past offsite disposal. Your purchasing agent may be able to tell you what was sent where and who carried it. Formerly owned sites may also be of concern, depending on how the contract of sale was written.

In the case of property to be sold by you, make sure to disclose to the buyer any past waste disposal and clarify this in the sales contract or you may be liable for abatement. For property bought by your company, if an environmental problem was disclosed to you before you purchased the property you could be directed to abate the problem. Even if the problem were not disclosed or if the vendor were to become unavailable, you may be compelled to help resolve the situation. Consult your legal people.

Track your wastes

A number of states—New Jersey, California, Wisconsin, Texas and others—already require a manifest system for offsite disposal. If you do not use a manifest, start a bound logbook containing at least the date, waste description, source, volume shipped, name of hauler and destination; or track the wastes, using shipping documents or purchase orders. Some plants will not pay the hauler/disposer until they get a letter back saying that the waste was disposed of in accordance with the purchase order.

Also, track the wastes disposed of on your own site. If plant activity warrants, set up a computer system to store the information and generate reports that will be required by Sect. 3002.

Assign responsibilities

Clear assignment of responsibility and authority should be made at both plant and headquarters levels. Make sure that the appropriate technical and legal people are involved in day-to-day decisions on storage, treatment and disposal. The time is past for these matters to be left solely in the hands of people without a proper waste-control background. Assign an attorney or qualified pollution-control person to follow development of RCRA regulations and keep the rest of the company informed. Some plants may handle sufficient types and volumes of wastes to justify one or more full-time professionals.

Transportation

Provide your truck drivers, shipping personnel and barge operators with response information in case of an emergency. Include phone numbers of company and agency people, properties of the waste, safety information, etc. (Do not forget that under the Clean Water Act (Sect. 311) you will eventually have to report harmful-quantity discharges of listed materials to the National Response Center, (800) 424-8802. Start reporting them now.)

Legislative and regulatory development

It is vital that you understand and get involved in the development of the laws and regulations at state, local and federal levels. 1979 will be a year of intense activity. Most of the administration and interpretation of the rules will eventually take place at the state level. So take part in this process through the Chamber of Commerce, State Manufacturers' Assn., area Chemical Industries Council, professional societies or any other active group. You may want to subscribe to the *Federal Register,* the *Bureau of National Affairs Environment Reporter* and the various newsletters that cover pollution control. Get on your state-agency mailing lists.

The states and EPA are looking for constructive comments, as are trade-association committees and task groups. The various government agencies do not have all the answers and are just as anxious as industry for the rules to be workable. Extensive input from industry at every stage is desired. Be aware of agency deadlines, submit your written comments within the proper time limits, and voice your comments at public meetings. Sect. 7006(1) requires that any challenge to a regulation be made within 90 days after promulgation, and prohibits any later challenge in an enforcement action.

There is another excellent reason to keep abreast of state rules. RCRA requires that state rules be at least as stringent as federal ones in order for the states to get authorization to receive notification under Sect. 3010. Be aware of both federal and state definitions of hazardous wastes because you will have to notify according to the most stringent.

Groundwater

Groundwater quality is one of the main concerns of this act. If you know or strongly suspect that you have a groundwater problem, start defining it now. Groundwater monitoring is a good first step. Depending on the nature of the problem, you may have to notify the agencies and nearby water users.

Groundwater problems are normally expensive and take a long time to resolve; decades are not unusual. It is not like air or water problems, which generally stop when the source is eliminated.

Permit application

Get a head start on your permit applications for operating hazardous-waste-management facilities by gathering data that will be required by Sect. 3005, including: site geology, hydrology and climate; qualifications of personnel; listing of nearby residences and commercial buildings; contingency plans, etc. This is a particularly burdensome task for land-based disposal operations.

Long-term actions

Here are some longer-term things you can do in legal, technical and administrative areas to make the transition required by the regulations.

Review offsite-disposal agreements

Many arrangements for pickup and disposal are handled by purchase orders that state "pick up trash bins as required" or are in equivalent language. This is totally inadequate. Deal only with competent organizations and spell out in detail what you will do and what they will do.

In addition to the usual commercial terms, a contract should address at least these questions:

1. Is a typical description of the wastes included? Remember that wastes vary from load to load. Include a disclaimer as to the nature of the wastes.

2. Does the contractor guarantee that his facilities, equipment and methods are compatible with the wastes?

3. Who is responsible for transportation of the wastes? When do title, ownership and risks transfer to the disposal contractor? Are the contractor and/or hauler held responsible for compliance with DOT rules?

4. Is it clearly stated that the contractor and his agents and employees will comply with applicable laws, and that all permits and licenses are in order?

5. Is there an "escape clause" in case things are not in order?

6. Is there a workmanship clause stating that the contractor will perform in a safe, efficient and lawful manner, using good operating, treatment and disposal practices?

7. Is there a good indemnification clause? That is, will you be held harmless if sued for personal injury, public rights, property damage, etc., because of the contractor's failure to act properly?

8. Does the contractor carry adequate insurance? This is critical.

Make sure new contracts are reviewed by your technical and legal people. Also review current arrangements your company has with outside firms for transport, storage and disposal.

Check out disposal site

It is not unusual to get drawn into a court case or get unwanted publicity as a result of improper actions on the part of a contractor. Send one of your environmental people to the sites you do business with. Most reliable outfits will be very willing to answer your questions and let you look around. Make your own judgments as to their competence, staffing, permits, problem history and general suitability for your purposes.

For instance: How close are the neighbors? Does the site sit on a swamp? How is the housekeeping? Does the operation have adequate technical staff and laboratory facilities or is it a one-man operation run from the cab of a truck? Look over the storage facilities; is there adequate containment? Is the site directly on the bank of a river? How about security? Is it possible for children or animals to wander into dangerous areas? Are there leachate collection systems? Are ponds lined or unlined? Where does runoff go? Are the facilities adequate to treat your particular waste? If the disposer is also the hauler, determine if your waste will be transferred to another site, and if so, check out that site.

Basically, you want your selection of a contractor to be proper. You may find yourself saddled with legal responsibility if this is not handled properly.

Good sources of information on a contractor's past problems are the state solid-waste and environmental agencies, and the state or county departments of health.

Get engineering people in early

Make sure your engineers get copies of pertinent regulations at each stage of development, particularly regarding Sect. 3004, which sets certain minimum standards for facilities that treat, store or dispose. The phone number of the Sect. 3004 EPA Desk Officer is (202) 755-9206. The regulations under RCRA go much deeper into saying how your facilities will be designed and operated than do any previous pollution-control regulations. It would be foolish to design a pond today without a look at these rules.

Get your line operators involved. There are many administrative and operating requirements in Sect. 3004, from training to response plans. Start to prepare these for your existing facilities.

Reduce your wastes

Limited offsite capacity for hazardous waste disposal, and design and operating requirements of RCRA, will appreciably increase the cost of disposal. The intent of Congress and these regulations is to foster conservation and recovery. Simple economics will shift the balance in process research and development toward higher material recoveries and lower waste volumes.

Advise your R&D and other technical people of the impact of the law. Direct some of their efforts to reducing waste volumes from existing processes, recovering or recycling accumulations of wastes, and developing new processes that minimize hazardous-waste-disposal problems and costs. Look at your raw materials; you may be able to afford more expensive ones and still come out ahead. There is a real opportunity here for creative process engineering.

Disposal capacity

Disposal capacity overall is the largest single unaddressed problem in RCRA. By EPA estimates, there are only about 100 commercial hazardous-waste facilities in this country, with most of industry's waste being disposed of onsite.

Plan for adequate capacity for your locations. Determine how much onsite capacity you have left. You may decide to provide additional storage capacity to avoid interfering with operations if ready disposal is not available or is uncertain.

Offsite capacity, at least for the short run, is expected to be seriously reduced under RCRA's impact. I know of a case where a company was looking for offsite disposal facilities for two wastes containing minor amounts of organics. One man spent three months contacting and checking out about 20 potential contractors before two were located who would definitely take the wastes.

You may be able to help a reputable disposal firm locate in your area by working with local groups. You may even decide to get into the business yourself.

The supply of disposal sites will be reduced because many on- and offsite facilities will not be able to upgrade to meet RCRA design standards, and citizen, community, media and legal pressures are forcing shutdown of existing sites as well as making it extremely difficult to open a new one. But demand for capacity is increasing because (a) the scope of Sect. 3001 defines many wastes as hazardous that previously were handled as nonhazardous, (b) testing costs will deter challenging the "hazardous" classification of many wastes and (c) hazardous-waste generation is growing at about 3% per year, according to EPA estimates. This adds up to a national problem that is rapidly getting critical.

Upgrade your own facilities

Most companies operate lagoons, incinerators, deepwells and other storage, treatment and disposal facilities. If the facilities do not meet the standards of Sect. 3004, you will have to negotiate a compliance schedule as part of your permit's requirements. Start emptying leaky ponds so they can be sealed and reused, or so the contents can be safely disposed of and the site properly secured.

Ask these questions: Are your incinerators equipped with proper effluent- and stack-sampling facilities? Do they have adequate controls and instrumentation and monitoring devices? Do you have data on the composition of your emissions and effluents? Do you have a solvent or acid pit behind the lab where you have been pouring discarded samples for years? And so on.

Get rid of accumulated wastes

Put your purchasing and sales departments to work. They have numerous contacts every day with people who use chemicals in their operations. If they know how much and what kind of waste you have, they may be able to find takers for free, or even for money. Do not forget dormant and off-grade stocks in your warehouses; you may be able to sell or recycle them.

Of course, such transactions are subject to the usual commercial quality-assurance problems, so check to see if they are covered by TSCA, Toxic Substance Control Act. If so, comply with all requirements. Check with your lawyers regarding the sales contract.

Segregate your wastes

If your inventory is thorough and detailed, you can begin segregating wastes according to hazardous and nonhazardous groupings. Take a good look at trash bins that are picked up by the local trash collector and taken to a sanitary landfill. They probably contain pallets, fiber drums, filter cartridges, bottles, bags, discarded laboratory samples and reagents, floor sweepings and other potentially contaminated materials. Set up separate trash bins or provide segregation in other ways. Educate your personnel and set up procedures so that hazardous wastes do not get into the nonhazardous systems.

Costs will increase

Operating cost budgets and capital plans should include added costs for waste disposal, even if it is done on your property. Costs will go up significantly unless you find you have a gold mine in some of your discarded materials. In your budgeting, include closure, post-closure monitoring and maintenance, and insurance costs.

Help customers

The products you sell often go to customers who are not as knowledgeable in the properties of the materials as you are. The formulations may be proprietary and the ingredients not spelled out in detail. Some portions of these products may end up as wastes. Product literature should make some mention of proper waste-handling practices. Be prepared to handle inquiries from customers on proper disposal. In extreme cases, it may be desirable to have waste materials shipped back to your plant where they can be handled with your own wastes.

Conclusion

As you review the proposed rules, I am sure you will think of other moves that can and should be made. Increased awareness of the developing regulations and their associated issues and problems should generate the kind of activity that will ease the impact of RCRA on industry and help assure that technically valid, practical and cost-effective regulations are promulgated.

The author

Mr. Richard Sobel is Director of Environmental Services for the Specialty Chemicals Div. of Allied Chemical Corp., P.O. Box 1087R, Morristown, NJ 07960. He is also chairman of the MCA's RCRA Task Group and of Allied Chemical's Solid Waste Committee. He joined Allied in 1950 and has held various technical and production positions. Mr. Sobel received a B.S. in Chemical Engineering from Cooper Union and an M.S. in Chemical Engineering from the University of Delaware.

Section 2
SOLID WASTE TREATMENT METHODS

Waste-sludge treatment in the CPI

Expedient sludge processing can reduce many industrial waste-disposal problems. This study focuses on thickening and dewatering equipment, presenting design and operational parameters, and plant and laboratory performance data.

Industries reviewed:

☐ Organic chemicals
☐ Petrochemicals
☐ Inorganic chemicals
☐ Fine and pharmaceutical chemicals
☐ Pulp and paper
☐ Mineral processing
☐ Secondary-metals processing

Processes covered:

☐ Thickening . . . gravity and flotation
☐ Dewatering . . . centrifugation, filtration and heat treatment
☐ Secondary treatment . . . aerobic and anaerobic digestion
☐ Chemical fixation
☐ Byproduct recovery

R. W. Okey, D. DiGregorio and E. G. Kominek, Envirotech Corp.

☐ The chemical process industries (CPI) produce a great volume and variety of waste solids, which are separated as sludge from process waters and aqueous wastes. The sludge must be treated and prepared for environmentally sound disposal. Here is a CPI-wide review of operational principles and design approaches for sludge-processing equipment and techniques. Volume reduction is the key to economical disposal and is the theme of this article.

A typical waste-processing flowsheet describing solids capture and concentration is shown in Fig. 1. Raw wastewater is treated in clarifiers specifically designed to provide a clear overflow, regardless of underflow concentration. Primary sludge is defined as the liquid waste containing suspended material that remains after primary clarification. The clear overflow is sent to wet-line bioconversion and secondary clarification units, which produce secondary sludge. The primary sludge is the main focus of this article because it possesses a unique character that differs from industry to industry. The secondary sludge is basically waste-biological cell tissue and is generally similar for all industries.

Primary sludges vary widely in specific gravity, chemical and biological stability, solubility, toxicity and particle size. However, some general classifications can be made. Table I lists primary sludges and waste solids generated by the CPI.

Table II lists thickening and dewatering performance data for sludges typically generated by certain industries. Though available cost data are insufficient to bracket the various industries, thickening and dewatering usually cost about $10-$50/ton of dry solids.

Originally published January 29, 1979.

Pretreatment

Most sludges require chemical conditioning prior to thickening and dewatering. Conditioners can be divided into two broad categories: inorganic coagulants, such as lime, hydrated aluminum sulfate (alum) and ferric salts; and organic flocculants, which have high molecular weight and are cationic, anionic or nonionic.

Inorganic coagulants have widespread use, and in most cases effectively capture colloids. They have the

Solid-wastes generated in the CPI		Table I
Origin	**Physical and chemical characteristics**	**Comments**
Coke manufacture	Coke and coal fines	Generally contained in scrubber wastes
Dyes and pigments	Reaction or raw-material sludge, highly variable	—
Fine-chemical and pharmaceutical	Raw-material solids, biological wastes	—
Inorganic chemicals	Insoluble salts, tailings, slimes	—
Metal processing (secondary)	Ash, scrubber wastes, metal hydroxide sludges	—
Petrochemical	Oily, greasy, asphaltic	Usually float
Plastic and rubber	Latex or plastic crumbs, often alum coagulated	—
Pulp and paper	Fibrous; some fine filter, often with lime or alum	—

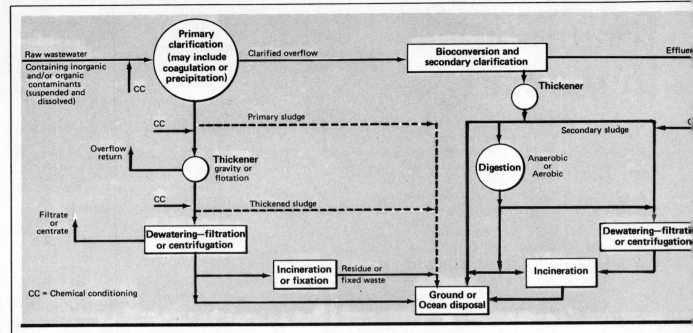

Typical waste-sludge processing flowsheet employed in the chemical process industries   Fig. 1

disadvantage of contributing substantially to the sludge volume, which frequently increases the overall dewatering problem.

Organic flocculants are specific in use and effective at low dosages. However, they do not scavenge colloids as well as the coagulants, and under conditions of very high or low pH undergo size degradation, with consequent loss of effectiveness.

Bench-scale tests are used to determine the optimum conditioning agent and procedure.

Heat treatment also is effective in preparing a sludge for thickening and dewatering, because it destroys the gelatinous components, which bind water. This method is discussed in the dewatering section.

Thickening

The purpose of thickening is to reduce sludge volume as much as possible prior to dewatering and/or disposal. Sludge-dewatering characteristics improve as solids concentration increases. Maximizing thickening results in cost and equipment performance benefits during dewatering operations.

Consider the dewatering of a calcium carbonate sludge by filtration. This type of sludge cannot be filtered effectively at feed concentrations of less than 6 or 7 weight %. At a concentration of approximately 10%, a filtration rate of 10–12 lb dry solids/h/ft² of filter area can be obtained. However, if the feed sludge is thick-

CPI waste-sludge thickening and dewatering performance data (summary of Table III, V, VI, and VII)   Table II

Industry	General characteristics of sludge	Concentration after thickening, % solids	Concentration after dewatering, % solids	Comments
Coke manufacture	Scrubber fines	20-25	40-50*	Vacuum filtration
Dye and pigment	Primary	7-10	20-30*	Vacuum filtration
Fine chemical and pharmaceutical	Biological	2-5*	15-25*	—
Inorganic chemical	Slimes, fines, tailings	10-30*	30-80*	—
Metals (secondary)	Metal fines, metal hydroxide	20-50	40-80	—
Petrochemical	Oily, greasy asphaltic	2-5	10-40	Thicken by flotation
Plastic and rubber	Latex or plastic crumbs, often alum-coagulated	7-8	15-25	—
Pulp and paper	Fibrous, fillers, often with lime or alum	2-50	20-55	—
		2-5	12-20	Combined primary and waste-activated, 30-50% with pressure filtration

*Estimated

Industry	Origin or nature of sludge	Feed-solids concentration, %	Loading lb/d /ft^2	Underflow solids, %	Polymer or coagulant type, concentration	Comments	Reference
Industrial performance data for thickening of waste sludge							**Table III**
Coke	Scrubber blowdown (coke and coal dust)	Unknown	Unknown	20-25	Betz 1115 0.25-0.50 mg/L	—	34
Dye and pigment	Primary sludges	Unknown	Unknown	7-10	Nalco 683 4-5 mg/L	Solids contact clarifier	34
Glass etching	Calcium fluoirde	Unknown	Unknown	15	Lime to pH 11.3	Solids contact clarifier	34
Latex production	Alum coagulated primary	Unknown	Unknown	Unknown	Alum	4-6 days settling in lagoon	6
Lead smelter	Scrubber wastewater	0.1-0.2	7.8	40-50	Unknown	—	34
Petrochemical	Oily general	Unknown	Unknown	2-5	Unknown	Flotation thickener	13
Pharmaceutical	General	Unknown	6	Unknown	Unknown	—	12
Pulp and paper	Mixed primary and secondary	0.3-0.35	3.1	3.3	Magnaflox	Primary to secondary > 20:1	34
Pulp and paper	Ground-wood or waste-activated	Unknown	Unknown	4.5 (float) 7-12	Unknown	Flotation disk-centrifuge	4
Pulp and paper	Deinking	1	2.2	3-5	Unknown	—	34
	Glassine	0.5	2.0	3-4	Unknown	—	
	Waste-activated	1.6	13.3	3-4	Unknown	—	
Pulp and paper integrated kraft	Biological waste	Unknown	34.1	2.5	Unknown	—	10
	Primary	Unknown	13.5	3.0	Unknown	—	
	Mixed primary and secondary	Unknown	34.0	3.0	Unknown	1:1 and 2:1 ratios	
PVC production	Chemical or primary	2.5	26.6	8	Unknown	Design values from bench scale	5
Rubber	Primary	Unknown	Unknown	15.4 (overflow)	Unknown	Flotation	7
Rubber, synthetic latex	Primary	144 mg/L	5.5 gal/min/ft^2	0.62-10.63 7.80 Avg	Alum 20-62 mg/L Cation 0.54 mg/L	Recycle rate 22-30%	8
	Secondary	129 mg/L	4.2 gal/min/ft^2	1.24-18.08 5.75 Avg	Anion 0.61-0.85 mg/L	Air to solids 0.1 lb/lb	
Steel (specialty)	Rollingmill and furnace cooling water, pickling rinse and cooling, tower blowdown, metal hydroxides	8-15	Unknown	25	Lime	—	11
Zinc electrolytic refinery	Reactor clarifier, gypsum from H$_2$SO$_4$ neutralization, metal hydroxides	Unknown	Unknown	Unknown	Unknown	Disposal to pond	9

ened to 20–22% dry solids, filtration rates can be as high as 50–60 lb dry solids/h/ft^2. At higher filtration rates, the size and cost of equipment is smaller.

Thickeners can concentrate slurries as dilute as a few hundred ppm to sludges of 5–15 weight % dry solids (or higher). Thickening rates depend on properties of the feed, feed solids concentration, retention time, feed concentration (underflow) required for dewatering, and overflow clarity required for disposal or recycle. Generally, thickening is accomplished by gravity sedimentation or flotation.

Sedimentation

A typical gravity thickener, Fig. 2, consists of a feedwell that baffles the slurry prior to entry into the sedimentation zone, and a rotating floor scraper that transports the thickened slurry to the centrally located underflow discharge-hopper. The feedwell serves as a flocculation zone when chemical conditioning is used.

The most common configuration for a gravity thickener is a circular tank. Side-water depths vary from 10 to 12 ft. Tank diameters vary from as small as 10–15 ft to as large as 600 ft. Floor slopes for thickeners are generally 10–15 deg.

Gravity-thickener design criteria are almost always established from experience or bench-scale tests. Traditionally, these tests are conducted using 2-L cylinders equipped with slow-turning rakes, and are interpreted by the Coe-Clevenger or the modified Kynch (Talmadge and Fitch) method. Each method has certain requirements that must be met, which involve selecting an operating point on a settling curve to calculate thickener sizing. These tests are also useful in predicting overflow clarity, underflow solids content, and the efficacy of chemical flocculants.

Gravity thickening of organic sludges (particularly waste-activated sludge) is complicated by anaerobic action. If the temperature is warm, bacteria in the sludge will decompose organic matter, releasing gases. This causes flotation problems, hinders compaction and creates noxious odors. Sedimentation-thickener loadings for organic sludges range from 4 to 12 lb/d/ft^2, while flotation thickening can handle loading rates of 36 to 48 lb/d/ft^2.

Flotation

Flotation thickening is used for slurries containing solids that float rather than settle, or have slow settling rates or poor compaction. Flotation keeps the system aerobic. A typical dissolved-air-flotation (DAF) thickener is shown in Fig. 3. Most flotation thickeners use recycle pressurization. Part of the flotator subnatant is pressurized. The incoming feed does not pass through the pressurization system but rather is mixed with pressurized recycle in the inlet-diffuser.

Flotation thickeners are rectangular or circular. Standard rectangular flotators vary in size from 100 ft^2 to as large as 1,800 ft^2. Rectangular flotators are usually used for small applications and where space considerations are paramount. Both types are equipped with surface skimmers and floor rakes. The surface skimmers remove float from the thickening tank to a float sump.

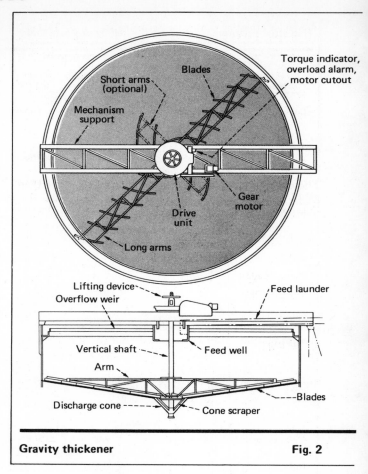

Gravity thickener   **Fig. 2**

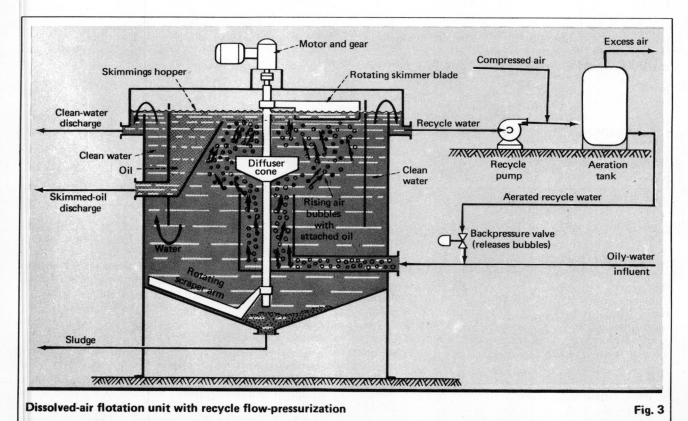

Dissolved-air flotation unit with recycle flow-pressurization   **Fig. 3**

Variables affecting centrifuge performance	Table IV

Machine variables	Process variables
Bowl diameter	Type of sludge
Bowl length	Loading rate
Differential speed	Chemical conditioning
Centrifugal force	
Beach length	
Beach angle	
Pool depth	

These skimmers should be of variable speed, 2–25 ft/min, to provide maximum flexibility. The floor rakes remove heavier solids that cannot be floated.

Flotation thickening may be evaluated by bench- or pilot-scale tests. Bench-scale apparatus is normally used to investigate flotation on a batch basis, predicting thickening performance in terms of float solids content and solids capture at various air/solids ratios. Bench-scale tests are also useful in evaluating the effect of chemical flotation aids on float solids and solids capture. Pilot-scale tests are useful for sizing as well as for predicting flotator performance. Bench and pilot devices are available from thickener manufacturers.

Table III lists industrial performance data for several sludge thickening operations.

Dewatering

The purpose of dewatering is to remove sufficient liquid from the thickened sludge to produce a cake that has optimal handling properties and solids content for subsequent processing or disposal. Dewatering is accomplished by mechanical methods, the most common being centrifugation and filtration, which include: vacuum, pressure, horizontal belt, belt press and precoat.

Flocculants are almost always used in dewatering industrial sludge to improve solids capture rate and cake discharge characteristics. Even though the cake may form well, the fine fraction usually contains an appreciable amount of material that tends to seal the cake. Flocculation coagulates the fine particles and helps prevent this. In a few cases, such as the filtration of water-softening sludge, no flocculation is required because the sludge does not contain a significant amount of fines. Determining the most economical chemical addition for a specific sludge should be based on bench-scale testing or pilot-plant studies.

Heat treatment of sludge prior to dewatering stabilizes and improves the dewatering characteristics of the sludge. It also destroys organics. It can be oxidative or nonoxidative, depending on the specific process involved. Waste sludge is treated at temperatures of 350–450°F and maintained at 150–250 psig for 20–30 min. Part of the sludge is solubilized, producing liquor having a COD, chemical oxygen demand, that can be as high as 4,000 mg/L.

Heat treatment reduces the specific resistance to filtration of a sludge by destroying the gelatinous components, which bind water. Dewatering improvement is similar for oxidative and nonoxidative systems. Sludges that gravity thicken to 2–3% total solids can be thickened to 10–15% total solids after heat treatment. Organic sludges that filter at 2–3 lb/h/ft² with chemical conditioners can be filtered without conditioners at 5–8 lb/h/ft² after heat treatment. Cake solids content is also improved by heat treatment.

Centrifugation

The solid-bowl centrifuge (Fig. 4) consists of a rotating bowl that concentrates and dewaters a slurry by separating the mixture into a cake and a dilute stream (centrate). The cake is transported within the bowl and discharged from it by a screw conveyor, which rotates within the bowl at a speed slightly less than the bowl

Industrial performance data for centrifugation of waste sludge							Table V
Industry	Origin or nature of sludge	Feed-solids concentration, %	Loading gal/min	Cake-solids concentration, %	Polymer or coagulant type, concentration	Comments	Reference
Chemicals	Primary	Unknown	Unknown	35	Polymer type unknown	Carried out in 40 x 60-in. machines, solids landfilled	21
Pulp and paper sulfite & NSSC*	Biological waste	1-2	Unknown	8-10	None	—	16
Pulp and paper						Solids capture	4
	Fine paper	3.5-12.0	56-120	16-35	Unknown	80-95%	
	Kraft	1.0- 4.5	95-320	15-40	Unknown	80-95%	
	Tissue	1.0- 4.0	20-38	21-42	Unknown	65-90%	
	Hardboard	2.0	43	30-35	Unknown	85%	
	Book	5.0-12.0	150	40-45	Unknown	85-92%	
PVC production	Chemical or primary	8	28	15-25	Unknown	—	5

*Neutral sulfite semi-chemical

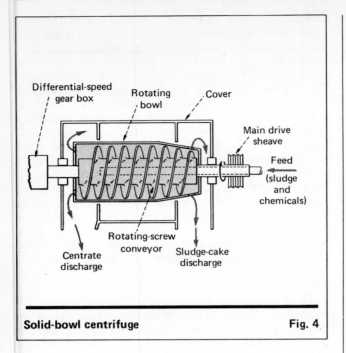

Solid-bowl centrifuge   **Fig. 4**

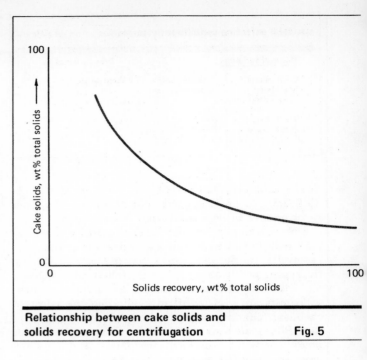

**Relationship between cake solids and
solids recovery for centrifugation**   **Fig. 5**

speed. The centrate overflows weirs on the opposite end
of the bowl.

There are numerous machine and process variables
that affect centrifuge performance. Table IV lists the
most important ones.

The capacity of a centrifuge is related to its size (bowl
diameter and length). During operation, a pool is
formed in the inner periphery of the bowl. The pool
volume is determined by the bowl diameter and length,
and pool depth. For a specific loading, solids capture
generally improves as pool volume increases.

If all variables remain constant, an increase in the
bowl diameter will increase pool volume and retention
time. Increased retention time permits a higher capture
of the smaller and/or lower-specific-gravity particles.
However, these fines decrease cake drainage characteristics, increasing the liquid content of the cake.

Increasing bowl length increases pool volume and
retention time. Generally, long-bowl machines
(length : dia. = 2–3 : 1) are more effective and have
greater capacity than short-bowl ones (length : dia. $\leq$
2 : 1).

The centrifugal force required to achieve a specific
level of performance for a given size of machine is
related to the characteristics of the sludge being dewatered. Most commercial machines are operated at

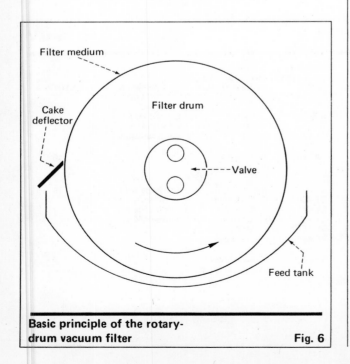

**Basic principle of the rotary-
drum vacuum filter**   **Fig. 6**

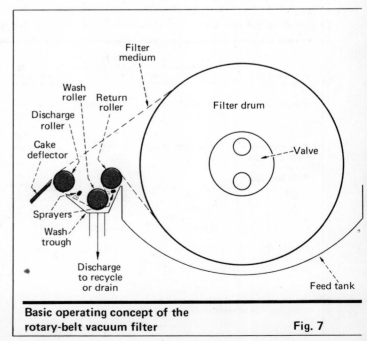

**Basic operating concept of the
rotary-belt vacuum filter**   **Fig. 7**

Industrial performance data for vacuum filtration of waste sludge							Table VI
Industry	Origin or nature of sludge	Feed-solids concentration, %	Loading lb/d/ft^2	Cake-solids concentration, %	Polymer or coagulant type, concentration	Comments	Reference
Aluminum extrusion	Clarifier, aluminum hydroxide	4-6	Unknown	15	Unknown	Roll discharge employed	34
Chemicals	Anhydrous sodium metasilicate manufacture	0.86-1.10	0.1-0.5 (gal/min/ft^2)	16.0-27.2	Unknown	Precoat filter	17
Coke manufacture	Scrubber blowdown (coke and coal dust)	20-25	150	Unknown	Unknown	Horizontal-belt filter	34
Ground wood pulping	Primary	Unknown	Unknown	24	2:1 Chipper fines added	—	15
Lead smelter	Thickened scrubber wastewater	40-50	18.7	80	Unknown	Agitated disk filter	34
Pharmaceutical	General	Unknown	2-biological 4-mixed primary and biological	Unknown	Up to 20 lb/ton	—	12
Pigmented paper	Primary	2.5-3.0	Unknown	30	Unknown	—	19
Pulp and paper	White water (W.W.)	1.33-4.70	1.7-13.4	23.3-33.0	Unknown	—	4
	Decoating and W.W.	5.85-10.02	2.1-11.0	34.6-42.9	Unknown	—	
	Boardmill	0.87-2.36	1.2-5.8	26.1-30.7	Unknown	—	
	Deinking and W.W.	5.87-7.15	3.1-10.0	31.4-36.4	Unknown	—	
	Felt mill	5.20-5.27	3.7-5.9	21.4-25.8	Unknown	—	
Pulp and paper	Combined primary and secondary	3.3	1-2	15.3	Unknown	—	34
Pulp and paper integrated kraft	Primary	2.6	Unknown	18-22	Unknown	—	14
Pulp and paper integrated kraft	Combined primary and secondary	1.38-2.70	2-13	12-18	FeCl$_3$ 5%	Ratio 1:1 and 2:1, formtime varied from 30-90 s, coil filter	10
Pulp and paper integrated kraft	Combined primary and secondary	1.63-1.90	2.4-5.5	12-13	None	Ratio 1:1 and 2:1, formtime varied from 30-90s	10
Pulp and paper integrated kraft	Combined primary and secondary	1.57-1.98	3.5-10	19-22	1:1 Bark fines added	Ratio 1:1 and 2:1 Rates do not include bark fines	10
Pulp and paper, kraft	Lime slurry from color removal process	16-26	17-51	44-53	None	Precoat and belt filters	20
Pulp and paper, sulfite and NSSC*	Waste activated	Unknown	Unknown	25-40	Unknown	Precoat filter	16
Steel	Pickling liquor and limestone	Unknown	22-68	51-67	Unknown	Lime Pilot operation	18
Steel	Limestone treated mill scale	Unknown	56	70	50% Excess of limestone	—	18
Steel (specialty)	Thickened rolling mill and furnace cooling wastes, cooling tower blowdown, metal hydroxides	25	17	40-60	Unknown	—	11

*Neutral sulfite semi-chemical

speeds sufficient to develop forces 1,000–3,000 times that of gravity. Increasing centrifugal force usually increases solids capture and cake solids content. However, cake solids content may remain constant or even decrease, depending on the quantity and characteristics of the additional solids captured.

The solids forced to the inner wall of the bowl are conveyed up the cone of the bowl, the beach, for discharge from the machine. The length and angle of the beach have a strong influence on centrifuge performance, particularly with sludges containing light, gelatinous solids (waste-activated or metal hydroxide sludges). As the sludge is conveyed along the beach (especially where contact with the pool ceases), additional moisture drains off prior to discharge. For most sludges, increasing dry-beach residence time will increase cake solids content.

As the cake is conveyed along the beach, it is subjected to centrifugal force acting perpendicular to the bowl and to a force acting parallel to the beach. The parallel force tends to push the solids down the beach into the pool. This "slippage" force is related to the centrifugal force and the beach angle.

For sludges consisting of light, gelatinous solids, the slippage force can be sufficient to cause conveyed solids to flow under the conveyer blades back into the pool. Solids capture deteriorates in this situation and centrifuge capacity is severely limited. Machines having small beach angles, or pools raised to slightly below the cake discharge point, can be used to overcome this problem.

The difference in rotational speed between the screw conveyor and the bowl is termed the differential speed. For a specific loading, it is advantageous to maintain the minimum differential speed necessary to convey all cake from the machine. Increasing differential speed above this level increases pool turbulence and speeds the cake conveying rate in the pool and on the beach. These factors generally reduce solids capture, especially in applications involving waste-activated, metal hydroxide and other difficult sludges.

The major process variables affecting performance are the type of sludge, the use of chemical flocculants, and the loading rate to the machine. For any application, loading rate and chemical conditioning can be varied to control centrifuge performance in terms of cake solids content and solids recovery. Fig. 5 shows the normal relationship between cake solids content and solids recovery. This relationship can be controlled by changing process conditions (loading rate) and machine conditions (pool volume and differential speed).

Centrifuges are best sized from pilot- or prototype-scale units operating at the plantsite under actual conditions. When this is not possible, performance may be estimated using batch laboratory centrifuges. Procedures for conducting laboratory tests are described in the literature [35]. Since these tests are batch and do not include provisions for continuous removal of settled solids from the pool, indirect methods must be used to judge whether solids can be successfully discharged from full-scale machines. This is essential. Generally, granular solids can be readily conveyed out of a full-scale centrifuge, whereas flocculated and/or compressible solids are significantly more difficult.

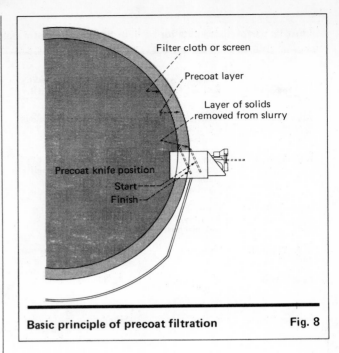

Basic principle of precoat filtration   **Fig. 8**

Table V lists performance data for a few industrial centrifugation operations.

Vacuum-drum filtration

The typical continuous vacuum filter consists of a rotating drum partially submerged in a vat containing slurry. The drum is divided into sections that can be separately evacuated by means of an automatic valve. The filter medium, which can be made of various natural and synthetic materials, overlies the face of the drum and supports the dewatering sludge.

The two most frequently used vacuum filters are the rotary drum (Fig. 6) and rotary belt (Fig. 7). Drum diameters vary from 3 to 12 ft; lengths vary from 3 to 40 ft.

Since the development of continuous filtration, the conventional vacuum-drum filter has been used in more sludge applications than any other filter type. This is mainly due to its operational flexibility, and ability to handle a wide range of slurries. The basic geometry of the drum filter permits the operator to vary the cycle time for cake formation, washing and dewatering. This minimizes inactive time, increases productivity/unit area and makes it easier to handle off-quality feeds. The conventional drum filter also permits discharge of thinner cake than is possible with disk, pan, horizontal and other filters.

The major disadvantage of vacuum-drum filters is progressive blinding of the filter medium, particularly with sludges containing extreme fines. Blinding generally results from: (1) plugging of the interstices within the filter medium by suspended solids in the feed, (2) chemical precipitation within the filter medium or (3) sealing of the surface of the medium caused by the shape of the feed solids.

Slow blinding occurs in the majority of existing applications, but several days to weeks may elapse before the filtration rate is severely reduced. To achieve a

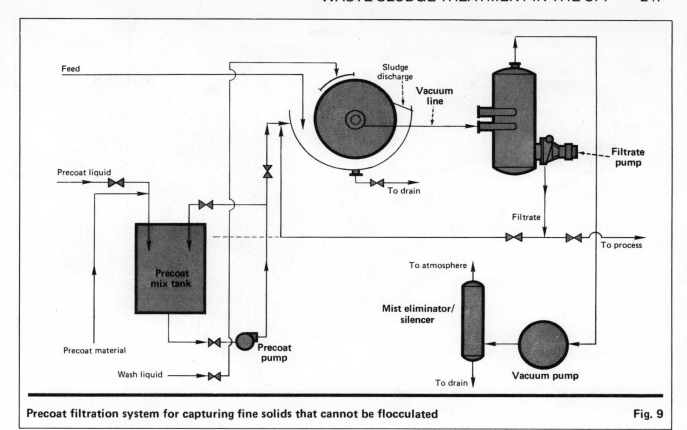

Precoat filtration system for capturing fine solids that cannot be flocculated   **Fig. 9**

balance between cloth life, operating time and capital cost, the design filtration rate must be 70–80% of the clean-medium rate. When blinding reaches a critical point, the medium must be replaced or rejuvenated with an acid or alkaline wash.

Another problem is poor cake-discharge. If a filter cake of minimum thickness and sufficient dryness is produced, the cake discharge will be practically complete. Generally, ¼-in. minimum thickness is desirable, although cakes of ⅛-in. can be discharged. However, if thin, slimy or moist cakes are encountered, cake discharge can be severely impaired. In addition, reduction in filter capacity, accelerated rate of blinding and difficulty in cake removal can occur if multifilament weaves of cotton and other natural fibers are used in the filter medium, because short fibers imbed in the cake.

Use of a rotary-belt-type drum filter eliminates or greatly reduces problems of blinding and cake thickness. The design (Fig. 7) consists of a sectionalized drum, in which the periphery of each section contains a soft rubber or synthetic strip that is slightly raised from the drum surface. The filter medium lies over these strips, sealing the vacuum side from the atmosphere. Since a typical filter operates at a 20–25 in. Hg vacuum, there is approximately 10–12 lb/in.^{2} of pressure providing sufficient sealing force. The filter medium is an endless belt that travels off the drum to a cake discharge roll and then to a wash chamber, where fluid is applied through high-pressure sprays to both sides of the cloth. Occasionally, an acid wash may be required to reduce blinding.

Cake discharge is accomplished by passing the filter medium over a small-diameter roller, which very abruptly changes the radius of curvature of the medium relative to the cake, causing the cake to break free from the cloth. Generally, a scraper blade is not required to loosen the cake from the medium but rather to deflect the cake to the discharge point of the filter. Cakes as thin as $\frac{1}{16}$ in. can be satisfactorily discharged.

Particulate sludges containing less than 50% of 200-mesh (74 microns) solids can be successfully handled by belt-type units, which can also handle most plating, steel mill, water softening, foundry, organic and pharmaceutical sludges, and even wet-air oxidation sludges.

Filtrate containing suspended-solids levels as low as 500 mg/L can be achieved with belt-type units, from feed concentrations as high as 10–20% solids. With other filters, such as precoat units, filtrate solids of only 25 mg/L or less are possible.

Vacuum-drum filtration may be evaluated from bench- or pilot-scale tests. Bench-scale apparatus procedures have been well developed and are described in the literature [36], [37]. The filter-leaf method is useful for predicting filter performance under a variety of operating conditions and for establishing medium type, bridging and submergence. Pilot-scale vacuum-drum filter tests are particularly useful in identifying and solving cake pickup and discharge problems.

Table VI lists performance data for several industrial vacuum-filtration operations.

Precoat filtration

Precoat vacuum filtration is used primarily for clarification purposes or for difficult filtering applications in

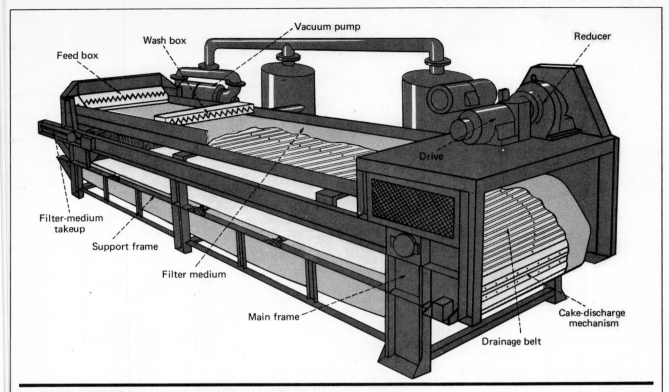

Horizontal-belt filter is best used with slurries containing granular solids   **Fig. 10**

which severe cake-discharge problems are expected. It is also used when feed characteristics are highly variable.

The precoat filter (Fig. 8) is similar to the rotary-drum vacuum filter. A cake of precoat material, such as diatomaceous earth or expanded perlite, is formed on the filter medium prior to sludge contact. Filtration proceeds continuously by shaving a portion of the precoat from the filter along with the filter cake. Shaving is accomplished by a sharp knife that removes 0.003–0.005 in. of precoat per drum revolution. As filtration continues, the knife advances toward the drum surface. The cake of precoat may last for several hours or several days, depending on operating conditions. A system employing a precoat unit is shown in Fig. 9.

For fine colloidal particles that will not settle or cannot be flocculated, the continuous precoat filter is almost the only solution. Precoat filters have been used successfully for clarifying slop oil, removing TNT fines from wastewater at ordnance plants, and dewatering oily or metal hydroxide sludges. Generally, the feed has a low solids concentration, and filtration rates can run as high as 40–50 gal/h/ft.[2] Precoat filter operating costs are usually higher than those for conventional drum or belt-type filters because of the precoat, which on the average is consumed at a rate of 10–15 lb of precoat/1,000 gal of filtrate.

Precoat-filter design criteria may be established from either bench- or pilot-scale tests. Bench-scale tests are similar to those employed for conventional vacuum filtration. A precoat material is initially formed on the filter leaf, and the slurry is filtered using the cake of precoat as the filter medium. Tests are useful in establishing filtration rate, solids capture and filter-cake solids content. Criteria such as precoat type and consumption (which serve to verify cake solids content) are best obtained from pilot-scale tests.

Horizontal-belt filtration

Horizontal-belt filters (Fig. 10) are best applied to slurries containing granular solids that form cake rapidly and have high dewatering rates. They are also used when cake-discharge problems are anticipated. A typical horizontal-belt filter includes a slurry flocculation unit that serves to distribute feed material across the width of the filter. A filter medium overlies the horizontal grids. Dewatering is achieved as the slurry is transported on the medium along the length of the filter. Vacuum, at controlled levels, may be applied at different zones along the filter.

These units vary in width from 5½ to 18 ft and in length from 16 to 110 ft. Filtration areas range from 10 to 1,200 ft.[2]

The horizontal, top-feed belt filter allows extensive washing and countercurrent staging for removal of objectionable solubles (such as mother liquor) from the cake. Continuous belt washing minimizes blinding. The horizontal-belt filter has been used primarily in industrial waste applications where an extremely wide range of particle sizes must be removed or where cake washing is a necessity.

Design criteria are determined by bench-top and pilot-scale testing in similar fashion to vacuum-drum filtration.

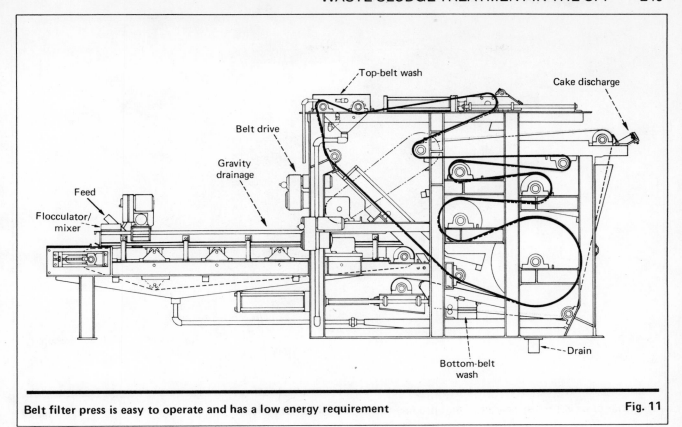

Top-belt wash

Cake discharge

Belt drive

Gravity drainage

Feed

Flocculator/ mixer

Bottom-belt wash

Drain

Belt filter press is easy to operate and has a low energy requirement

Fig. 11

Belt-press filtration

The belt press (Fig. 11) is widely used in Europe but has only recently been applied to sludge dewatering in the U.S. A typical belt press includes a slurry flocculator that assists in distributing feed across the width of the unit. An adjacent drainage zone removes liquid from the flocculated slurry by gravity or vacuum. Moisture is removed from the drained cake by mechanical and shear forces that are exerted as the cake is sandwiched between two endless belts and transported around several rollers of various diameters. Pressure is applied to the cake by a combination of increasing belt tension and decreasing roller diameters. Scrapers are used to continuously discharge the cake.

The advantages of the belt press over other mechanical dewatering methods include low energy requirements and operating simplicity. The belt press overcomes cake pickup problems experienced in vacuum filtration of sludges that are difficult to dewater. It is ideally suited for operation with polymer conditioning. This eliminates handling problems associated with ferric chloride and/or lime chemical conditioners, which may be required for vacuum or pressure filtration of difficult sludges.

There are numerous belt-press configurations. Units are available with belts up to 10 ft wide. Lengths vary depending on the configuration.

The most widely accepted method of sizing belt presses and predicting their performance is to run pilot-scale tests. Manufacturers of these devices have units available for determining capacity, solids capture, cake solids and polymer conditioning. Until more oper-

ating experience is available on belt presses, pilot tests should be performed prior to specifying and sizing.

Pressure filtration

Recessed-plate or plate-and-frame types of filter presses consist of a series of rectangular plates supported face-to-face in a vertical position. Filter cloth is fitted over the face of each plate. Hydraulic rams or powered screws are used to hold the plates together during dewatering. A typical pressure filter is shown in Fig. 12.

Filter pressing is a batch operation in which chemically conditioned sludge is pumped into the space between the plates. Pressures of 100–250 psi are applied for periods of one to several hours. The solids in the slurry are retained by the filter cloth and gradually fill the space inside the plate while liquid is forced through the filter cloth. At the termination of the dewatering period, the plates are separated and the sludge cake removed.

Pressure filtration can produce cake of higher solids content, containing 5–20 percentage points less moisture, than other filtration processes. It is frequently employed where low cake moisture is required. However, it is a batch process and may require cake breaking and storage facilities.

In applications where very hydrous, cellulose pulps are encountered (pulp-and-paper and municipal sludges), pressure filtration is frequently used. Sludges containing water and oil are also good feeds. The pressures are usually sufficient to prevent blinding of the filter cloth by the oil.

Design criteria for sizing pressure filters are generally

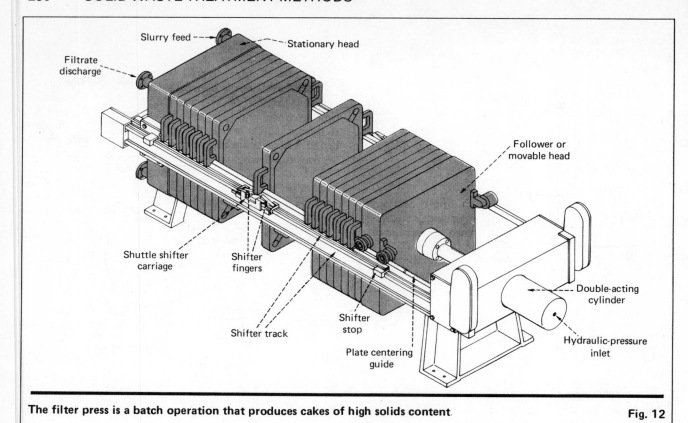

The filter press is a batch operation that produces cakes of high solids content.

Fig. 12

obtained from pilot-scale tests. Attempts to use specific resistance to filtration, as measured by laboratory-scale apparatus, have generally been unsuccessful. Pilot-scale apparatus varies in size from several square inches to small prototype units of about 100 in². The results obtained from these devices correlate well with full-scale yields. These units are also very useful in identifying proper filter media and in measuring the effects on performance of cake thickness, time, pressure and chemical conditioning.

Table VII lists performance data for several industrial pressure-filtration operations.

Reverse osmosis and ultrafiltration

The pressure-driven membrane processes—reverse osmosis and ultrafiltration—have found only limited use in dewatering sludges because they cannot tolerate large concentrations of suspended material without serious reduction in flux. For example, ultrafiltration systems can concentrate solids to only 3–5%. However, these processes are useful when it is necessary to separate and concentrate a colloidal or dissolved species from a slurry. Toxic materials present in a liquid stream at very low concentration can be concentrated to 1–3% solids in pressure-driven membrane processes without any severe penalty other than an increase in required system pressure, resulting from the increasing osmotic pressure. Usually, the molecular weight of the species to be separated and concentrated is high enough that the pressure increase is insignificant. The process chosen depends on the size of the species to be concentrated. For ionic species, reverse-osmosis is used, and for or-

ganic species ultrafiltration is used. The concentrated liquid or viscous material can then be coincinerated with sludge.

Secondary sludge treatment

Waste biological sludge is generated in the secondary stages of waste treatment (Fig. 1). Organic chemical, petrochemical and pulp and paper wastes usually require biological treatment for BOD, biochemical oxygen demand, reduction in compliance with EPA's BPTCA (Best Possible Technology Currently Available) standards. This is accomplished in aerated lagoons, contact-media units and activated-sludge-treatment units. This discussion will consider only the latter.

The sludge production of activated-sludge processes is shown in Table VIII. The data are typical and are based on feeds consisting of less than 0.2 lb inerts/lb BOD. The amount of inerts in the feed is critical. The sludge produced has a concentration of 0.5% to 1.5% solids after discharge from a secondary clarifier. Thickening is essential to reduce this volume prior to dewatering operations.

If primary treatment produces a sludge volume that is much larger than that from secondary treatment, the sludges can be mixed and dewatered together. However, they are often kept separate for better control in dewatering and disposal.

Biological sludge stabilization, which is often practiced prior to dewatering or disposal, can be aerobic or anaerobic. Anaerobic digestion has the advantages of lower power requirements and potential recovery of methane.

Aerobic-digestion basins have sludge retention times of 10–20 days, with feeds of 5–10 lb of solids/100 ft³. Up to 40% of the volatile solids are destroyed.

Anaerobic digestion reduces organic sludges to methane, carbon dioxide, ammonia and hydrogen sulfide. The methane can be recovered and used for power generation or other purposes. Digestion is temperature dependent. In the mesophilic temperature range, 50–95°F, the mixed cultures of bacteria require 20–55 days for digestion. In the thermophilic range, 100–140°F, digestion time is reduced to 15–20 days [38].

Chemical fixation

Chemical fixation is a process for binding hazardous waste in a chemical matrix that is impervious to water penetration. This minimizes the hazard of chemicals being leached into groundwater.

One acceptable procedure involves fixing the toxic substance in a silicate matrix [25]. Presumably, this process causes the formation of metal silicates, or the direct occlusion of metals in SiO_2, depending on the pH. Fixation is acceptable for many heavy metals and certain inorganic anions such as arsenates. To date, fixation has not been very effective with toxic organics. Table IX lists sludge types that have been chemically fixed.

Taconite tailings derived from the beneficiation of iron ore have been successfully fixed using a carbonate bonding process [1]. The waste is mixed with water and lime hydrate, and then reacted with a gas rich in CO_2 to form a calcite crystal matrix. This procedure may be broadly applicable to fine inorganic solids that cannot be contained when released to the environment.

The moisture content of sludge has a substantial impact on the cost of fixation. Increasing solids content from 30% to 50% reduced cost in one application from $30 to $10/ton of dry solids.

Tests are required to demonstrate feasibility and cost effectiveness of fixation because of the unpredictable variation in sludge compositions.

Byproduct recovery

Byproduct recovery can be a significant factor in sludge management [2]. Waste tars, spent catalysts and

---

Industrial performance data for pressure filtration of waste sludge   Table VII

Industry	Origin or nature of sludge	Feed-solids concentration, %	Loading lb/h /ft²	Cake-solids concentration, %	Polymer or coagulant type, concentration	Comments	Reference
Chemical	Biological and chemical	1.7	Unknown	37-38	5% $FeCl_3$	Max pressure 120 psi 75-90 min cycle time	22
Chemical	Primary and waste activated	4.5	147,000 lb/d on 5,800 ft²	32	Lime and $FeCl_3$ CaO-20% $FeCl_3$-2.5%	66.6% Biological 33.3% Chemical conditioners	34
Glass etching	Reactor-clarifier, calcium fluoride	15	15	40	Unknown	Belt filter	34
Petrochemical	General Oily	2-5%	Unknown	10-40	$2-$10/ton Types unknown	>99% solids capture, gravity-belt press	13
Pulp and paper	Board-mill	Unknown	—	27.5-50.3	Unknown	100-900 psi 1-10 min	4
	Deinking	Unknown	—	38.5-64-3	Unknown		
Pulp and paper, bleached kraft	Primary	Unknown	50 tons/d	35	2.5 lb/ton	Belt press twin wire	23
Pulp and paper, integrated kraft	Primary	18-22	Unknown	37-40	Unknown	V-press, dried to 70-85%	14
Pulp and paper, integrated kraft	Combined primary and secondary	19-22	Unknown	33.5	1:1 Bark fines added	Batch press	10
Pulp and paper, sulfite and NSSC*	Secondary	1-2%	0.4 to 3.6	30-50 filter press 20 belt press 7 ultrafilter	Unknown 10-20 lb/ton None	>99% solids capture	16
Pulp and paper, sulfite and NSSC*	Primary	Unknown	10-25 gal/min/ft²	30-40	0-5 lb/ton	—	16
Steel	Oily chemical	Unknown	Unknown	50-80	Unknown	—	22

*Neutral sulfite semi-chemical

Production from activated-sludge processes   Table VIII

Process	Sludge retention time, d	Waste sludge lb/lb BOD removed
High rate	2-4	0.5 -0.7
Conventional	4-8	0.4 -0.6
Extended aeration	30	0.15-0.13

*Based on a feed containing less than 0.2 lb inerts/lb BOD.

Types of sludges that have been chemically fixated   Table IX

Oily wastes and tars
Mining tailings
Paper fines
Strong acids or alkalis
Dissolved metals
Nonvolatile organics
Resins
Phosphate wastes
Chemical emulsions

other materials from petrochemical processing are often sold for metal and other valuable component recovery. Carbide lime and acetylene process byproducts are used as sludge conditioners. Waste aluminum and iron salts are used as flocculants.

In many cases, particularly in the secondary-metals industry, thickeners and filters may be employed for product recovery and waste management. Consider the recovery of zinc from sludge produced during rayon manufacturing [3]. Starting with a mixed sludge containing zinc hydroxide and cellulose at about 3% solids, the following steps can be taken to obtain a reusable product:

1. Heat-treat the sludge at 280–300°F to increase filterability.
2. Filter the sludge in a press to about 35% solids.
3. Digest the sludge cake with H_2SO_4 (to produce a 25–30% $ZnSO_4$ solution).

Ultimate disposal techniques other than incineration for waste sludge   Table X

Industry	Origin and nature of sludge	Treatment prior to disposal	Rate	Solids concentration, %	Disposal technique	Comments	Reference
Chemical	Aerobic digestor biological waste	Aerobic digestion	1 million gal/d	Unknown	Ground disposal	15-day digestion	24
Chemical	General chemical, primary	Concentration* after separation of toxics	Unknown	35	Landfill	—	21
Chemical	Waste activated	Bioconversion	Unknown	Unknown	Deep-well disposal	—	21
Latex production	Alum coagulated primary	Thickening	Unknown	Unknown	Landfill	—	6
Metal processing (secondary)	Hydroxide formation	Concentration	Unknown	Unknown	Fixation	Costs $50-$100/ton, dependent on solids content	25
Organic chemical	Waste activated and flyash	None	Unknown	3(activated waste)	Lagoon	95% Flyash 5% Waste activated	28
Organic chemical	Aerobic industrial fermentation	None	8,000 to 152,000 gal/d	0.7 to 5.5	Ground disposal	100 acres in 1-acre plots	26
Paint	Solvent recovery, latex washout	None	Unknown	Variable	Landfill	—	1
Paint	Oil and water-based paint	Sedimentation and concentration	Unknown	Unknown	Lagoon	Conditioned with polymer	31
Pesticides	All types	Concentration	Variable	Unknown	Contained burial	—	32
Pharmaceutical	General	Concentration and bioconversion	Unknown	Unknown	Landfill	—	12
Plastics	All sources All wastes	None	Unknown	Variable	Landfill	—	7
PVC production	Chemical or primary	Concentration	23 tons/d	15-25	Sanitary landfill	—	5
PVC and rubber intermediate	Ferric coagulated primary	Clarification	Unknown	Unknown	Pit disposal	—	27
Rubber	All sources	None	Unknown	Variable	Lagoon or landfill	—	7
Synthetic fibers	Polyester manutacture, primary and waste activated	Clarification and bioconversion	Unknown	Unknown	Lagoon	—	29
Uranium processing	Sands and slimes	Leach extraction filtration	Plant flow	Variable	Tailings pond	—	30

*Thickening and/or dewatering

4. Refilter the sludge in a filter press (to remove $CaSO_4$ and organics).

5. Oxidize ferrous iron with H_2O_2; adjust pH to 4.5.

6. Remove ferric iron on a rotary precoat filter.

This is only one example of the many waste-recovery processes that are now or will soon be in use. With the cost of disposal on the rise, recovery and reuse is a potential profit area and an engineering and process design challenge.

Disposal

Though methods of sludge disposal are beyond the scope of this article, Table X lists examples of industrial disposal techniques other than incineration.

References

1. Carbonate Bonding of Taconite Tailings, USEPA, Office of Research and Development, EPA-670/2-74-001, Jan., 1974.

2. Burd, R. S., A Study of Sludge Handling and Disposal, U.S. Dept. of the Interior, Federal Water Pollution Control Administration, Office of Research and Development, WP-20-4, May, 1968.

3. Iammartino, N. R., Wastewater Clean-up Processes Tackle Inorganic Pollutants, *Chem. Eng.*, Sept. 13, 1976, p. 118.

4. State-of-the-Art Review of Pulp and Paper Waste Treatment, USEPA, Office of Research and Monitoring, EPA-R2-73-184, 1973.

5. Wastewater Treatment Facilities for a Polyvinyl Chloride Production Plant, USEPA, Water Pollution Control Research Series, 12020 DJI, June, 1971.

6. Putting the Closed Loop into Practice, *Environ. Sci. Technol.*, Vol. 6, No. 13, Dec. 1972, p. 1,072.

7. Fluidized-Bed Incineration of Selected Carbonaceous Industrial Wastes, USEPA, Water Pollution Control Research Series, 12120 FYF, March 1972.

8. Air Flotation—Biological Oxidation of Synthetic Rubber and Latex Wastewater, USEPA, Environmental Protection Technology Series, EPA-660/2-73-018, Nov., 1973.

9. Environmental Considerations and the Modern Electrolytic Zinc Refinery, *Min. Eng.*, Vol. 29, No. 11, Nov. 1977, p. 31.

10. Kehrberger, G. J., Mulligan, T. J., South, W. D., and Djordjevic, B., Thickening and Dewatering Characteristics of Kraft Mill Sludges from a High-Purity Oxygen Treatment System, *Tappi*, Vol. 57, 1974, p. 119.

11. Wills, Jr., Robert H., Crucible Waste Treatment, *Clear Waters*, J. of N.Y. WPCA, Vol. 6, No. 4, Dec. 1976, p. 15.

12. Interim Final Effluent Guidelines for The Pharmaceutical Manufacturing Industry, USEPA, 440/1—75/060 Group II, 1976.

13. Grove, George W., Use Gravity Belt Filtration for Sludge Disposal, API Special Report, *Hydrocarbon Process.*, May 1975, p. 82.

14. Methods of Pulp and Paper Mill Sludge Utilization and Disposal, USEPA, Office of Research and Monitoring, EPA-R2-73-232, 1973.

15. Bishop, Fred W., and Drew, A. E., Disposal of Hydrous Sludges From a Paper Mill, *Tappi*, Vol. 60, No. 11, Nov. 1971, p. 1830.

16. Miner, R. A., Marshall, D. W., and Gellman, I. L., A Pilot Investigation of Secondary Sludge Dewatering Alternatives, National Council of the Paper Industry for Air and Stream Improvements Inc., USEPA, EPA-600/2-78-014, Feb., 1978.

17. Turkki, E. V., Hildebrand, A. S., and Nemerow, N. L., Removal of Fine Solids from Chemical Waste Stream Through Precoat Rotary Vacuum Filtration, *Proc.*, 30th Purdue, Ind., Waste Conf., 1977, p. 122.

18. Limestone Treatment of Rinse Waters from Hydrochloric Acid Pickling of Steel, USEPA, Water Quality Office, 12010 DUL, Feb. 1971.

19. Sludge Material Recovery System for Manufacturers of Pigmented Papers, USEPA, Waste Pollution Control Research Series, 12040 FES, July 1971.

20. Color Removal from Kraft Pulp Mill Effluents by Massive Lime Treatment, USEPA, Office of Research and Monitoring, EPA-R2-73-086, Feb. 1973.

21. Gossett, J. W., How Dow Chemical Deals with Diverse Wastes, *Chem. Process.*, Vol. 31, No. 1, Jan. 1968, p. 20.

22. Kellogg, Stephen R., and Weston, Roy F., Treatment of Various Industrial Sludges by Pressure Filtration, The Eleventh Mid-Atlantic Regional Meeting of ACS, University of Delaware, E-6158, April 1977.

23. Keener, P. M., and Metzger, L. R., Startup and Operating Experience with a Twin-Wire Moving Belt-Press for Primary Sludge, *Tappi*, Vol. 60, No. 9, Sept. 1977, p. 120.

24. Howard, J. W., Poduska, R. A., and Walls, W. V., Upgrading of Industrial Wastewater Treatment Facilities at Tennessee Eastman Co., 1977 Water & Wastewater Equipment Mfrs. Assn. Industrial Conference, Atlanta, Ga., 1977.

25. Landreth, R. E., and Mahlock, J. L., Chemical Fixation of Wastes, *Ind. Water Eng.*, Vol. 14, No. 4, July/Aug. 1977, p. 16.

26. Woodley, Richard A., "Spray Irrigation of Organic Chemical Wastes," Commercial Solvents Corp., Terre Haute, Ind., 1977.

27. Kemp, C. E., BFG Chemical Upgrade Treatment, *Water and Sewage Works*, Vol. 119, Nov. 1972, p. 94.

28. Kumke, G. W., Hall, J. F., and Oeben, R. W., Conversion to Activated Sludge at Union Carbide's Institute Plant, *JWPCF*, Vol. 40, No. 8, Part I, Aug. 1968, p. 1,408.

29. Reuse of Chemical Fiber Plant Wastewater and Cooling Water Blowdown, USEPA, Water Pollution Control Research Series, 12090 EUX, Oct. 1970.

30. State-of-the-Art Uranium Mining, Milling, and Refining Industry, USEPA, Office of Research and Development, EPA-670/2-74-030, March 1974.

31. Waterborne Wastes of the Paint and Inorganic Pigments Industries, USEPA, Office of Research and Development, EPA-670/2-74-030, March, 1974.

32. The Pesticide Manufacturing Industry—Current Waste Treatment and Disposal Practices, USEPA, Water Pollution Control Research Series, 12020 FYE, Jan. 1972.

33. "Ultimate Disposal of Liquid Wastes by Chemical Fixation," Chem Fix Div., Environmental Sciences Inc., 1977.

34. Data from Envirotech Corp., Eimco Machinery Products Div., Salt Lake City, Utah.

35. Vesiland, P. A., "Treatment and Disposal of Wastewater Sludges," Ann Arbor Science, Mich., 1974.

36. Perry, R. H., et al., "Chemical Engineers' Handbook," 5th ed., McGraw-Hill, New York, 1973, p. 19–60.

37. Nelson, P. A., and Dahlstrom, D. A., Moisture Content Correlation of Rotary Vacuum Filter Cake, *Chem. Eng. Progress*, Vol. 53, No. 7, July 1957, p. 320.

38. "Waste Water Treatment Plant Design," Manual of Practice No. 8, WPCF and ASCE, Lancaster Press, Pa., 1977.

The authors

OKEY   KOMINEK   DIGREGORIO

Robert Okey is Director of Technical Services for Envirotech Corp., Eimco Process Machinery Div., 669 West Second South, P.O. Box 300, Salt Lake City, UT 84110. (801) 521-2000. Previously, he was the water-and-waste staff technologist for an engineering and construction firm, a consultant, and Eastern Washington State District Engineer. He has taught at Seattle University and the University of Southern California. He has a B.S. in Agricultural Engineering from Iowa State College and a B.S. in Civil Engineering and M.S. in Sanitary Engineering from the University of Washington, Seattle. Mr. Okey has authored many articles on wastewater treatment and contributed to the Water Pollution Control Federation Manuals of Practice. He is a member of WPCF, AIChE and ASCE, and is a registered Professional Engineer in eleven states.

Edward G. Kominek is the Manager of Carbon System Operations for Envirotech Corp., Eimco Process Machinery Div. During his 35 years in water- and waste-treatment activities, his major experience has been in process design and research and development. He has received B.S. and M.B.A. degrees from the University of Chicago and is a registered professional engineer in Ohio, Illinois and Arizona. Mr. Kominek has authored several papers on water and waste treatment. He is a member of AIChE, WPCF, AWWA, API and others.

David DiGregorio is Director of Sanitary Engineering Research and Development for Envirotech Corp., Eimco Process Machinery Div. His responsibilities include the development and application of unit processes for wastewater treatment. Before joining Eimco in 1971, he worked in research and development at Dorr-Oliver. Mr. DiGregorio has a B.S. in Civil Engineering from the University of Massachusetts and an M.S. from Cornell University, Ithaca, New York. He has authored papers on waste treatment and solids processing, worked on two EPA-sponsored research contract studies and contributed to the WPCF Manuals of Practice. He is a member of WPCF and a registered Professional Engineer in Utah.

Equipment for dewatering waste streams

Develop your product parameters and feed conditions—
then choose among gravity thickners, centrifuges, gas
flotation, vacuum filters and filter presses for dewatering.

Richard R. Evans and *Richard S. Millward*, Dorr-Oliver Inc.

☐ The objective of dewatering is to achieve a concentration of wastes that will yield the minimum cost for subsequent sludge handling. To select the best dewatering technique, one must know the available choices to evaluate them in terms of the specific problem.

Before considering ultimate disposal of wastes, it is important to check for any recoverable material that might be worth recycling. A process waste stream may be thought of as not having enough product value to financially justify installing recovery facilities, until environmental factors alter the economics.

As an example, a paper-mill effluent might contain some valuable long fibers, as well as short fibers and other components. The effluent might be thickened in a primary clarifier, dewatered by vacuum filtration, and further dewatered in a continuous press, in preparation for incineration in a hog-fuel furnace. Or, the flowsheet might first use a Dutch State Mines (DSM) screen* to capture and return the valuable long fibers to the paper-making process, while the screen underflow continues through the above dewatering steps.

If the dewatered solids cannot be reused, they are disposed of either in the ocean (a rapidly decreasing option) or on land. Disposal volume must often be minimized, as by incineration, pyrolysis, wet-air oxidation, or aerobic or anaerobic stabilization.

Today's high energy costs require that mechanical dewatering be performed before any thermal processing. If the wastes are to be incinerated, certain parameters must be determined before selecting the dewatering process(es). Such parameters might be moisture content, volatile-solids content and calorific value of the volatile solids. For other ultimate disposal methods, appropriate parameters should be defined.

Sludges encountered in the CPI

Three general categories of waste sludges occur in the chemical process industries (CPI): primary solids, chemical precipitates and biological solids.

*A special screen with 50–350 micron apertures used for classification and dewatering.

Originally published October 6, 1975.

In petroleum refining, primary solids from American Petroleum Institute (API) separator bottoms include corrosion products, carbon, catalyst fines, chemical precipitates, sand and silt—all of which are oil-coated. In the pulp and paper industry, primary sludge solids contain fibers and insoluble constituents; while in the organic-chemicals field, primary solids may be unwanted byproducts, used processing aids (such as filter aids or catalysts), precipitated metals or off-specification products. Process engineers are best able to define the facility's primary solids, because these solids are a direct output of material balances.

A second class of sludges consists of chemical precipitates generated during treatment of liquid waste streams, as it is prepared for acceptable discharge to the environment. Precipitated metals are a major constituent of this class, such as aluminum hydroxide from wastes containing aluminum chloride in organic streams. Metals may also be precipitated during the purification of raw materials, such as ferrous hydroxide from the production of titanium dioxide.

Full view of a gravity thickener in operation

Biological sludges, which constitute a third class, vary widely in characteristics, insofar as they affect dewatering. The thickening characteristic of the sludge is often described by the sludge-volume index (SVI), which is the volume in milliliters occupied by one gram of biological sludge after the aerated liquor settles for 30 minutes. With the activated-sludge process, SVI values can range from 50–350. This wide variation in settling characteristic is usually found from one type of substrate to another, and is not found within a properly designed and well-operated facility.

Types of dewatering

Since all dewatering equipment is affected by the volume to be processed per unit time, the thickening of solids is the first step to be performed. The major emphasis is on the solids fraction, with secondary consideration given to the liquid fraction. The clarity of the liquid fraction is not critical, assuming the solids capture is taken into account in the overall material balance (for the purposes of equipment sizing), and the effects of the solids recycle to the process are fully recognized and evaluated. Design criteria for thickening should be developed by normal test procedures, provided typical feedstock is available.

An illustrative material, relatively new to many process engineers in the CPI, is waste-activated sludge (WAS). WAS, having an SVI of 100, might be discharged from the liquid-processing clarifier with a 0.5% suspended-solids concentration. This material could be thickened by sedimentation to about 2.5% solids; by dissolved-gas flotation to 4.0–4.5% solids; or to 4.5–5.0% solids by nozzle-disk centrifugation. This thickening rejects about 90% of the water originally contained in the solids, significantly decreasing the volume of material to be processed in subsequent steps.

Gravity thickeners are generally constructed in circular basins, allowing continuous withdrawal of solids from the center cone, while the arms of the thickener mechanism continuously feed it (F/1 and F/2).

A recent development in gravity-thickener equipment* solves many of the old problems automatically. The rake arm has a dual-axis hinge, which allows a universal-joint action whenever an overload is encountered. Only the rake blades and the streamlined arm are mounted on—to contact the dense sludge—thus cutting resistance to a minimum.

Gravity thickeners have been successfully applied in CPI processes, but not in the area of waste-activated sludge, in which large tankage is required—350–500 ft²/[(tons of solids)(d)]. Problems with anaerobic decomposition (and subsequent gassing), and gas flotation of solids, occur with WAS, particularly when the process streams and ambient conditions are in the range of 25–35°C, which leads to accelerated bacterial action.

A typical flowsheet of an activated-sludge system, using a disk-nozzle centrifuge thickener, is shown in F/3. The settled activated sludge is taken from the bottom of the final clarification unit and recycled to the aeration system. The waste stream is hydrocyclone-degritted and screened before being pumped to the centrifuge for thickening.

*Called CableTorq, Dorr-Oliver Inc., Stamford, Conn.

The size of this system depends on the total weight of solids to be removed and the expected feed concentration from the clarifier. A finite percentage of solids recovery across the centrifuge is not critical, as long as the required solids wastage is maintained, and the desired thickened concentration is achieved.

The centrate is returned to the aeration system as part of the activated-sludge recycle. (In a typical system, with an 85% capture across the centrifuge, the centrate-solids returned to the basin are less than 0.5% of the total solids returned to aeration each day.)

Full-scale experience shows that a recirculating load of fines has not been developed. This can be understood due to the small proportion of centrate fines in the total solids recycle, and to the well-recognized flocculating capacity of the polymeric materials produced by the activated-sludge microorganisms.

For satisfactory separations with nozzle-disk machines, no chemicals are needed. A typical performance curve of this type of equipment is shown in F/4.

Centrifugal thickeners have operational flexibility due to their extremely low holdup time. Unlike gravity thickeners, they can be used with activated sludge without the potential of anaerobic decomposition.

In dissolved-gas flotation, part of the clarified liquid from the flotation unit is recycled and saturated with gas (usually air) at approximately 60 psig. The sudden release of gas forms extremely fine bubbles in close contact with the flocculent solids. This decreases the effective gravity of the solids enough to cause an accelerated separation of the solids upward from the liquid.

Flotation thickening does not require significant waste-stream pretreatment, such as degritting or screening, but a polyelectrolyte is needed for reasonable design loadings, measured in tons/[(d)(ft²)] solids per flotation-unit area. An economic comparison of flotation thickening versus nozzle-disk centrifuge ($/ton of solids) is provided in the accompanying table.

Dewatering filters

Most chemical engineers are familiar with rotary-drum filters. References 1–4 can be consulted to pursue vacuum filtration and practice in depth.

The same basic considerations that underlie the choice of chemical-process equipment for environmental problems, also dictate the selection of dewatering filter devices [1]. Filtration is an appropriate dewatering method if the solids that build up in a test unit measure 0.05–5 cm/s.

If filtration is used, the filtrate flow must be maintained through the filter media over a period of time. Problems arise with ultrafines (which may foul the media, hampering filtrate flow); unstable solutions that produce precipitates in the media; or with blinding materials that prevent the building of cake of sufficient thickness for discharge.

A belt-type filter, where the filter medium is continuously sprayed by a backwash before it returns to the filtration position, solves the blinding problem. But the amount of liquid used in washing must be properly calculated because it can be considerable. (The solids recycled in this wash stream must also be included in the overall material balance.)

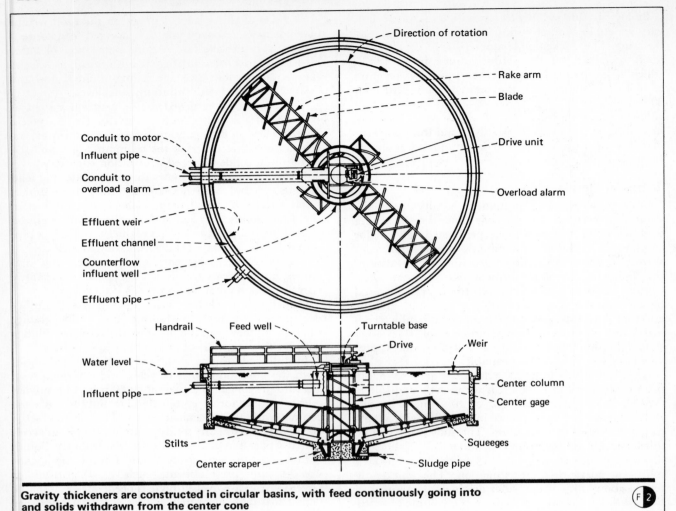

Gravity thickeners are constructed in circular basins, with feed continuously going into and solids withdrawn from the center cone   F 2

Flotation thickening versus nozzle-disk centrifuge thickening of waste-activated sludge

Basis for calculations: capital costs as of July 1975, amortized at 15% and 15 years; chemicals for flotation: $4.50/ton; solids product and loading: 30 lb/(day)(ft²); labor: $5/h; power: 3¢/kWh; feed-sludge composition: 0.8% solids.

Solids product, tons/day	Flotation thickening, $/ton			Nozzle-disk centrifuge, $/ton		
	Operation and maintenance	Amortized capital	Total	Operation and maintenance	Amortized capital	Total
12	10.20	10.86	21.06	8.61	8.45	17.06
24	9.22	8.15	17.37	8.39	8.39	16.76
36	8.65	7.26	15.91	7.48	8.67	16.15
48	8.27	6.70	14.97	5.61	6.50	12.11

Note: The apparent anomaly at the 36 ton/day level is caused by a change in centrifuge size, from a 30-in to a 36-in bowl diameter.

A series of metallic springs is another approach to filter-medium maintenance. The springs are essentially nonfouling, and are effective in dewatering solutions that contain fibers. The medium is returned to a clean condition with each revolution of the filter. Care should be exercised when there is a significant fraction of fines, because some of the fines will be captured by the deposited fibers on the medium.

The precoat filter is a modification, used when the solids to be captured are extremely fine, or when the feed contains components that will blind and smear the filter medium. The precoat presents a continuously renewed filter medium, because the previously deposited solids are removed by the cutting knife.

The automatic, batch filter-press, used for a considerable time in Europe, has a wide variety of environmental applications. Although not a continuous device, it produces a lower-moisture cake. As such, it has stirred interest as a pretreatment means to produce a cake for incineration. (The combustion energy of the organics is sufficient to incinerate the cake; no auxiliary energy is needed.)

Europeans generally produce the cake for landfill, commonly using a fly-ash conditioning agent to maintain cake porosity in slow-filtering materials. If preparation of a suitable feed for incineration is the objective, more attention must be given to the volatile solids-to-solids ratio in the cake. Polymeric conditioning, with minimal addition of inorganic chemicals or ash, will probably be desirable.

The continuous, horizontal-belt press or vertical press is presently popular in Europe, and installations are becoming more common in the U.S. Performance has not yet been documented, but it is expected to be between that of a rotary-drum filter and a batch filter press, in terms of cake-moisture content. Continuous operation is desirable in some instances, assuming the moisture content of the product is consistent with the downstream process.

Centrifuges

The long cylindrical-conical, solid-bowl centrifuge has become popular for dewatering solids encountered in wastewater treatment. This is particularly true when the solids are such that filtration is only marginally acceptable, due to cloth-blinding potential or inability

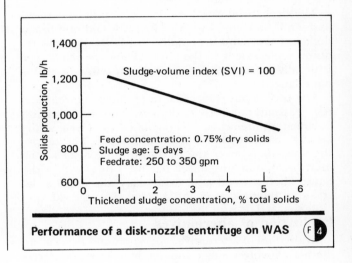

An activated-sludge system using a centrifuge   F 3

Performance of a disk-nozzle centrifuge on WAS   F 4

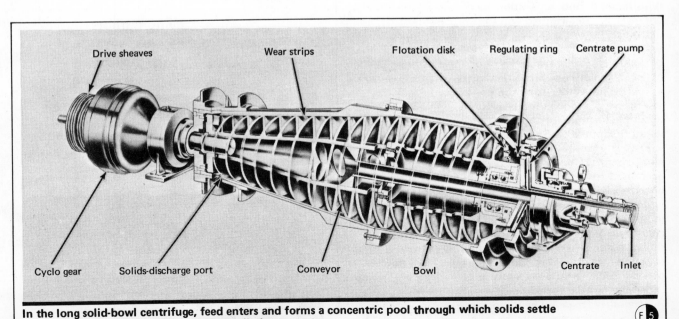

In the long solid-bowl centrifuge, feed enters and forms a concentric pool through which solids settle to the outer wall and are removed at the discharge ports   F 5

to build a cake thick enough for discharge purposes.

A cross-section of this type of centrifuge is shown in F/5. The feed—brought to the interior through a stationary feed-tube—enters the machine and forms a concentric pool through which the solids settle to the outer wall. Solids are continuously removed by the scroll conveyor across the drying beach, to the discharge ports at the small end of the conical section. The liquid flows countercurrently through the cylindrical section, to the overflow dam of variable elevation, which affords pool-volume regulation.

The major process variables are the liquid and solids feedrates, and chemical conditioning of the sludges. If the feedrate is increased, the detention time within the bowl decreases, reducing the clarification and drying capacity of the machine.

Polymers may be added to the sludge feed to help capture fine solids, improve dewatering, and optimize operation of the machine. The centrifuge has a certain solids-handling capacity for specific machine parameters and for different sludges.

The mechanical variables are pool depth, bowl speed and scroll differential speed. The pitch and number of flights of the scroll are another variable, but this requires replacement of the scroll. Thus, the scroll type is usually established for a specific application and is not changed.

The pool depth is specified to optimize the clarification capacity and cake dryness, and to minimize polymer requirements; it is regulated by a weir at the centrate outlet of the cylindrical section. The deeper the pool, the greater the clarification capacity of the unit, because of the longer retention time within the bowl. The increased pool depth may also cause a decrease in the dryness of the dewatered-sludge cake.

Evaluation of alternatives

Evaluation of alternative separation and dewatering methods involves the optimum matching of process requirements and separation techniques. The process engineer has determined what his process must do, and he must find out what the available liquid-solids separation techniques can do. The entire process may be modified to obtain the best or most economical fit between available separation alternatives and the over-all performance needs of the process.

This art of tradeoff analysis is too complex to reduce to simple formulas or to a computer-type program, but it can be approached systematically by drawing upon:

Practice—Find out what has been used on the same service before. Gravity thickeners, for example, have not been used to any extent on waste-activated sludge, but nozzle-disk centrifuges and dissolved-air flotation have.

Analogous Service—For instance, a mixture of hydrated metal oxides from lime precipitation of an organic-chemicals plant effluent, prior to biological treatment, should behave similarly to hydrated metal oxides from treatment of metal-finishing wastes. Therefore, the process engineer should look into sludge dewatering experience in the metal-finishing industry.

Application Profile—The importance of various parameters to the process must be analyzed. Most can be measured in terms of cost. However, there are factors whose cost cannot be assigned objectively. One example would be using a closed centrifuge instead of an open filter for separating sewage solids, because greater cleanliness and odor control is afforded by the centrifuge. Perhaps the most important of such cost factors are the process cost penalties associated with the degree of separation that can be accomplished, and the resulting level of recycling of solids to the process. Other factors might include the value attached to space, mechanical simplicity, reliability, holdup of materials in the process, controlability, maintainability, etc.

Unit-Operation Capability—This involves knowledge of the costs, characteristics, capabilities and limitation of the liquid-solids separation techniques. Whenever possible, the fitness of the unit operation in the particular application should be confirmed by appropriate test work. Incorporation of "typical design values" has been avoided in this article, because generalizing with the number of variables at hand invites disaster. Although test work may not be easily accomplished on new processes (because of lack of feedstocks), environmental considerations must be fully evaluated when the chemical-process development goes to the pilot-plant stage.

Engineers have available the experience of manufacturers of separation equipment. But first, the engineers should develop their objectives (such as product parameters of solids in the liquid phase, and liquids in the solid phase) and define feed conditions, including solids concentration and density, particle size and mother-liquor characteristics. On such basic data, a selection of dewatering equipment can be made.

References

1. Tiller, F/M., Bench-Scale Design of SLS Systems, *Chem. Eng.*, Apr. 29, 1974, p. 117.
2. "Process Design Manual for Sludge Treatment and Disposal," EPA 625/1-74-006, U.S. Environmental Protection Agency-Technical Transfer, Oct. 1974.
3. Silverblatt, C.E., others, Batch, Continuous Processes for Cake Filtration, *Chem. Eng.*, Apr. 29, 1974, p. 127.
4. "The Application of Chemical Engineering to the Treatment of Sewage and Industrial Liquid Effluents," I. Chem. Eng. Services for the Institution of Chemical Engineers, London, England, Apr. 1975.

The authors

Richard R. Evans is Director of Technology, Environmental, Dorr-Oliver Inc., 77 Havemeyer Lane, Stamford, CT 06904, where he has held several management positions, including manager of industrial water projects and senior industrial-water-wastes engineer. Before, he has with Infilco for 18 years. He has a B.S. degree in chemical engineering from Ohio State University and is a member of AIChE, the Water Pollution Control Federation and the International Assn. of Water Pollution Research.

Richard S. Millward is Product Manager, Environmental, with Dorr-Oliver Inc., where he started as an applications engineer. His experience also includes management and sales positions with Vacumite Corp., which is an industrial filtration equipment supplier. He has a B.S. degree in civil engineering from Bucknell University and has authored a dozen technical papers in the environmental control area.

Sludge drying—one way to waste-reduction

Though controversial because of its heavy energy demands, sludge drying is nonetheless getting attention from municipalities and industries concerned that they can't meet local land-disposal restrictions.

☐ Amid all the uncertainty over public and governmental attitudes toward the final disposal of waste material, industrial and municipal producers of waste sludge are turning more and more often to drastic measures—including drying.

That's strong medicine under most circumstances. Not only does drying (from typically 15% to 90%-by-weight solids) take a lot of energy, but once it is completed, processors are often left with a major marketing problem—selling the stuff to skeptical fertilizer consumers. But to many a sludge producer, the energy/marketing uncertainty is outweighed by the environmental outlook: More-drastic sludge treatment is coming into favor at regulatory agencies, since it cuts down on waste-disposal headaches. In particular, it reduces the need for landfill.

Drying is one of four overall strategies for sludge treatment. In order of processing severity these are:

Pasteurization. One such method—composting—involves holding wet sludge in a tank or pond, allowing biological action enough time to eat up any pathogens present. The sludge can then be spread over cultivated fields. In some cases, pasteurization is accomplished by a mild heat treatment or by exposure of sludge to ionizing radiation.

Dewatering. In this method, water is removed by vacuum filters, centrifuges or filter presses. Solids content of the treated sludge ranges between 25 and 60%, depending on the severity of treatment. Often, chemical or thermal conditioning is applied to improve mechanical dewatering methods.

Drying. The main subject of this article, drying can take place in rotary, fluid-bed, multi-effect and milling-type equipment. Heat may be applied directly—usually in the form of exhaust gases from a combustion source—or indirectly (typically as steam). Solids content reaches about 90%. However, some processes employ simple air drying—a method that cannot achieve anything like 90% solids, but is drying nonetheless.

Incineration. The ultimate method of disposal, incineration is often used in conjunction with dewatering and drying to produce energy for those methods. Some processors are moving to pyrolysis as an alternative to incineration.

Whatever the disposal process, or combination of processes, the aim is to produce a waste material suitable for landfill, fertilizer or recycling. The ultimate disposition has a deciding effect on the type of process chosen. Drying can produce a fertilizer, provided there are no toxic metals present in the waste, and that odor can be controlled. If fertilizer production is not possible, or simply not worth the marketing trouble, then another method may be chosen, though many of the alternatives employ drying in part.

Besides its environmental advantage, sludge drying adds a lot of flexibility to the waste-disposer's bag of tricks. It's part and parcel of process schemes leading up to incineration or pyrolysis. It reduces volume and weight. And because nitrogen content isn't eaten up, as it is in composting, drying provides a salable product (if no toxic materials are present).

But drying is far from a perfect solution to the waste problem. For one thing, companies or municipalities that commit themselves to this alternative have to be prepared to handle an incessant inventory buildup, in case the market for fertilizer goes flat. The

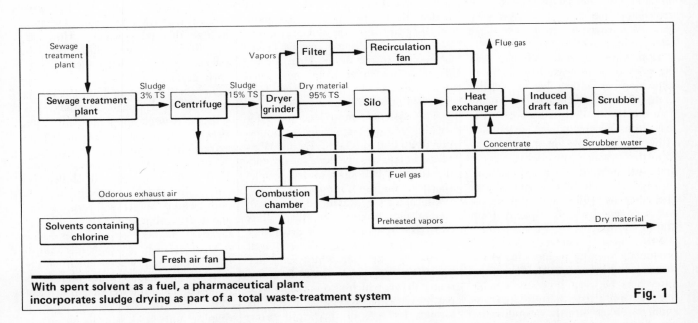

With spent solvent as a fuel, a pharmaceutical plant incorporates sludge drying as part of a total waste-treatment system

Fig. 1

Originally published June 4, 1979.

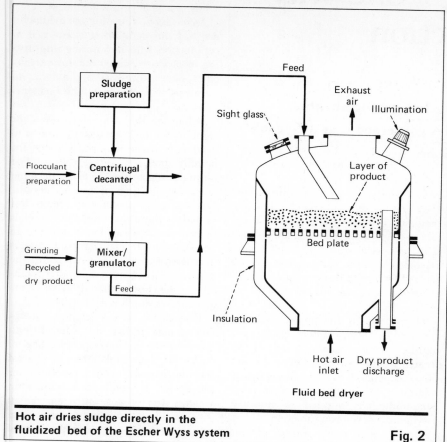

Fluid bed dryer

Hot air dries sludge directly in the fluidized bed of the Escher Wyss system   **Fig. 2**

supply of sludge would always be present, no matter what the economic conditions.

Certain processing problems also may crop up. Dried sludge may pose a difficult odor-control problem. It could ignite in direct-heated dryers. Its nitrogen content may be too low anyway (at least 5% nitrogen is necessary to make effective fertilizer).

VENERABLE PRACTICE—Sludge drying goes back at least fifty years in some localities. Milwaukee first tried it in the 1920s, and Chicago began using it in the 1940s. Both cities take waste heat from nearby power generators. However, both now are shifting away from this practice, since waste heat is more valuable in other applications. Chicago plans to stop drying sludge altogether by 1985, and Milwaukee will soon cut back its capacity from 200 to 130 tons/d.

Most new processes take heat produced either by incinerating dried sludge or by burning fuel. The major differences between processes involve the methods by which heat is applied—either directly or indirectly, at what temperature, and in what combination with other treatment methods.

DIRECT HEAT—Escher Wyss Ltd. (Zurich) exposes ground, dewatered sludge to direct heat in a fluid-bed dryer. Dewatering is accomplished in a centrifugal decanter, with the aid of added flocculant. In a subsequent mixer/granulator, the sludge is mixed with some recycle dry product before being sent to the dryer. Heat for the dryer comes either from incinerated dry sludge or from purchased fuel. The bed operates at about 600°C.

This system is most economical for treating sewage from a local population of 20,000 to 50,000, Escher Wyss says. Designs are available for producing 100 to 1,000 kg/h of dry product (particle size up to 5 mm).

Chicago's CE Raymond, a unit of Combustion Engineering, Inc., offers two basic systems: a flash dryer that treats sludge in a cage mill, sending it on to a cyclone from which some dry product is recycled to the mill; and a rotary dryer, which also operates with partial recycling. Raymond designs its systems for any of three processing modes—complete incineration of prod-

uct, partial incineration plus partial output of soil conditioner, or conditioner production only. The technology has been in operation in cities such as Chicago and Houston for many years.

Ecological Services Products, Inc. (Dunedin, Fla.) also offers a rotary-dryer design, with prior filter-press dewatering. The complete system, from filter to screw conveyor and into the rotary dryer, followed by cyclone separation of dust and offgases, and scrubbing of the exhaust, aims at yielding a bagged product suitable as a soil conditioner. Systems have been installed in Florida and Georgia.

In Italy, Item S.p.A. (Milan) is now in the pilot-plant stage of development with a "drying tunnel." Its essential aim is to dry and pulverize at the same time (as does Raymond's cage mill and some other techniques) by forcing sludge and direct air through a long chamber at velocities up to 50 m/s.

Organic Recycling, Inc. (West Chester, Pa.), a unit of UOP Inc., has tested a "toroidal dryer" at a municipal waste-treatment plant in Washington, D.C., albeit without great success. While not yet available commercially (the firm intends to produce an improved design in 1980 or later), the system takes dewatered sludge (20% solids), admitting it with some recycled product along the outer circumference of the 16-ft device. Air preheated from 500 to 1,200°F takes the moisture content down to 6%, according to original design objectives. The process's chief advantage, claims Organic, is high thermal efficiency. However, the firm had some difficulty with internal ignition during the method's early development.

In another system that hasn't yet made it to commercialization, Resources Conservation Co. (Renton, Wash.) combines a novel water-extraction step with more-prosaic rotary drying. Dewatered sludge (about 15% solids) is mixed with chilled triethylamine (TEA)—a solvent miscible with water at temperatures below about 20°C, but staying in a distinct layer at higher temperatures. The mixture is then fed to a mechanical dewatering unit where much of the solvent, with bound water and oil, is removed. Then the wet cake is dried at 120-150°C. Because TEA evaporation requires only 133 Btu/lb, drying is faster and less expensive.

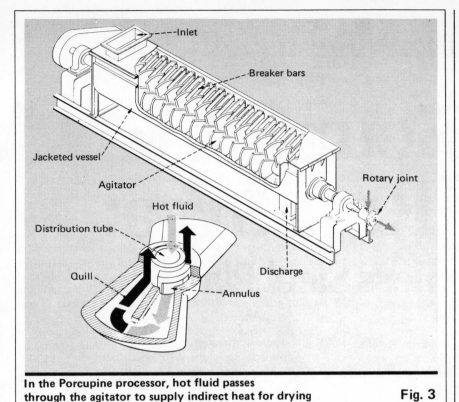

In the Porcupine processor, hot fluid passes through the agitator to supply indirect heat for drying   **Fig. 3**

Calling its process "B.E.S.T." (for Basic Extractive Sludge Treatment), Resources Conservation has failed so far to sell it to its prime prospect, the nearby city of Portland, Ore. Portland opted instead for a landfill alternative, avoiding altogether the question of sludge drying. The firm does not consider its route useful in industrial sludge treatment, because wastes from these sources are too variable. So for now, the process still awaits commercialization at some municipality.

INDIRECT HEATING—Indirect heating via a heat-exchange fluid—steam, water, hot oil, Dowtherm or the like—negates any worry about ignition hazards, but is not as thermally efficient. The solids content of sludge from such systems never gets much above 60 or 70%, mostly because the heat transfer coefficient decreases rapidly as the material dries.

The Environmental Engineering Div. of Von Roll Ltd. (Zurich) offers both direct and indirect designs. Its indirect system employs a thin-film dryer built by Luwa Corp. (its U.S. subsidiary located in Charlotte, N.C.). In this vessel, sludge is spread thinly over the internal cylindrical surface while heat is applied from the outside. Dry material is scraped off by rotating blades, and then falls to the bottom.

Von Roll's direct system, which gets solids up to 90%, is a dryer/grinder in which sludge and hot gases flow in parallel fashion. A system built for Bayer, at its Eberfeld pharmaceutical plant near Düsseldorf, West Germany, puts centrifuged sludge (15% solids) through the grinder, which is heated from combustion gases derived from several sources: sewer gas, chlorinated solvents, and oil underflow from an exhaust-gas condenser.

Another system for drying sludge is the Porcupine, offered by Bethlehem Corp. (Easton, Pa.). Resembling a grinder more than anything else, the unit mixes and grinds sludge by means of rotating blades that are heated internally with steam. The blades also propel the sludge along the unit's length, discharging the dry product at the far end. The major advantage Bethlehem claims for this system, compared to a thin-film dryer, for example, is higher surface area per unit mass of sludge processed at any one time.

EVAPORATORS—Another alternative is to employ multi-effect evaporators to remove water.

At the Gibraltar, Pa., dyestuff plant of Crompton & Knowles Corp. (Fair Lawn, N.J.), a multi-stage evaporator is treating wastes containing 10% organic salts. Vapor is passed through an activated-carbon adsorption system. The sludge is suitable as landfill.

Foster Wheeler Corp. (Livingston, N.J.) and Dehydro-Tech Corp. (East Hanover, N.J.) offer the Carver-Greenfield process of multiple-effect evaporation. This system, designed chiefly for food-processing applications (e.g., meat rendering), is under study by ITT Rayonier for treating pulpmill wastes at the firm's Port Angeles, Wash., facility. Dried waste would be used as boiler fuel, or, if approved by the U.S. Food and Drug Administration, it may find use as animal feed.

Another firm that may make animal feed via this method is Adolph Coors, at its Golden, Colo., brewery. Eli Lilly Co. uses the process on pharmaceutical sludge now being pyrolyzed.

ULTIMATE DISPOSITION—The pyrolysis option for getting rid of sludge altogether can replace incineration. Though it seems easier, incineration may force costly scrubbing measures to prevent air pollution. Furthermore, heavy metals may show up in the ash and cause land pollution problems anyway—particularly in the case of chromium ions that become oxidized to the harmful +6 valence level.

Mindful of this problem, Japan's NGK Insulators, Ltd. (Nagoya) is developing a system of indirect drying followed by sludge pyrolysis. The dryer closely resembles a Porcupine unit, if in fact it isn't one. The first commercial installation of NGK's process, rated at 60 metric tons/d, will be completed later this year.

In any case, there will probably always be a place for incineration. Under a one-year contract with the U.S. Dept. of Energy, Battelle Memorial Institute (Columbus, Ohio) is investigating a dual fluidizing-bed system of drying and incineration. In the first bed, hot sand fluidizes incoming sludge and imparts heat to the system, generating steam that is piped to another part of the waste-treatment facility. The dried material and sand are then transported to the incineration bed, from which hot sand is recycled back to the first one. Some garbage is incinerated along with the sludge. Ideally, refuse and sludge provide all the energy for the process.

Reginald I. Berry

Increasing sludge-digester efficiency

Using sodium bicarbonate instead of lime for pH adjustments in sewage-sludge digesters yields better control and increased production of byproduct methane.

Nicholas R. Barber and Carl W. Dale,
Church & Dwight Co.

☐ Even when highly sophisticated wastewater treatment plants become commonplace, bacterial action still will probably account for a significant part of the operation, particularly in the earlier stages of treatment.

Bacterial action, while not 100% effective in terms of total waste conversion, is still the most economical way to convert most organic wastes into harmless, even useful, products. The food necessary for bacterial growth—the waste itself—is in ample supply. But the environment in which the bugs live must be carefully controlled to keep the overall treatment process running smoothly.

Aerobic or anaerobic?

Both types of bacteria are necessary to complete nature's cleanup cycle of "dead" biomass. Aerobic bacteria work faster and seem to be more easily controlled; thus they have become the standard choice for most waste-treatment systems.

Some modern plants use both bacterial types. While the anaerobic system is one of the oldest wastewater treatment methods, it has traditionally been viewed as an unpredictable operation. This has hampered its development even though it is an extremely efficient process from an engineering standpoint—confinement of two-thirds of the organic solids to less than 1% of the flow volume.

For years, some plants with anaerobic digesters have collected the methane produced and have used it for heating digesters, other equipment, or plant buildings. Fluctuations in methane production are both expected and tolerated, but that, too, may change. Methane represents conservation of much of the energy originally present in raw waste, while aerobic processes produce oxides, such as CO_2 and water, having little or no chemical energy content. Energy released by this oxidation is lost to the atmosphere or is used by the bacteria.

The waste-conversion chemistry is very much the same in all plants, and for discussion purposes can be divided into the two categories of aerobic and anaerobic. Let us look at the aerobics first.

Aerobic activated-sludge processes

Conventional activated-sludge plants remove 90 to 95% of the BOD but less than 50% of the nitrogen. While BOD removal greatly reduces oxygen demand in receiving streams, discharge of soluble ammonium ions represents a potential oxygen demand in its own right. This nitrogenous oxygen demand (NOD) is exerted by autotrophic nitrifying bacteria as they oxidize ammonia to nitrate:

$$NH_4 + 2 HCO_3^- + 2O_2 \rightarrow NO_3^- + 2 CO_2 + 3 H_2O$$
$$14 \qquad 100 \qquad 64 \qquad 14$$
$$\text{(as CaCO}_3)$$

Conversion of 14 parts of nitrogen from ammonia to nitrate consumes 100 parts of alkalinity (expressed as $CaCO_3$), a ratio of 100:14 or 7.14:1. Formation of 1 mg/L of nitrogen in the nitrate form from NH_3 may destroy more than 7 mg/L of alkalinity. Twice as much alkalinity is destroyed as was formed in the initial hydrolysis of amino groups to ammonia.

In the above reaction, bicarbonate is converted to carbon dioxide. The pH of an aerobic system depends on the ratio of dissolved CO_2 (H_2CO_3) and HCO_3^-:

$$pH = pK_1 + \log \frac{[HCO_3^-]}{[CO_2]}$$

During nitrification the H_2CO_3 level increases through the conversion of bicarbonate and also through oxidation of organic material. When wastewaters are low in alkalinity, the resulting pH drop can cripple the nitrifying bacteria with resulting inhibition of ammonia oxidation. This effect will become noticeable below a pH of 7, and a pH below 6.5 is seriously inhibitory.

Lime or sodium bicarbonate?

Traditionally, lime has been the means of pH control in activated-sludge or trickling-filter systems, although other alkaline materials—including soda ash, caustic soda, and ammonia—have been used with some success. On the surface, lime appears to be a good choice; it is less expensive than most other materials in terms of cost per pH-unit shift. But it is a "harsh" chemical with no buffering capacity. It is easy to overdose, causing pH to rise so high that a new set of problems arises.

Dosing treatment systems with nonbuffering chemicals can give rise to other problems. Lime must be diluted and added slowly, with good mixing, to prevent buildup of local hot spots of high pH that can halt bacterial growth. The strong caustic nature of lime also dictates careful mixing and handling. Construction materials must be chosen for the mixing and dispensing equipment that will not be corrosively attacked. Lime treatment also tends to encrust equipment with scale that must be removed from time to time. Scale formation is likely to occur above pH 6.3 due to reaction of calcium with natural bicarbonate to form insoluble calcium carbonate—a reaction that reduces alkalinity still further.

Originally published July 17, 1978.

Because the soluble components of digester liquor may ultimately gain access to the environment, use of ammonia for alkalinity or pH control is not recommended and often not permitted. This and most other agents for pH control can also irritate eyes, skin, and respiratory tracts.

Virtually all of the problems mentioned can be eliminated by use of sodium bicarbonate ($NaHCO_3$) for alkalinity and pH control. It is the only nonhazardous chemical commonly used in waste treatment plants for these purposes. Sodium bicarbonate is a common constituent of all aqueous systems and is in human and animal diets. Its nontoxic nature obviates the need for special handling or feeding equipment. While sodium bicarbonate is more expensive on a per-pound basis, it has other desirable properties that may make it less expensive compared to lime, in terms of overall plant operation.

Before we get into these unusual properties, however, let us review the chemistry of anaerobic-sludge digesters and the problems that can arise from improper pH and alkalinity control.

Anaerobics: methane from biomass

While anaerobic bacteria work more slowly, the advantage lies in the better separation of water and solids obtained after the sludge has been digested and stabilized. Better sludge compaction greatly reduces the cost of hauling and disposal costs, and generated methane is an increasingly timely and important by-product.

In anaerobic digestion, microbial action is a balanced process involving two groups of bacteria, the acid formers and the methane formers. In the first stage, through action of the acid formers, there is a breakdown of fats, proteins and carbohydrates by hydrolization, producing the short-chain fatty acids (acetic, propionic, butyric, valeric). Natural bicarbonate alkalinity is destroyed and replaced by volatile-acid alkalinity, (VAA) according to the following equation.

$$HCO_3^- + HAc \rightarrow CO_2 + H_2O + Ac^-$$

In systems where a significant fraction of bicarbonate has been replaced by acetate, total alkalinity concentrations will remain essentially constant although its neutral buffering capacity has been much reduced. The second-stage methane formers are far more sensitive to their physical and chemical milieu. They perform the most important step in anaerobic digestion, the conversion of the fatty acids to carbon dioxide and methane gas. Methane formation is the slowest reaction in the overall degradation process and is the rate-controlling step. Shock increases in toxic substances or changes in pH or temperature inhibit methane-forming bacteria metabolism, resulting in an accumulation of fatty acids. A vicious cycle develops because an excess of fatty acids will cause an imbalance in the digestion process by dropping the pH below the 6.8 to 7.2 range, optimum for methane formation. At pH values lower than 6.5, methane formers may not grow at all.

pH control: trickier for anaerobics

In a closed anaerobic digester, gases in the volume above the liquor are CO_2 and methane, both formed by bacterial action. The amount of CO_2 dissolved in the liquor depends on the partial pressure of CO_2 in the gas mixture, and the pH of the liquor, in turn, depends on the alkalinity (HCO_3^-) and the dissolved CO_2.

Anaerobic digester atmospheres contain 30 to 50% CO_2. A system producing 38% CO_2 and 62% CH_4 will contain 500 mg/L of CO_2 in solution in the digester liquor. Since pK_1 for H_2CO_3 is about 6.3, the molar ratio of bicarbonate to soluble CO_2 must be 5:1 to guarantee a pH of 7.0. This means that bicarb alkalinity must be at least $5.6 \times 10^{-2}M$ or 2,800 mg/L as $CaCO_3$, regardless of the volatile acids or acetate concentration, and this bicarbonate concentration would represent the minimum requirement providing only marginal operational confidence. The concentrations of bicarbonate alkalinity required to maintain a pH of 7.0 are shown in the table.

CO_2 content in gas, %	Bicarbonate alkalinity required, mg/L as $CaCO_3$
25	1,875
30	2,250
35	2,625
40	3,000
45	3,375
50	3,750

Regular gas analyses, used with this table to establish minimum alkalinity requirements, can be helpful in avoiding problems. If gas analysis is not undertaken, a minimum concentration of about 2,500 mg/L alkalinity is recommended as a safeguard against contingencies such as overloading, heat loss, toxic ions, or introduction of acid or alkaline materials.

Some sludge digesters become so retarded they are producing little or no methane or CO_2 by the time recovery procedures are begun. Raising the pH is essential, but there can be problems here. Alkaline agents such as lime, caustic soda ash, or ammonia cannot directly increase bicarbonate alkalinity without first reacting with soluble CO_2:

$$NaOH + CO_2 \rightarrow NaHCO_3$$
$$Na_2CO_3 + H_2O + CO_2 \rightarrow 2\ NaHCO_3$$
$$Ca(OH)_2 + CO_2 \rightarrow Ca(HCO_3)_2$$

As a result, CO_2 removed from solution by these reactions must be replenished by CO_2 from the gas phase to reestablish equilibrium—which creates a partial vacuum. This places great stress on tank structures and may allow atmospheric oxygen to be sucked into the tank atmosphere. In an anaerobic system, this can be toxic for the bacteria. Also, the resulting air-methane mixture poses the threat of explosion.

In addition to the advantages outlined for sodium bicarbonate in aerobic systems, $NaHCO_3$ is the one natural buffer that can eliminate the toxicity and vacuum problems of pH adjustment in anaerobic digesters. It can directly shift the equilibrium to a neutral pH value without disturbing the physical and chemical balance of delicate biological systems—particularly the extremely sensitive methane-forming bacteria.

At many plants using anaerobic digesters, alkalinity concentrations run dangerously close to the marginal levels required for pH control. Increasing bicarbonate levels above minimal requirements not only serves as protection against upsets, but also improves methane production and sludge compaction.

Alkalinity controls methane production

Increasing bicarbonate levels in digesters may accomplish both improved treatment efficiency and increased energy production simultaneously. Rate of gas production is linearly proportional to the bicarbonate alkalinity. Each additional gram per liter of bicarbonate alkalinity (up to 5 g/L) can increase methane production by 500 Btu/(lb volatile solids)(d) or 126,000 cal/(g v.s.)(d).

The experiment that generated these data employed four 20-liter, plastic, pilot-plant digesters with seed sludge obtained from a sewage-treatment plant. Two of the mini-digesters were fed different rates of $NaHCO_3$ alone, one was given a mixture of lime and $NaHCO_3$, and one was treated with lime alone. When bicarbonate alkalinity was intentionally decreased, the digester system stability deteriorated. A "retarded" digester was established, which was characterized by low pH and low gas yields, accompanied by a rapid increase in volatile acids and high CO_2 concentrations in the evolved gas.

All four digesters were slowly nursed back to health, the recovery taking 30 to 40 days, with careful analysis to determine optimum dosage. Due to problems of calcium carbonate precipitation above pH 6.5 and creation of a vacuum from addition of lime, pH adjustment was best achieved by lime addition until pH was 6.3 to 6.5, followed by $NaHCO_3$ for a pH of 7.0 to 7.2. Recommended minimum alkalinity was 2,500 mg/L. Once this minimum is established, bicarbonate alkalinity will be recycled continually in a digester according to:

$$R\text{-}COOH \ NH_4(HCO_3) \rightarrow R\text{-}COONH_4 + H_2O + CO_2$$
$$R\text{-}COONH_4 + H_2O \rightarrow CH_4 + NH_4(HCO_3)$$

Bicarbonate alkalinity in the more desirable range of 2,500 to 5,000 mg/L gives more buffer capacity, and much larger increases in volatile acids can be handled with minimum ph drop. This provides a good safety factor and allows time for control if an upset should occur.

Daily addition of 840 mg/L sodium bicarbonate will give a sodium level of 230 mg/L, which will also stimulate digester activity. Sodium ion can be tolerated by methane-forming bacteria at concentrations up to 8,000 mg/L with little cation toxicity. In general, monovalent cations are less toxic than the diavalent ions.

An ounce of prevention . . .

After the pH of the digester has been restored, therapeutic dosages of $NaHCO_3$ can be used to maintain alkalinity reserves.

The daily dose of alkalinity required for any desired steady-state concentration is given by:
$$D_d = D_{max} (1 - e^{-\lambda})$$
Where:
D_d = daily dose of alkalinity to achieve D_{max}, mg/L
D_{max} = desired increase in alkalinity to reach steady-state, mg/L
λ = displacement constant, day^{-1} (reciprocal of average detention time in days)
Sample calculation:
Given: Digester detention time = 30 days
λ = 0.033 day^{-1}
pH = 6.8
Alkalinity, current = 1,700 mg/L

Alkalinity, desired = 3,500 mg/L
$D_d = D_{max} (1 - e^{-\lambda}) = (3,500 - 1,700) (1 - e^{-0.033})$,
= $1,800(1 - e^{-0.033})$ $D_d = 49$ mg/L as $CaCO_3$

Improved settling characteristics

Rapid sludge settling is desirable in almost every stage of wastewater treatment. This is particularly true in small, activated aerobic-sludge plants, which often operate at higher than average sludge-return rates. Initial settling rate determines the efficiency of secondary settling tanks.

Colloidal materials in wastewater are usually negatively charged, and salts of polyvalent cations, such as alum and ferric chloride, are often effective coagulants. Because these salts are moderately strong acids, their use may reduce bicarbonate alkalinity and/or pH values below the range required for optimum flocculation. Each mg/L of alum added destroys about 0.5 mg/L of alkalinity as $CaCO_3$.

While lime or other nonbuffering alkaline materials can be used to adjust pH to compensate for alum, there is a good reason—beyond pH adjustment—why $NaHCO_3$ should be given serious consideration. Sodium bicarbonate is a flocculant in its own right, and when it is used with alum or ferric chloride there is synergistic action. Many experiments with alum coagulation of turbid waters have shown that both pH and alkalinity must be considered for optimal flocculation.

In other chemical treatments sometimes used in wastewater plants, the buffering property of bicarbonate can control and maintain pH values near the optimum for the particular reaction used. Such treatments include peroxide oxidation of H_2S for odor control and breakpoint chlorination of nitrogenous materials—an alternative way of decreasing NOD, which has traditionally been accomplished by nitrification-denitrification systems.

Consider bottom-line costs

Sodium bicarbonate is more expensive than lime but when overall costs are considered, bicarbonate may be preferable. This is particularly true in treatment plants with flows less than five million gal/d or those plants with predominantly low pH wastes. The five million gal/d flow is an economical breakpoint based on cost of metering and mixing equipment—necessary for lime systems—storage space, and manpower.

The authors

Nicholas R. Barber is manager of chemical applications, Church & Dwight Co., Inc., P. O. Box 369, Piscataway, NJ 08854. Previously he worked for Inselek, Inc., RCA Corp., Cities Service Oil Co. and Mobil Oil Corp. He holds a B.A. in chemistry from Rutgers University. He is currently working toward his MBA at Rutgers University Graduate School of Business.

Carl W. Dale is a project administrator for research and development at Church & Dwight Co. He previously worked for Public Service Electric & Gas Co. (Harrison, N.J.) and at Foster D. Snell, Inc. He holds a B.A. in chemistry from the University of Connecticut and an M.S. in environmental science from Rutgers University. He is a member of the New Jersey (North and South sections), New York and Pennsylvania Water Pollution Control Assn., and the American Water Works Assn.

Converting solid wastes and residues to fuel

Solid wastes and residues can be an important energy resource in process plants. Here are design, economics, and process options of the commercial thermal-conversion techniques for recovering this energy.

Jerry Jones, SRI International

☐ Organic solid wastes and residues* (municipal refuse, tires, sludges, waste plastics and packaging materials, and agricultural and forestry residues) do not offer a complete solution to our energy supply problems, but can be an

*Residues have some positive value (often quite low), as opposed to solid wastes, which represent a disposal cost. In this article, all residues will be referred to as solid wastes. Efforts are also underway to photosynthetically produce feedstocks for energy production. Such biomass materials have the potential to make a much greater contribution to energy supplies than waste materials or residues.

important energy resource in some regions and for specific processing plants. Energy can be recovered from solid wastes and residues by numerous thermal routes as well as by biochemical conversion.

Fig. 1 summarizes the options now available. This article describes the thermal-conversion processes that fall into the categories of pyrolysis, gasification and liquefaction and does not discuss solid-waste incineration or

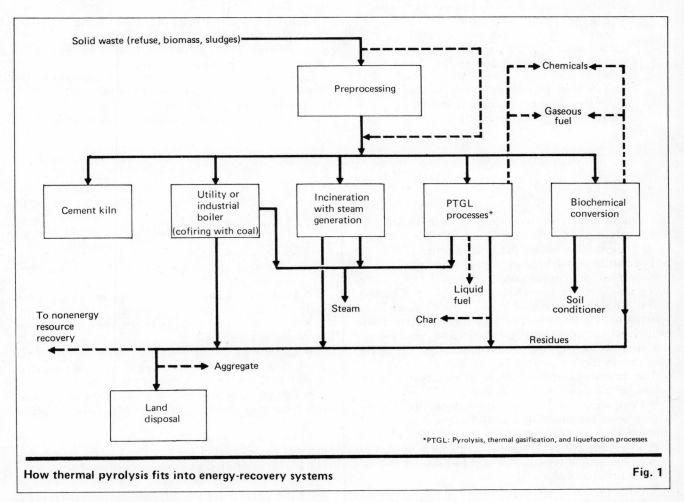

How thermal pyrolysis fits into energy-recovery systems

*PTGL: Pyrolysis, thermal gasification, and liquefaction processes

Fig. 1

Originally published January 2, 1978.

Solid-waste incineration. Solid material is completely combusted in a furnace to produce a hot flue gas. The principal way to recover energy is to transfer heat from the gas via a steam boiler. In waterwall incinerators, boiler tubes surround the interior walls of the furnace.

Destructive distillation or pyrolysis (e.g., metallurgical-coke production in vertical-flued ovens). Carbonaceous materials are thermally decomposed into fuels (gases, liquids and char) by indirect heat transfer, with no injection of oxygen, steam or hot CO_2 into the reactor (see Box 2, reactions 1, 2, 3, and 4). Currently, many authors define pyrolysis somewhat differently. By their definition, a pyrolysis process may be autothermic: that is, some of the heat for thermal decomposition is directly transferred from hot gases that are the products of partial combustion of char or gases. This latter definition is not rigorously correct because there is actually a partial-combustion (or starved-air-combustion) zone in the reactor (see Reaction 10).

Gasification. A carbonaceous material is reacted with oxygen, steam or hot CO_2 to produce fuel gas (see Reactions 5, 6, 7 and 8 for examples of char gasification). Gasification reactors may also include a pyrolysis zone in which char is produced. Heat is usually generated within the reactor by partial

combustion of the char and is transferred directly from hot combustion-product gases to the solid waste (see Reaction 10).

Liquefaction processes. These yield greater quantities of organic liquids relative to fuel gases. Pyrolysis processes may be designed to produce large quantities of liquids and can therefore be categorized as liquefaction (see Reactions 1 and 2 for examples). Other liquefication processes use solvent hydrogenation, with or without catalysts. In a third type of liquefaction process synthesis gas ($CO + H_2$) is generated and converted catalytically to methanol, polymer gasoline, or other light organic liquids.

Slagging reactor. The reactor temperature exceeds the ash melting point, so that the ash from the reactor can be removed as a molten slag.

Direct heating or firing. Heat is transferred directly from hot combustion-product gases to solid wastes.

Indirect heating or firing. Fuel combustion takes place in a separate chamber, with heat transfer from the hot combustion-product gases to the solid wastes by means of convection and radiation through the reactor wall or fire-tube walls separating the two chambers, or by means of a recirculating heat carrier (hot ceramic balls, hot sand, hot char, molten metal, or molten salts).

---

biochemical conversion (see Boxes 1 and 2: Definitions, and Pyrolysis and Gasification Reactions). We will refer to all processes that involve pyrolysis, thermal gasification or liquefaction of solid wastes as PTGL processes.

Some PTGL processes are designed to discharge product fuel gases directly into the combustion chamber of a steam boiler. Because this type of process reactor (sometimes referred to as a starved-air or partial-combustion chamber) involves solid-waste pyrolysis and gasification, processes of this type will be included in our discussion even though the fuel produced is not stored.

The primary purpose here is to classify the PTGL process options, and to illustrate that:

■ Comparisons of PTGL process options should begin with process fundamentals (solids-flow mechanism, bed characteristics, reactor type, etc.).

■ The many process options available must be considered on the basis of type of recovered fuel desired and the waste material available.

■ PTGL processes are a possible but not necessarily simple solution to energy recovery from solid wastes.

■ Second-generation PTGL technologies now emerging will require more research and development work before becoming commercial.

PTGL process features

Potential advantages of the PTGL processes, relative to other solid-waste options, generally cited by proponents of PTGL technologies are:

■ Production of a storable fuel as opposed to only steam generation from incineration of wastes.

■ Recovery of char that can be converted into activated carbon or synthesis gas.

■ Lower costs for air pollution control because of the much lower gas volumes requiring cleaning, compared to those for solid-waste combustion in incinerators.

■ Production of a residue (a glassy aggregate or char) that is environmentally more acceptable than ash from an incinerator.

Many PTGL processes offer some or all of these advantages, but certain problems may develop. For example, a highly concentrated, organically contaminated wastewater stream is sometimes produced. This stream forms in processes where high-moisture-content waste is gasified, the moisture-laden fuel gas cooled, and the moisture condensed (the water comes from the feed material plus water of reaction). Water-soluble oxygenated organics (see Reactions 1 and 2) leave the system in the liquid

---

Pyrolysis and gasification reactions   Box II

1.   Carbonaceous solids $\xrightarrow{\text{Heat}}$ High-and moderate-mol. wt. organic liquids (tars and oils, some aromatics) + char + Low-mol. wt. organic liquids (many organic acids and aromatics)

$+ CH_4 + H_2 + H_2O + CO + CO_2$

$+ NH_3 + H_2S + COS + HCN$ Amounts depend on nitrogen and sulfur content in feedstock

2.   Organic liquids $\xrightarrow{\text{Heat}}$ Aromatic organics $+ \dfrac{\text{low-mol. wt.}}{\text{organic liquids}} +$ char $+ CH_4 + H_2 + H_2O + CO + CO_2$

$+ NH_3 + H_2S + COS + HCN$ Amounts depend on nitrogen and sulfur content in feedstock

Pyrolysis gas-phase equilibria

3. $CH_4 + H_2O \rightleftharpoons CO + 3H_2$ (endothermic)*

4. $CO + H_2O \rightleftharpoons CO_2 + H_2$ (slightly exothermic)

Char gasification

5. $C + 2H_2O \rightarrow CO_2 + 2H_2$
6. $C + H_2O \rightarrow CO + H_2$ (endothermic)†
7. $C + CO_2 \rightarrow 2CO$
8. $C + 2H_2 \rightarrow CH_4$ (exothermic)‡

Combustion for gasification heat source

9. $C + O_2 \rightarrow CO_2$ (complete combustion)
10. $nC + n/2\, O_2 \rightarrow nCO$ (partial combustion)¶   (exothermic)

*Overall methane yield decreases greatly with increasing temperature because of this reaction.

† High-energy-consuming reactions.

‡At atmospheric pressure, the equilibrium conversion to methane is quite low and decreases as temperature rises from $1,000°F$ to $1,500°F$. High-pressure operation is required to achieve a significant degree of hydrogenation of char or liquid products.

¶ Also known as starved-air combustion.

---

phase, and a wastewater treatment process is required. The extent of this problem varies from process to process, and depends on the raw-material characteristics and reactor type and design. This problem is avoidable and the technique for prevention will be examined.

Another problem is emission of fine particulates. In slagging reactors having very high temperature zones ($>2,000°F$), salt vaporization and subsequent condensation in the gas stream cause submicrometer particles to form. This potential problem will be discussed in more detail later.

Basic process characteristics

Process options can best be compared by classifying them on the basis of solids-flow mechanism, bed characteristics, reactor type (see below), heat-transfer method, evolved gas flow, temperature profiles, operating pressure, and the use of catalysts.* The relationships among these

*The U.S. Dept. of Energy is currently sponsoring research into high-pressure and catalytic liquefaction and gasification of residues. Until recently, most solid-waste-recovery processes were operated at 1 atm. Likewise, the use of catalysts has been limited.

process characteristics—and their impact on product types, product yields, and environmental-control needs—are important to any comparison of processes. For instance, experience with demonstration plants, and analyses of alternative systems presented in the literature, clearly indicate that environmental control needs vary from process to process and represent a large fraction of the system investment and operating costs.

Feedstock pretreatment

The amount of preprocessing is determined by technical and economic design considerations and varies for different waste materials and types of process reactors. Size reduction may be necessary to (1) ensure troublefree reactor-feeder operation, (2) prevent bridging in certain types of reactors, or (3) increase heat-transfer efficiency within a reactor. Size reduction will also aid in metals-recovery and separation of organic fractions from municipal refuse. Densification (cubing or pelletizing) of wastes such as cotton-gin trash or rice straw will increase reactor throughput. Drying is another preliminary

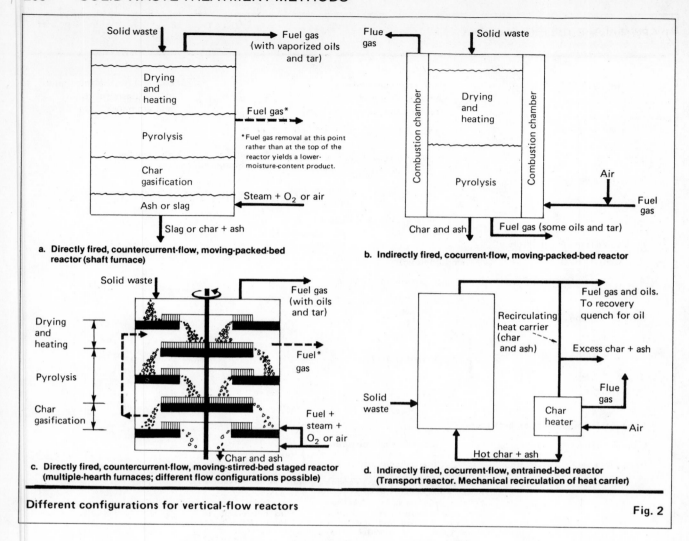

a. **Directly fired, countercurrent-flow, moving-packed-bed reactor (shaft furnace)**

b. **Indirectly fired, cocurrent-flow, moving-packed-bed reactor**

c. **Directly fired, countercurrent-flow, moving-stirred-bed staged reactor (multiple-hearth furnaces; different flow configurations possible)**

d. **Indirectly fired, cocurrent-flow, entrained-bed reactor (Transport reactor. Mechanical recirculation of heat carrier)**

Different configurations for vertical-flow reactors

Fig. 2

processing operation often used to treat high-moisture-content feedstocks and will be discussed later.

Reactor types and characteristics

Classification of processes by solids-flow direction and bed conditions in the reactor appears to be the best approach. Figs. 2 through 5 are simple sketches of the basic reactors.

A number of the reactors that have been drawn are arbitrarily divided into different zones. Only the indirectly-heated reactors without steam or oxygen injection may be described as pyrolysis reactors. The actual length of the solids-drying and heating zone will be a function of feed moisture content and heat-transfer rate. All reactors that contain both gasification and pyrolysis zones are usually directly fired (or autothermic) and may include steam injection for char gasification.

Table I lists the basic classes of reactors now being demonstrated or under development and identifies specific characteristics. Table II lists several R&D programs that use electrically heated reactors. For commercial applications, electric heating will probably not be economical but this type of reactor is more easily controlled during experimentation.

Predicting product yield and composition

To predict product yield and composition is an extremely complex task, and experimentation may be the only way to accurately correlate individual reactor output with feed composition, feed particle-size characteristics, and reactor operating conditions. Even for a relatively simple case where the physical and chemical characteristics of a dry feedstock are assumed constant, selection of the operating conditions to achieve desired product distribution (gases, oil, and char) and characteristics is not a simple matter. For example, in a packed-bed reactor (see Fig. 2a), the stream flows that must be adjusted include:
- Solid-waste feedrate to top of bed.
- Air or oxygen feed to char gasification zone.
- Steam feed to char gasification zone.
- Char removal rate from bottom of bed.

Once these flowrates are set at levels that allow steady-state operation, certain reactor operating conditions (unique to the flowrates selected) will result:
- Bed depth.
- Temperature profiles (solids and gases).
- Residence times (solids and gases).
- Product yield and composition (oil, gas, char).

Presently, it is difficult to predict the proper operating

Basic types of pyrolysis, thermal gasification and liquefaction (PTGL) reactors new demonstrated or under development   Table I

| Solids flow and bed conditions | Typical reactor vessels | Heat transfer | Relative direction of gas flow | Examples of processes, developers, R&D programs | Feedstock | Main products | |
						Fuels or char materials	Steam
I Vertical-flow reactors							
A. Moving packed bed (gravity solids flow; also called fixed bed)	Refractory-lined shaft furnace	Direct	Countercurrent	✓ Forest Fuels Mfg., Inc. (Antrim, (N.H.)	FAR‡		X
				Battelle Northwest (Richland, Wash.)	Refuse	X or	X
				American Thermogen (location unknown)	Refuse		X
				✓ Andco/Torrax Process (Buffalo, N.Y.)	Refuse		X
				H. F. Funk Process* (Murray Hill, N.J.)	Refuse	X	
				✓ Tech-Air Corp/Georgia Inst. Tech. (Atlanta, Ga.)	FAR	X	
				✓ Union Carbide Purox Process (Tonawanda, N.Y.)	Refuse, FAR	X	
				✓ Motala Pyrogas (Sweden)	Refuse	X	
				Urban Research & Development (E. Granby, Conn.)	Refuse	X	
				Wilwardco, Inc. (San Jose, Calif.)	FAR, sludge	X	
			Cocurrent	U. of California (Davis, Calif.)	FAR	X	
			Crossflow	Foster Wheeler Power Products (London, England)	Refuse, tires	X	
	Refractory or metal retort	Indirect—wall	Cocurrent	Destrugas Process (Denmark)	Refuse	X	
			Crossflow	Koppelman Process (Encino, Calif.)	FAR	X	
B. Moving stirred staged bed (gravity solids flow)	Refractory-lined multiple-hearth furnace	Direct	Countercurrent	✓ BSP/Envirotech (Belmont, Calif.)	Sludge, refuse	X	X
				✓ Nichols Research & Engr. (Belle Mead, N.J.)	Sludge, wood	X	X
				Garrett Energy Research & Engr. (Claremont, Calif.)	Manure	X	
				Hercules/Black, Crow & Eidsness (Gainesville, Fla.)	Refuse	X	
C. Moving entrained bed (may include mechanical bed transport)	Refractory-lined tubular reactor	Indirect by RHC†	Cocurrent	✓ Occidental Petroleum Co./Garrett Flash Pyrolysis Process (LaVerne, Calif.)	Refuse	X	
II Fluidized reactors	Refractory-lined or metal-walled vessel	Direct	—	Copeland Systems Inc. (Oak Brook, Ill.)	Sludges	X	
				Coors Brewing Co./U. of Missouri (Rolla, Mo.)	Refuse, FAR	X	
				Energy Resources Co. (Erco) (Cambridge, Mass.)	Refuse, FAR	X	
				Hercules/Black Crow & Eidsness (Gainesville, Fla.)	Refuse	X	
		Indirect by RHC		Bailie Process/Wheelabrator Incin. Inc. (Pittsburgh, Pa.)	Refuse	X	
				A. D. Little Inc./Combustion Equipment Assoc. (Cambridge, Mass./New York, N.Y.)	Refuse	X	
III Horizontal- and inclined-flow reactors	Rotary kiln or calciner						
A. Tumbling solids bed	Refractory-lined reactor	Direct	Countercurrent	✓ Devco Management Inc. (New York, N.Y.)	Refuse		X
				✓ Monsanto Landgard/City of Baltimore, Md.	Refuse	X	X
				Watson Energy Systems (Los Angeles, Calif.)	Refuse		X
	Metal retort in firebox	Indirect—wall	Countercurrent or cocurrent	Ecology Recycling Unlimited, Inc. (Santa Fe Springs, Calif.)	Refuse	X	
				Pyrolenergy System/Arcalon (Amsterdam)	Refuse, FAR	X	
				Pan American Resources, Inc. (West Covina, Calif.)	Refuse, FAR	X	
				Kobe Steel (Japan)	Tires	X	
				JPL/Orange County, Calif. (Fountain Valley, Calif.)	Sludge	X	
				Rust Engineering (Birmingham, Ala.)	Refuse, sludge	X	
	Metal retort	Indirect by RHC	Cocurrent	Tosco Corp/Goodyear Tire and Rubber (Los Angeles, Calif./Akron, Ohio)	Tires	X	
B. Agitated solids bed	Metal retort (mixing conveyor)	Indirect—wall or fire tubes	—	✓ Deco Energy Co. (Irvine, Calif.)	Tires	X	
				✓ Enterprise Co. (Santa Ana, Calif.)	Refuse	X	
				Kemp Reduction Corp. (Santa Barbara, Calif.)	Refuse, FAR	X	
	Refractory chamber (vibrating conveyor)	Indirect—fire tubes	Cocurrent	✓ PyroSol (Redwood City, Calif.)	Fluff from scrapped autos		X
C. Static solids bed	Metal chamber & conveyor belt	Indirect—fire tubes	Crossflow	Thermex, Inc. (Hayward, Calif.)	Tires	X	
IV Molten metal or salt beds							
A. Floating solids bed (horizontal flow)	Moving molten-lead hearth	Indirect by RHC	—	Michigan Tech. U. (Houghton, Mich.) (Puretec Pyrolysis System)	Refuse, FAR	X	
B. Mixed molten-salt bed (various possible flow schemes)	Vertical shaft or mixed bed	Indirect by RHC	—	Battelle Northwest (Richland, Wash.)	Refuse	X	
				Anti-Pollution Systems, Inc. (Pleasantville, N.J.)	Refuse, sludge	X	
V Multiple-reactor systems							
A. Combined entrained-bed/static-bed reactor system	Tubular metal retort and	Indirect-wall	Cocurrent	U. of California (Berkeley, Calif.)	Pulping liquor	X	
	static-hearth refractory chamber	Direct	—				
B. Combined moving packed-bed/entrained-bed reactor	Vertical shaft	Direct	Countercurrent	Battelle Columbus Laboratories* (Columbus, Ohio)	Paper, biomass	X	
	Vertical shaft (char gasification)	Direct	Cocurrent				
C. Combined mechanically conveyed static-solids-bed/moving packed-bed reactor	Traveling-grate refractory chamber	Direct	Countercurrent	Mansfield Carbon Products, Inc. (Gallatin, Tenn.)	Refuse	X	
	Refractory-lined shaft furnace	Direct	Countercurrent				

* Pressure above atmospheric.   † Recirculating heat carrier.   ‡ Forestry and/or argricultural residues   ✓ At commercial or demonstration scale.

Electrically heated reactors in the R&D stage   **Table II**

Typical reactor vessels	Examples of processes, developers, R&D programs	Feedstock	Main products	
			Fuels or char materials	Steam
Tubular metal retort[f]	Wright-Malta Corp*,[†] (supercritical pressure) (Ballston Spa, N.Y.)	FAR[‡], refuse	X	X (electric power)
Back-mix reactor	Dept. of Energy Wood to Oil Process Development Unit *,[†] Bechtel Corp., (Albany, Ore.)	FAR	X	
Tubular quartz reactor[f]	Princeton U. (Princeton, N.J.)	Manure	X	
Metal retort[f]	Dept. of Energy/Pittsburgh Energy Research Center (Pittsburgh, Pa.) (Formerly Bureau of Mines)	Refuse	X	
Metal retort[f]	Stanford U. (Palo Alto, Cal.)	Refuse	X	
Metal retort[f]	U. of Southern California (Los Angeles, Cal.)	Refuse, sludge	X	
Metal retort[f]	New York U. (New York, N.Y.)	Refuse	X	
Glass vessel[f] (Molten-salt bed)	U. of Tennessee (Knoxville, Tenn.)	Tires, plastics, FAR	X	
Entrained-bed reactor[f] Fluidized-bed reactor Cyclonic burner[f] Moving packed bed[f]	Texas Tech. U. (Lubbock, Tex.) (All reactors are autothermic)	Manure	X	

* Pressure above atmospheric.   † Sodium carbonate catalyst   ‡ Forestry and/or agricultural residues.   f Bench-scale equipment.

conditions that achieve specific product yields for most heterogeneous waste materials having variable moisture and ash content. Empirical methods must be used for design.

Even though a quantitative analysis of the many PTGL process options is a complex subject, the general rules outlined below may be useful:

■ The yield of condensable organics (oils and tars) is maximized when solids are rapidly heated to a moderate temperature level (~1,000°F), and the product gas and vapors immediately quenched. Liquid organics are highly oxygenated if produced from a cellulosic feedstock (CH_2O). Such liquid products, therefore, are partially miscible with water and may contain corrosive organic acids. The reactors shown in Figs. 2d, 3a, and 3b are examples of systems that are capable of producing high oil yields.

■ Noncondensable-fuel-gas yield (CO, H_2, CH_4) is maximized when the evolved gases and char are held at high temperatures (1,400°F) for relatively long residence times. Cocurrent flow of solids and gases allows this condition to exist and may minimize the yield of condensable organics. In cocurrent-flow systems having wet solids feed, moisture driven off in the drying zone flows through the pyrolysis zone. This allows the steam-carbon reactions (see Box 2, Reactions 5 and 6) to proceed without steam injection and favors high fuel-gas yield (Figs. 4b and 4c are examples of reactors that may have relatively high gas yields. Operation of a fluidized-bed reactor at high temperature can also give a high gas yield (see Fig. 3).

Indirectly heated reactors yield a higher-heating-value fuel gas (500 to 800 Btu/scf) than directly fired, air-blown reactors. The latter type produces a fuel gas in the range of 100 to 150 Btu/scf. Oxygen blowing will yield a gas of moderate heat content (250 to 350 Btu/scf). Net energy production, however, may be lower for the indirectly heated reactor system as a result of economic limitations on the use of very large heat-transfer surfaces.

Certain trace components in waste or residues can present design problems relative to corrosion and pollution control. For instance, polyvinyl chloride (PVC) in municipal refuse will cause HCl to form in the product gas of a system. Source separation of PVC can minimize this problem. The reducing atmosphere in the reactor can also yield NH_3, HCN, H_2S from the sulfur and nitrogen compounds in the feed, and these products will be partitioned between the gas phase and liquid phase. The

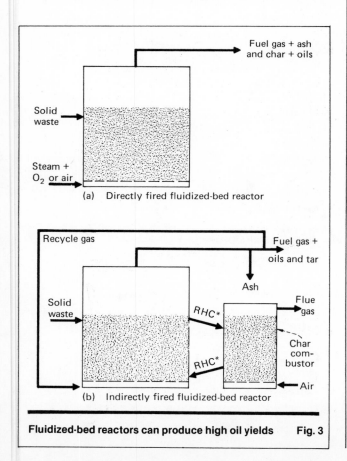

(a)   Directly fired fluidized-bed reactor

(b)   Indirectly fired fluidized-bed reactor

Fluidized-bed reactors can produce high oil yields   **Fig. 3**

concentration of heavy metals, such as cadmium and mercury, in the feedstock must also be taken into account. Systems must be included to remove these compounds from the air and water effluent streams.

■ The temperature profiles in the drying, pyrolysis and gasification zones are extremely important. Water pollution potential is the greatest with countercurrent flow reactors wherever product gases leave the drying zone at a low temperature (500°F), and water is condensed out of the fuel gas. Low temperature results in minimal thermal cracking of high-molecular-weight organic vapors. This condition does yield a somewhat higher overall thermal efficiency for the process, however. (See Figs. 2a and 2c for reactors that can operate under these conditions.)

■ The condensable organic materials in the product fuel gas may include low-molecular-weight oxygenated organics (acids, ketones and aldehydes), oxygenated aromatics, heavy aromatic oils, and tars. The volume of wastewater condensed from product fuel gas may be in excess of 75 gal/ton of municipal refuse ($>$ 25% moisture in feed), unless the refuse is pre-dried, which is costly. The high water content of many agricultural and forestry residue feedstocks may lead to even greater volumes of wastewater if these feedstocks are not dried prior to processing.

■ Only directly fired, moving-packed-bed reactors with countercurrent flow are generally operated in a slagging mode (see Fig. 2a). Slag flow-properties will be determined by feed characteristics such as glass composition and reactor temperature. Submicrometer inorganic particles may be formed by salt condensation from the vapor phase in such systems, and these particles may not be completely removed in the packed bed.

Design considerations for specific products

Solid wastes now being considered for feedstocks include municipal refuse, tires, sludges, plastic wastes, forestry-product wastes, agricultural residues such as rice hulls, corn husks, wheat and rice straw, and cattle manure. The feedstock characteristics (C:H:O ratio, cellulose content, lignin content, ash content and composition, water content, particle size, particle flowability) must be considered when selecting a process to produce a fuel gas, oil, or char. It is beyond the scope of this article to discuss this subject in detail. As an example, however, we will discuss a design option for cases where an oil product is desired from a cellulosic (CH_2O) feedstock.

In such a system, the product vapor from pyrolysis of solids will contain oxygenated organic molecules. To maximize the oil yield, it is necessary to minimize the thermal cracking of reactive radicals that are oil precursors. Such cracking can be minimized by immediate and rapid quenching to below the gas dewpoint by means of a cool-oil quench stream, but this technique also condenses the water vapor in the gas stream. As noted earlier, this water comes mainly from the moisture in the solid feed, but also forms as a reaction product. It may be preferable to pre-dry the feed to minimize wastewater treatment problems such as oxygenated-oil removal. However, the costs and possible complications from other environmental problems (odor and particulate control) must be considered before such a decision is made. Some water may also be desirable in the oil to serve as a viscosity modifier.

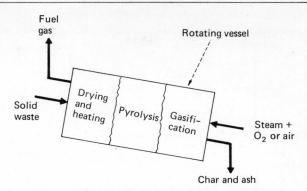

a. Directly fired, countercurrent-flow, tumbling-solids-bed reactor (refractory-lined rotary kiln)

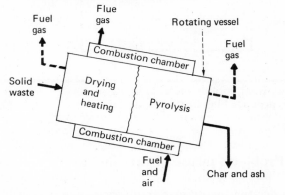

b. Indirectly fired, tumbling-solids-bed reactor; cocurrent or countercurrent flow (metal-shell rotary calciner)

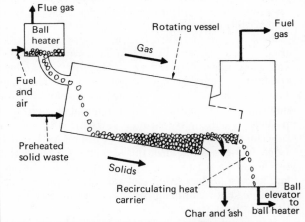

c. Indirectly fired, tumbling-solids-bed reactor(refractory-lined rotary kiln or metal retort)

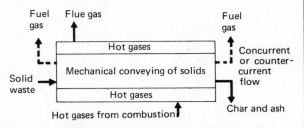

d. Indirectly heated, mechanically conveyed tunnel reactor with agitated or static-solids bed

(Note: Solids conveyed by traveling pan conveyor or screw conveyer)

Horizontal-flow reactors can have high gas yields   Fig. 4

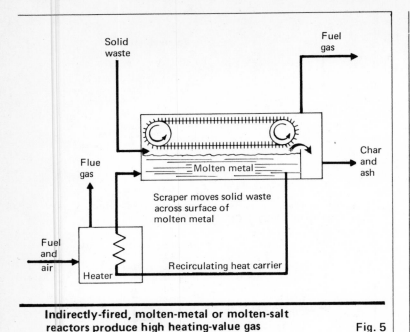

Indirectly-fired, molten-metal or molten-salt reactors produce high heating-value gas   Fig. 5

Problems in use of raw fuel gas

Reactor selection and operation will depend to a great extent on the form of recovered energy desired from a specific feedstock. Immediate combustion of reactor product gas to generate steam eliminates water pollution resulting from condensation of water from a fuel gas loaded with liquid organics.

It is important, however, to ensure against simply trading an air pollution problem for a water pollution one. Inorganic particulate matter in the fuel gas going to a combustion chamber may vaporize in the hot chamber (about 2,000°F), causing submicrometer particles to form downstream by inorganic-salt condensation. These small particles readily pass through low-efficiency particulate collectors. Where immediate combustion of particulate-laden fuel gas is planned, the problem can be minimized by choosing a reactor that will produce the lowest particulate loading in the exiting fuel gas (for example, a packed bed instead of a tumbling bed, or a stirred bed instead of a fluidized bed).

Reactor choice may not completely solve the problem, though. Submicrometer inorganic particles can also form in the bottom of a high-temperature, directly fired packed-bed reactor. Such particulate-laden streams will not be completely cleaned by passing through a coarse bed of refuse. Cleaning the fuel gas before combustion will minimize this problem; however, such hot-gas cleaning is not a simple process. More work is required to develop an energy-efficient hot-gas-cleaning process.

Recovered char

Char recovery is desirable because of the material's potential value as a feedstock for activated-carbon produc-

tion or as a low-quality adsorbent. Char is also a natural binder for water-soluble inorganics, reducing the leaching potential of certain constituents from buried PTGL-reactor residues in landfills, and is a storable solid fuel. However, the practicality of char removal depends on char quantity and characteristics—which in turn are functions of feedstock conditions, process type and operating conditions—and local demand.

First- and second-generation processes

Pyrolysis is applicable to a wide range of materials. Before 1977, only processes using directly fired shaft furnaces and directly fired rotary kilns were being demonstrated in the U.S. for municipal refuse applications with capacities of over 100 tons/day.* Some directly fired reactors are still new relative to producing fuels from wastes. The directly fired multiple hearth furnace, which has long been used for sludge incineration, roasting, and activated-carbon and charcoal production (from pyrolysis of coal and wood wastes), has recently been extensively tested for pyrolysis and gasification of refuse, sludge and manure.

Pyrolysis reactors using indirect heat transfer have not been demonstrated to the same extent as directly fired shaft furnaces and directly fired rotary kilns. They may therefore be classified as second-generation processes. These reactors can yield a higher-Btu-content product gas than can the directly fired or first-generation process reactors, but they present some design challenges.

As second-generation processes develop, problems that are not of great concern in refractory-lined, directly fired reactors must be addressed. With heat transfer by conduction through a surface such as a wall, for instance, the reactor should ideally have a metal shell with oxidizing conditions on the fire side and reducing conditions (with H_2S and H_2 present) on the other side, requiring the use of expensive, high-nickel alloys such as Inconel. The designer of an indirectly heated reactor may instead select recirculating nongaseous heat carriers that will allow use of a refractory-lined vessel. By such a choice, however, the designer exchanges the problem of using an expensive construction material for the potential operating problem and expense of a recirculating solid or molten heat carrier.

*Two of the most important current industrial uses of these types of reactors are as blast furnaces (iron-ore reduction in shaft furnaces) and as cement kilns (directly fired refractory-lined rotary kilns).

The author

Jerry Jones is manager of the Environmental Control Group of the Chemical Engineering Laboratory at SRI International, Menlo Park, CA 94025, where he is engaged in process development, process economics and market evaluation studies. His areas of interest include environmental control processes, resource recovery, and energy production from wastes and nonconventional feedstocks. He holds B.S. and M.E. degrees in chemical engineering from Cornell University and an M.S. in environmental engineering from Stanford University. He is a registered professional engineer in California and Illinois and a member of AIChE, ACS, and the Water Pollution Control Federation.

Section 3
SOLID WASTE DISPOSAL METHODS

Solid-waste disposal: Landfarming
Solid-waste disposal: Deepwell injection
Solid-waste disposal: Incineration
Solid-waste disposal: Landfilling
Solid-waste disposal: Solidification
Disposal of flue-gas-cleaning wastes
How sludge characteristics affect incinerator design
Incineration of industrial wastes
How to burn salty sludges
Incineration's role in ultimate disposal of process wastes
Ultimate disposal of spilled hazardous materials

Solid-waste disposal: Landfarming

The upper zone of soil is an environment where biological, chemical and physical processes function continuously to modify and shunt compounds into the natural biogeochemical pathways. Landfarming uses these processes to effectively and safely treat and dispose of certain biodegradable industrial wastes.

R. L. Huddleston, *Continental Oil Co.*

☐ Use of the upper soil zone to manage industrial wastes, both hazardous and nonhazardous, has been referred to as land spreading [1], land application [2], sludge farming [3], landfarming [4], land disposal [5] and soil cultivation [6]. Landfarming is the designation officially recognized by the U.S. Environmental Protection Agency. It is the most appropriate because it implies an orderly process characterized by planning and post-application care, both of which are vital to proper waste management. Landfarming must be viewed, planned and managed with the same degree of care and attention given any other technical process operation.

Landfarming of waste in the upper soil zone is not a new concept. It has been practiced, though sparingly, by the petroleum industry for at least 25 years in the disposal of oily solid wastes without any reported environmental problems or operational difficulties. In the last few years, management of biodegradable petroleum wastes by landfarming has dramatically increased. At present, probably half of all petroleum oily wastes are being landfarmed, and major generators of other degradable wastes (pharmaceutical and organic-chemical plants [13]) are using or considering use of the method.

The major advantages of landfarming as a solid-waste treatment/disposal method are:

1. Effectiveness at a comparatively reasonable cost.
2. Relative environmental safety.
3. Use of natural processes that recycle the waste.
4. Relative process simplicity, requiring no dependence on high-maintenance or failure-prone equipment.
5. Possible improvement of soil structure and fertility.

Landfarming is a relatively simple operation, but the

Originally published February 26, 1979.

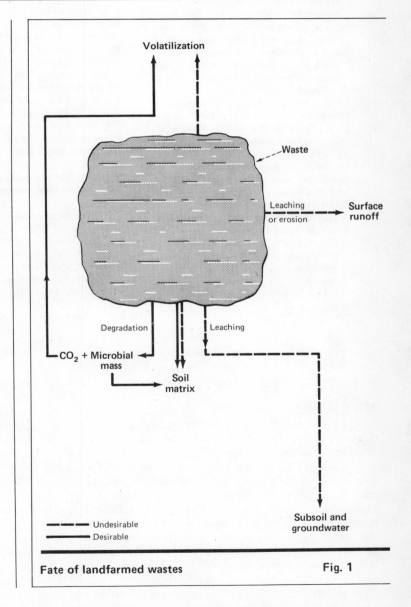

Fate of landfarmed wastes   Fig. 1

275

Sect. 250.41(46)—Landfarming of a Waste means application of waste onto land and/or incorporation into the surface soil, including the use of such waste as a fertilizer or soil conditioner. Synonyms include land application, land cultivation, land irrigation, land spreading, soilfarming, and soil incorporation.

Sect. 250.45-5—Landfarms

(a) Hazardous waste not amenable to landfarming.

1. Ignitable waste
2. Reactive waste
3. Volatile waste
4. Waste which is incompatible when mixed

[Note: Exceptions are allowed. (Where exceptions are allowed in these regulations, the burden of proof is on the owner/operator.)]

(b) General requirements

1. A landfarm shall be located, designed, constructed and operated to prevent direct contact between the treated area and navigable water.

2. A landfarm shall be located, designed, constructed, and operated to minimize erosion, landslides and slumping in the treated area.

3. A landfarm shall be located, designed, constructed and operated so that the treated area is at least 1.5 meters (5 feet) above the historical high-water table. [Note: exceptions allowed]

4. The treated area of a landfarm shall be at least 150 meters (500 feet) from any functioning public or private water supply or livestock water supply. Note: exceptions allowed if:

(i) No direct contact will occur between the treated area of the landfarm and any functioning public or private water supply or livestock water supply;

(ii) No migration of hazardous constituents from the soil in the treated area of the landfill to any public or private water supply or livestock water supply will occur; and

(iii) A soil monitoring system as specified in Sect.

250.45-5(e) has been installed and is being adequately maintained.

5. A landfarm shall be located on an area that has fine grained soils (i.e., more than half the soil particles are less than 73 microns in size) which are of one of the following types, as defined by the Unified Soil Classification System (ASTM Standard D 2487-69): OH-organic clays of medium to high plasticity; CH-inorganic clays of high plasticity, fat clays; MH-inorganic silts, micaceous or diatomaceous fine sandy or silty soils, elastic silts; CL-inorganic clays of low to medium plasticity, gravelly clays, sandy clays, silty clays, lean clays; OL-organic silts and organic silt-clays of low plasticity. [Note: exceptions allowed]

(c) Site preparation

1. Surface slopes of a landfarm shall be less than 5%, to minimize erosion in the treated area by waste or surface runoff, but greater than 0% to prevent the waste or water from ponding or standing for periods that will cause the treated area to become anaerobic. [Note: exceptions allowed]

2. Caves, wells (other than active monitoring wells) and other direct connections to the subsurface environment within the treated area of a landfarm, or within 30 m (100 ft) thereof, shall be sealed.

3. Soil pH in the zone of incorporation shall be equal to or greater than 6.5 [Note: exceptions allowed]

(d) Waste application and incorporation

1. Waste application and incorporation practices shall prevent the zone of incorporation from becoming anaerobic.

2. Waste shall not be applied to the soil when it is saturated with water. [Note: exceptions allowed]

3. Waste shall not be applied to the soil when the soil temperature is less than or equal to 0°C.

4. The pH of the soil-waste mixture in the zone of incorporation shall be equal to or greater than 6.5 and maintained until the time of facility closure. [Note: exceptions allowed]

5. Supplemental nitrogen and phosphorus added to the soil of the treated area, for the purpose of increasing the rate of waste biodegradation, shall not exceed the rates of application recommended for agricultural purposes by the U.S. Dept. of Agriculture or Agricultural Extension Service.

(e) Soil monitoring

1. Background soil conditions shall be determined by taking one soil core per acre in the area to be treated. The depth of the soil core shall be three times the depth of the zone of incorporation or 30 centimeters (12 inches), whichever is greater. The bottom one third of the soil core shall be quantitatively analyzed for those constituents known or expected to be in the waste which make it hazardous. At new facilities, soil cores shall be taken and analyzed prior to beginning operation. At existing facilities, background soil cores shall be taken and analyzed within six months after the effective date of these regulations.

2. Soil conditions in the treated area of a landfarm shall be determined by taking one soil core per acre, semiannually. The depth of the soil core shall be three times the depth of the zone of incorporation or 30 centimeters (12 inches), whichever is greater. The bottom one third of the soil core shall be quantitatively analyzed for the constituents in the waste which make it hazardous. [Note: exceptions allowed]

3. If soil monitoring shows that the concentration of a hazardous constituent in the bottom one third of the soil core has significantly exceeded the background levels established in accordance with paragraph (e)(1), the owner/operator shall:

(i) Notify the Regional Administrator within seven days;

(ii) Determine, by soil monitoring, the areal extent of vertical contaminant migration in the soil; and

(iii) Discontinue all landfarming in the contaminated area, as determined in (ii), until corrective measures can be taken.

---

principal components, waste and soil, are *not* simple. Their complexities must be appreciated if landfarming is to be acceptably practiced.

How landfarming works

Soil contains an array of microbial "factories" that constantly recycle organic and inorganic matter by breaking them down into forms required by higher forms of life. This natural recycling represents the most important mechanism of the landfarm process and is carried out by microorganisms that are present in very high populations in typical soil, as shown in Table I. It is apparent from the population distribution pattern in the table that the upper soil zone contains, by far, the largest microbial populations. For this reason, landfarming is restricted to the soil plow-zone, 6–8 in. or 15–20 cm.

Wastes added to the soil environment are subject to one or more of the following processes: decomposition, leaching of water-soluble components, volatilization, and incorporation into the soil matrix. Fig. 1 describes the typical fate of landfarmed waste.

Direct assimilation of the waste into the soil matrix is undesirable (and may prevent the landfarming of the waste) if it causes soil toxicity or soil structure breakdown. However, it can be innocuous, or beneficial if the waste constituents increase soil fertility or water-holding capacity. Maximization of the degradation processes is essential. Waste loss by volatilization, erosion or leaching must be minimized to avoid adverse environmental effects.

When biodegradable waste is added to the soil, it is attacked by microflora (bacteria, yeasts, molds, etc.) and decomposed. Many biochemical steps are involved, which always result in some of the carbon being converted to carbon dioxide. The remainder of the carbon is assimilated into microbial cell mass, but only if appropriate amounts of available nitrogen and phosphate (the only form of available phosphorus) are in the soil. If necessary, these inorganic nutrients can be supplied by adding ordinary lawn or farm fertilizer.

Waste conversion to cell mass is desirable, because a larger cell mass results in faster waste decomposition,

(f) Growth of food-chain crops

Food-chain crops shall not be grown on the treated area of a landfarm.

(g) Closure

1. A landfarm shall be designed and operated so that, by the time of closure, the soil of the treated area(s):

(i) Is returned to its pre-existing condition, as established in paragraph (e)(1) if the facility began operation after promulgation of this requirement (i.e., a new facility).

(ii) Is returned to equivalent pre-existing condition, as determined by soil analysis of similar local soils that have not had hazardous waste applied to them, if the facility began operation prior to the promulgation of this requirement (i.e., an existing facility). Soil analysis of similar local soils shall not be required at existing facilities if background soil data are available and those data establish background conditions for the treated area(s).

2. Soil of the treated area(s) of a new or existing facility that does not comply with paragraph (g)(1)(i) or (ii), respectively, shall be analyzed to determine whether it meets the characteristics of a hazardous waste as defined in Subpart A [43 Fed. Reg. 58954]. In the event the soil is determined to be a hazardous waste, it shall be removed and managed as a hazardous waste in accordance with all applicable requirements of this Part. [Note: exceptions allowed]

Definitions
Sect. 250.13(a) Ignitable waste

1. Definition—A solid waste is a hazardous waste if a representative sample of the waste:

(i) Is a liquid and has a flash point less than 60°C (140°F) determined by the method cited below or an equivalent method, or

(ii) Is not a liquid and is liable to cause fires through friction, absorption of moisture, spontaneous chemical changes, or retained heat from manufacturing or processing, or, when ignited, burns so vigorously and persistently as to create a hazard during its management, or

(iii) Is an ignitable compressed gas as defined in 49 CFR 173.300(b), or

(iv) Is an oxidizer as defined in 49 CFR 173.151.

2. Identification method

(i) Flash point of liquids shall be determined by a Pensky-Martens Closed Cup Tester, using the protocol specified in ASTM Standard D-93-72, or the Setaflash Closed Tester using the protocol specified in ASTM standard D-3278-73 or any other equivalent method as defined in this Subpart.

(ii) Ignitable gases shall be determined by the methods described in 49 CFR 173.300.

Sect. 250.13(c) Reactive waste

1. Definition—A solid waste is a hazardous waste if a representative sample of the waste:

(i) Is normally unstable and readily undergoes violent chemical change without detonating; reacts violently with water, forms potentially explosive mixtures with water, or generates toxic gases, vapors, or fumes when mixed with water; or is a cyanide or sulfide bearing waste which can generate toxic gases, vapors, or fumes when exposed to mild acidic or basic conditions.

(ii) Is capable of detonation or explosive reaction but requires a strong initiating source or which must be heated under confinement before initiation can take place, or which reacts explosively with water.

(iii) Is readily capable of detonation or of explosive decomposition or reaction at normal temperatures and pressures.

(iv) Is a forbidden explosive as defined in 49 CFR 173.51, a Class A explosive as defined in 49 CFR 173.53, or a Class B explosive as defined in 49 CFR 173.58.

Note—Such waste includes pyrophoric substances, explosives, autopolymerizable material and oxidizing agents. If it is not apparent whether a waste is a hazardous waste using this description, then the methods cited below or equivalent methods can be used to determine whether the waste is hazardous waste.

2. Identification method

(i) Thermally unstable waste can be identified using the Explosion Temperature Test cited in Appendix II of this Subpart [43 *Fed. Reg.* 58962] (waste for which explosion, ignition or decomposition occurs at 125°C after 5 minutes is classed as hazardous waste).

(ii) Waste unstable to mechanical shock can be identified using the Bureau of Explosives impact apparatus and the tests cited in 49 CFR 173.53(b), (c), (d), or (f), as appropriate.

Sect. 250.41(45) Incompatible Waste

means a waste unsuitable for commingling with another waste or material, because the commingling might result in:

(i) Generation of extreme heat or pressure,

(ii) Fire,

(iii) Explosion or violent reaction,

(iv) Formation of substances which are shock sensitive, friction sensitive, or otherwise have the potential of reacting violently,

(v) Formation of toxic (as defined in Subpart A) dusts, mists, fumes, gases, or other chemicals, and

(vi) Volatilization of ignitable or toxic chemicals due to heat generation, in such a manner that the likelihood of contamination of groundwater, or escape of the substances into the environment, is increased, or

(vii) Any other reactions which might result in not meeting the Air Human Health and Environmental Standard. (See Appendix I [43 *Fed. Reg* 59017] for more details.)

Sect. 250.41(98) Volatile Waste

means waste with a true vapor pressure of greater than 78 mm Hg at 25°C.

---

and conversion leads to waste organic matter being "fixed" into the soil as natural organic matter, which may enhance soil structure and fertility.

Landfarmed waste almost always contains constituents that are not biodegradable, such as: metals, salts, coke, dust, sand, etc. The landfarm acts as a repository for these.

Wastes suitable for landfarming

Wastes to be landfarmed should contain organic constituents that are susceptible to biodegradation and not subject to significant leaching while the degradation process proceeds. Petroleum oily wastes and waste treatment biosolids are good examples. Some organic chemical-plant wastes may also meet these criteria.

Biopersistence is a good indicator of the landfarm suitability of a waste. Of course, the higher the biopersistence, the less suitable the waste. Biopersistence can be assessed using relatively simple laboratory biodegradation tests [8].

Suitable wastes should not contain extremely high salt or metal concentrations. Even though soil has the capacity to bind cations, and most heavy metals are water insoluble at pH values above 7, landfarming is not an effective treatment/disposal method for inorganic substances. At pH values below 7, the leachability of metals markedly increases.

Microbial distribution in soil [7]					Table I
	Organisms/gram of soil				
Depth (cm)	Aerobic bacteria	Anaerobic bacteria	Actinomycetes	Fungi	Algae
3-8	7,800,000	1,950,000	2,080,000	119,000	25,000
20-25	1,800,000	379,000	245,000	50,000	5,000
35-40	472,000	98,000	49,000	14,000	500
57-75	10,000	1,000	5,000	6,000	100
135-145	1,000	400	—	3,000	—

Factors controlling biodegradation   Table II

- Composition (structure) of waste
- Contact between waste and soil microorganisms
- Presence of adequate oxygen
- Temperature
- pH
- Presence of available inorganic nutrients
- Moisture content

**Soil constituents and
characteristics to test for   Table III**

Arsenic	Copper	Selenium
Barium	Lead	Silver
Cadmium	Mercury	Sodium
Chromium	Nitrate	Zinc

- pH
- Important constituents in the waste (such as: oil, specific organics, pesticides or radioactive materials)
- General soil type
- Soil cation-exchange capacity

**Core sampling to a depth of 8-12 ft
helps establish site suitability   Fig. 2**

Also, wastes should not contain hazardous components at levels that might cause contamination of air, subsoil or groundwater. Highly reactive, toxic or radioactive wastes should not be landfarmed.

A waste may be unacceptable for landfarming because a particular constituent causes environmental or safety problems. In some cases, pretreatment of the waste will modify or remove this constituent. For example, heat treatment will remove pathogens that would ordinarily prevent landfarming. Pretreatment is specific to a particular waste and in most cases the technology required will have to be developed. However, there are pretreatment methods for waste volume reduction and for recovery of valuable waste constituents. These should be used whenever possible.

Factors controlling biodegradation

The rate and extent of biodegradation of waste in the soil are strongly influenced by many chemical and physical factors. Some of the most important are listed in Table II. The biodegradation rate primarily depends on the chemical structure of the waste, which cannot be readily controlled.

Waste surface area and biodegradation rate are generally directly related. The degree of physical contact between waste and soil microflora can be readily controlled. For flowable wastes, several successive cultivations of waste and soil (after waste application) will accomplish the necessary contact. For solid or semisolid wastes, some mechanical means must be used to shred, pulverize or otherwise maximize waste surface area prior to incorporation into the soil.

The presence of adequate free oxygen in the soil is essential to effective biodegradation. Aerobic biodegradation of organic matter is much more rapid and complete than is anaerobic degradation, and some substances, such as saturated hydrocarbons, are not significantly biodegradable under anaerobic conditions. Though the soil plow-zone is aerobic under normal conditions, it quickly becomes anaerobic if water-saturated or overloaded with biodegradable organic matter. Adequate drainage and proper waste loading will prevent anaerobic conditions.

The biodegradation rate in the soil decreases as temperature decreases, and under frozen conditions biodegradation essentially stops. However, soil temperature is impractical to control, and is not considered a dominant factor in landfarming.

Soil pH strongly influences microbial activity and should be maintained between 7 and 9. The pH is easily controlled by adding inexpensive agricultural products such as lime. The water solubility of heavy metals increases drastically at low pH values, and this situation must be avoided.

As mentioned, available nitrogen and/or phosphate in the soil is essential to achieve the maximum biodegradation rate. These inorganics are easily supplied by common agricultural fertilizers. The amounts required are dependent on such factors as available nitrogen and phosphate in the waste, waste biodegradation rate, rate of waste application, and fertilizer persistence in the landfarm soil. Though these factors vary from waste to waste, there are a few guidelines that may be used to estimate initial fertilizer quantities:

1. For readily biodegradable organic matter, such as certain pharmaceutical waste, as much as 1 part of *nitrogen*/25 parts of waste *carbon* added to the soil may be required to obtain a maximum biodegradation rate. For such waste, many soils will not require phosphate addition, but if needed this should not exceed 1 to 10% of the nitrogen required.

2. For less-rapidly-biodegradable waste, such as oily waste, nitrogen and phosphate requirements are smaller—probably 1 to 10% of those required for readily degradable matter.

3. Note that if fertilization is excessive, unnecessary levels of ammonium, nitrate and phosphate will be present in the soil-waste zone, and nitrate contamination of groundwater may occur.

Soil moisture content affects the rate of waste biodegradation. Microbial activity in the soil is optimum at soil moisture concentrations between 50 and 80% of the soil water-holding capacity [7]. Depending on the soil type, such values correspond to about 6 to 22 wt. % of water in the soil [14]. As a rule of thumb, soil water-concentrations that support plant growth are also appropriate for microbial degradation. Artificial control of soil moisture is expensive and has not proved justifiable even in arid climates such as Montana's.

Operational steps

The following is a step-by-step description of the landfarming procedure.

Site selection—The site chosen for landfarming should be cultivatable with ordinary farm equipment, unencumbered with surface or subsurface utilities and reasonably isolated from residential or public-use areas. The general subsurface geology should be known and not include caverns, vertical faults, wells or other features that could provide direct access to groundwater. Groundwater should be no closer than four feet to the surface, even if the soil has a high clay content, unless some unusual feature, such as an impenetrable barrier, can function to prohibit contamination.

The ground surface should be flat enough to prevent erosion problems, yet slightly sloped to prevent ponding. The site should not be subject to flooding and must be compatible with some method of containing and/or treating surface-water runoff, if necessary.

After these initial requirements have been met, representative core samples should be taken to a depth of 8–12 ft. Such cores can be obtained as shown in Fig. 2.

The cores should be sectioned (uppermost, mid-depth and deepest) and separately analyzed both for substances that are restricted by U.S. drinking water standards and for potentially hazardous substances that are or might be in the waste. Also, certain soil-characterization tests should be made on these core sections. A list of important soil constituents and characteristics that should be tested for is given in Table III. Core data not only help establish site suitability, but provide a description of the site prior to waste addition for future reference.

Site preparation—The entire site should be fenced to prevent unauthorized entry, and engineered to adequately handle surface runoff as previously mentioned. The specific area to receive wastes should be graded, if

Landfarm waste-oil loss rates [4,6,11,12]			Table IV
Type of waste-oil and operator	initial % oil in soil†	grams oil lost/ kg soil/ yr	barrels oil lost/ acre/yr
Refinery oily waste			
Conoco—Billings, Mont.	1	7	51
Conoco—Ponca City, Okla.	5	9	66
Shell—Houston, Tex.	10	165	1,205
Petroleum crude oil			
Conoco—Ponca City, Okla.	5	17	124
Sun*	1	14 (±2)	102
Waste lube oil			
Sun*	2	18 (±5)	131
New Zealand	2	27	197
	5	38	277
	7	69	504
	9	454	3,314
Waste vacuum-pump oil			
Union Carbide—Oak Ridge, Tenn.	7	171	1,248
No. 6 fuel oil			
Shell—Houston, Tex.	10	311	2,270
Sun*	2	20 (±7)	146

Note: All data are approximate
*Combined data from three locations: Marcus Hook, Pa., Tulsa, Okla., and Corpus Cristi, Tex.
†Immediately after application

necessary, and have a uniform surface for even waste application. The area should be cultivated by disking, plowing or rototilling just before waste application.

The soil plow-zone pH should be adjusted, if necessary, to be between 7 and 9.

Waste analysis—Wastes to be landfarmed should be tested for all the constituents listed in Table IV except,

Waste is mixed into the plow-zone by disk or rototiller Fig. 3

This manual core sampler
assists post-waste-addition testing   Fig. 4

of course, soil type and cation-exchange capacity. Also, the wastes' biopersistence should be determined.

Waste application—The waste should be distributed uniformly. Flowable wastes can be applied by a tank truck equipped with a distribution arm or manifold. Solid or semisolid waste can be applied by truck and spread with a tractor.

Some liquid wastes are best applied directly to the subsoil, in order to reduce odor problems, for example. Equipment is available that will lift the soil surface, squirt in the waste and replace the surface.

The amount of waste to be landfarmed should be based on predetermined contents criteria, such as organics, oil, salt and metals. In most cases, the only way to safely select the criteria is to pilot landfarm the specific waste in question. Many such tests have been conducted with refinery and petroleum-production oily wastes [4], [6], [10], [11]. It has been found that oily-waste additions resulting in 5 to 10 wt.% oil in the plow-zone soil can be effectively handled as a single landfarm addition step. These wastes typically contain 10 to 50% oil, 30 to 80% water and 5 to 50% solids. After several additions, oily-waste content in the soil can be as high as 25 wt. % or higher.

Soil-waste blending—Waste can be effectively mixed into the soil plow-zone using a disk or rototiller (Fig. 3). Usually about six cultivations are required to adequately blend waste and soil.

Some landfarm operators let fluid wastes partially dry prior to blending them into the soil, claiming that such practice results in more-effective soil-waste mixing [4]. Other landfarm managers proceed to cultivate waste into the soil shortly after its application.

If fertilizer is required, it should be applied at this point. Again, soil-waste pH should be tested and adjusted if found to be outside the 7 to 9 range.

Post waste-addition care—Periodic soil-waste cultivation generally assists waste biodegradation by facilitating soil aeration and soil-waste contact. The optimum cultivation frequency depends on waste and soil characteristics. For oily wastes, cultivation every four to eight weeks, except during the winter, appears

beneficial. Some operators till even more frequently [3]. Soil pH should be measured at least quarterly.

The landfarm area should be periodically sampled to ensure that waste degradation proceeds as expected and to judge when waste reapplication can be made. In addition, core samples extending below the plow-zone should be taken to detect any adverse waste leaching that could ultimately contaminate groundwater. These tests should be performed at least annually. The core sampler shown in Fig. 4 is an excellent device for sampling to about 30 in. of soil depth.

Waste-oil loss rates and costs

Industrial waste-oil loss rates accomplished by landfarming are shown in Table IV. Typically, the cost of landfarming oily waste is about $16.50/ton of waste or $2.75/bbl.

References

1. Dotson, G. K., Dean, R. B., Kenner, B. A., and Cooke, W. B., Land Spreading—A Conserving and Non-Polluting Method of Disposing of Oily Wastes, *Proc.*, Fifth Intl. Water Poll. Res. Conf., Vol. 1, Pergamon Press, 1971, Sect. 11.36, pp. 1–15.
2. Johnson, J. B., and Connor, L. J., Economic and Regulatory Aspects of Land Application of Wastes to Agricultural Lands, "Land as a Waste Management Alternative," Ann Arbor Science Pub., Inc., Ann Arbor, Mich., 1977, pp. 29–44.
3. Lewis, R. S., Sludge Farming of Refinery Wastes as Practiced at Exxon's Bayway Refining and Chemical Plant, *Proc.*, Natl. Conf. on Disposal of Residues on Land, Information Transfer, Inc., Rockville, Md., 1977, pp. 87–92.
4. Huddleston, R. L., and Cresswell, L. W., The Disposal of Oily Wastes by Land Farming, *Proc.*, Petrol. Refiners Wastewater Forum, Sponsored by EPA, Amer. Pet. Inst., Natl. Petrol. Refiners Assn. and U. of Tulsa, Jan. 1976.
5. Adamczyk, A. F., Land Disposal of Food Processing Wastes in New York State, "Land as a Waste Management Alternative," Ann Arbor Science Pub. Inc., Ann Arbor, Mich., 1977, pp. 743–756.
6. Kincannon, C. B., Oily Waste Disposal by Soil Cultivation Process, EPA-P-2-72-110, Dec. 1972.
7. Alexander, M., "Introduction to Soil Microbiology," John Wiley and Sons, Inc., New York, 1961.
8. Ryan, J. A., Factors Affecting Plant Uptake of Heavy Metals from Land Application of Residuals, *Proc.*, Natl. Conf. on Disposal of Residues on Land, Information Transfer, Inc., Rockville, Md., 1977, pp. 98–105.
9. Bartha, R., and Pramer, D., Features of a Flask and Method for Measuring the Persistence and Biological Effects of Pesticides in Soil, *Soil Sci.*, Vol. 100, No. 1, 1965, pp. 68-70.
10. Huddleston, R. L., and Meyers, J. D., Treatment of Refinery Oily Wastes by Land Farming, presented to the Natl. AIChE Meeting, Philadelphia, Pa., June, 1978.
11. Raymond, R. L., Hudson, J. O., and Jamison, V. W., Final Report on Cleanup of Oil in Soil by Biodegradation, Project 05 21.3 and 21.4 of the Committee on Environmental Affairs of the Amer. Pet. Inst., Mar. 23, 1975.
12. Francke, H. C., and Clark, F. E., Disposal of Oily Wastes by Microbial Assimilation, National Technical Information Service, Springfield, Va., NTIS-Y-1934, May 1974.
13. Nelson, D., Laboratory and Field-Scale Investigation of Mycelial Waste Decomposition in the Soil Environment, presented to the Soc. for Industrial Microbiology Meeting, Michigan State U., East Lansing, Aug. 1977.
14. Guidelines Support Document, prepared for U.S. EPA by Environmental Research and Technology, Inc., Concord, Mass., Feb. 1978.

The author

Robert L. Huddleston is Group Leader of the Environmental Research Group in the Research and Development Dept. of Continental Oil Co., P.O. Box 1267, Ponca City, OK 74601. He holds B.S. and M.S. degrees in microbiology from the University of Oklahoma. Mr. Huddleston has published over 35 research papers, been awarded two patents and he has served on several American Petroleum Inst. and Soap and Detergent Assn. technical study committees and is a Past President of the Soc. for Industrial Microbiology.

Solid-waste disposal: Deepwell injection

Underground injection of liquid waste into a permeable rock formation isolated from potable water and mineral-bearing strata is in some cases the only safe method for disposing of certain indestructible, nonconvertible or otherwise hard-to-treat hazardous substances.

Marvin E. Smith, Subsurface Disposal Corp.

□ Deepwell disposal of industrial wastes has been practiced for over 25 years and today is a well-developed, carefully controlled, environmentally acceptable method of liquid waste disposal.

The primary concern in the planning, construction and operation of a deepwell disposal system is the complete protection of underground potable water, which by U.S. EPA definition includes waters with up to 10,000 ppm total dissolved solids, TDS. Of secondary concern is the protection of oil, gas and other valuable minerals. Development of a deepwell system is a multidisciplinary effort requiring geological, engineering, chemical, biological and legal expertise, none of which should be ignored.

A deepwell disposal system consists of a disposal zone, well (Fig. 1) and surface facility for pretreating the waste liquids. The disposal zone must be located below potable-water aquifers, and isolated from them by thick, relatively impermeable and fracture-resistant strata such as shale, limestone or dolomite. The zone must totally contain the waste liquids and have no other utility value. Deepwell disposal-zone depths vary from a few hundred to several thousand feet. A survey of well depths conducted for EPA in 1974 is shown in Table I.

Disposal wells are constructed using oil- and gas-industry-proven technology, incorporating special adaptions for problems unique to waste injection.

The key to a successful deepwell system is *compatibility* of the waste liquids with materials of construction, formation fluid and the formation itself.

Suitable wastes

All types of waste from a wide range of industries are being injected into deepwells. These wastes vary from relatively innocuous to highly toxic, from practically inert to highly reactive.

Industrial wastes represent an almost infinite variety of compounds, both organic and inorganic, with varying mixtures and a wide range of component concentrations within the mixtures. They may be true solutions, emulsions, liquid-solid mixtures, acids and bases, in fluids that may be excessively hot or cold and possibly subject to further reactions with changes in temperature and/or pressure and atmospheric exposure.

Basically, an industrial waste stream is deepwell injected only after all reasonable alternative disposal methods have been evaluated and found less desirable in terms of environmental protection and dependability. Chemicals found in waste streams that may have to be deepwell injected can be divided into organic and inorganic types. They may be liquids, gases or solids, the gases and solids being either dissolved in the liquids or simply carried along by them.

The organic chemicals include: acids, such as maleic, formic, adipic, cresylic, salicylic and acetic; alcohols, such as methanol, tertiary butanol, phenol and isopropanol; solvents, such as acetone, toluene, xylene, formaldehyde, ethylbenzene, benzaldehyde and methyl ethyl ketone; and other compounds, such as sodium naphthenate, sodium cresylate, calcium and sodium acetate, and large molecular structures such as styrene polymers and various polymeric resins.

Inorganic chemicals include: acids, such as sulfuric, hydrochloric and phosphoric; bases, such as sodium hydroxide; and salts, such as sodium chloride, sodium sulfate, sodium sulfide, sodium carbonate, sodium sulfite, arsenic sulfide, ammonium bisulfate, sodium bromide and calcium carbonate.

Every waste stream has unique characteristics that may highly complicate or make impossible a surface disposal method. A stream containing chlorinated hydrocarbons (for example, trichlorobenzene) and byproducts is a candidate for deepwell disposal because these organics are difficult to destroy by conventional

Originally published April 9, 1979.

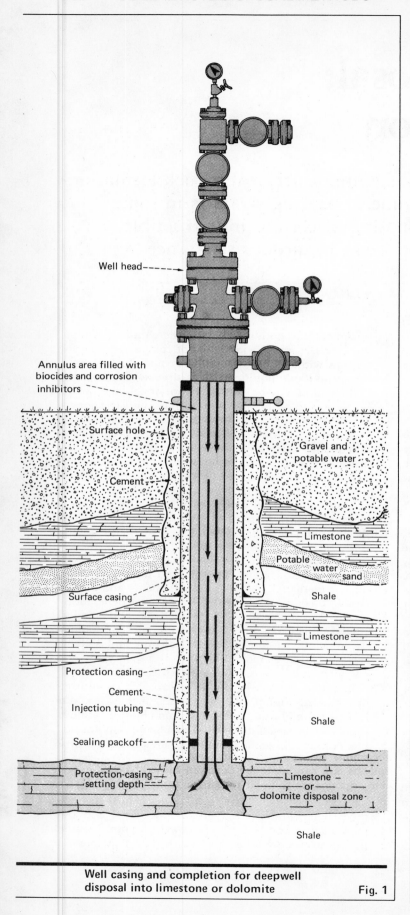

Well head

Annulus area filled with
biocides and corrosion
inhibitors

Surface hole

Gravel and
potable water

Cement

Limestone

Potable water sand

Surface casing

Shale

Limestone

Protection casing

Cement

Injection tubing

Shale

Sealing packoff

Protection-casing
setting depth

Limestone
— or —
dolomite disposal zone

Shale

**Well casing and completion for deepwell
disposal into limestone or dolomite**

Fig. 1

Underground-Injection-Control Program

With the passage of the Safe Water Drinking Act in December 1974, Congress conferred on EPA regulatory jurisdiction over injection wells through the Underground-Injection-Control Program, UIC. The Act called for complete prohibition of any underground injection after December 1977 unless authorized by permit, and required compliance with specified inspection, monitoring, recordkeeping and reporting regulations. EPA published the proposed regulations for UIC in the Aug. 31, 1978, *Federal Register.* However, after reviewing the industrial response to the proposals, EPA decided to rewrite them.

The reproposed regulations are due for publication in March 1979. Sixty days are allowed for public comments, four to six months for review of these, and final promulgation should be accomplished by the end of 1979.

Initially, 22 states (as determined by need) will be required to establish UIC programs. They have 270 days after the promulgation of regulations to do so, with a time extension available. If a state fails to qualify, EPA will implement the program.

methods. Such a stream might also contain large quantities of water and inorganic salts. Although incineration might be an acceptable method of disposal, a very large amount of energy is required to oxidize the waste stream, and the resulting scrubber operation would increase the waste volume by a factor of 4.5.

Many organics, for example ethylbenzene, either survive treatment or result in unacceptably high BOD, COD or TDS levels in the effluent. Others, such as formaldehyde, have potential odor problems and are produced in excessive concentrations. Some organics, such as *tert*-butyl alcohol, are biologically nondegradable, while others, such as ketone, are not amenable to activated-carbon adsorption.

Another group of chemicals difficult to decompose is the inorganic salts. Many sodium salts, such as sodium sulfate, inhibit treatment methods used to reduce organics concentration. The salts can be incinerated, but would then result in an additional scrubber wastestream and create a problem in solid-waste disposal. If no scrubber is used, compounds such as sodium bromide will result in toxic gases (e.g., hydrogen bromide) being emitted into the atmosphere.

The largest users of deepwell systems are the chemical, petrochemical and pharmaceutical companies, which operate about half of the active wells. Oil refineries and natural-gas plants are second, with the metals industry third. Some other users are the food, and pulp and paper industries. Table II shows an industrial classification of injection wells that was prepared for EPA in 1977.

Preinjection treatment

The purpose of pretreatment is to prevent plugging of the disposal zone formation, and damage to equip-

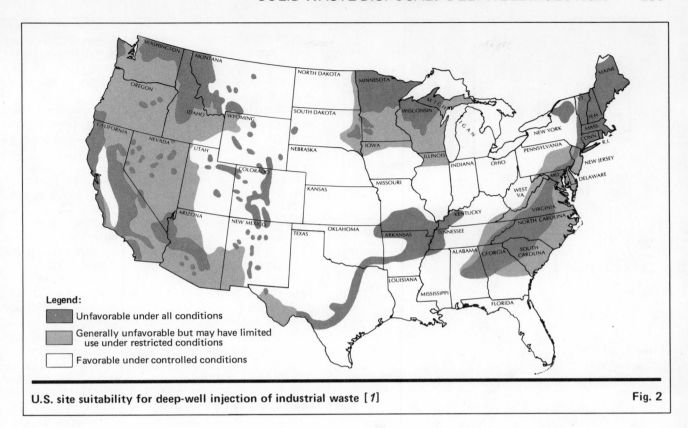

Legend:
- ■ Unfavorable under all conditions
- ■ Generally unfavorable but may have limited use under restricted conditions
- □ Favorable under controlled conditions

U.S. site suitability for deep-well injection of industrial waste [1]   Fig. 2

ment. Unless the zone is cavernous, it is highly desirable to remove by filtration and/or clarification all solids greater than one micron from the waste liquid, because accumulation of solid can plug the formation.

In many cases, dissolved waste constituents will react with formation fluid, causing precipitates that will eventually plug the formation and reduce disposal zone capacity. Such reactions can be categorized as listed below [1]:

■ Precipitation of alkaline earth metals (calcium, barium, strontium and magnesium) as relatively insoluble carbonates, sulfates, orthophosphates, fluorides and hydroxides.

■ Precipitation of metals such as iron, cadmium, zinc and chromium as insoluble carbonates, hydroxides, orthophosphates and sulfides.

■ Precipitation of products of oxidation-reduction reactions.

Waste constituents that are objectionable must be removed or neutralized through chemical treatment and liquid/solid separation. The solids and sludges that accumulate from pretreatment are periodically disposed of by sanitary landfilling.

Another method used to prevent precipitates from forming is to first inject a buffer solution into the well to serve as a barrier between the two incompatibles, the waste liquid and formation fluid. This buffer is determined from lab testing and should be the most economic fluid compatible with each of the other fluids. Mild strengths of salt water solution are most frequently found adequate. This technique is recognized as an art rather than a science, and the size of the buffer zone is usually a tradeoff between cost and estimated life of

well effectiveness. A buffer-zone volume may be several million gallons.

Well site selection

Enough geologic information is available in the U.S. to define areas in which deepwell injection is feasible and may be performed safely; areas where injection is possible but may produce detrimental effects to the environment; and areas where injection is impossible, marginal or will definitely have detrimental environmental effects. Fig. 2 presents a general mapping of U.S. site suitability for deepwell injection.

The feasibility of a well site within an acceptable area may be determined by using as disposal zone guidelines the ten criteria [1] of: (1) uniformity, (2) large areal extent, (3) substantial thickness, (4) high porosity and

Deepwell completion depths (259 wells) [1]		Table I
Depth, ft	No. wells	Percentage
0–1,000	20	7.7
1,001–2,000	56	21.6
2,001–3,000	33	12.7
3,001–4,000	34	13.1
4,001–5,000	39	15.1
5,001–6,000	44	17.0
6,001–7,000	18	7.0
7,001–8,000	12	4.6
8,001+	3	1.2

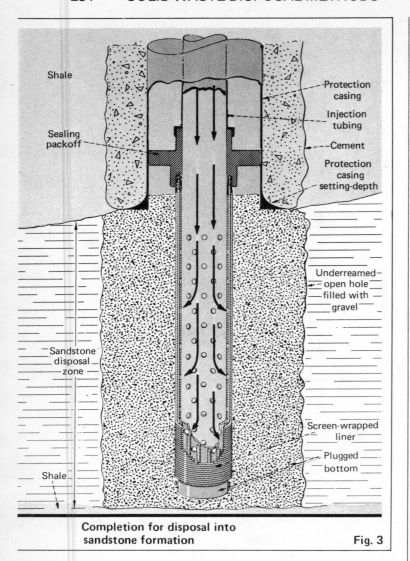

Completion for disposal into sandstone formation — Fig. 3

(Labels in figure: Shale; Sealing packoff; Protection casing; Injection tubing; Cement; Protection casing setting-depth; Underreamed-open hole filled with gravel; Sandstone disposal zone; Screen-wrapped liner; Plugged bottom; Shale)

Industrial classification of injection wells (267 wells) [7]   Table II

	Wells	Percentage
Mining (9.4%)		
Metal	2	0.7
Coal	1	0.4
Oil and gas extraction	17	6.4
Nonmetallic	5	1.9
Manufacturing (80.5%)		
Food	6	2.2
Paper	3	1.1
Chemicals and allied products	131	49.1
Petroleum refining	51	19.1
Stone and concrete	1	0.4
Primary metals	16	6.0
Fabricated metals	3	1.1
Machinery—except electronics	1	0.4
Photographics	3	1.1
Others (10.1%)	27	10.1

permeability, (5) low pressure, (6) saline aquifer, (7) separation from potable-water horizons, (8) adequate overlying and underlying aquicludes, (9) no poorly plugged wells nearby, and (10) compatibility between the mineralogy and fluids of the reservoir and the injected wastes.

A reservoir so chosen will be able to safely contain any waste that may be injected into it, provided the injected volume does not exceed the available volume of the reservoir, and injection pressures do not exceed critical formation pressures.

Oil-industry experience has shown that nearly all types of rocks may, under favorable circumstances, have sufficient porosity and permeability to yield or accept large amounts of fluid. However, sedimentary rocks— e.g. sandstone, limestone and dolomite—are the most likely to have the geologic characteristics suitable for waste injection wells, especially when deposited in a marine environment. Most limestone, dolomite, and limestone-dolomite formations are nonporous and nonproductive; however, highly fractured formations are excellent candidates for waste disposal.

The oil and gas industry has already explored and frequently developed those natural resources in practically every area where deepwell injection systems are feasible. A large number of electrical logs* are available from which to select the best disposal zones. Prospective zones are mapped from log data to verify that they meet all criteria. The depositional character of the zones can be determined from log data. It would be difficult to justify budgeting an exploratory venture while not possessing the foreknowledge, provided by log data, that such an effort was reasonably assured of being successful.

In general, the formation selected for the disposal zone will be of sandstone, limestone or dolomite and will never be free of natural contaminants. Selection of the best zone is usually simplified by a few basic considerations. First, hydrological research of the area will determine the maximum depth of waters that must be protected. As stated, the zone chosen must be at some greater depth, with sufficient thickness of impermeable formations to assure complete isolation.

Second, since greater depth means higher drilling costs, higher injection pressures, and more potential problems, the disposal zone chosen will be the shallowest that has the necessary thickness, permeability and porosity adequate for the particular waste. Should there be more than one such zone, it is generally considered a reasonable tradeoff to acquire an alternate zone in the same wellbore, in anticipation of the time when the primary completion is no longer serviceable.

Drilling and completion

The design of a given system must be based upon the type of disposal zone, its permeability and storage capacity, the volume of waste to be injected, the rate at which it is to be injected, and the type of waste.

In the drilling and completion of a disposal well, the techniques employed are the same as for drilling an oil or gas well. However, certain specifics get special atten-

---

*Recorded measurement of conductivities and resistivities down length of an uncased borehole; gives complete record of formations penetrated.

tion. Surface casing must be set below the base of the potable water and cemented full-length back to the surface (see Fig. 1). The protection casing, which is set at or through the disposal zone, must also be cemented back to the surface. This procedure assures total isolation and protection of potable water zones.

All materials of construction that may come in contact with the waste liquid must be selected based upon compatibility. This includes flow lines, valves, pumps, storage tanks, filters, wellhead valves, downhole tubing string, packer, screen-wrapped liner, casing and cementing. The annulus space in the well between the injection tubing and protection casing is filled with biocides and corrosion inhibitors for metals protection.

The drilling of the well down to the disposal zone, but not into it, is accomplished with conventional drilling fluids, which usually contain large quantities of clays, gels and barite materials. Prior to reaching the disposal zone, the fluid in the wellbore is displaced with a true solution containing no solids, so that the disposal zone cannot be contaminated with drilling fluid components. The true solution is usually a brine conditioned with a polymer to have the thixotropic properties necessary in a drilling fluid. The brine concentration is adjusted for fluid density in order to keep disposal-zone pressures under control at all times.

The disposal zone is usually cored, either conventionally or by taking side-wall samples, and these formation samples are analyzed in the laboratory to obtain as much information as possible about the zone's ability to accept the waste liquids.

It may be determined that some form of acid-stimulation pretreatment of the formation would be helpful. When this is done, the waste products of reaction are usually flushed back out of the well, so that no potentially damaging contaminants are left in the formation. Then, the completed well is usually backflowed to assure total cleanup of any possible residue from drilling. A few thousand barrels of clean formation water are collected and are, in turn, reinjected into the well under controlled pressure and rate conditions. From this, a reasonable assessment of the well's ability to accept liquids may be made. The injection rate for a well is fairly constant. However, from well to well rates vary from 75 to 1,000 gal/min, with 250 to 300 gal/min as a median rate.

Frequently, the waste disposal liquids and the formation fluid may not be compatible. In such instances, a buffer zone is created, as already discussed.

If the disposal zone is a limestone or dolomite formation, its primary mechanism for receiving waste liquids will be vugular-type porosity, plus fractures and fissures. For such formations, the well is usually completed with either perforated casing through the disposal zone—or with casing set and cemented at the top of the zone, after which the zone is drilled and left uncased. Such formations are usually highly consolidated and there is little concern for the formation rubblizing.

Sometimes, these types of disposal zones are of cavernous proportions, and in such instances the requirement for surface removal of the most minute particles from the waste is not nearly so important. (However, this type of well is the exception.)

Distribution of injection wells by state [1]		Table III
EPA Region	Wells	No. operating
Region II		
New York	4	1
Region III		
Pennsylvania	9	0
West Virginia	7	6
Region IV		
Alabama	5	2
Florida	10	4
Kentucky	3	2
Mississippi	2	1
North Carolina	4	0
Tennessee	4	2
Region V		
Illinois	8	4
Indiana	13	11
Michigan	34	21
Ohio	10	6
Region VI		
Arkansas	1	1
Louisiana	85	52
New Mexico	1	1
Oklahoma	15	10
Texas	124	57
Region VII		
Iowa	1	0
Kansas	30	21
Region VIII		
Colorado	2	0
Wyoming	1	1
Region IX		
California	5	4
Hawaii	4	1
Nevada	1	1
Total	383	209

If the disposal zone is a sandstone formation, its primary mechanism for receiving waste liquids will be the interconnecting void spaces present throughout the compacted sand bed. For sandstone, completion is more complicated.

Sand particles are consolidated because of mineralized cementation impurities. However, these cementing agents are frequently altered in some manner when contacted by waste liquids, and an unconsolidated condition results, in which sand particles may tend to move or shift. This, in turn, allows migration of fines and impurities, often resulting in a plugging action. Such a well is usually completed with a gravel-packed screen-wrapped-liner, as shown in Fig. 3.

This type of completion is accomplished by setting and cementing casing at the top of the disposal zone, drilling through the zone, and then underreaming the open-hole section to increase its diameter. By this method, the wellbore formation facing, which is in first contact with the waste liquids, will be substantially increased in area, and the tendency toward plugging up will be proportionately reduced. A liner is set in this open-hole section, and the annulus space between liner and open hole is packed with a predetermined-size gravel.

The liner may be slotted or perforated. It is externally wrapped with a special wire to ensure that the opening between successive wraps is regulated to a particular mesh, so that sand particles capable of passing through the liner are greatly restricted. The gravel tends to keep unconsolidated sand and fines from shifting and making their way back into the wellbore.

No deepwell system can be designed and installed with the expectation of a successful operating life unless the effects of all the parameters related to the reservoir, the waste and the physical dimensions of the well are incorporated into the design.

Monitoring

The deepwell system is instrumented to continuously record injection pressures and rates. Any unusual variation in either of these is the first alert to review the entire system for irregularities.

The annulus space between the injection tubing and protection casing has a pressure monitor installed. Any variation will alert the well operator to the possibility of a tubing or casing leak.

However, such problems are the exception, not the rule. Properly maintained disposal systems should experience many years of successful operation with minimal upkeep. The most outstanding example is a well more than twenty years old that can still accept waste liquid on a vacuum.

Permits

Presently, all governmental control is still at the state level; however, this is soon to change (see p. 108).

In most states, the essential components of the permit application include:

1. Justification. As a general rule, a well will not be permitted if there are other ways of treating the waste(s). (Economics are not considered in the making of this decision.)

2. Selection of well site. Usually within the confines of the existing plant property.

3. A complete hydrologic assessment of the potable water (including mapping) within a minimum 2½-mile radius.

4. A complete geologic assessment of the chosen disposal zone, including isopach and structure maps, E-W and N-S cross-sections, and a map of surface features.

5. A detailed engineering design of the well construction, including appropriate drawings.

6. A complete chemical analysis of each waste stream, including volumes.

7. A description of the process(es) from which the waste(s) are generated (trade secrets may be protected).

8. A design of the surface facility for preconditioning the waste fluids before subsurface injection.

9. An identification of every wellbore ever drilled within 2½ miles of the proposed disposal well location, an establishment of each well's present condition, and a determination as to whether the well might serve as an access from the disposal zone to the potable zones.

Note: All wells that present a possible danger must be properly corrected before a permit will be issued.

10. Disposal-zone calculations that forecast fluid and pressure frontal movement away from the disposal wellbore with respect to time and cumulative volume injected.

The permit application is completed and submitted to the appropriate state regulatory agency, which will examine it in detail and evaluate it with respect to any and all activities already ongoing or planned in the area of interest.

Assuming everything is found in order, a public hearing will be set, at which time all interested parties may give testimony or comment on the merits of the proposed project. If the regulatory agency finds that all available information weighs sufficiently in favor of the applicant, a permit will be granted.

Table III presents the distribution by state of the 383 permitted injection wells in existence as of 1977, and the operating status of the wells.

Costs

Several factors determine the economics of a deepwell system. The major ones are [1]: depth of disposal zone, proposed rate of injection, preinjection treatment required, formation injectivity, power costs, labor costs, and contract drilling costs.

Of course, costs vary with each operation, but in general, installation of a deepwell disposal system costs about three quarters of a million to a million dollars.

For a well that is one mile deep and has an injection rate of 6 million gal/mo (140 gal/min), maintenance costs are about $60,000/yr/active well and energy costs run about 2¢/gal or less, depending on utility rates. If neutralization of the waste is required prior to injection, this could cost as much as $150,000/yr/active well.

References

1. Reeder, L. R., and Associates, Review and Assessment of Deepwell Injection of Hazardous Waste, U.S. EPA, EPA/600/2-77/029A, June 1977, pp. 2, 42, 44, 48, 80 and 110.
2. Warner, D. L., et al., An Introduction to the Technology of Subsurface Wastewater Injection, U.S. EPA, EPA/600/2-77/240, Dec. 1977.
3. Galley, J. E., Subsurface Disposal in Geologic Basins—A Study of Reservoir Strata, Amer. Assn. Petrol. Geol. Memoir 10, 1968, pp. 1–10.
4. Warner, D. L., Subsurface Disposal of Liquid Industrial Wastes by Deepwell Injection, Amer. Assn. Petrol. Geol. Memoir 10, 1968, pp. 11–20.
5. Van Everdingen, A. F., Fluid Mechanics of Deepwell Disposal, Amer. Assn. Petrol. Geol. Memoir 10, 1968, pp. 32–42.
6. Donaldson, E. C., Subsurface Disposal of Industrial Wastes in the United States, U.S. Bur. Mines Inform. Circ. No. 8212, 1964.
7. Hubbert, M. K., and Willis, D. G., Mechanics of Hydraulic Fracturing, Amer. Assn. Petrol. Geol. Memoir 18, 1971.

The author

Marvin E. Smith is Technical Services Manager for Subsurface Disposal Corp., 5555 West Loop South, Bellaire, TX 77401. Telephone: 713-666-8158. Prior to joining Subsurface, he had an extensive career in all phases of oil and gas operations, including drilling, production and management. Mr. Smith received a B.S. in chemical engineering from the University of Missouri at Rolla in 1950. He is a member of API, AIME, ASME, NACE and the International Assn. of Drilling Contractors. He is a technical advisor to the EPA on its Underground Injection Control Program and is a registered engineer in Texas.

Solid-waste disposal: Incineration

Thermal decomposition greatly reduces waste volume, destroys toxic organics and affords waste heat recovery. Here are selection criteria and operating parameters for the major incineration processes.

David A. Hitchcock, BSP Div., Envirotech Corp.

☐ Incineration is an engineered process that uses thermal decomposition via oxidation to convert a waste to a less bulky, toxic or noxious material. The principal products of incineration from a volume standpoint are carbon dioxide, water and ash, while the products of primary concern due to their environmental effects are compounds containing sulfur, nitrogen, halogens and heavy metals (mercury, arsenic, selenium, lead and cadmium).

If the gaseous combustion products of incineration contain undesirable compounds, secondary treatment such as afterburning, scrubbing or filtration is required so as to lower concentrations to acceptable levels prior to atmospheric release. The solid and liquid effluents from the secondary treatment processes may require treatment prior to ultimate disposal. The variables having the greatest effect on the complete oxidation of the wastes are waste combustibility, residence time, flame temperature and turbulence in the reaction zone.

Generally, only organic wastes are candidates for incineration, although some inorganics can be thermally degraded. In the chemical process industries, incineration is most frequently used to dispose of tarry and gummy petroleum and plastic intermediate wastes, and general refuse. The combustibility of the waste is an important factor in determining the applicable incineration process. Waste combustibility is characterized by flammability limits, flash point, ignition temperature and autoignition temperature. In general, the lower these values the less severe the oxidation environment required (lower temperature and less excess oxygen). Table I lists basic waste-data considerations in the selection of an incineration system.

Incineration offers the benefits of reducing waste volume, completely eliminating harmful bacterial and viral constituents, destroying toxic organic compounds and affording the opportunity for waste heat recovery.

There are several types or categories of incinerators available today that may be used to thermally decompose chemical-process-industries wastes. These include the:

1. Multiple-hearth furnace.
2. Fluidized-bed incinerator.
3. Liquid-waste incinerator.
4. Waste-gas flare.
5. Direct-flame incinerator.
6. Catalytic combustor.
7. Rotary kiln.
8. Wet-air oxidation unit.
9. Molten-salt incinerator.
10. Multiple-chamber incinerator.
11. Ship-mounted incinerator.

Each of these devices has advantages and disadvantages that must be evaluated prior to final process selection. The major features and applications for each class of device will be identified and some of the operating characteristics and criteria presented.

Table II lists residence times and operating temperature ranges for the various incineration processes.

A matrix matching a simplified classification of waste against the different incineration processes is presented in Table III. The classification is by no means complete, and technical thermal criteria must be considered in any application. However, the matrix offers a broad picture of the types of wastes that can be handled by the various thermal systems discussed. In general, most equipment manufacturers maintain application test facilities, which are available for engineering design data-collection and process feasibility test-work.

Originally published May 21, 1979.

U.S. Resource Conservation and Recovery Act
proposed regulations for incineration (43 *Fed. Reg.* 59008)

Sect. 250.45-1 Incineration

(a) An owner/operator of an incinerator shall comply with the requirements of this Section when burning hazardous waste.

(b) Trial burns.

(1) The owner/operator shall conduct a trial burn for each hazardous waste which is significantly different in physical and chemical characteristics from any previously demonstrated under equivalent conditions. The trial burn shall include as a minimum the following determinations:

(i) An analysis of the hazardous waste for concentrations of halogens and principal hazardous components;

(ii) An analysis of the ash residues and scrubber effluent for the principal hazardous components;

(iii) An analysis of the exhaust gas for the concentrations of the principal hazardous components, hydrogen halides, CO, CO_2, O_2, and total particulates;

(iv) An identification of sources of fugitive emissions and their means of control;

(v) A measurement of combustion temperature and computation of residence time;

(vi) A computation of combustion efficiency and destruction efficiency;

(ii) A computation of scrubber efficiency in removing halogens;

(2) The results from each trial burn shall be submitted to the Regional Administrator.

(c) Monitoring. The owner/operator shall monitor and record the following in each trial burn and each operational burn:

(1) Combustion temperature;

(2) Carbon monoxide and oxygen concentrations in the exhaust gas on a continuous basis, and

(3) The rate of hazardous waste, fuel and excess air fed to the combustion system at regular intervals of no longer than 15 min.

(d) Combustion criteria.

(1) The incinerator shall operate at greater than 1,000°C combustion temperature, greater than 2 s retention time, and greater than 2% excess oxygen during incineration of hazardous waste, unless the waste is hazardous because it contains halogenated aromatic hydrocarbons, in which case the incinerator shall operate at greater than 1,200°C combustion temperature, greater than 2 s retention time, and greater than 3% excess oxygen during incineration of the hazardous waste.

(2) The incinerator shall be operated at a combustion efficiency equal to or greater than 99.9%, as defined in the following equation:

$$CE = \frac{C_{co_2}}{C_{co_2} + C_{co}} \times 100$$

Where:

CE = combustion efficiency

C_{co_2} = concentration of CO_2 in exhaust gas

C_{co} = concentration of CO in exhaust gas

Incinerators that burn waste that is hazardous only because it is listed in Sect. 250.14(b)(1) Sources generating hazardous waste, are exempt from this requirement.

Note to (d)(1) and (2)—Incinerators may operate at other conditions of temperature, retention time and combustion efficiency if the facility owner/operator can demonstrate that an equivalent degree of combustion will be provided.

(3) The incinerator shall be operated with a functioning device to automatically cut off waste feed to the incinerator when significant changes occur in flame combustion temperature, excess air or scrubber water pressure.

(e) Destruction and emission control criteria.

(1) The incinerator shall be designed, constructed, and operated to maintain a destruction efficiency of 99.99%, as defined in the following equation:

$$DE = \frac{W_{in} - W_{out}}{W_{in}} \times 100$$

Where:

DE = destruction efficiency

W_{in} = mass feed rate of principal toxic components of waste going into the incinerator (g/min)

W_{out} = mass emissions rate of principal toxic components in waste in the incinerator combustion zone (g/min)

Incinerators that burn waste that is hazardous only because it is listed in Sect. 250.14(b)(1) are exempt from this requirement.

(2) An incinerator used to thermally degrade hazardous waste containing more than 0.5% halogens shall be equipped with emission control equipment capable of removing 99% of the halogens from the exhaust gases.

(3) The incinerator shall be operated in a manner that assures that emissions of particulate matter do not exceed 270 mg/dry stand. m³ (0.12 gr/dry stand. ft³) at zero excess air. Compliance with this requirement may be achieved by having particulate emissions which, when corrected to 12% CO_2 by the formula below, are less than 180 mg/stand. m³ (0.08 gr/dry stand. ft³).

$$PE_c = \frac{PE_m C_s}{1.5 \ C_m}$$

Where:

Pe_c = corrected particulate emissions, mg/m³ (gr/dscf)

PE_m = measured particulate emissions, mg/m³ (gr/dscf)

C_s = stoichiometric CO_2 conc., ppm

C_m = measured CO conc., ppm

(4) The incinerator shall be designed, constructed and operated so that fugitive emissions of unburned hazardous waste and combustion products are controlled.

The multiple-hearth, fluidized-bed and liquid-waste incinerators can be operated under normal incineration and in a pyrolysis (oxygen starved) mode. Before reviewing the incineration processes, let us first discuss pyrolysis.

Pyrolysis (air-starved) incineration

Normal incineration requires 40–100% excess air over the stoichiometric value. Pyrolysis is theoretically a zero-air indirect-heat process similar to distillation or cracking. However, in practical application it is an air-starved process in that combustion is occurring with air levels less than the stoichiometric requirement for combustion. In pyrolysis, waste organic compounds are distilled or vaporized to form combustible gas, which is discharged from the furnace. Heat for the process is provided by the partial combustion of the pyrolysis gas within the furnace and by the combustion of elemental carbon. The unoxidized portion of the combustible gas

may be used as fuel in an external combustion chamber, with the resulting energy recovered by conventional waste-heat-boiler technology. Fixed carbon levels in the furnace ash are higher for pyrolysis than for normal incineration.

Pyrolysis is normally employed when the waste material has high calorific content. Autogenous sludges with a high calorific value (containing a gross-Btu-to-moisture ratio greater than 3,500 Btu/lb of water) are best processed in oxygen-starved conditions.

If the feed material is capable of releasing heat greater than 6,000 Btu/lb of contained water, pyrolysis is a virtual necessity to prevent the ash fusion and resulting clinker problems common in normal incineration, due to the abnormally high combustion temperatures encountered.

Nonautogenous sludges (having a gross-Btu-to-moisture ratio less than 3,000 Btu/lb of water) are often combined with high calorific organic materials to raise

Basic data considerations in selecting an incineration system [12] Table I	
Type(s) of waste	Liquid, solid, gas or mixtures
Ultimate analysis	Carbon, hydrogen, oxygen, nitrogen, water, sulfur and ash
Metals	Calcium, sodium, copper, vanadium, etc.
Halogens	Bromides, chlorides, fluorides
Heating value	Btu/lb
Solids	Size, form and quantity
Liquids	Viscosity as a function of temperature, specific gravity and impurities
Gases	Density and impurities
Special characteristics	Toxicity and corrosiveness, other unusual features
Disposal rates	Peak, average, minimum (present and future)

Operating parameters for incineration processes Table II		
	Temperature range	Residence time
Multiple hearth	600–1,000°F (drying zone) 1,400–1,800°F (incineration)	0.25-1.5h
Fluidized bed	1,400–1,800°F	Liquids and gases—seconds Solids—longer
Liquid incinerator	1,200–3,000°F	0.1-2s
Direct flame	1,000–1,500°F	0.3-0.5s
Catalytic combustor	600–1,000°F (1,500°F maximum)	1s
Rotary kiln	1,500–3,000°F	Liquids and gases—seconds Solids—hours
Wet-air oxidation	300–550°F (1,500 psig)	10-30 min
Molten salt	1,500–1,800°F	0.75s
Multiple chamber	1,500–1,800°F	Gases—seconds Solids—minutes
Pyrolysis	900–1,500°F	12-15 min

the energy release level above 3,500 Btu/lb of water. Typical high calorific waste supplements include shredded paper or wood, unusable oily wastes, and spent filter aids from organic filtration processes.

Normal incineration is used when a specific requirement exists for low levels of fixed carbon in the furnace ash, and when the higher temperatures do not create ash fusion problems.

Multiple-hearth furnace

Fig. 1 shows a multiple-hearth-furnace incineration system. Such furnaces range from 6 to 25 ft dia. and 12 to 75 ft high. The diameter and number of hearths are dependent on the waste feed, the required processing time and the type of thermal processing employed. Normal incineration usually requires a minimum of six hearths, while pyrolysis applications require a greater number.

Normally, sludge or other waste material enters the furnace by dropping through a feed port located in the furnace top. Rabble arms and teeth, attached to a vertically positioned center shaft, rotate counterclockwise to spiral the sludge across the hearths and through the furnace. The waste drops from hearth to hearth through passages alternately located either along the periphery of the hearth or adjacent to the central shaft (as shown in Fig. 1). Although the rabble arms and teeth all rotate in the same direction, additional agitation of the waste (back rabbling) is accomplished by reversing the angles of the rabble teeth. Waste retention time is controlled by the design of the rabble tooth pattern and the rotational speed of the central shaft.

Burners and combustion air ports are located in the walls of the furnace, the exact positioning being determined by evaluation of the waste's combustion requirements. Each hearth contains temperature sensors and controllers, and temperature is adjusted by burner and air port modulation.

The hearths are made of refractory, and the central shaft is cast iron and often insulated with castable refractories. The rabble arms and teeth are alloy cast-

ings. Construction materials vary in grade to suit waste requirements.

Cleaning of the multiple-hearth-furnace exhaust gas is usually accomplished by passing the hot gas from the furnace or from a waste heat boiler through a precooler, where it is cooled to the adiabatic saturation temperature by spraying fine water droplets into the hot gas stream. Normally, the adiabatic saturation temperature ranges from 170 to 190°F, depending on the water vapor content of the gas. The cooled gas then passes through a venturi throat into which additional water is sprayed. A 20 to 35-in. water column pressure-drop in the venturi throat provides the energy needed to collect fine particulates on the water droplets in the gas stream.

The gas stream with entrained water then enters a particle disengagement unit, such as a cyclonic scrubber or an impingement-plate scrubber if additional cooling is required. The latter usually consists of two or three impingement plates that pass the gas countercurrent to cold water. The gas is now subcooled (approximately 120°F) and significantly stripped of water vapor, greatly reducing total volume. After passing through an induced-draft fan, the effluent gas is discharged to the atmosphere.

Multiple-hearth furnaces operating in pyrolysis or oxygen-starved modes can handle feed materials with heat-release potentials greater than 25,000 Btu/lb of water and still maintain internal furnace temperatures between 1,200 and 1,500°F. Temperatures inside the external combustion chamber where the pyrolysis gases are burned can approach 3,000°F, greatly increasing the temperature driving force for energy recovery.

A large petrochemical company in the South recently installed two very large multiple-hearth furnaces for the disposal of waste biological sludge and chemical manufacturing residues. Each furnace, 25 ft 9 in. OD with eight hearths, incinerates approximately 25,000 lb/h of waste material, and releases approximately 50 million Btu/h. Earlier attempts using different types of equipment to incinerate the waste encountered significant problems because high concentrations of chloride and

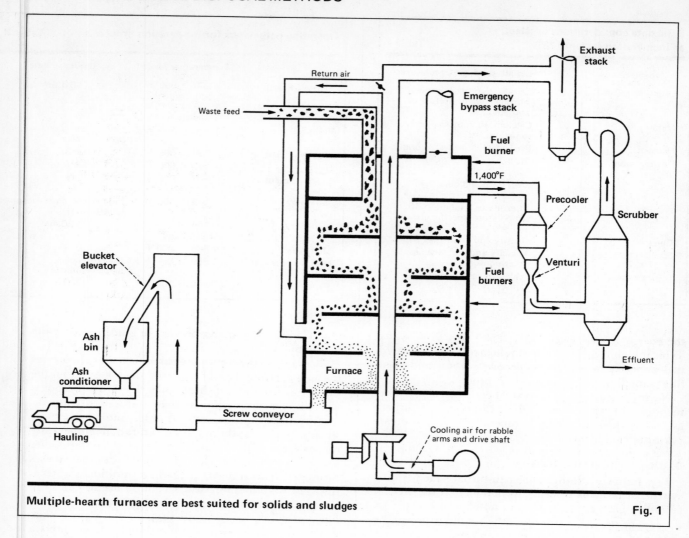

Multiple-hearth furnaces are best suited for solids and sludges   Fig. 1

calcium salts in the feed sludges produced low melting eutectic compounds. The waste compounds are now incinerated in a pyrolysis mode, with temperatures in the pyrolysis zone held below fusion levels of the ash compounds.

Fluidized bed

The fluidized-bed system shown in Fig. 2 is commonly used for incineration. The fluidized bed is a simple device consisting of a refractory-lined vessel con-

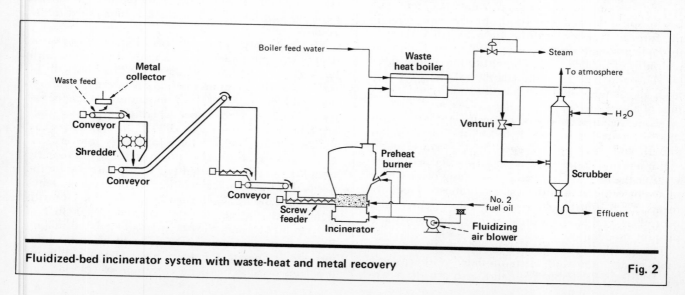

Fluidized-bed incinerator system with waste-heat and metal recovery   Fig. 2

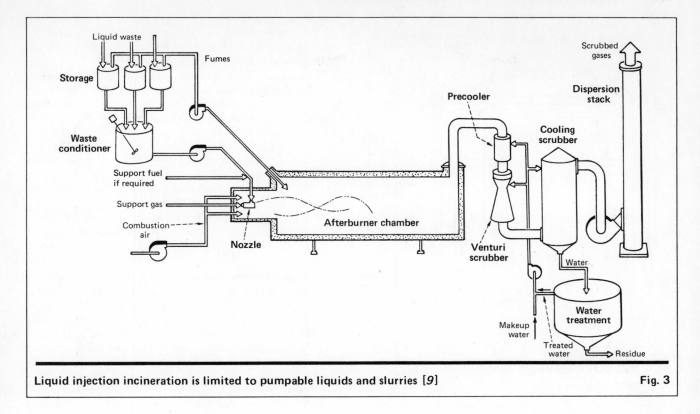

Liquid injection incineration is limited to pumpable liquids and slurries [9]   Fig. 3

taining inert granular material. Gases are blown through this material at a rate sufficiently high to cause the bed to expand and act as a theoretical fluid. The fluidizing gases are injected through nozzles that permit flow up into the bed but restrict downflow of the material.

Normally, bed design restricts combustion to the immediate area of the bed. This maintains the "free-board" area above the bed for separating the inert particles from the rising gases and for minor combustion of devolatilized components. The hot gases leave

Matrix for matching waste type with incineration processes   Table III

Waste type	Rotary kiln*	Multiple hearth*	Fluidized bed*	Liquid incinerator	Catalytic combustor	Multiple-chamber incinerator	Wet-air oxidation	Molten-salt incinerator
Solids								
Granular homogeneous	X	X	X					
Irregular bulky (Pallets, etc.)	X					X		
Low melting point (Tars, etc.)	X		X	If material can be melted and pumped				
Organic compounds with fusible ash constituents	X	X						X
Gases								
Organic vapor laden				X	X			
Liquids								
High organic strength aqueous wastes, often toxic	If equipped with auxiliary liquid injection nozzles			X			X	
Organic liquids	If equipped with auxiliary liquid injection nozzles		X	X				X
Solids/liquids								
Waste contains halogenated aromatic compounds (2,200 °F minimum)	X		X	If liquid				X
Aqueous organic sludges	Provided waste does not become sticky upon drying	X	X				X	

*Suitable for pyrolysis operation

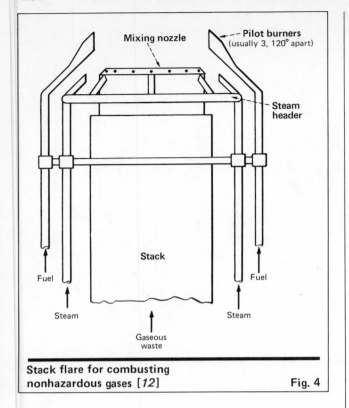

**Stack flare for combusting
nonhazardous gases [12]**   **Fig. 4**

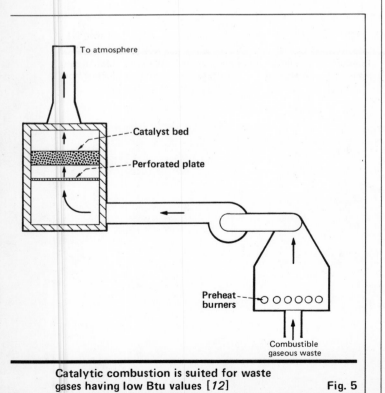

**Catalytic combustion is suited for waste
gases having low Btu values [12]**   **Fig. 5**

the fluidized bed and enter heat-recovery or gas-cleaning devices that are similar to those used with multiple-hearth furnaces and other incinerators.

Sludge or other waste feed enters the bed through nozzles located either above or within the bed. Preheat-ing of the bed to startup temperatures is accomplished by a burner located above and impinging down on the bed. If support fuel is required, it is injected through nozzles within the bed.

Because of the extremely intimate contact between combustion gases and the waste being burned, excess air for normal incineration is usually limited to approximately 40% above the stoichiometric air requirement for combustion of the waste.

Fluidized beds are subject to problems caused by low ash-fusion temperatures. These can be avoided by keeping operating temperatures below the ash fusion level or by adding chemicals that raise the fusion temperature of the ash to an acceptable level.

Experimental work is being done to permit the application of pyrolysis technology to fluidized beds.

Liquid-waste incinerators

The liquid-waste incinerator is probably the most flexible, and certainly the most laborfree, type of incinerator. The general feed requirement of such an incinerator is that the feed waste act as a liquid and exhibit a viscosity less than 10,000 SSU, the maximum allowable for pumping. Some wastes are solids at room temperature, but melt when heated and are pumpable and have viscosities that allow atomization in the liquid-waste burner. As is the case in any combustion process, effective liquid-waste incineration requires that the waste be relatively homogeneous with a reasonably constant fuel or heating value. Liquid-waste incinerators find applications ranging from complete combustion of noncombustibles, such as contaminated water, to combustion of totally organic compounds, such as waste solvents.

A typical liquid-waste incineration system is shown in Fig. 3. The heart of the system is the waste atomization device or nozzle (burner). Because a liquid combustion device is essentially a suspension burner, efficient and complete combustion is obtained only if the waste is adequately divided or atomized and mixed with the oxygen source. Atomization is usually achieved either mechanically using rotary cup or pressure atomization systems, or via gas fluid nozzles using high-pressure air or steam. For proper atomization in most burner nozzles, the liquid waste should be treated to obtain a maximum viscosity of approximately 750 SSU. Highly viscous materials are usually heated to sufficient temperature to achieve this viscosity. Other methods of lowering viscosity include production of one- or two-phase emulsions, or solubilizing the waste in a lower-viscosity liquid. The burner nozzle is mounted at one end of the refractory-lined chamber and exhaust gases exit from the other end to gas-cleaning equipment.

Liquid-waste incinerator systems are equipped with waste storage and blending tanks to ensure a reasonably steady and homogeneous waste flow. The tank system is equipped to burn the liquid-waste fumes.

Though the system in Fig. 3 is designed in a horizontal configuration, many systems are built with the combustion chamber and burner oriented on the vertical axis. Normal combustion-chamber sizing is based upon a heat release of approximately 25,000 Btu/ft³ although some highly turbulent vortex designs use heat-release rates approaching 100,000 Btu/ft³ [13]. Generally, a good design requires that the flame not impinge upon the refractory wall and that the unit operate at temperatures below the ash fusion point. Liquid-waste incinerators operate at temperatures ranging between 1,500 and 3,000°F, depending on the requirements of the process.

Gas incinerators

Waste gases are incinerated using three different approaches. Probably the most common method is the **direct-flame flare** (Fig. 4) used in most petroleum refineries. This device is effective when the waste gas is not classified as a hazardous waste and has sufficient hydrocarbon content to act as a self-supporting fuel requiring only an oxygen source. Flares have reasonable combustion efficiencies but are unable to maintain hazardous waste at combustion temperatures for a long enough time to ensure complete destruction.

Flares are located on the ground or are elevated. The tower or elevated flare is the most commonly used due to its inherent safety benefits. It is essentially an open pipe through which the gases are passed and combusted using ambient oxygen. Steam is often employed as an atomization medium to promote complete and smokeless combustion. Pilot burners are normally mounted on the flare for ignition purposes.

Waste gases containing particulates or other constituents requiring scrubbing are often incinerated in a **direct-flame incinerator.** If the waste gas does not contain sufficient hydrocarbon content to support self-

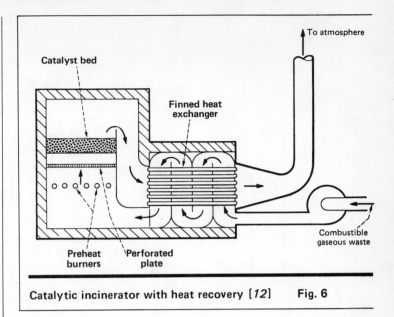

Catalytic incinerator with heat recovery [12]   **Fig. 6**

combustion, auxiliary burners and fuel are required. The waste gas can be directed to a liquid incinerator or other combustion device, depending on its volume and combustible content. Direct-flame incineration requires temperatures of 1,000–1,500°F to ensure complete combustion.

Complete combustion can be produced with 40% excess air levels in contrast to the 75% excess levels

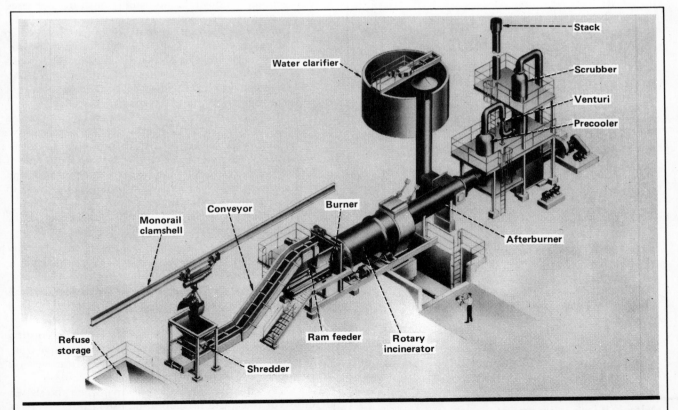

This rotary-kiln incineration system for irregular solid-wastes can also be used as a liquid incinerator   **Fig. 7**

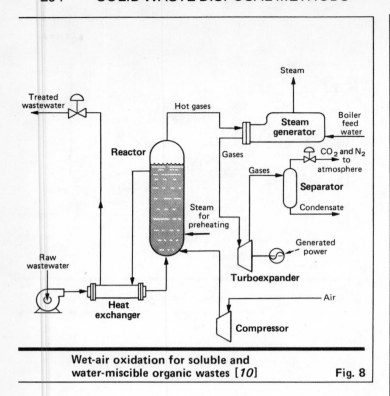

Wet-air oxidation for soluble and water-miscible organic wastes [10]   **Fig. 8**

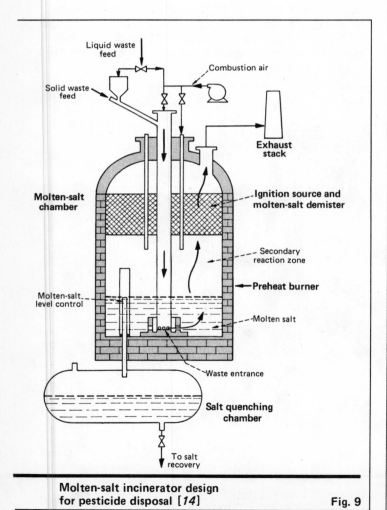

Molten-salt incinerator design for pesticide disposal [14]   **Fig. 9**

required in multiple-hearth furnaces. This greatly improves operating economy, because the reduced amount of air sharply decreases heat loss.

The third type of waste-gas combustion device is the **catalytic combustor,** shown in Fig. 5. Many waste gases consist of weak mixtures of organic compounds and air, such as fumes from paint-drying ovens and printing operations, which do not contain sufficient fuel value to enable self-combustion. Prior to the development of catalytic combustion, these gases often required direct-flame incineration temperatures approaching 2,000°F in order to achieve burner stability. The catalytic-incineration units can carry out thermal incineration of such gases at much lower temperatures, about 600–1,000°F, and therefore are much more economical to operate [12].

Catalytic incineration is limited to a maximum operating temperature of approximately 1,500°F to avoid burnout of the catalyst, which is usually made of platinum or rhenium. The engineer must be aware of the fuel value of the waste gas so that the heat of reaction of combustion is not sufficient to raise temperatures in excess of the catalyst burnout temperature-range.

Generally, catalytic incinerators are considered for operation with waste containing hydrocarbon levels that are less than 25% of the lower explosive limit. When the waste gas contains sufficient heating value to cause concern about catalyst burnout, the gas may be diluted by atmospheric air to ensure operating temperatures within the operating limits of the catalyst. However, the waste gas usually contains combustible materials at levels far below those required to support autogenous combustion and usually must be preheated to the catalytic reaction temperatures.

Catalytic-combustion systems often produce clean heated gas as a product and are well suited for waste-heat recovery units, as shown in Fig. 6. Such units significantly reduce the preheat fuel requirement. Alternatively, heat-recovery devices for the production of steam or heated ambient air may be installed.

In addition to the maximum temperature limitation of most catalysts, catalytic combustors are also sensitive to poisons, such as heavy metals, phosphates, arsenic compounds, and elemental and compounded forms of halogens and sulfur. Fouling agents such as alumina and silica dusts, iron oxides and some silicons tend to erode or foul the catalyst.

Rotary kilns

The rotary kiln is a cylindrical, horizontal, refractory-lined shell that is mounted at a slight incline. Rotation of the shell causes mixing of the waste with the combustion air. The length-to-diameter ratio of the combustion chamber normally varies between 2/1 and 10/1, and the speed of rotation is normally in the range of 1 to 5 rpm. Combustion temperatures vary according to the waste being incinerated, but normally range from 1,500 to 3,000°F. Residence times vary from several seconds to hours, depending on the waste.

Rotary kilns are especially effective when the size or nature of the waste precludes the use of other types of incineration equipment. Specifically, waste materials such as glass bottles, cardboard boxes, discarded wooden packing-cases, paper and other unmanageable solid wastes are often co-incinerated in rotary kilns. Kilns are usually designed for batch feeding, which allows for a highly flexible installation. However, continuous feeding is possible if a reasonably homogeneous feedstream is available.

Since rotary kilns are normally totally refractory lined with no exposed metallic materials of construction, they resist higher incineration temperatures better than do most other types of incineration equipment. Temperatures approaching 3,000°F, although certainly unusual, are within the capabilities of rotary equipment. Of course, high-temperature incineration (usually greater than 1,600°F) is subject to the same ash-fusion problems encountered in other types of waste incinerators.

The rotary-kiln incineration system shown in Fig. 7 was installed to meet the varying requirements of a large pharmaceutical manufacturer. The unit is fired by conventional, as well as liquid-waste burners, which enables the installation to double as a liquid-waste incinerator. Waste is delivered to the facility by dump trucks and deposited in a refuse storage area. The wastes are transferred to the shredder by a monorail clamshell and delivered to a ram feeder for introduction to the 1,600°F rotary incinerator.

The exhaust gases from the facility pass through an afterburner chamber to ensure complete destruction of organic compounds, and then through a precooler, venturi scrubber and packed column prior to discharge to the atmosphere.

Wet-air oxidation

Wet-air oxidation (Fig. 8) is a unique process that has been successfully applied to aqueous solutions containing oxidizable compounds in relatively low concentrations (from a standard combustion viewpoint). The process operates on the principle that the rate of oxidation of organic compounds is significantly increased at higher pressures. Thus, by pressurizing an aqueous organic waste (pressures approach 1,500 psi), heating it to an appropriate temperature and then introducing atmospheric oxygen, an incomplete liquid-phase oxidation reaction is produced, which destroyed most of the organic compounds.

The process exhibits varying levels of combustion efficiency, depending upon the characteristics of the waste. Wet-air oxidation is often used as a pretreatment step to destroy toxic compounds before conventional biological wastewater treatment. Oxidation efficiencies ranging between 60 and 100% are reported. The heat released in the oxidation process in the reactor is recovered in a countercurrent heat exchanger. The process becomes thermally self-sufficient when the chemical oxygen demand of the influent waste reaches a level of 20,000 to 30,000 mg/L [11].

Coke-plant waste and amiben waste are typical wastes being treated by wet-air oxidation processes [7].

Molten-salt incinerators

Molten-salt incinerators (Fig. 9) have recently been developed to pilot-plant and demonstration scale for incineration of organic-waste compounds. Usually the

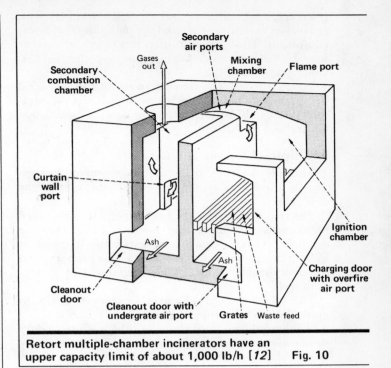

Retort multiple-chamber incinerators have an upper capacity limit of about 1,000 lb/h [12]   Fig. 10

molten-salt bath is composed of approximately 90% sodium carbonate and 10% sodium sulfate and is designed for operation in the range of 1,500 to 1,800°F. Substitution of other salts, such as potassium carbonate, allows for even lower incineration temperatures. The use of reactive salts, such as the eutectic mixtures NaOH-KOH and Li_2CO_3–Na_2CO_3–K_2CO_3, produces the additional benefit of entrapping potentially toxic or objectionable offgas constituents, such as heavy metals

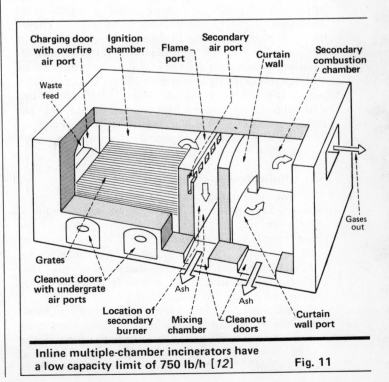

Inline multiple-chamber incinerators have a low capacity limit of 750 lb/h [12]   Fig. 11

(mercury, lead, cadmium, arsenic, selenium) [15]. This reduces or eliminates the need for pollution-abatement equipment. The spent salt often can be regenerated or may be land-disposed. Startup and support fuels include gas, oil and coal.

Wastes such as free-flowing powders and shredded materials may be directly fed to molten-salt incinerators. Waste liquids may be sprayed into the combustion air and fed to the unit.

Multiple-chamber incinerators

Multiple-chamber incinerators are generally classified as the retort type (Fig. 10) or the inline type (Fig. 11).

The retort multiple-chamber incinerator is identified by an arrangement of the chambers that forces the combustion gases to make a 90-deg change in direction in both horizontal and vertical axes. The primary- and secondary-reaction chambers are separated by a common wall.

The inline multiple-chamber incinerator is distinguished by an intermediate secondary-burner/mixing zone followed by a third chamber, which is a secondary combustion chamber. The combustion gases only change direction in the vertical plane.

Typically incinerated in multiple-chamber incinerators are plastic wastes, such as polyvinyl chloride and epoxy, acrylic and phenolic resins, and plant refuse consisting of wood, garbage, paper and rubber [12]. Generally, the multiple-chamber incinerators are considered unsuited for flowable materials, such as sludges, liquids and vapors.

Multiple-chamber incinerators are more labor intensive than other incineration equipment because of the extreme variation in the form of feed waste and the special handling that this requires.

Retort-type incinerators are considered to have an upper capacity limit of approximately 1,000 lb/h, above which effective turbulence in the mixing chamber is reduced, resulting in incomplete combustion of the vaporized waste components. Inline multiple-chamber incinerators are not well suited to applications with a waste capacity of less than 750 lb/h, due to flame propagation restrictions in these small-size units. Although the upper limit for use of inline units has not been formally established, rates in the range of 2,000 lb/h are common. Specially designed units with considerably larger capacities are possible.

Ship-mounted incinerators

To dispose of organics or organometallics that cannot be burned in conventional incinerators, specialty incineration services are available. However, most of the companies that offer custom incineration are now faced with increasing U.S. EPA scrutiny of their effluents. One viable alternative may be incineration ships that burn hazardous and toxic wastes (especially highly chlorinated organics) on the high seas.

Shell Chemical Co. has burned much of its chlorinated waste aboard the *Vulcanus* (a ship owned by Ocean Combustion Services, a subsidiary of the Hansa Line, a West German shipping company) in the Gulf of Mexico under a limited permit from EPA. EPA closely scrutinized the operation both prior to actual incineration and during the process to evaluate environmental effects. In the operation (for which a valid permit still exists), the chlorinated wastes were incinerated at a flame temperature of approximately 2,400°F. The 2,000 to 2,200°F stack gases generally contained 25 to 75 ppm carbon monoxide, 9.0 to 12.5% oxygen, and 5.2 to 6.2% hydrochloric acid, and less than 200 ppm of chlorine, all of which were within the permit specifications as reported by EPA [6]. Incineration efficiencies were reported to be in excess of 99.9%.

RCRA regulations

The U.S. Resource Conservation and Recovery Act (RCRA) proposed regulations, as written by EPA, are shown on p. 186. These proposed rules include specifications on residence time, combustion temperature, excess air, destruction efficiency, combustion efficiency, scrubber efficiency and automatic feed cutoffs.

EPA believes that incineration should be actively encouraged because it reduces or eliminates the volume and/or toxicity of waste to be land-disposed.

References

1. Lewis, C. R., and others, Incineration of Industrial Wastes, *Chem. Eng.*, Vol. 83, Oct. 18, 1976, pp. 115–121.
2. Ocean Incineration Anew, *Environ. Sci. and Technol.*, Vol. 11, March 1977, pp. 236–237.
3. Brockway, M. E. and Cheremisinoff, P. M., Understanding Incinerators, *Plant Eng.*, Vol. 32, Apr. 27, 1978, pp. 103-106 and May 11, 1978, pp. 171–174.
4. Becker, K. P. and Wall, C. J., Incinerate Refinery Waste in a Fluid Bed, *Hydrocarbon Process.*, Vol. 54, Oct. 1975, pp. 88–93.
5. Dunn, K. S., Incineration's Role in Ultimate Disposal of Process Wastes, *Chem. Eng.*, Vol. 83, Oct. 6, 1975, pp. 141–150.
6. Ricci, L. J., Offshore Incineration Gets Limited U.S. Backing, *Chem. Eng.*, Vol. 83, Jan. 5, 1976, pp. 86–88.
7. Stevens, J. I., Crumpler, E. P., and Shih, C. C., Thermal Destruction of Chemical Wastes, 71st Annual Meeting AIChE, Nov. 14, 1978.
8. Sebastian, F. P. and Cardinal, P. J., Solid Waste Disposal, *Chem. Eng.*, Oct. 14, 1968, pp. 112–117.
9. Hydro Systems for Liquid and Fume Incineration, BSP Div., Envirotech Corp., Belmont, Calif., Bulletin ESI-300-3-73 5M.
10. Wet Air Oxidation for the Chemical Process Industry, Zimpro Inc., Rothschild, Wis., Bulletin No. 1220.01, Sept. 1977.
11. BSP Rotary Process Equipment, BSP Div. Envirotech Corp., Belmont, Calif., Bulletin No. BSP-02-429-5M, Sept. 1977.
12. Ottinger, R. S., et al., Recommended Methods of Reduction, Neutralization, Recovery or Disposal of Hazardous Waste, TRW Systems Group, EPA Report PB-224 582, Aug. 1973.
13. Scurlock, A. C., et al., Incineration in Hazardous Waste Management, U.S. EPA Office of Solid Waste Management Programs, PB-261 049, 1975.
14. Yosim, S. J., et al., U.S. Patent 3,845,190, Oct. 1974, assigned to Rockwell International Corp.
15. Powers, P. W., How to Dispose of Toxic Substances and Industrial Wastes, Noyes Data Corp., Park Ridge, N.J., 1976.

The author

David A. Hitchcock is Manager, Industrial Market Development for the BSP Div. of Envirotech Corp., One Davis Drive, Belmont, CA 94002. Telephone: 415-592-4060. He is responsible for directing the application of thermal products and identifying new-product needs in processing and waste treatment. Mr. Hitchcock holds a B.S. degree in chemical engineering and a B.S. degree in business administration from Oregon State University and is a registered Professional Engineer in the State of Colorado. He is a member of the Inst. for Briquetting and Agglomeration.

Solid-waste disposal: Landfilling

Landfill disposal of both hazardous and nonhazardous waste has become more difficult and expensive than ever. This article outlines siting, designing and operational procedures for hazardous-waste landfills, incorporating concepts and constraints posed by the Resource Conservation and Recovery Act (RCRA).

William A. Duvel, Jr., Ph.D., Environmental Research and Technology, Inc.

☐ Proposed RCRA regulations will completely change the chemical process industries' land disposal practices. All landfill sites will have to meet the Subtitle D sanitary landfill criteria (which outlaw open dumps). And if a site handles hazardous waste, it will have to meet the additional, more-stringent requirements of Subtitle C (see page 230).

Basically, RCRA requires that landfills be designed, constructed and operated so that discharges are minimized or do not occur. Hazardous-waste landfilling is a very site-specific operation, requiring comprehensive preplanning to avoid costly or irreparable circumstances—such as purchasing and preparing a site and not being able to use it. Ultimately, the success of a landfill depends on the operating personnel.

Fundamental considerations

Before planning the design of a landfill or any hazardous-waste disposal facility, three fundamentals must be considered:

1. The disposal system should not interfere with process-plant operation. Disposal is essentially a service function. Good preplanning requires that:

■ The disposal site (or sites) must be large enough to accommodate all the waste to be generated at a plant. It is advisable to secure sufficient, properly located property that allows for future site development.

■ Sufficient equipment redundancy or emergency/contingency bypass plans must be provided to ensure that waste can be accepted at all times.

■ Surge capacity and equalization facilities must be built in to allow for changes in waste quantity and quality due to production and process variations.

■ The disposal system should be operable under all weather conditions. If this is impractical, sufficient storage must be provided to accommodate waste until operations can be continued.

2. The disposal area must be structurally sound, environmentally acceptable, and permanent. The landfill must not be subject to flooding, slumping, sliding or other similar external or internal displacement phenomena. It must not cause water, air or noise pollution, or endanger public health.

3. The disposal system must conform to all applicable legal and regulatory requirements, namely RCRA regulations (which include the federal permit program) and any additional or more-stringent state and local requirements.

Waste volume and characteristics

It is absolutely essential to determine (or, at least, estimate) waste volume, production rate, and physical, chemical and engineering properties. This information is required by RCRA and is necessary for calculating spatial requirements, estimating cover requirements (if any), determining personnel and equipment needs, and planning the sequence and method of filling.

It is especially important to define explosive, corrosive, reactive, toxic and volatile hazards associated with the waste, as defined by RCRA Sect. 3001 proposed regulations (43 *Fed. Reg.,* Dec. 18, 1978, p. 58954). These hazards are generally of an immediate nature and are relatively easily determined. Sect. 3004 proposed regulations prohibit, unless within specified exceptions, landfill disposal of ignitable, reactive and volatile wastes, and bulk liquids, semisolids and sludges.

Wastes must not be incompatible, that is, become explosive, corrosive, reactive or toxic, or release hazardous fumes or gases when mixed. (A table of potentially incompatible wastes is provided in Appendix I of the

Originally published July 2, 1979.

Sect. 250.45-2 Landfills.
(a) Site Selection.

1. A landfill shall be located, designed, constructed, and operated to prevent direct contact between the landfill and navigable water.

2. A landfill shall be located, designed, and constructed so that the bottom of its liner system or natural in-place soil barrier is at least 1.5 m (5 ft) above the historical high water table.*

3. A landfill shall be at least 150 m (500 ft) from any functioning public or private water supply or livestock water supply.*

(b) Construction and operation.

1. A landfill shall be located, designed, constructed, and operated to minimize erosion, landslides, and slumping.

2. A landfill shall be located, designed, constructed, and operated so that its liner system or natural inplace soil barrier is compatible with all of the waste to be landfilled.

3. The exact location of each hazardous waste and the dimensions of each cell with respect to permanently surveyed bench marks shall be recorded. The contents of each cell shall also be recorded.

4. Waste, containerized or non-containerized, that is incompatible (see Appendix I) shall be disposed of in separate landfill cells.

5. Each container of liquid hazardous waste shall be surrounded by an amount of sorbent inert material capable of absorbing all of the liquid contents.

6. The following hazardous waste shall not be disposed in a landfill:*
(i) Ignitable waste;
(ii) Reactive waste;
(iii) Volatile waste;
(iv) Bulk liquids, semi-solids, and sludges.

7. Diversion structures (e.g., dikes, drainage ditches) shall be constructed such that surface water runoff will be prevented from entering the landfill.*

8. Surface water which has been in contact with the active portions of a landfill shall be collected and treated or disposed of as a hazardous waste in accordance with requirements in this Subpart unless it is analyzed and found not to be hazardous waste as identified or listed in Subpart A or it is collected and discharged into a navigable water in compliance with a NPDES permit issued under the Clean Water Act.

9. Where gases are generated within the landfill, a gas collection and control system shall be installed to control their vertical and horizontal escape.*

10. A minimum of 15 cm (6 in.) of cover material shall be applied daily on active portions of a landfill. Active portions which will not have additional waste placed on them for at least one week shall be covered with 30 cm (12 in.) of cover material.*

11. In areas where evaporation exceeds precipitation by 20 in. or more and where natural geologic conditions allow, a landfill shall have a natural in-place soil barrier on the entire bottom and sides. This barrier shall be at least 3 m (10 ft) in thickness and consist of natural in-place soil which has a permeability $\leq 10^{-7}$ cm/s and meets the requirements of (b)(14).*

12. An owner/operator of a landfill using the design in paragraph (b)(11) or any similar design which does not have a leachate collection system shall demonstrate to the Regional Administrator, at the time a permit is issued pursuant to Subpart E, that liquids will not accumulate in the landfill to the extent that they may be discharged to the surface or to groundwater.

13. In areas where climatic and natural geologic conditions do not allow meeting the requirements of (b)(11), a landfill shall have one of the following liner systems covering the entire bottom and sides:
(i) *Design I.* The liner system shall have a slope of at least 1 percent at all points and be connected at all low points to one or more leachate collection sumps, (which meet the specifications in paragraph (b)(17)), so that leachate formed in the landfill will flow by gravity into the leachate collection sump(s) from which the leachate can be removed and treated or disposed of as specified herein. The liner system shall consist of:

(A) A soil liner which is at least 1.5 m (5 ft) in thickness and composed of natural in-place soil or emplaced soil which has a permeability $\leq 10^{-7}$ cm/s and meets the requirements of (b)(14); and

(B) A leachate collection and removal system overlying the soil liner which is at least 30 cm (12 in.) in thickness and composed of permeable soil capable of permitting leachate to move rapidly through the system and into the leachate collection sump(s).

(iii) *Design II.* The liner system shall have a slope of at least 1 percent at all points and be connected at all low points to one or more leachate collection sumps, so that leachate formed in the landfill will flow by gravity into the leachate collection sump(s) from which the leachate can be removed and treated or disposed of as specified herein. The liner system shall consist of:

(A) A leachate detection and removal system, placed on the natural base of the landfill, which shall consist of a minimum of 15 cm (6 in.) of permeable soil capable of permitting leachate to move rapidly through the system and into the collection sumps;

(B) A membrane liner system overlying the leachate detection and removal system composed of a 15 cm (6 in.) layer of clean permeable sand or soil overlaid with a synthetic membrane liner which meets the specifications in (b)(17) and which is overlaid with a 15 cm (6 in.) layer of clean permeable sand or soil;

(C) A soil liner overlying the membrane liner system which is at least 1 m (3 ft) in thickness and composed of soil which has a permeability $\leq 10^{-7}$ cm/s and meets the requirements of (b)(14); and

(D) A leachate collection and removal system overlying the soil liner which is at least 30 cm (12 in.) in thickness and composed of permeable soil capable of permitting leachate to move rapidly through the system and into the leachate collection sumps.*

14. The soils used in a soil liner or natural inplace soil barrier shall meet the following minimum criteria:
(i) Be classified under the Unified Soil Classification System CL, CH, SC and OH (ASTM Standard D2487-69),
(ii) Allow greater than 30 percent passage through a no. 200 sieve (ASTM Test D1140),
(iii) Have a liquid limit equal to or greater than 30 units (ASTM Test D423),
(iv) Have plasticity greater than or equal to 15 units (ASTM Test D424),

(v) Have a pH of 7.0 or higher (see Appendix IV), and
(vi) Have a permeability not adversely affected by anticipated waste.*

15. A synthetic membrane liner shall meet the following minimum criteria:
(i) Be of adequate strength and thickness (minimum 20 mils) to insure mechanical integrity;
(ii) Be compatible with the waste to be landfilled;
(iii) Be resistant to attack from soil bacteria and fungus;
(iv) Have ample weather resistance to withstand the stress of extreme heat, freezing, and thawing;
(v) Have adequate tensile strength to elongate sufficiently and withstand the stress of installation and/or use of machinery and equipment;
(vi) Be of uniform thickness, free from thin spots, cracks, tears, blisters, and foreign particles;
(vii) Be placed on a stable base; and
(viii) Have a permeability $\leq 10^{-12}$ cm/s or its equivalent.

16. A landfill overlying an underground drinking water source shall have a groundwater monitoring system and a leachate monitoring system as specified in § 250.43-8.

17. A leachate collection sump (as required in (b)(13)) shall be designed and constructed:
(i) Of materials both compatible with and impermeable to the leachate formed in the landfill;
(ii) So that the sump is accessible for removal of leachate if the sump pump becomes inoperative and/or the stand pipe becomes damaged; and
(iii) With a volume $\geq$ three-months expected volume of leachate but no less than 1,000 gal.

18. The owner/operator shall remove leachate from the collection sump as frequently as necessary to maintain gravity flow in the collection and removal system and shall check the collection sump at least monthly to assure compliance with this requirement.

19. Landfill liner systems and natural in-place soil barriers shall not be placed over earth materials exhibiting a permeability of greater than 10^{-4} cm/s.

(c) Closure.

1. At closure, the owner/operator of a landfill shall place a final cover over the landfill. This final cover shall consist of at least 15 cm (6 in.) of soil with a permeability $\leq 10^{-7}$ cm/s which meets the criteria of (b)(14), underlying 45 cm (18 in.) of soil capable of supporting indigenous vegetation. The top 15 cm (6 in.) shall be topsoil.

2. Where trees or other deep-rooted vegetation are to be planted on the completed landfill, the final cover shall consist of the 15 cm (6 in.) soil layer specified in paragraph (c)(1) underlying at least 1 meter (3 ft) of soil capable of supporting the deep-rooted vegetation and indigenous vegetation.*

3. The final grade of the final cover shall not exceed 33%. Where final grades exceed 10%, horizontal terraces shall be constructed. Terraces shall be of sufficient width and height to withstand a 24-h, 25-yr storm. A terrace shall be placed at every 10 ft of rise in elevation when the slope is less than 20% and at every 20 ft when the slope is greater than 20%.*

(d) Post-closure care.

1. During the post-closure period, which shall con-

*Exceptions allowed.

tinue at the landfill for a period of at least 20 years (see § 250.43-7), the owner-operator of the landfill:

(i) Shall maintain the soil integrity, slope, and vegetative cover of the final cover and all diversion and drainage structures;

(ii) Shall maintain the groundwater and leachate monitoring systems and collect and analyze samples from these systems in the manner and frequency specified in § 250.43-8;

(iii) Shall maintain surveyed bench marks;

(iv) Shall maintain and monitor the gas collection and control system where such a system is installed to control the vertical and horizontal escape of gases; and

(v) Shall restrict access to the landfill as appropriate for its post-closure use.*

2. No buildings intended for habitation shall be constructed over a landfill where radioactive waste as listed in Subpart A has been disposed.

Sect. 250.43-1 General site selection.

(a) Facilities shall not be located in an active fault zone.

(b) In accordance with Executive Order 11988, "Floodplain Management," a facility shall not be located in a "regulatory floodway" as adopted by communities participating in the National Flood Insurance Program (NFIP) managed by the Federal Insurance Administration (FIA) of the U.S. Department of Housing and Urban Development. In cases where regulatory floodways have not been designated by the FIA, the owner/operator shall obtain an analysis, using FIA-approved methods, to determine whether the facility is located within a non-regulatory floodway (i.e., a floodway which is currently not regulated by the FIA). A facility shall not be located in an area determined by the analysis to be a regulatory floodway.

(c) In accordance with Executive Order 11988, "Floodplain Management," a facility shall not be located in a "coastal high hazard area" as defined on a Flood Insurance Rate Map (FIRM) by the FIA. In cases where a coastal high hazard area has not been designated by the FIA, the facility owner/operator shall obtain an analysis, using FIA-approved methods, to determine whether the facility is located within a coastal high hazard area. A facility shall not be located in an area determined by the analysis to be a coastal high hazard area.*

(d) In accordance with Executive Order 11988, "Floodplain Management," a facility shall not be located in a 500-year floodplain.

(e) In accordance with Executive Order 11990, "Protection of Wetlands," a facility shall not be located in a wetland.*

(f) A facility shall not be located so as to be likely to jeopardize the continued existence of Endangered and Threatened Species as listed pursuant to the Endangered Species Act of 1973 (16 U.S.C., 1530 et seq.) in 50 CFR; nor result in the destruction or adverse modification of their Critical Habitat as contained in 50 CFR Part 17, Subpart F: Critical Habitat, 1760 et seq.*

(g) A facility shall not be located in the recharge zone of a sole source aquifer designated pursuant to Section 1424(e) of the Safe Drinking Water Act (Pub. L. 93-523).*

(h) Active portions of a facility shall be located a minimum of 60 m (200 ft) from the property line.*

Sect. 250.43-8 Groundwater and leachate monitoring.

An owner/operator of a landfill or surface impoundment facility shall install, maintain and operate a Groundwater Monitoring System and a Leachate Monitoring System as specified in this Section and shall comply with the Sampling and Analysis and the Recordkeeping and Reporting requirements of this Section.

(a) Groundwater monitoring system.

1. A Groundwater Monitoring System shall consist of a minimum of four (4) monitoring wells meeting the following specifications:

(i) At least one well shall be located in an area hydraulically upgradient from the active portion of the facility so as to yield samples representative of the background quality of the groundwater which flows under the facility.

(ii) A minimum of 3 monitoring wells shall be installed hydraulically downgradient of the active portion of the facility and shall be sunk to different depths in order to detect any leachate which has migrated into groundwater(s) underlying the facility property. Each well shall be constructed to draw samples from the depths where the facility owner/operator can demonstrate that contamination is most likely to occur.

(iii) At least one of the 3 wells specified in (ii) shall be located immediately adjacent to the active portion of the facility. The other wells shall be located within the property line of the facility to provide the greatest opportunity for interception of any leachate that migrates into groundwater(s) underlying the facility.

2. All monitoring wells shall be cased, and the annular space shall be backfilled with an impermeable material in order to prevent surface water from entering the well bore and interaquifer water exchange.*

(b) Leachate monitoring system.

1. A Leachate Monitoring System shall be installed within the zone of aeration underlying the facility without drilling through the bottom and side liners or soil barriers of the landfill or surface impoundment and shall be designed to collect samples in the zone of aeration between the bottom of the liner or soil barrier and the top of the water table.

(c) Sampling and analysis.

1. The background level of the quality of both the groundwater and the water in the zone of aeration underlying the facility shall be established by conducting the comprehensive analysis specified in paragraph (c)(6) on samples collected from the Groundwater Monitoring and Leachate Monitoring Systems on a monthly basis for at least one year. For a new facility, comprehensive analysis of monthly samples shall begin at least 3 months prior to the treatment, storage or disposal of any hazardous waste at the facility.*

2. After the background level has been established pursuant to (c)(1), samples shall be taken from the Groundwater Monitoring System at least once a year and analyzed pursuant to the requirements of (c)(6) and, in addition, samples shall be taken from the Groundwater Monitoring System on the following frequency and analyzed pursuant to the requirements of (c)(5):

(i) Semi-annually, if the groundwater flow rate ranges

between 25 and 50 m/year (82 and 164 ft/year) or

(ii) Quarterly, if the groundwater flow rate is greater than 50 m/year (164 ft/year).

3. After the background level has been established pursuant to (c)(1), samples shall be taken from the Leachate Monitoring System at least once a year and analyzed in accordance with the requirements of (c)(6) and, in addition, samples shall be taken from the Leachate Monitoring System at least once each quarter and analyzed in accordance with the requirements of (c)(5).*

4. If after the background levels are established pursuant to (c)(1), the analysis of samples taken pursuant to (c)(2) or (c)(3) shows that the quality of groundwater or water in the aeration zone significantly differs, as determined by the Student's t, single-tailed test at the 95% confidence level, from the background quality of these waters, the owner/operator shall:

(i) Notify the Regional Administrator within 7 days after such a finding;

(ii) Determine, if possible, the cause of the difference in quality (e.g., the result of a spill, a design failure, an improper operating procedure); and

(iii) Determine the extent of groundwater contamination or the potential for groundwater contamination and discontinue operation of the facility until the Regional Administrator determines what actions are to be taken.

5. A minimum analysis shall quantify the following characteristics of the sample:

(i) Specific conductivity, mho/cm at 25°C;

(ii) pH;

(iii) Concentration of chloride, mg/L;

(iv) Concentration of total dissolved solids, mg/L;

(v) Concentration of dissolved organic carbon, mg/L;

(vi) The concentrations of the principal hazardous constituents, or indicators thereof, found in the largest quantity in the hazardous waste disposed of in the facility, mg/L.

6. A comprehensive analysis shall quantify the following characteristics of the sample:*

(i) Those characteristics listed in paragraph (5); and

(ii) The concentrations of the contaminants and the levels of the properties listed in Appendix II, except radioactivity levels if the facility does not treat, store, or dispose of waste containing radioactive substances.

(iii) Concentration of beryllium, mg/L.

(iv) Concentration of nickel, mg/L.

(v) Concentration of cyanide, mg/L.

(vi) Concentration of phenolic compounds (as phenol) mg/L.

(vii) Presence of organic constituents as determined by a scanning by gas chromatography.

(d) Recordkeeping and reporting.

1. An owner/operator of a facility shall forward to the Regional Administrator at the end of each reporting quarter two copies of the monitoring data developed pursuant to the requirements of paragraphs (c)(2) and (c)(3) during the reporting quarter.

2. An owner/operator of a facility shall be required to retain, for a minimum of 3 years, all records of monitoring and analytical activities and data, including all original strip chart recordings and instrumentation, calibration, and maintenance records.

Properties of landfilled waste that influence long-term stability	Table I

Moisture content, %
Solids content, % by weight
Dry bulk density, lb/ft^3
Permeability, cm/s
Compression index
Shear strength
Unconfined compressive strength
Specific gravity of solids
Particle size analysis
Liquid limit
Plastic limit

proposed regulations for Sect. 3004, p. 59018.

From an engineering perspective, the important question is whether the waste material can be moved conveniently with conventional earthmoving equipment, such as bulldozer, grader or loader. If not, the properties of the waste should be changed. Such operations include: dewatering, mixing with dry sorbent (such as flyash or soil), chemical fixation, or some combination of these. Though waste need not be completely dry when placed, it must be workable and stable after being dried and/or mixed with other materials onsite. A normally convenient test for evaluating overall workability is the slump test (ASTM test C-143-74) used for concrete mixes in construction projects.

Landfilled waste must have long-term stability. In general, it is advisable to determine the properties listed in Table I. Most cases do not require all of these, but this type of information is usually useful in determining compaction capability, ability to withstand loads, and slope stability.

Site selection criteria

The two general instances of site selection are: (1) a new disposal facility in conjunction with a new plant, and (2) a new or expanded disposal facility in conjunction with an existing plant. For new plants, disposal site criteria are factored into the overall plant-siting analysis, for example, availability of water, raw materials and transportation. For existing plants, siting is essentially predetermined because conditions are fixed.

There are four broad criteria categories that influence site selection: engineering; environmental; legal, regulatory and political; and economic. Some criteria are fixed, while others are flexible. Fixed criteria indicate fatal flaws that absolutely eliminate sites. Flexible criteria can be applied more loosely and are used to rank alternative sites.

Table II lists generalized siting criteria. Fixed criteria specifically mentioned in RCRA-proposed regulations are listed in Table III. Exceptions are allowed, but these are very limited and will not apply for most situations.

The proposed regulations also include some hidden siting criteria. Section 250.45-2(b)(19), for example, requires that soils of less than 1×10^{-4} cm/s permeability be located under the landfill.

The proposed regulations apply equally to new and existing disposal facilities, with no grandfather provision provided. Thus, many existing facilities may become illegal overnight because, by their very location, they cannot meet the siting criteria.

Particular attention should be given to securing public and political acceptance for prospective sites. Siting new hazardous-waste disposal facilities is a highly sensitive and speculative venture, because of public concern over past practices such as at Love Canal (New York) and Valley of the Drums (Kentucky).

Preliminary site investigation

Landfill design begins during the siting study. Typically, the designer should review topographic, geologic and soil survey maps, aerial photographs, local experience, and individual mapping of potential sites. Most of these sources are readily available for assisting field investigation and developing a conceptual site plan.

Ordinarily, an enlarged U.S. Geological Survey 7½-min topographic map of the site is suitable for preliminary planning and conceptual design. However, mapping of the disposal site further aids planning and design, helps in the preparation of construction drawings, and is required for permit applications.

General site maps should be prepared at a scale of 1 in. = 400 ft or larger, depending on site size. Cross-sections and detailed design drawings should be prepared at a scale of 1 in. = 100 ft or larger, with 2-ft or 5-ft contour intervals, as appropriate. The site map should show all relevant features on or near the site, as listed in Table IV.

A site visit should be conducted to verify conditions described in the background literature and to observe additional site details. Site reconnaissance includes walking the entire site, noting vegetative cover, soil types, rock outcrops, springs and seeps, water-flow quantities, slope conditions and any other features necessary for a thorough evaluation. Observations should be recorded on the site maps.

Conceptual design

The conceptual design is simply the overall plan for development of the site. The two most important limiting design factors, as required by RCRA, are leachate containment by a natural or constructed liner, and control of surface water on and around the site.

Landfill development (filling) is determined by local topography, depth to groundwater, and natural clay formations that act as liners. Taking maximum advantage of topographic configurations minimizes excavation and grading.

Many landfill options are available. Fig. 1 shows a typical natural lined landfill in flat terrain. Though all details of surface and groundwater management are not shown, note that the constructed liner is vertically keyed into the naturally occurring impermeable strata, creating a bowl that isolates the area under the fill from contact with groundwater. The fill can be started on the original grade, or the intervening soil between the surface and impermeable strata can be excavated to provide more storage (as shown in Fig. 1).

Fig. 2 shows a typical natural lined landfill in hilly terrain, with the fill started on the original grade. The fill is built up against the hill, which forms one side of the containment structure.

A third possibility is shown in Fig. 3, in which a valley is used to form three sides of the containment basin. Since valleys are natural avenues for surface runoff and may have springs or seeps along the sides, control of surface and ground water is difficult. It is particularly important to route surface water around the fill and to collect and divert groundwater. Site preparation in valleys is quite complex.

Subsurface investigation

There are a variety of drilling and excavating techniques available for determining subsurface soil and rock strata location and thickness, properties of low permeable strata, and groundwater depth and quality.

Test pits are an inexpensive and rapid way of obtaining soil and rock samples over a large area. They are typically dug with a backhoe to a maximum depth of 15–20 ft. Sampling to greater depth is achieved using test borings, which are more expensive. These borings also allow for onsite testing of soils, measuring of groundwater depth and quality, and installation of monitoring wells. Determination of the number of wells, and their depth, location and method of drilling, is highly site-specific.

Site development plan

The site development plan is the detailed presentation of the conceptual design, including site preparation, startup, interim stages and final closure. The method of development should be illustrated in a series of phase drawings that include: a base map showing existing topography and initial site improvements; interim maps and/or cross-sections showing site topography, excavations and other changes at various stages of filling; detailed drawings of roads, drainage ditches, the leachate collection system, constructed liners, monitoring wells and other related facilities; and a map showing the final planned contours, and facilities to be located on the fill.

The number and complexity of drawings is a function of the size, complexity and duration of the filling operation. In general, the drawings should be sufficiently detailed so that operators will be able to understand the methods to be used. Although permit requirements under RCRA Sect. 3010 have not been fully defined, very detailed and complete drawings and specifications will be required.

The method of filling depends on site topography and waste characteristics. If the waste is reasonably dry and immediately workable, it is spread in the specified area in thin horizontal layers of 6-in. to 2-ft thickness (depending on compaction characteristics) and compacted with heavy equipment. Successive layering is continued until a reasonable working height of 10–20 ft is reached. A set of layers is termed a lift, and lifts are repeatedly made until the specified area is filled. Fig. 4 shows a typical section of this type of development.

Dry waste can also be worked as at municipal landfills. The waste is brought in by trucks across the top of the fill and dumped at the working face, where it is spread and compacted along the slope face. Develop-

---

Hazardous-waste disposal-site selection criteria   Table II

Criterion	Comment
Engineering	
Physical size	Should be large enough to accommodate waste for life of production facility.
Proximity	Should be as close as possible to production facility to minimize transport cost. Must be 500 ft from nearest water supply and 200 ft from property line.
Access	Should be all-weather, have adequate width and load capacity, with minimum traffic congestion.
Topography	Should minimize earth-moving, take advantage of natural conditions.
Geology	Must avoid areas with earthquakes, slides, faults, underlying mines, sinkholes, and solution cavities.
Soils	Should have 5-ft minimum soil between base of fill and high groundwater table, natural clay liner or clay available for liner, and final cover material available.
Environmental	
Surface water	Must be outside 100-yr floodplain; should be outside 500-yr floodplain. No direct contact with navigable water. Must avoid wetlands.
Groundwater	Base of fill must be 5 ft above high groundwater table, must avoid sole-source aquifer, should avoid areas of ground water recharge.
Air	Locate to minimize fugitive emissions and odor impacts.
Terrestrial and aquatic ecology	Must avoid unique habitat areas (important to propagation of rare and endangered species), and wetlands.
Noise	Should minimize truck traffic and equipment operation noise.
Land use	Should avoid populated areas and areas of conflicting land use (parks, scenic areas, etc.).
Cultural resources	Must avoid areas of unique archaeological, historical and paleontological interest.
Legal/public	
Legal/regulatory	Consider: RCRA and state requirements for permits; do not neglect local zoning ordinances, erosion control permits and other similar documentation.
Public/political	Must gain local acceptance from elected officials and local interest groups (Sportsman Club, local Sierra Club, Environmental Defense Fund, etc.).
Economic	
Property acquisition	Consider actual land cost plus related costs such as realtors' and attorneys' fees.
Site development	Consider excavation, grading, liner, new roads and other development costs.
Annual costs	Consider high fixed annual cost and low operating cost vs. low fixed annual cost and high operating cost. Many landfill operations are low-capital, high-operating-cost systems.
Salvage value	Do not consider salvage value, because hazardous-waste disposal site will probably not be an asset.

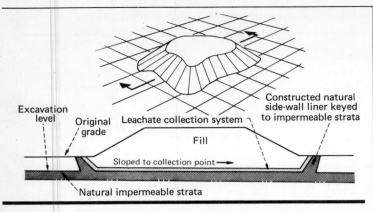

Landfill built below original grade in flat terrain   Fig. 1

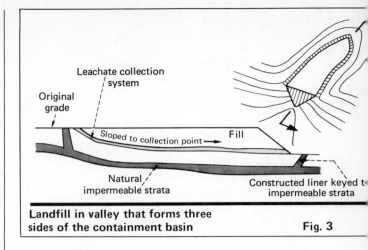

Landfill in valley that forms three sides of the containment basin   Fig. 3

ment and filling is similar to that shown in Fig. 4.

If the waste is wet and not immediately workable, alternate work areas are established. For example, one area receives wet waste, which is allowed to dry while another is worked.

Containerized waste, in 55-gal drums, for example, is placed by forklift or front-end loader into rows. The proposed regulations require each drum containing liquid to be surrounded by enough sorbent material to absorb the contents if the container should leak or burst. Therefore, disposal in this fashion requires very careful placement and covering. Of course, compaction of the drums is not desirable because of the danger of crushing them and thereby releasing the contents.

The proposed regulations require a daily natural-soil cover of 6 in. for encapsulating the waste in cells. These cells provide containment, prevent exposure to the elements, prevent fires, and eliminate the attraction of birds and rodents. Such cover may not be necessary in all cases—for example, some inorganic residues and slags are not subject to wind erosion, are not flammable and are not attractive to rodents, birds or other wildlife.

Also, incompatible materials handled at the same landfill should be placed in separate areas, eliminating the potential hazards created by mixing.

Landfills in flat terrain are built either from the bottom up or by starting at one end and working to-

ward the other. Fig. 5 shows these alternative methods of development. Generally, it is preferable to start at one end and work toward the other, because this minimizes waste exposure to the elements and may help reduce leachate.

As shown in Fig. 6, landfills on sloped terrain or in valleys are built from the upper to the lower elevation, or vice versa. Generally, the former is preferable for hazardous-waste disposal, because this method does not impound surface water, thus minimizing the potential for generating leachate.

Management of surface water

Appropriate temporary and permanent structures must be provided to control surface runoff. The site development plan must include provision for managing precipitation falling directly on the fill and for runoff flowing onto the site.

Runoff is usually captured in a diversion channel and routed completely around the site. Channel size, configuration and construction materials are generally a function of site-specific conditions. Typical construction materials are natural soil lined with grass, asphalt or rubble, and half-sections of concrete or asphalt. Split corrugated-metal pipe is popular for interim drainage. It can be placed quickly and relocated conveniently.

The proposed RCRA regulations indicate that all sur-

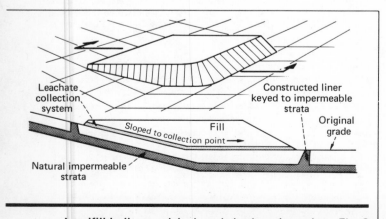

Landfill built on original grade in sloped terrain   Fig. 2

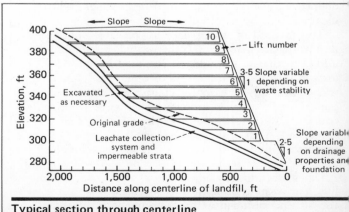

Typical section through centerline of valley landfill, showing lift details   Fig. 4

Prohibitive siting criteria specified in RCRA proposed regulations   Table III

In active fault zones
Within the 100-yr floodplain
Within a coastal high-hazard area*
Within a 500-yr floodplain*
Within wetlands*
In critical habitats*
In the recharge zone of a sole-source aquifer*
Within 200 ft of the property line of the facility*
Within 500 ft of any functioning public, private or livestock water supply*
In direct contact with navigable waters*
Within 5 ft of the historical high-water table*

*Exceptions allowed

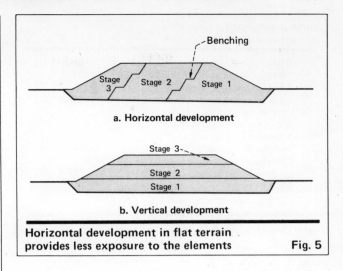

a. Horizontal development

b. Vertical development

Horizontal development in flat terrain provides less exposure to the elements   Fig. 5

face runoff must be diverted from the site. However, in many cases it does not seem practical to design a drainage system for extreme meteorological events. It may be appropriate to use a 25 or 50-year storm design to calculate channel sizes. The final promulgated regulations will probably determine storm design conditions.

The fill should be sloped and graded so that rainwater is not impounded. Drainage facilities should be sloped to allow maximum collection without erosion. Acceptable grades vary, depending on the type of material over which the water will flow. For easily erodable materials, grades as flat as 0.5–1% are appropriate.

Important features to include on site maps   Table IV

Contour lines
Property lines
Rights-of-way (power lines, rail lines, pipelines, etc.)
Access roads and roads on landfill
Weighing facilities (if any)
Existing and proposed utilities (above or below ground)
Lifts
Discharge points of groundwater
Direction of groundwater flow
Wells
Springs
Swamps
Sinkholes
Public water supplies
Fire ponds
Soil or cover stockpiles
Quarries
Sand and gravel pits
Burrow areas
Fire hydrants
Underground and surface mines
Mining spoilpiles
Gas and/or oil wells
Water towers
Buildings
Fences
Diversion ditches
Leachate collection and treatment system
Groundwater monitoring points

Runoff down steep slopes must be contained in paved ditches, rock channels or closed conduits to prevent erosion losses.

Runoff should be channeled to sedimentation ponds for collecting soil and other eroded material. The completed surfaces of the landfill should be sloped at least 2% to ensure runoff of direct precipitation into the collection system. The proposed regulations indicate that surface runoff that has been in contact with the fill will be considered as hazardous and must be treated as such until it has been determined otherwise. An NPDES permit is required for discharge of this surface runoff.

The proposed regulations require benching (terracing) at 10-ft intervals on side slopes of less than 5:1, and at 20-ft intervals on steeper slopes (3:1 maximum).

Groundwater protection

Keeping surface water from entering the fill helps protect groundwater. This is accomplished by proper siting, controlling and diverting surface runoff, prohibiting placement of excessively wet waste, and using

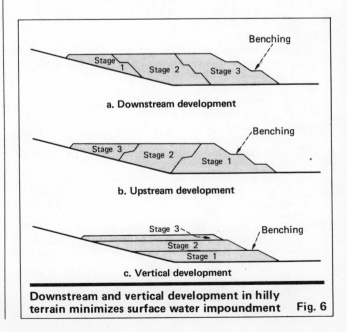

a. Downstream development

b. Upstream development

c. Vertical development

Downstream and vertical development in hilly terrain minimizes surface water impoundment   Fig. 6

Criteria influencing landfill equipment selection   Table V

Quantity and characteristics of waste being handled
Distance waste must be moved on site
Site cleaning requirements
Auxiliary tasks such as road and diversion-channel maintenance
Variation in waste quantities
Variation in weather conditions
Maintenance needs
Standby or equipment backup needs
Equipment reliability with age
Availability of parts and maintenance
Capital and operating costs
Operator comfort and safety

---

cover materials, such as soil and vegetation.

Additional protection is provided by using impermeable liners. The proposed regulations require the use of liners on the bottom and sides at all hazardous-waste landfills. They allow the use of natural inplace liners, liners constructed from natural materials, and synthetic membrane liners. The latter can be used only with a natural-liner backup. All liners must be compatible with the waste placed in the fill.

Areas having 20 in. or more of evaporation in comparison to precipitation per year must have a liner provided along the bottom and sides of the landfill, equivalent to 10 ft of natural soil having 10^{-7} cm/s

---

Factors to consider in estimating landfill costs   Table VI

I Capital requirements
1. Facility investment
   a. Direct investment
      Piping and underdrain
      Foundations and structural
      Electrical
      Instrumentation and monitoring
      Excavation and site preparation
      Buildings
      Services
      Roads
      Earthmoving equipment
      Liner
      Other equipment
   b. Indirect investment
      Engineering
      Contractor expenses and fees
   c. Contingencies
2. Other capital requirements
   a. Prepaid royalties
   b. Allowance for startup and modification
   c. Inventory capital
   d. Land
II Operating expenses
1. Raw materials (consumables, like gasoline)
2. Raw materials (nonconsumables, like drainage pipe)
3. Operating labor
4. Maintenance
5. Land preparation
6. Utilities
7. Overhead

(about 0.1 ft/yr) permeability, if geologic conditions allow. Where conditions do not allow a natural liner to be formed, one must be constructed. Two alternatives are given in the proposed regulations, and equivalent liner systems will be permitted.

RCRA Design I (Fig. 7) consists of two elements: a leachate collection system composed of permeable soil at least 12 in. thick lying immediately under the fill, and a soil liner under the leachate collection system composed of at least 5 ft of soil having a permeability no greater than 10^{-7} cm/s. The liner must be sloped to one or more collection points so that leachate can be drawn out.

RCRA Design II (Fig. 8) consists of a more complicated sandwich of materials, incorporating both a natural liner and a membrane liner. The sandwich consists of (from top to bottom): a leachate collection system (minimum 12 in. of permeable soil); a soil liner (minimum 3 ft thick, maximum 10^{-7} cm/s permeability); minimum 6 in. of clean sand; a synthetic-membrane liner; minimum 6 in. of clean sand; and a second leachate collection system. Given the highly complex nature of this system and the very high cost of installing two liners and two leachate collection systems, it is doubtful whether Design II could ever be economically justified.

These requirements and the attendant specifications for liner materials have come under vigorous attack by many different organizations and companies faced with the prospect of installing these expensive systems. While there is likely to be some change in specific requirements, it is not likely that EPA will abandon the need for a liner. Thus, it is reasonable to assume that a liner will be required at all future hazardous-waste disposal sites.

Gas collection

For wastes expected to generate gases, a venting system must be provided within the landfill. Biodegradable wastes are likely to produce methane, carbon dioxide, hydrogen sulfide and other potentially hazardous and/or malodorous gases.

Auxiliary facilities

The features already described are the primary factors in landfill design. However, several auxiliary points deserve attention:

■ Access roads—Provide all-weather access that accommodates anticipated traffic and avoids congestion and road hazards.

■ Security—Limit access to authorized personnel by providing a chain link fence (or other physical barrier) and locked gate. Prevent intrusion by wild animals, domestic livestock, children, bike-riders, hikers, and others. See Sect. 250.43-2 Security.

■ Buildings—Provide shelter for equipment maintenance and storage; and restrooms for workers (mobile trailers are adaptable for employee facilities).

■ Utilities—Provide with power, water, radio and/or telephones.

Landfill operation

Good practice dictates preparing a written landfill operating plan to accompany the design maps and drawings. The plan should include: operating instructions, traffic routing, recording and monitoring procedures, periodic maintenance schedules, contingency/emergency plans, and safety procedures.

The number of people required at a site is dependent on the operating size, the hours of operation, the type and size of equipment, and other parameters. The day-to-day success and safety of the landfill operation depends on the equipment operator and his supervisor. The instructions detailed in the landfill design should be interpreted and explained to the operating personnel so that they fully understand what is expected. The proposed RCRA regulations require formal training and documentation of the qualifications of the operating personnel, and annual review of training, especially that related to emergency and contingency procedures.

Equipment selection is important and can be confusing. Both track and rubber-tired spreading equipment (bulldozers, loaders and graders) are popular and versatile. Separate compaction equipment (roller, tamping foot or vibratory) are sometimes used to increase compaction to gain maximum material strength and area utilization. Table V lists principal selection criteria.

The landfill may be operated in wet or freezing weather, depending on location. In cold areas, equipment must be winterized, and particular care taken in driving over frozen soil or fill so that equipment is not damaged. Snow cover should be removed and not incorporated in the fill.

In wet weather, wet wastes will not dry out, equipment may become bogged down, and trucks and other site vehicles will carry mud onto approach roads. Onsite roads often become rutted and slippery, creating difficult and sometimes hazardous driving conditions. Under such circumstances it is often advisable to suspend operations temporarily.

During dry periods, dust may become a problem. Dust is generated from drying of access roads and sloughing of mud from transport vehicles. It also develops at the working face of the landfill, due to equipment activity. Hazardous-waste dust, such as that from asbestos wastes, must be strictly controlled. Nuisance dusts (nonhazardous) can be controlled by spraying with water or other dust suppressants.

RCRA-proposed regulations require contingency and emergency procedures in the event of a discharge or release of hazardous waste. As a minimum, these procedures must follow the provisions of the Spill Prevention, Control and Countermeasures Plan of the Clean Water Act. The contingency/emergency plan must include specific communication and notification procedures and equipment, arrangements with local authorities (fire and police) for notification and assistance, evacuation plans, etc.

Recordkeeping is another important aspect of the landfill operation. The proposed regulations use a philosophy of monitoring hazardous wastes from cradle to grave. This includes a requirement for recording the exact location of each hazardous waste in the landfill with respect to permanently surveyed markers. It is very important, therefore, to accurately monitor and record exactly where and when each type of waste material is placed into the fill.

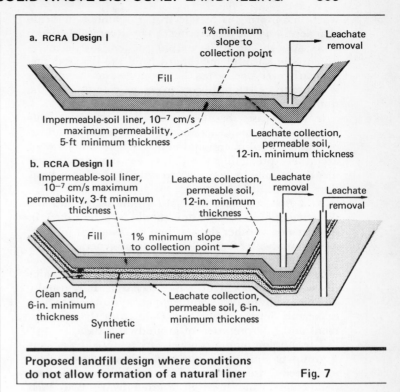

Proposed landfill design where conditions do not allow formation of a natural liner   Fig. 7

Groundwater monitoring

A hazardous-waste landfill should be fully monitored on a continuous basis so as to: protect the owner/operator against unfounded accusations, claims or suits; provide warning against improper practices or malfunctions; satisfy regulatory requirements.

The monitoring program should be initiated before disposal begins so that a natural background data-base is established, and should continue throughout the landfilling operation and for some time after site closure. The proposed regulations require monitoring for 20 years after closure, but exceptions are allowed if the owner can demonstrate that monitoring can be safely discontinued sooner. Monitoring time after closure depends on the degree of hazard associated with the waste, depth to groundwater, groundwater use and historical problems. Be aware that some groundwater problems take years to develop.

As a minimum, monitoring should include sampling and analysis of water from the monitoring wells, the runoff system and the leachate collection system. Periodic visual observation of general site conditions—looking for evidence of erosion, seepage, slumping and other similar phenomena—is also necessary. Groundwater and leachate monitoring are by far the most important. As noted previously, the proposed RCRA regulations require monitoring in (a) the aeration zone (leachate monitoring), drawing samples from the leachate collection system under the landfill, and (b) the saturated zone (groundwater monitoring), drawing samples from the groundwater adjacent to the fill.

The requirement for monitoring in the aeration zone has come under heavy attack. Opponents claim that aeration-zone monitoring methods are not sufficiently established to provide reliable data, and that groundwater monitoring is as effective. Proponents indicate that aeration zone monitoring can be accomplished and is an appropriate early warning system for detecting potentially harmful migrations from the disposal site.

Nearly everyone agrees that groundwater monitoring is appropriate. Controversy arises in determining the number, location and depth of wells.

In many cases, because of vertical and lateral diffusion and mixing, leachate migration into groundwater has a three-dimensional plume. There is no way of accurately predicting the plume location and movement, so it may be necessary to sample at various locations and depths.

As a first step, the depth to groundwater and the direction of groundwater flow should be determined. In the absence of other data, it may be assumed initially that groundwater flow follows the same general contour as the surface of the land. However, there are numerous exceptions to this rule of thumb, especially near major rivers or other areas of active recharge. It is usually advantageous to use a competent geohydrologist to help make this determination and to assist in the development and interpretation of groundwater data.

The simplest monitoring network required by RCRA is a group of four wells, one hydraulically up-gradient and three down-gradient of the landfill in a line perpendicular to the direction of groundwater flow. Well spacing and depth should be determined by local conditions. Additional wells may be necessary as the landfill is developed.

A well can be designed to sample an entire aquifer. This usually eliminates the possibility of missing a leachate plume. However, dilution could conceal the presence of leachate. If such a situation is suspected, wells should be provided in clusters to sample water at various depths in the aquifer. Samples are obtained by bailing, pumping or suction, depending on the type of well.

The frequency of sampling depends on site conditions. As a minimum, quarterly sampling is usually adequate to detect changes in groundwater conditions. The proposed regulations require monthly sampling for one year to establish background data, and then semiannual sampling if the water flowrate is 82–164 ft/yr, and quarterly sampling if the flowrate is greater than that. More-frequent sampling may be appropriate if groundwater contamination is suspected.

Parameters to look for in the groundwater samples vary with the situation. The proposed regulations require monitoring for specific conductance, pH, chloride, total dissolved solids, dissolved organic carbon, and principal hazardous substances in the waste. A more comprehensive analysis includes: all of the above plus all of the parameters listed in the primary and secondary drinking-water standards (except radioactivity) plus testing for beryllium, nickel, cyanide and phenolic compounds. And a gas chromatographic scan must be conducted.

Closure and postclosure

Closure and postclosure requirements are mandated in the proposed regulations. On closure, the landfill must be graded to maximize runoff and minimize erosion. The upper surface must be sealed with a 6-in. impermeable cap of clay of no greater than 10^{-7} cm/s permeability, and then covered with 18 in. of natural soil that is revegetated with shallow-rooted plants. Three feet of soil cover is required for deep-rooted vegetation.

Ideally, the landfill should be graded and contoured throughout its active life with a view toward closure. As various sections or areas are completed, they should be closed by covering and revegetation. This procedure minimizes wind and water erosion and leachate potential. If available onsite, cover material should be stockpiled as the landfill is being constructed so that it will be available as needed later on.

Cost

There are no textbook values indicating the cost of building and operating a hazardous-waste landfill, nor are cost curves available for rapidly estimating disposal costs. Costs vary with site conditions, waste characteristics and particular situations.

Generally, disposal costs for relatively simple systems using natural soil lining and requiring no leachate monitoring system are about $5–10/dry ton of waste disposed. More-complicated systems, requiring a constructed liner or synthetic-membrane liner, may be expected to cost from $10–50/dry ton.

Estimating the disposal cost for a particular facility is a relatively straightforward exercise in engineering economics. The major costs are earthmoving and the liner. The steps in developing the cost estimate are the same as those for any construction project. Table VI lists the principal factors to consider when estimating landfill capital and operating costs.

References

1. Engineering and Design Manual—Coal Refuse Disposal Facilities, U.S. Dept. of the Interior, Mining Enforcement and Safety Administration, 1975.
2. State Decision-Makers' Guide for Hazardous Waste Management, U.S. EPA SW-612, 1977.
3. Sanitary Landfill, Amer. Soc. of Civil Engineers, New York, ASCE—Manuals and Reports on Engineering Practice No. 39, 1976.
4. Noble, G., "Sanitary Landfill Design Handbook," Technomic Pub. Co., Inc., Westport, Conn., 1976.
5. Procedures Manual for Groundwater Monitoring at Solid Waste Disposal Facilities, EPA-530/SW-611, 1977.
6. Residual Management by Land Disposal, EPA-600/9-76-015, 1976.
7. Cedergren, H. R., "Seepage, Drainage, and Flow Nets," John Wiley & Sons, New York, 1977.
8. Kays, W. K., "Construction of Linings for Reservoirs, Tanks and Pollution Control Activities," John Wiley & Sons, New York, 1977.
9. Michael Baker, Jr., Inc., FGD Sludge Disposal Manual, Electric Power Research Institute, Inc., Palo Alto, Calif., FP-977 Project 786-1, 1979.

The author

William A. Duvel, Jr., is a Senior Program Manager at Environmental Research & Technology, Inc. (ERT), 696 Virginia Road, Concord, MA 01742. He is responsible for development and supervision of major environmental programs. He has had over ten years of experience as a consultant in wastewater and solid waste. Recently, he conducted a study for the Mfg. Chem. Assn., evaluating technical aspects of RCRA Sect. 3004 proposed regulations.

Dr. Duvel holds a B.S. in biology from Tufts University, and an M.S. and Ph.D. in environmental science from Rutgers University. He is a member of Amer. Soc. of Civil Engs., Amer. Water Resources Assn., and Water Pollution Control Federation.

Solid-waste disposal: Solidification

These processes can convert a hazardous waste into a nonhazardous substance. Of the five basic technologies, one can usually be customized to handle any specific toxic waste.

Robert B. Pojasek, Energy Resources Co.

☐ The term solidification collectively defines disposal technologies that fixate or encapsulate waste in a solid matrix end-product. Fixation techniques chemically and physically bind the waste with a solidification agent. Encapsulation methods physically surround the waste within the agent. Both techniques reduce waste permeability and produce an end-product having significant shear strength.

Solidification has been practiced in Europe and Japan for many years, but until recently it was too expensive to supplant less-environmentally-acceptable disposal methods in the U.S., being limited in application to radioactive wastes. However, the hazardous-waste-disposal requirements of the Resource Conservation and Recovery Act (RCRA) have spurred increased interest in solidification techniques because they can turn a hazardous waste into a substance that is considered nonhazardous. Before discussing this point further, let us identify the types of solidification processes and some commercial examples.

The U.S. Army Engineers' Waterways Experiment Station at Vicksburg, Miss., evaluated solidification processes for the EPA and grouped them according to the principal additive as:

- Silicate and cement-based.
- Lime-based.
- Thermoplastic-based.
- Organic polymer-based.
- Encapsulation techniques.

Silicate and cement-based processes

These generally use portland cement and other additives, such as fly ash, to form a monolithic, rocklike end-product. Most wet wastes can be mixed directly, though most vendors prefer to work with a thickened or pastelike sludge. Processing can be batch or continuous, with the end-product usually placed in the disposal site before curing.

Cement-based techniques have proven successful on many sludges generated by the precipitation of heavy metals. The high pH of the cement mixture tends to keep the metals in the form of insoluble hydroxide or carbonate salts. Metal ions may also be taken up into the cement matrix. Waste components such as asbestos, latex, metal filings and plastic can increase the strength and stability of the end-product. However, certain inorganic and organic compounds often hinder setting and curing. Special additives solve retarding and swelling problems, and increase end-product impermeability and shear strength.

Chemfix, Inc. offers a proprietary process (U.S. Patent No. 3,837,872) based on the reactions between soluble silicates and silicate setting agents (such as portland cement, lime or gypsum) that react in a controlled manner with the waste to provide a product of high stability having a rigid, friable structure. Metals in the waste are chemically fixed as silicates and hydroxides, and are protected from resolubilization by being incorporated in the impermeable matrix.

Handling and disposal is done at the generator's site by mobile units having flowrates of 300–500 gpm. The end-product is not placed in a secured landfill. To date, this process has been the most widely used technology in the U.S. for chemical waste disposal via solidification.

Stablex Corp. uses a proprietary Sealosafe technology (U.K. Patent No. 1,485,625), with the solidification process based primarily on portland cement and fly ash. The end-product has the consistency of synthetic rock. It is as impermeable as impervious clay, and ten times more impermeable than concrete. Metals are chemically fixed within an aluminosilicate matrix.

Most wastes handled by Stablex are processed at regional treatment and disposal facilities. Typically, the solidified product is used for land reclamation, a siting advantage for the disposal area.

Lime-based processes

Lime-based methods depend on the reaction of lime with fine-grained siliceous material and water to pro-

Originally published August 13, 1979.

Technical comparison of solidification processes   **Table I**

Process	Advantages	Disadvantages
Cement-based	1. Additives are available at a reasonable price. 2. Cement mixing and handling techniques are well developed. 3. Processing equipment is readily available. 4. Processing is reasonably tolerant of chemical variations in sludges. 5. The strength and permeability of the end-product can be varied by controlling the amount of cement added.	1. Low-strength cement-waste mixtures are often vulnerable to acidic leaching solutions. Extreme conditions can result in decomposition of the fixed material and accelerated leaching of the contaminants. 2. Pretreatment, more-expensive cement types, or costly additives may be necessary for stabilization of wastes containing impurities that affect the setting and curing of cement. 3. Cement and other additives add considerably to weight and bulk of waste.
Lime-based	1. The additives are generally very inexpensive and widely available. 2. Equipment required for processing is simple to operate and widely available. 3. Chemistry of pozzolanic reactions is well known.	1. Lime and other additives add to weight and bulk of waste. 2. Stabilized sludges are vulnerable to acidic solutions and to curing and setting problems associated with inorganic contaminants in the waste.
Thermoplastic	1. Contaminant migration rates are generally lower than for most other techniques. 2. End-product is fairly resistant to most aqueous solutions. 3. Thermoplastic materials adhere well to incorporated materials.	1. Expensive equipment and skilled labor are generally required. 2. Sludges containing contaminants that volatilize at low temperatures must be processed carefully. 3. Thermoplastic materials are flammable. 4. Wet sludges must be dried before they can be mixed with the thermoplastic material.
Organic polymer	1. Only small quantities of additives are usually required to cause the mixture to set. 2. Techniques can be applied to either wet or dry sludges. 3. End-product has a low density as compared to other fixation techniques.	1. Contaminants are trapped in only a loose resin-matrix end-product. 2. Catalysts used in the urea-formaldehyde process are strongly acidic. Most metals are extremely soluble at low pH and can escape in water not trapped in the mass during the polymerization process. 3. Some organic polymers are biodegradable. 4. End-product is generally placed in a container before disposal.
Encapsulation	1. Very soluble contaminants are totally isolated from the environment. 2. Usually no secondary container is required, because the coating materials are strong and chemically inert.	1. Materials used are often expensive. 2. Techniques generally require specialized equipment and heat treatment to form the jackets. 3. The sludge has to be dried before the process can be applied. 4. Certain jacket materials are flammable.

duce a hardened material known as pozzolanic concrete. The two most common additives are fly ash and cement-kiln dust. Other additives are used to enhance either material strength or impermeability. Handling is similar to that for cement-based techniques, though setting time is slower, and the end-product may require compaction during placement.

Lime-based techniques suffer from the same setting and curing problems as do cement-based ones. They are better suited for inorganic than for organic wastes. Decomposition of organic material after curing can result in increased permeability and decreased shear strength.

I.U. Conversion Systems, Inc. has a Poz-o-Tec process (U.S. Patent No. 3,785,840) that involves the precise mixing of waste products, fly ash and lime-rich materials. These additives are generally inexpensive and widely available. The end-product can be engineered to the customers' requirements. A landfill can be designed to support heavy construction, or the process can be adapted to produce road-bed aggregate, artificial reefs, mine reclamation fill, impermeable liners or capping materials.

Thermoplastic-based processes

Thermoplastic techniques mix bitumen, paraffin or polyethylene with dried waste at temperatures usually above 100°C. The mixture solidifies during cooling and is usually containerized and/or thermoplastically coated before disposal. These processes require special equipment and are relatively expensive. Organic chemicals that act as solvents toward the binder cannot be processed. Nor can wastes containing high concentrations of strong oxidizing salts, such as nitrates, chlorates and perchlorates.

This is a physical process, with the thermoplastic material adhering to the incorporated waste. The end-product resists most aqueous solutions. Contaminant migration rates are generally lower than for other solidification processes.

Werner and Pfleiderer Corp. has a process for thermoplastic binding of nuclear wastes. The firm presently uses a nonproprietary, commercially available oxidized asphalt binder. The waste stream and heated binder is mixed in a specially designed twin-screw extruder. Water content is vaporized and removed, providing considerable volume reduction for many wastes.

Matching waste types with solidification processes		Table II
Process	**Major wastes treated**	**Untreatable wastes**
Cement-based	Toxic inorganics Stack-gas-scrubber sludges	Organics Toxic anions
Lime-based	Toxic inorganics Stack-gas-scrubber sludges	Organics Toxic anions
Thermoplastic	Toxic inorganics	Organics Strong oxidizers
Organic polymer	Toxic inorganics	Acidic materials Organics Strong oxidizers
Encapsulation	Toxic and soluble inorganics	Strong oxidizers

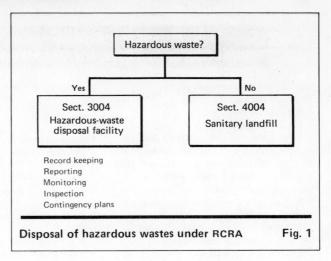

Disposal of hazardous wastes under RCRA   Fig. 1

Organic polymer processes

In these, a monomer is thoroughly mixed with the waste. A catalyst is then added to form the polymer, which is containerized prior to disposal. As with the thermoplastic binders, the polymer typically does not chemically fix the waste.

The most common polymer technique is the urea-formaldehyde process developed for nuclear waste handling. A problem may arise with contaminated water remaining after the material has solidified. However, recent developments by urea-formaldehyde users (Teledyne Corp. and ANEFCO) and chemical suppliers have alleviated this problem.

Recently, Washington State University and Dow Chemical Co. have developed a number of polyester and polyvinyl resins for use in waste solidification.

Encapsulation processes

Encapsulation is often used to describe any process in which waste particles are coated with a binder. Under an EPA contract, TRW, Inc. developed a process that uses a small amount of polybutadiene to form wastes into cubes. A layer of high-density polyethylene is then fused onto the entire surface of the cube. This seamless encapsulating jacket has exceptional resistance to water and environmental deterioration and can withstand great stress. While the cost is quite high, this method may provide adequate containment for small volumes of highly toxic materials. Stabatrol, Inc. is perfecting a vaulting technique for physically encapsulating wastes on a larger scale.

Table I compares the advantages and disadvantages of the solidification processes [1].

Wastes suitable for solidification

Each waste must be tested for solidification potential. Some wastes can be solidified directly, while others require pretreatment. Wastes that have been solidified include: steelmill, plating and lead-smelting acid wastes, cooling-tower blowdown, mine tailings, dye liquors, spent catalysts, effluent-treatment sludges, incinerator ashes, food-production sludges, spent oxides, sulfur residues and stack-gas-scrubber sludges. Table II

presents a general matchup of waste type with solidification process [1].

In most industrial sludges, toxic metals are precipitated as amorphous hydroxides that are insoluble at an elevated pH (9 to 11). By controlling pH, toxic cations can be contained. If the precipitates are made to crystallize rather than remain as gels or amorphous compounds, additional immobilization is achieved.

Anions are typically more difficult to chemically bind into insoluble compounds than are cations. Sulfates and chlorides, the two most common anions in industrial wastes, produce few insoluble compounds. Most successful attempts at solidification of anionic materials involve physical encapsulation and isolation of the salts by insoluble materials.

Most unincineratable wastes are potential candidates for solidification. Materials that are not readily solidified include oils, greases, organic solvents, paint solvents and plastic components.

Traces of certain chemical substances will accelerate or retard the solidification setting process. If setting is accelerated, the process equipment may foul. If significantly retarded, the end-product may fail to solidify. Other waste components may alter end-product characteristics. For example, high concentrations of sulfate may cause many cement-based formulations to swell excessively when exposed to water.

When is a waste hazardous?

RCRA Subtitle C, Sect. 3001—Identification and Listing of Hazardous Waste, classifies a waste as hazardous if it is on one of the lists or is ignitable, corrosive, reactive or toxic. Toxicity is determined by an extraction test. The leachate produced is analyzed in accordance with the regulations, and the results compared to certain performance standards. The current proposed standards are ten times the concentration of the parameters listed in EPA's Primary Interim Drinking Water Regulations. If any of these parameter concentrations is exceeded, the waste is hazardous and must be disposed of in accordance with Sect. 3004 (see Fig. 1).

For solidified waste, a Structural Integrity Procedure precedes the extraction test. Should the waste fracture or chip during the test, it must be ground to a $\frac{3}{8}$-in.

U.S. Resource Conservation and Recovery Act *proposed* extraction procedure for determining toxicity (43 *Fed. Reg.* 58956)

Sect 250-13 Hazardous-waste Characteristics
(d) Toxic waste.

(1) Definition—A solid waste is hazardous waste if the extract obtained from applying the Extraction Procedure cited below to a representative sample of the waste has concentrations of a contaminant that exceed the specified levels. The contaminants are arsenic, barium, cadmium, chromium, lead, mercury, selenium, silver, endrin, lindane, methoxychlor, toxaphene, 2,4-D and 2,4,5-TP Silvex.

(2) Identification method. (i) Extraction procedure. (A) Take a representative sample (minimum size 100 g) of the waste to be tested and separate it into its component phases using either the filtration method or the centrifugation method described in this section. Reserve the liquid fraction under refrigeration at 1-5°C (34-41°F) for use as described in paragraph (F) of this section.

(1) Filtration Method. [Not included]

(2) Centrifugation Method [Not included]

(B) Take the solid portion obtained in paragraph (i), and prepare it for extraction by either grinding it to pass through a 9.5-mm (⅜ in.) standard sieve or by subjecting it to the following structural integrity procedure.

Structural Integrity Procedure

Equipment: Compaction tester having a 1.25-in.-dia. hammer weighing 0.73 lb. and having a free fall of 6 in.

Procedure: 1. Fill the sample holder with the material to be tested. If the waste sample is a monolithic block, then cut out a representive sample from the block having the dimensions of a 1.3-in.-dia. × 2.8-in. cylinder.

2. Place the sample holder into the compaction tester and apply 15 hammer blows to the sample.

3. Remove the now compacted sample from the holder and transfer it to the extraction apparatus.

(C) Take the solid material from paragraph (B), weigh it and place it in an extractor. A suitable extractor will not only prevent stratification of sample and extraction fluid but also ensure that all sample surfaces are continuously brought into contact with well-mixed extraction fluid.

(D) Add to the extractor a weight of deionized water equal to 16 times the weight of solid material added to the extractor. This includes any water used in transferring the solid material to the extractor.

(E) Begin agitation and adjust the pH of the solution to 5.0±0.2, using 0.5N acetic acid. Hold the pH at 5.0±0.2 and continue agitation for 24±0.5 hr. If more than 4 mL of acid for each g of solid is required to hold the pH at 5, then once 4 mL of acid per g has been added, complete the 24-hr extraction without adding any additional acid. Maintain the extractant at 20–40°C (68–104°F) during extraction.

(F) At the end of the 24-hr extraction period, separate the material in the extractor into solid and liquid phases as in paragraph (A). Adjust the volume of the resulting liquid phase with deionized water so that its volume is 20 times that occupied by a quantity of water at 4°C equal in weight to the initial quantity of solid material charged to the extractor (e.g., for an initial weight of 1 g, dilute to 20 mL). Combine this solution with the original liquid phase from paragraph (A). This combined liquid, and any precipitate which may later form, is the Extraction Procedure Extract. [For contaminant analysis procedures, see *Fed. Reg.* 58957.]

---

mesh prior to extraction. If the waste does not break down, it is leached as a solid. If it passes the extraction test, the waste is not hazardous and may be disposed of in a Sect. 4004 landfill. (The proposed Extraction Procedure is shown in the box above. The RCRA final promulgated rules may contain some changes but the basic procedure will probably be similar.)

Therefore, though a waste may be toxic and thus hazardous prior to solidification, it will no longer be considered hazardous if after solidification the leachate extracted is not toxic. Of course, the RCRA regulations are still not promulgated, and no wastes have yet been tested. However, some precedents do exist.

The National Sanitation Foundation in Ann Arbor, Mich., under a contract with the Michigan Dept. of Natural Resources, conducted extraction tests for inorganics on Stablex Corp.'s solidification of five typical automotive plating wastes containing arsenic, cadmium, mercury, lead and nitrate. Results showed that the leachate did not contain an amount of any of the regulated chemicals greater than the maximum contaminant level proposed for hazardous waste disposal. In addition, the study reported that organics present in the waste did not hamper control of inorganics [2].

The Univ. of Louisville and Combustion Engineering, Inc. tested an adaptation of the I.U. Conversion Poz-o-Tec process for Louisville Gas and Electric Co. and concluded that the end-product had acceptable physical and environmental properties [3, 4]. Chemfix has obtained for its clients permits, for one-time use of its process on a specific waste, from more than a dozen states—including Texas and Illinois, two of the strictest. In addition, EPA has conducted tests on various commercial processes [5, 6].

Do it yourself or contract for it?

Table III is a list of commercially available materials that when properly mixed with a waste will provide solidification. Often, additives are required to enhance hardness or water resistance.

In-house solidification allows the waste generator to have more knowledge and control over the process. However, it takes time to determine the proper solidification agent for each waste stream and to optimize the formulation and handling parameters. Design and shakedown of the processing is also time-consuming. Independent laboratories can verify the properties of

---

Readily available products for solidifying waste   Table III

Generic solidification agents

Asphalt
Grout
Cement
Epoxy grout
Urea-formaldehyde
Pozzolanic cements
Plasticized sulfur

Portland cement applications

1. Type I is the normal cement used in construction.
2. Type II is recommended for use in the presence of moderate sulfate concentrations (150-1,500 mg/kg).
3. Type III has a high early strength and is recommended for use where a rapid set is required.
4. Type IV is recommended for use in large-mass concrete work due to its low heat of hydration.
5. Type V is a special low-alumina, sulfate-resistant cement recommended for use in the presence of high sulfate concentrations (>1,500 mg/kg).

U.S. companies offering solidification services	Table IV
ANEFCO	White Plains, N.Y.
Chemfix, Inc.	Kenner, La.
Chem Nuclear Systems, Inc.	Bellevue, Wash.
Chemonazone Div., Ionization International, Inc.	Chicago, Ill.
Chemical Waste Management, Inc.	Calumet City, Ill.
Dow Chemical Co.	Midland, Mich.
Dravo Corp.	Pittsburgh, Pa.
Environmental Technology Corp.	Pittsburgh, Pa.
Industrial Resources, Inc.	Chicago, Ill.
IU Conversion Systems, Inc.	Horsham, Pa.
John Sexton Landfill Contractors, Inc.	Oakbrook, Ill.
Manchak Environment, Inc.	Santa Barbara, Calif.
Nuclear Sources and Services, Inc.	Houston, Tex.
Solidtec, Inc.	Morrow, Ga.
Stabatrol, Inc.	Norristown, Pa.
Stablex Corp.	Radnor, Pa.
Stock Equipment Co.	Chagrin Falls, Ohio
System Technology Corp.	Zenia, Ohio
Teledyne Corp.	Timonium, Md.
Todd Shipyards Corp.	Galveston, Tex.
TRW, Inc.	Cleveland, Ohio
United Nuclear Industries, Inc.	Richland, Wash.
Washington State University	Pullman, Wash.
Wehran Engineering Corp.	Middletown, N.Y.
Werner and Pfleiderer Corp.	Waldwick, N.J.

the end-product so that the necessary regulatory approvals can be obtained.

Many U.S. companies offer solidification services. Most have had valuable experience handling and processing different waste streams. Services available vary from problem identification to design, construction and operation of a turnkey facility. Table IV lists companies contacted by the author that handle inquiries from industry.

Some vendors regard their process as proprietary. Others emphasize the sale of equipment and services, revealing the process chemistry to clients. Since some additives are considered proprietary, these are sold to the client at a price that includes a royalty charge.

As with other disposal methods, the claims of the vendor should be substantiated before contracting for the services. It is best to have several process vendors examine the feasibility of solidifying a particular waste. Because the end-product can be disposed outside of a secure landfill, leaching and physical testing evaluations should be obtained for peace of mind regarding future liability, and for regulatory acceptance. Independent laboratories familiar with the process used and with leaching protocols in general can provide this service.

Costs and comments

It is difficult to compare the costs of various processes. Each application is quite different and the process must be designed to satisfy the requirements of both the generator and the waste type.

An onsite facility can be either turnkey or mobile. Pricing of turnkey facilities is a function of scale and depends on the retrofitting and waste pretreatment needed. Mobile facilities are often owned and operated by a contractor. The generator pays for the service as well as the process. Mobile units may be worth purchasing for cooperatives and trade organizations that service small waste generators.

If a waste is treated at a regional facility, there will be the extra expense of transportation. However, pretreatment is often accomplished with another waste, and economy of scale usually applies.

A generator who wishes to have complete control over the waste solidification sequence can devise a generic system using a commercially available solidification agent. After some initial testing and independent verification, the generator will have an efficient controllable process. However, this alternative is not for everyone. Testing requirements may be costly if initial attempts to solidify are unsatisfactory.

References

1. Pojasek, R., ed., "Toxic and Hazardous Waste Disposal," Vol. 1 and 2, Ann Arbor Science Publishers Inc., Ann Arbor, Mich., 1979.
2. Leachate Testing of Hazardous Chemicals From Stabilized Automotive Wastes, National Sanitation Foundation, Ann Arbor, Mich.
3. Univ. of Louisville and Combustion Engineering, Inc., Environmental Effect of FGD Disposal: A Laboratory/Field Landfill Demonstration, EPA Symposium on Flue Gas Desulfurization, Las Vegas, March 1979.
4. Louisville Gas and Electric Co. and Univ. of Louisville, Physical Properties of FGC Waste Deposits at the Cane Run Plant of Louisville Gas and Electric Co., EPA Symposium on Flue Gas Desulfurization, Las Vegas, March 1979.
5. Aerospace Corp., Disposal of Flue Gas Cleaning Wastes, EPA Shawnee Field Evaluation, EPA-600/7-78-024, Feb. 1978.
6. Radian Corp., Potential Groundwater Contamination Resulting from the Disposal of Flue Gas Cleaning Wastes, Flyash and SO_2 Scrubber Sludge, 71st Annual Air Pollution Control Assn. Meeting, Houston, June 1978.

The author

Robert B. Pojasek is Laboratory Director for the Environmental Div. of Energy Resources Co., 185 Alewife Brook Parkway, Cambridge, MA 02138. His current projects include disposal of hazardous chemical wastes and potable-water source quality. An Executive Board Member of ACS, he organized and chaired its International Symposium on Disposal of Hazardous Waste, at Honolulu in April.

Dr. Pojasek has a B.S. in chemistry from Rutgers U. and a Ph.D. in chemistry from the U. of Massachusetts. He is also a member of the Water Pollution Control Federation and the New England Water Pollution Control Assn.

Disposal of flue-gas-cleaning wastes

The projected increase in use of coal for power will byproduce large quantities of ash and sulfur compounds from the cleaning of flue gases. Here is a summary of current technology for disposing of those byproducts.*

Julian W. Jones,† Industrial Environmental Research Laboratory, U.S. EPA

☐ Coal-burning power plants in the U.S. produce large quantities of wastes that accumulate in ash-collection devices or flue-gas desulfurization (FGD) systems. Since coal use is being encouraged, and since FGD systems are often necessary to reduce sulfur oxide emissions, the rate of coal-ash and FGD-waste production is expected to grow rapidly.

Coal-ash production by electric utilities is expected to reach 65 million metric tons/yr, including over 45 million metric tons of fly ash, by 1980[1]. U.S. electric-utility commitments to FGD currently total 46,425 MW of electrical generating capacity[2]. Assuming

that all of these plants will burn a typical high-sulfur Eastern coal and use limestone scrubbing, approximately 24 million metric tons/yr (dry) of waste will be produced by 1985, when all these committed plants are onstream.

Although extensive utilization of coal ash and FGD waste is practiced elsewhere,** less than 20% of the coal ash produced in the U.S. is currently used[1], and there have been no full-scale commitments by U.S. utilities to produce FGD gypsum for sale. This situation may change because of technological innovations or overriding economic concerns. However, since disposal is the most attractive choice for many utilities, the environmental, technological and economic effects of various disposal options will have to be addressed.

Disposal of coal ash, either by ponding or landfill, has been practiced in the U.S. for many years, and ash ponds at power plants overflow to rivers in many locations. Until recently, such practice was not of major environmental concern, probably because of the rela-

* Mention of tradenames, commercial products, or commercial processes does not constitute endorsement or recommendation for use.

† The author wishes to express his appreciation to Jerome Rossoff of The Aerospace Corp. and Dr. Richard R. Lunt of Arthur D. Little, Inc. for their assistance in preparation of this paper.

** In Great Britain, the Central Electric Generating Board (CEGB) for at least the past ten years has produced and marketed the bulk of 10–11 million metric tons of coal ash annually[3]. In Japan, over 1 million metric tons of gypsum were made, primarily for use in wallboard and portland cement, by FGD processes in 1975[4].

Originally published February 14, 1977.

of FGD waste disposal also involves disposal of the associated ash.

Studies of ash disposal alone have been undertaken by the U.S. Environmental Protection Agency[5] (EPA) and the Electric Power Research Institute[6] (EPRI); and the Tennessee Valley Authority (TVA) is currently conducting additional studies under EPA sponsorship[7]. This article will not address these studies, but will rather be limited to studies for the disposal of FGD wastes. Since these wastes do contain some fly ash, they will hereafter be referred to as FGC (flue gas cleaning) wastes.

Current FGC waste-disposal practices

Most of the U.S. utility companies' installations of nonregenerable (e.g., lime/limestone) scrubbing systems have used disposal methods similar to those for ash disposal—e.g., ponding. The earliest systems employed unlined disposal ponds, but because of concerns about possible effects on groundwater quality, low-permeability clay-lined ponds have been provided at some of the more recent installations. In addition, chemical treatment (fixation) of the FGC wastes has been chosen by some utility companies because of (1) expected improvement in structural properties and (2) resistance to leaching [8].

Federal regulations or guidelines for disposal of FGC wastes have not yet been issued. However, the Federal Water Pollution Control Act Amendments of 1972 and the Safe Drinking Water Act of 1974 both indicate that the desire of Congress is to eliminate or minimize chemical pollutant discharges into surface waters and groundwaters[9]. In addition, the Resource Conservation and Recovery Act of 1976, signed into law last October, now requires EPA to establish regulations or guidelines for disposal of wastes from air pollution control devices[10].

Recognizing the need for better definition of approaches to meet this legislative challenge, EPA in 1972 initiated a major program of research and development in the area of FGC waste disposal, as a follow-on to modest efforts initiated in the late 1960s in support of the limestone scrubbing program. The primary objectives of the 1972 program were to better quantify any potential environmental problems associated with FGC waste disposal, to assess current FGC waste-disposal technologies, and to recommend alternate disposal approaches.

Initial efforts under the program indicated that disposal techniques that minimized or eliminated the potential for water pollution were required. At that time, the best available techniques were landfilling of chemically fixed FGC waste, or disposal of untreated waste in ponds lined with low-permeability clay, or plastic or rubber. Detailed results of this program have been described[11,12].

Current EPA R&D

In late 1974, plans were formulated to greatly expand the R&D on FGC waste as part of EPA's Energy Research Program. These efforts included improved quantification of environmental parameters and were also aimed at reducing costs, investigating a broader

tively low solubility of the chief chemical constituents in most coal ashes. However, from the time that FGD systems were introduced, there was major concern about disposal of the wastes or "sludges," because of: the large amount of occluded water in these wastes; the wastes' quite variable physical and chemical properties; and the soluble and slightly soluble constituents in the water.

As a result of this concern, considerable research and development has been undertaken by governmental and private organizations. The major thrust of these efforts has centered on FGD wastes. However, since such wastes contain varying amounts of fly ash, depending on whether the ash is collected separately or simultaneously with the sulfur oxides in the flue gas, any study

range of alternative disposal strategies, and encouraging byproduct usage. The program, entitled "FGC Waste and Water Program," currently consists of the 19 projects listed in Table 1. Of these, three ongoing projects* have produced most of the results summarized below.

Pond disposal of FGC wastes

Historically, the most frequently used alternative for disposal of FGC wastes (including coal ash) has been ponding. In most cases, the slurried solids are pumped. The slurry comes either from the bottom outlet of a clarifier or directly from the scrubber, with the pond acting as the settler/clarifier. Clarified liquor is normally returned to the scrubber system.

Two major environmental considerations are inherent to ponding FGC wastes: (1) the water-pollution potential of soluble materials, and (2) the land-degradation potential of nonsettling, or physically unstable, waste solids.

The water pollution potential of FGC wastes is primarily dependent on the chemical properties of the wastes but is also affected by the physical properties (e.g., permeability) that affect the rate of leachate generation. These properties vary widely for different power-plant and scrubber operations; important variables include the sulfur and ash content of the coal, the type of scrubber system (lime, limestone, dual alkali), the amount of unreacted alkali in the waste, the amount of ash in the waste, the degree of dewatering of the waste, and the degree of oxidation of sulfites to sulfates. Pollution potential can be attributed to soluble trace metals, chemical oxygen demand (COD) from the sulfites, excessive total dissolved solids (TDS), and excessive levels of other major chemical species (sulfate, chloride, calcium, magnesium, sodium).

An example of the pollution potential from ponding untreated FGC wastes is shown in Table II, which compares some of the results of leachate analyses from EPA field-disposal tests [13] (currently underway) with the EPA proposed National Interim Drinking Water Regulations[14] and the 1962 U.S. Public Health Service Drinking Water Standards, both conservative criteria.

The concentrations in these analyses come very near to the maximum observed in the leachate from untreated wastes of the pond and, as expected, reflect the concentrations of the constituents in the raw waste liquor. After reaching these maximums, these concentrations have tended to diminish, indicating a flushing out of the solubles in the waste. Laboratory data have shown essentially complete flushing of solubles after five pore-volumes of leachate have been generated[15]. The long-range pollution potential, after this initial flushing takes place, is represented primarily by the solubility of calcium sulfate (gypsum). This indicates that some form of control must be considered to limit

* (1) FGC-Waste Characterization, Disposal Evaluation, and Transfer of FGC-Waste-Disposal Technology (the Aerospace Corp.)
(2) Shawnee FGD Waste Disposal Field Evaluation (TVA/The Aerospace Corp.)
(3) Evaluation of Alternative FGD Waste Disposal Sites (A.D. Little, Inc.)
The TVA and Aerospace Corp. efforts in the first two projects address primarily ponding and landfill of FGC wastes; the A.D. Little efforts focus primarily on the feasibility of mine and ocean disposal.

Comparing leachate with drinking water			Table II
Constituent	Selected pond a* leachate analyses, mg/l	National interim drinking water regulations, mg/l	1962 U.S. public health service drinking water standards, mg/l
Arsenic	0.08	0.05	0.05
Mercury	0.0067	0.002	0.002
Selenium	0.013	0.01	0.01
Lead	0.062	0.05	0.05
Sulfate	980	—	250
Chloride	3,500	—	250

*Contains untreated wastes from a lime-scrubbing FGD system

the entry of major species and trace elements into public water supplies.

If selected properly, the composition of the bottom and sides (or dikes) of a pond can reduce the pollution potential of FGC wastes by (1) attenuating the pollutant migration into the ground via ion exchange or adsorption, and/or (2) greatly reducing the rate of pollutant migration because of very low permeability (e.g., 10^{-6} cm/s). Such bottoms and sides may consist of the naturally occurring soil that makes up the disposal site, or of a natural liner material, such as placed and compacted clay, or of a synthetic liner. Although impermeable linings can alleviate the pollution potential of the operating FGC waste pond, the retention of water is of concern in reclaiming the disposal site when the pond is retired.

Physical instability of untreated FGC wastes is primarily caused by small platelet or needle-like crystals of calcium sulfite hemihydrate, which settle very slowly and trap water even when filtered. Pond-settling untreated sulfite-rich FGC waste solids typically results in a concentrated slurry of less than 50% solids[9]. At this concentration, these wastes exhibit "thixotropic" behavior†, so that dewatering to about 65–70% solids is necessary in most cases to achieve physical stability, and optimum compaction is achieved at about 80% solids[16]. Several approaches (e.g., vacuum filtration) may be taken to increase the final percentage of solids to greater than 50%.

However, typical FGC wastes are generally very difficult to pump at greater than 50% solids[17]. Since the pond is used as the primary dewatering mechanism, only those approaches that would greatly improve the settling properties of the solids are applicable. Examples include: process modifications to increase the size of the sulfite crystals, oxidation to produce gypsum,** and chemical treatment (e.g., the Dravo Corp. proprietary Synearth process).

Decreasing the moisture in FGC wastes can make possible future reclamation of the pond. However, if a hydraulic head, created by decanted waste liquor plus rainwater, is permitted to remain indefinitely on top of

† FGC wastes are not truly thixotropic materials, the viscosity of which depends not only on the rate of shear but the time over which the shear force is applied (i.e., they become stable again after the shear force is removed); FGC wastes generally remain "liquified" once rendered to that state, unless further dewatered.
** FGC wastes from low-sulfur Western coal-fired plants primarily contain gypsum as the reaction product.

the wastes, soluble constituents will be forced into the ground water below, if the pond is not reclaimed.

Ponding costs can vary considerably, depending on the technical design of the disposal system and on site-specific factors such as labor costs and costs for delivery of liner materials. Preliminary estimates by The Aerospace Corp. have indicated ponding costs varying from $5.50 to $10 per metric ton of dry solids, excluding pond reclamation costs[16].

Landfill disposal of FGC wastes

Coal ash has generally been dispoed of as landfill, where ponding is impractical. Although landfill disposal has also occurred with FGC wastes (e.g., at Duquesne Light Co.'s Phillips Station), the primary motive has been to avoid the land-deterioration and water-pollution problems associated with ponding.

For landfill disposal, dewatered or treated FGC wastes are hauled to the disposal site, or waste slurries are pumped to the site for dewatering or treatment there. The dewatered wastes are placed and compacted. Although sanitary landfill procedures (i.e., daily covering) are not necessary, exposure to rainfall should be at a minimum so as to reduce leaching and (in the case of untreated wastes) possible reslurrying. The disposal site should be designed so that a minimum of runoff water enters it. Since some runoff will probably be generated around the site, trenches for collection and recycle or treatment should be provided.

Finally, FGC wastes must have enough physical stability to be handled by earth-moving equipment for landfill disposal. This means dewatering the material to at least 50% solids and possibly as high as 65–70% solids by mechanical, thermal[9] or chemical means, or a combination of these.

There are four methods for improving dewatering: (1) adding dry solids, (2) improving the dewatering equipment, (3) controlling crystal size; and (4) oxidizing to obtain gypsum. They follow approximately that order of increasing complexity. (The above statement applies to medium- and high-sulfur applications; FGC wastes from low-sulfur Western-coal-fired plants primarily contain gypsum as the reaction product.)

One of the easiest ways to dewater FGC wastes is by adding dry solids, such as fly ash. Gravity settling followed by vacuum filtration results in a maximum of only 50–60% solids with many FGC wastes; but if an equivalent amount of dry fly ash is added to a 55% solids material, an increase to about 70% solids is obtained.

Another approach improves the gravity-settling equipment so that it can produce 65–70% solids after vacuum filtration. This possibility is being examined in an EPA-sponsored study at Auburn University (Grant No. R804531-01-0), where improved gravity-settler design studies, as well as filtration and pumpability tests, will be conducted.

Another approach increases the size of calcium sulfite hemihydrate crystals. Radian Corp. has developed a process (patents applied for) that controls crystal size through the use of mechanical equipment in conjunction with lime/limestone scrubbing.

One outgrowth of such scrubbing is strong evidence that dual-alkali technology offers another approach through which crystal size can be controlled. In this case, solids are produced in a special reactor outside the scrubber loop[18].

Increasing the reaction-product crystal size is also one of the purposes of oxidizing FGC waste to gypsum, since gypsum crystals are typically much larger and thicker than sulfite crystals, and therefore settle quicker and trap less water.

Still another way to "dewater" FGC wastes is by one of several proprietary chemical-treatment or "fixation" processes offered by firms such as IU Conversion Systems, Inc., Dravo Corp., and Chemfix, Inc. Whereas all these companies' processes are designed to produce a physically stable material with reduced solubility, the processes and products are quite different. Each has treated FGC wastes from large (10 MW equivalent) EPA lime/limestone pilot scrubbers at TVA's Shawnee Steam Plant. Preliminary results from field disposal tests on the treated materials have recently been reported[13,15].

In general, the results show physical stability, greatly increased compressive strength, lower permeability, and about 50% lower concentrations of major solubles* (e.g., chloride and sulfate) in the leachate, when compared to untreated wastes from the same pilot scrubbers. Total costs for treatment and disposal of the Shawnee-type waste (high-sulfur coal, fly ash collected in the scrubber) were estimated at $8.10–$12.70 per (dry) metric ton of waste[13,15].

Mine disposal of FGC wastes

The concept of coal-mine disposal of FGC waste has greatly interested engineers in the flue-gas-desulfurization industry for many years. The attraction was created by the existence of railroad transportation between the coal mine and the power plant, and the need for material to fill the void left by mining of the coal. In addition, many plants may not have sufficient land area for on-site disposal.

However, studies for disposal of FGC waste in mines (e.g., the U.S. Bureau of Mines Contract S0144135 with Radian Corp. to examine the use of these wastes to prevent underground coal-mine subsidence, and the current EPA Contract 68-03-2334 with Arthur D. Little, Inc.) have only recently been undertaken.

The contract with A. D. Little is designed in part to assess the technical, environmental and economic factors associated with mine disposal of FGC wastes through engineering studies as well as laboratory and pilot-scale experiments. The preliminary assessment phase of the effort has essentially been completed, and some initial conclusions are possible.

An initial review identifies four general categories of mines as the most promising candidates: active surface-area coal mines; active underground coal mines; inactive or mined-out portions of lead/zinc mines; and inactive or mined-out portions of active underground limestone mines. Each category was reviewed with regard to: the alternatives for placement; the physical properties of FGC wastes that would be suitable; the

---

* Trace-element concentrations have not consistently shown a comparable decrease.

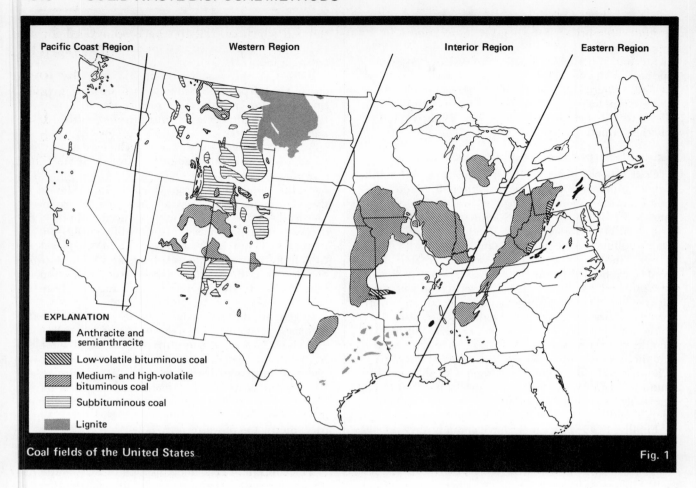

Coal fields of the United States

EXPLANATION

■ Anthracite and semianthracite

▨ Low-volatile bituminous coal

▨ Medium- and high-volatile bituminous coal

▤ Subbituminous coal

▦ Lignite

Pacific Coast Region   Western Region   Interior Region   Eastern Region

Fig. 1

environmental/technical (e.g., operational) impacts; the capacities; and the availability and accessibility (via transportation systems) for FGC waste disposal. As a result of this review, the following mines were determined most promising:

1. Active Interior* surface-area coal mines.

2. Active Eastern* and Interior* room-and-pillar underground coal mines.

In general, Interior surface-area coal mines appear to be more promising than Western (Rocky Mountain and Pacific Coast) surface-area coal mines. However, surface-area mines both in the Interior and the West were considered much more promising than Eastern surface contour mines, because of the latter's low capacity for FGC wastes and the difficulty for waste placement. The problem with the Western surface-area mines is the greater potential for groundwater contamination, relative to Interior mines, because of a lower net annual rainfall, resulting in less dilution of leachate. In addition, FGC wastes from Western plants burning low-sulfur coal normally contain a higher concentration of dissolved salts.

Individual Interior surface-area mines have substantial capacity for receiving FGC wastes, and disposal is considered technically feasible within existing mining operations. The wastes must be dewatered to the extent necessary for landfill operations, so that they can be

* Refers to one of four U.S. "provinces" in which coal fields occur i.e., Eastern (including Appalachia), Interior (between Rocky Mountains and Appalachia), Rocky Mountain, and Pacific Coast.

dumped into a mined-out strip (which can be adjacent to one being mined) and covered with overburden. Placing FGC waste in the mine void should assist in returning the terrain to its original elevation.

The principal environmental impact anticipated from this disposal method is an increase in total dissolved solids (TDS) in waters that are recharged by leachate from the disposal site. This impact may be lessened by placing part of the overburden in the mined-out strip prior to placing the FGC waste, thereby elevating the waste above the groundwater table. In addition, dilution to acceptable TDS levels can be encouraged by maintaining a suitable distance between the disposal site and the nearest stream, or by assuring that the receiving streams have a sufficiently high flowrate. Although more expensive, chemical treatment of the FGC wastes should also lower the TDS level in the leachate, as well as reduce the quantity of leachate, because of a lower permeability of the treated material.

Disposal of FGC wastes in a surface lignite mine is currently planned for Unit 2 of the Milton R. Young Station of Minnkota Power Cooperative in Center, N.D. [19], which will generate about 450 MW at full load. The flue gas will be scrubbed using alkaline ash (collected by the electrostatic precipitators of Units 1 and 2) as the absorbent. Disposal of treated FGC wastes in a local surface mine was originally considered by an Ohio utility for two new 400-MW units, but was dropped in favor of a disposal site available closer to the power plant[20]. A Montana utility also considered surface mine disposal of FGC wastes, but decided to use a settling pond closer to the plant.[21] Other units planned for this plant may cause reconsideration of the surface mine[18].

Among the underground mines studied, Eastern and Interior coal mines using conventional room-and-pillar methods were determined to be promising. Mined-out portions of room-and-pillar mines can be bulkheaded and filled with FGC waste by pumping it down from the surface through holes in the roof of the mine. Since long-wall mining involves roof-collapse in the mined-out area, placing FGC waste could substantially complicate the mining operation.

Placement of either treated or untreated FGC wastes' in room-and-pillar underground mines can reduce their acid drainage to groundwater, since filling the voids in these mines would reduce oxidation of pyritic sulfur and the formation of acid. Residual calcium carbonate in the FGC wastes would also partially neutralize existing acid.

In addition to these benefits, certain treated wastes could reduce long-term subsidence of the room-and-pillar mines by providing lateral pillar and roof support. However, untreated wastes, on the other hand, would not be expected to provide sufficient strength for subsidence control unless substantial settling of the solids (to about 65–70% solids) occurred after placement.

Contamination of groundwater in the mine would be most likely to occur if untreated FGC wastes were used. The water quality of nearby streams would be affected also. Such effects must be viewed relative to the amount of dilution and/or soil attenuation anticipated between the mine and the stream, as well as relative to the alternative of acid mine drainage (in the case where FGC wastes are not placed) affecting the stream quality. In other words, use of room-and-pillar coal mines for FGC waste disposal must be decided according to the hydrogeology of the site.

Underground room-and-pillar lead/zinc mines in the U.S. interior possess vast capacity for FGC waste disposal; both active and inactive mines appear feasible from operational and capacity standpoints. However, they are not linked to electric utilities by a transportation system, so the economics of FGC disposal in lead/zinc mines are probably not favorable. However, underground room-and-pillar limestone mines in the U.S. interior that are located above the water table appear to be feasible sites for FGC waste disposal from a technical and environmental standpoint. Still, these dry mines are limited in number and capacity and may be viable in only a few situations.

Ocean disposal of FGC wastes

Ocean disposal of various types of wastes has been taking place for many years. With increasing awareness of the adverse effects of much of this activity, recent efforts have been undertaken to reverse the degradation of the ocean environment. The Marine Protection Research and Sanctuaries Act of 1972 (PL92-532) is the major federal legislative action in this area. Section 2(b) of the Act states, "The Congress declares that it is the policy of the United States to regulate the dumping

Chemical treatment of FGC wastes has shown physical stability, greatly increased compressive strength, lower permeability, and about 50% lower concentrations of major solubles (e.g., chloride and sulfate) in the leachate, when compared to untreated wastes.

of all types of materials into ocean waters and to prevent or strictly limit the dumping into ocean waters of any material *that would adversely affect human health, welfare, or amenities, or the marine environment, ecological systems, or economic potentialities.*" [Italics are author's.]

The italicized portion of the quotation represents one of the major criteria by which methods for the ocean disposal of FGC wastes are being judged in the Arthur D. Little, Inc. studies for EPA. These studies were undertaken because it was apparent that if ocean disposal of FGC wastes proved environmentally acceptable, power plants currently burning low-sulfur oil in the northeastern U.S. could switch to high-sulfur coal. Many plants in the Northeast cannot presently burn coal because of a lack of land-based FGC-waste diposal sites; however, many of these plants do have access to the ocean. Another reason for undertaking the studies was the recognition that the major chemical constituents in FGC waste (i.e., chlorides, sulfates, calcium) are found in relatively high concentrations in seawater, so that the impact of adding these constituents would probably not be very significant.

An examination of the physical and chemical characteristics of FGC wastes has identified the following environmental concerns relating to ocean disposal:

1. Ocean-floor sedimentation—The particle (crystal) sizes of most untreated FGC wastes are much smaller than the coarse-grained sand particles of the ocean floor, which are most conducive to the marine life found on the Continental Shelf. Therefore, deposition of these wastes in large quantities may have an undesirable effect in the relatively shallow Continental Shelf area, particularly if disposal results in "paving" of the ocean floor.

2. Solids suspended in the water column—Settling of disposed FGC wastes toward the ocean floor through the water column, as well as resuspension of solids from the ocean floor, would expose various types of marine life to elevated concentrations of suspended sediments. The effects of sediments vary from minimal to serious, depending on the organism involved.

3. Sulfite-rich wastes—Many FGC wastes contain significant quantities of calcium sulfite hemihydrate, presenting a potential problem of sulfite toxicity to marine life and a potential chemical oxygen demand (COD) problem if oxidation to sulfate occurs. The sulfite toxicity depends on the degree of waste ingestion by the marine life; the COD depends on the kinetics of both sulfite dissolution and sulfite oxidation to sulfate.

The actual impacts of sulfite toxicity, dissolution and oxidation will require further investigation.

4. Trace contamination in wastes—Untreated and treated FGC wastes may contain several trace elements in concentrations in excess of acceptable levels for the marine environment. These elements include mercury, zinc, selenium, cadmium and nickel (among others). Significant dilution of the wastes with seawater (i.e., dispersion) may be necessary for ocean disposal to be acceptable.

In examining these environmental concerns, various methods of transportation and disposal of the FGC wastes were examined, including surface craft (e.g., bottom-dump barge and slurry dispersion) and pipeline (outfall). Various chemical and physical forms of the FGC wastes were also considered—i.e., sulfite-rich wastes, sulfate-rich wastes, and treated wastes in both "soil-like" and "brick-like" forms. Both Continental Shelf and deep ocean disposal of the wastes were examined from environmental, technical and economic standpoints.

As a result of the current study, several preliminary conclusions were drawn. Until more-definitive data are available, it would appear that disposal of sulfite-rich FGC wastes on the Continental Shelf or in the deep ocean would not be advisable. In addition, it appears that all soil-like FGC wastes—sulfite or sulfate, treated or untreated—should not be disposed of via quick-dumping surface craft or pipeline (outfall) on the Continental Shelf. Several other options appear more promising—all via surface craft:

1. Dispersed disposal of sulfate-rich FGC wastes on the Continental Shelf.

2. Concentrated disposal of treated, brick-like FGC wastes on the Continental Shelf.

3. Dispersed disposal of sulfate-rich FGC wastes in the deep ocean.

4. Concentrated disoposal of both sulfate-rich and treated FGC wastes in the deep ocean.

Conclusions

Vast quantities of FGC wastes will be produced by coal-burning power plants in the mid-1980s. Although opportunities for utilization of these wastes will undoubtably exist, disposal of FGC wastes will be practiced at most of the plants. Because of environmental concerns related to disposal of these wastes, several alternatives are being studied, including variations within each alternative. The alternatives are ponding, landfill disposal, mine disposal, and (possibly) ocean disposal. Ponding and landfill disposal are both current practices, but improvements in environmental protection and in costs associated with these methods are emerging through a combination of EPA and private-industry efforts. Mine disposal, particularly coal-mine disposal, of FGC wastes is being considered by private industry and, according to EPA studies to date, appears quite promising. Ocean disposal is currently under study and, although a degree of uncertainty exists regarding its viability, several approaches appear promising. All of these alternatives are being studied to encourage the environmentally acceptable use of domestic coal for power production.

References

1. Faber, J.H., National Ash Assn., "U.S. Overview of Ash Production and Utilization," presented at Fourth International Ash Utilization Symp., St. Louis, Mar. 24–25, 1976.

2. PEDCo-Environmental Specialists, Inc., Summary Report—Flue Gas Desulfurization Systems, Sept.–Oct., 1976, prepared under EPA Contract 68-02-1321, Dec. 1, 1976.

3. Dent, J.K., Central Electric Generating Board (CEGB), London, U.K., "England Overview," presented at Fourth International Ash Utilization Symp., St. Louis, Mar. 24–25, 1976.

4. Ando, J., Chuo University, Tokyo, "Status of Flue Gas Desulfurization and Simultaneous Removal of SO_2 and NO_x in Japan," presented at Sixth EPA Symp. on Flue Gas Desulfurization, New Orleans, Mar. 8–11, 1976.

5. Phillips, N. P. and Wells, R. M., Radian Corp., Solid Waste Disposal Final Report, EPA-650/2-74-033, (NTIS No. PB 233-144/AS), May 1974.

6. Holland, W. F. et al., Radian Corp., The Environmental Effects of Trace Elements in the Pond Disposal of Ash and Flue Gas Desulfurization Sludge, prepared for Electric Power Research Institute, Sept. 1975.

7. Interagency Agreement EPA-IAG-D5-E-721 (TV-41967A) between the U.S. Environmental Protection Agency and the Tennessee Valley Authority, May 7, 1975, Subagreement No. 19, "Flyash Characterization and Disposal," June 17, 1975.

8. Crowe, J. L. and Elder, H. W., Tennessee Valley Authority, "Status and Plans for Waste Disposal from Utility Applications of Flue Gas Desulfurization Systems," presented at Sixth EPA Symposium on Flue Gas Desulfurization, New Orleans, Mar. 8–11, 1976.

9. Princiotta, F. T., U.S. Environmental Protection Agency, Chairman, Sulfur Oxide Throwaway Sludge Evaluation Panel (SOTSEP), Vol. I: Final Report—Executive Summary, EPA-650/2-75-010a (NTIS No. PB 242-618/AS), Apr. 1975.

10. Resource recovery and Conservation Act of 1976, also known as Amendment to the Solid Waste Disposal Act (42 U.S.C. 3251 and following), Sec. 1004, para. 27, signed into law Oct. 21, 1976.

11. Jones, J. W. and Stern, R. D., U.S. Environmental Protection Agency, "Waste Products from Throwaway Flue Gas Cleaning Processes—Ecologically Sound Treatment and Disposal," in Proceedings: Flue Gas Desulfurization Symposium—1973, EPA-650/2-73-038 (NTIS No. PB 230-901/AS), Dec. 1973.

12. Jones, J. W., U.S. Environmental Protection Agency, "Environmentally Acceptable Disposal of Flue Gas Desulfurization Sludges: The EPA Research and Development Program," in Proceedings: Symposium on Flue Gas Desulfurization—Atlanta, Nov. 1974, Vol. II, EPA 650/2-74-126b (NTIS No. PB 242-573/AS), Dec. 1974.

13. Fling, R. B. et al., The Aerospace Corp., Disposal of Flue Gas Cleaning Wastes: EPA Shawnee Field Evaluation—Initial Report, EPA-600/2-76-070 (NTIS No. PB 251-876/AS), Mar. 1976.

14. U.S. Environmental Protection Agency, "National Interim Primary Drinking Water Regulations," Fed. Reg. (40-141), Dec. 24, 1975.

15. Rossoff, J. and Rossi, R. C., The Aerospace Corp. "Flue Gas Cleaning Waste Disposal—EPA Shawnee Field Evlauation," presented at Sixth EPA Symposium on Flue Gas Desulfurization, New Orleans, Mar. 8–11, 1976.

16. Rossoff, J. et al., The Aerospace Corp., "Disposal of By-Products from Non-Regenerable Flue Gas Desulfurization Systems: A Status Report," in Proceedings: Symp. on Flue Gas Desulfurization, Atlanta, Nov. 1974, Vol. I, EPA 650/2-74-126a (NTIS No. PB 242-572/AS), Dec. 1974.

17. Rossoff, J. and Rossi, R. C., The Aerospace Corp., Disposal of By-Products from Non-Regenerable Flue Gas Desulfurization Systems: Initial Report, EPA-650/2-74-037a (NTIS No. PB 237-114/AS), May 1974.

18. Lunt, R. R., Arthur D. Little, Inc., Personal Communication, May 1976.

19. Murad, F. Y., Combustion Equipment Associates, Hillier, L.V., Minnkota Power Cooperative, Inc., and Kilpatrick, E.R., Minnesota Power and Light Co., "Boiler Flue Gas Desulfurization by Fly Ash Alkali," presented at Fourth International Ash Utilization Symp., St Louis, Mar. 24–25, 1976.

20. Rossoff, J., The Aerospace Corp., Personal Communication, May 1976.

21. Crowe, J. L. and Elder, H. W., Tennessee Valley Authority, "Status and Plans for Waste Disposal From Utility Applications of Flue Gas Desulfurization Systems," presented at Sixth EPA Symp. on Flue Gas Desulfurization, New Orleans, Mar. 8–11, 1976.

The author

Julian W. Jones is currently responsible for all EPA research and development concerned with disposal and utilization of waste from power-plant flue-gas cleaning systems. He has spent his professional career managing a variety of government-sponsored R&D projects, from Apollo spacecraft propulsion (with NASA) to lime/limestone scrubbing for powerplant SO_2 control (with EPA). An author of numerous technical articles, he holds a B.S. (Magna Cum Laude) and M.S. in chemical engineering from the University of South Carolina.

How sludge characteristics affect incinerator design

The composition and properties of various types of sludges must be known before selecting dewatering equipment, developing design criteria for an incinerator, deciding whether sludge pretreatment is necessary, and developing procedures for air-pollution control of the incinerator flue gas and for handling the resulting ash and effluents.

R. G. Novak, J. J. Cudahy, M. B. Denove, R. L. Standifer, and W. E. Wass,
Hydroscience, Inc., Knoxville, Tenn.

☐ The most economical approach to sludge* disposal generally involves dumping the thickened or dewatered sludge on the land or in the ocean. For various reasons—including sludge toxicity, or the unavailability or high cost of land near the treatment site—incineration is, at times, recommended prior to land disposal of the inert ash.

This article will discuss the relationships between sludge characteristics and the design considerations for a sludge-incineration system. Among the design factors covered are: chemical conditioning, sludge dewatering and sludge incineration.

Conditioning and dewatering

Waste-activated sludge prior to thickening is primarily composed of water (99–99.5% by weight). A portion of the free water not bound within the sludge mass can be removed by gravity settling. The remaining free water, which occupies the voids around the settled sludge mass, can be removed by mechanical dewatering. However, this does not remove the water physically bound within the solids.

The amount of bound water remaining in a dewatered waste-activated sludge is approximately six times greater than the amount of incinerable solids in the cake. For other organic sludges, the percentage of bound water will vary as a function of the sludge constituents.

Since significant quantities of energy, usually as purchased auxiliary fuel, are required to evaporate any water remaining in the sludge prior to incineration,

*In this paper, sludge is defined as waste-activated and/or industrial organic sludges containing incinerable organic components, which have varying amounts of water and dissolved inorganic salts.

dewatering is important for removing the free water and water trapped within the floc, and, thus, for minimizing auxiliary fuel costs. The sludge-dewatering characteristics vary, and are a function of the raw-waste properties and the operating conditions of the treatment process.

Chemical conditioning with inorganic salts plus lime or polyelectrolytes is generally required. Laboratory and pilot-scale dewatering studies are recommended to develop design criteria for the dewatering equipment, and to determine the moisture content of the dewatered sludge, which has a direct influence on size of the incineration system, and auxiliary-fuel costs.

Various types of mechanical-dewatering equipment are listed in Table I. The need for prior thickening, typical chemical-conditioning requirements, and a range of the solids content of the dewatered cake are noted. This summary is developed for a typical waste-activated sludge primarily composed of biological solids (microorganisms).

A pressure filter generally produces the driest cake, but is the most costly dewatering alternative. However, auxiliary-fuel costs will be less, due to the lower moisture content of the dewatered cake. This is true even though significantly larger cake quantities exist, due to the high dosage of inorganic conditioning-chemicals.

Since incinerator size is a function of the water content of the sludge, the pressure filter will require a smaller incineration system than is needed in the other dewatering alternatives. But labor and maintenance costs are higher than for other methods.

The belt press and DCG-MRP offer the lowest capital and labor costs but require higher auxiliary-fuel costs.

Originally published May 9, 1977.

319

Performance of dewatering equipment on waste-activated sludge   Table I

Dewatering unit	Need for sludge thickening	Installed capital cost	Utility, labor and maintenance costs	Chemical-conditioning requirements	Chemical-conditioning costs, $/ton	Total dry solids content in dewatered cake	Lb of water remaining in cake/lb dry sludge solids[5]
Dual-cell gravity (DCG) plus Multi-roller press (MRP)	Not essential	Low	Low	Polymer 12-15	36-45	10-13% (11.5)[6] →	7.69[2]
Belt filter press	Not essential	Low	Least	Polymer 12-15	36-45	10-13% (11.5) →	7.69[2]
Centrifuge	Recommended	Higher	Higher	Polymer 15-18	45-54	10-12% (11) →	8.09[2]
Vacuum filtration	Absolute requirement	Higher	Higher	$FeCl_3$ 150-200 Lime 450-600	50-54	12-15%[1] (13.5) →	8.64[3]
Recessed-plate pressure filter	Absolute requirement	Highest	Highest	$FeCl_3$ 150 Lime 500-900	42-54	23-30%[1] (26.5) →	4.02[4]

[1] Includes the effect of adding a high percentage of lime and ferric chloride to the sludge solids.
[2] Assumes insignificant fraction of dissolved solids in the sludge cake.
[3] Assumes that conditioning chemicals or reaction products in the cake are 35% of the dry sludge solids (DSS).
[4] Assumes that the conditioning chemicals or reaction products in the cake are 45% of the DSS.
[5] DSS are defined as original biological solids.
[6] Numbers in parentheses indicate an average value; they are used to calculate water remaining in cake per pound of dry sludge solids.

This equipment is extremely simple to operate and maintain.

The centrifuge and vacuum filter are slightly more costly to operate, install and maintain than the belt press and DCG-MRP systems. The moisture content of the dewatered cake is basically the same for all of the methods except the pressure filter.

Results obtained from Example I   Table II

Wet cake summary		
Moisture content	70% (wt)	
Total dry solids	30%	
Volatile solids		14.6%
Nonvolatile solids		3.6%
$CaCl_2$ (total dissolved solids)		1.0%
Conditioning chemicals: $Ca(OH)_2$ and $Fe(OH)_3$		10.8%
		30.0%

Cake ash summary[1]	Wet cake, %	Ash, %
Nonvolatile solids	3.6	23.4
$CaCl_2$ (total dissolved solids)	1.0	6.5
Chemical solids	10.8	70.1
	15.4	100.0

[1] Does not consider hydroxide dissociation.

Many factors must be considered before selecting a dewatering system; there is no obvious answer.

Example 1 (see also Table II) illustrates the effect of conditioning chemicals on the quantity and composition of a sludge, and the ash produced from incineration of this sludge.

During some pilot-plant work, in order to obtain a satisfactory filter cake containing 30% dry cake solids, 0.6 lb of hydrated lime/lb dry sludge solids (DSS) and 0.32 lb ferric chloride/lb DSS were required. Over 75% of the ash resulting from incineration of this wet cake comes from the filter conditioning chemicals. The effect of the dissolved $CaCl_2$ formed by the reaction of the lime and $FeCl_3$ can also be significant, especially on the fusion point of the incineration ash.

Sludge incinerators

The multiple-hearth and rotary-hearth incinerators are commonly used for the incineration of many types of industrial sludges that have been conditioned and dewatered.

Fig. 1 is a cross-section of a six-hearth, counterflow incinerator. Dewatered sludge enters the top (or No. 1) hearth and is conveyed to the lower hearths by rotating plows. Combustion- and cooling-air enters the lower hearths and flows upward, counter to the sludge flow. In this particular unit, the drying zone occupies the top three hearths; the burning zone occupies hearths No. 4 and 5, and the cooling zone occupies hearth No. 6.

The flue gases are drawn through the incinerator and

gas-cleaning system by an induced-draft fan. Ash is discharged from the bottom shelf into an ash quench-trough. Since the top three hearths are used for drying, the temperature of the exhaust flue gas is lowered to about 800°F or less, as the hot combustion gases (at about 1,600°F) sweep over the wet, cold sludge and perform useful drying work. While this exchange of heat evaporates an important percentage of sludge moisture, it does not raise the sludge temperature much above the adiabatic saturation temperature of the combustion gas (170–180°F). Thus, the moisture content of the sludge on all of the drying hearths is high enough, so that the heat-transfer rate to the wet sludge, rather than the evaporation rate, is the controlling factor.

A temperature record of the flue gas and sludge on the various hearths illustrates this. At hearth No. 1, the sludge temperature is 100°F, and the flue-gas exit temperature is 700°F. At hearth No. 2, the flue-gas temperature is 1,000°F, and the sludge temperature is 150°F. At the last drying hearth (No. 3), the flue-gas temperature is about 1,150°F, and the sludge temperature is only 180°F. It is not until the sludge gets to the burning hearth that it reaches combustion temperatures, as in hearths No. 4 and 5. Typically, the moisture content of the sludge is about 40% when it discharges to the first burning hearth.

Although sludge temperatures in the drying hearths are relatively low, the vapor pressure of water and volatile organics at temperatures of 150 to 180°F is significant, and potential emission and odor problems must be considered.

For a municipal-sludge incineration process, experience has shown that afterburning is not usually necessary. However, for an industrial sludge, attention must be given to the possible evaporation and steam distillation of odorous and environmentally sensitive organic compounds in the drying section of a multiple-hearth incinerator. At temperatures of 150–180°F, significant vapor pressures of the sludge liquids can be reached, especially if the sludge contains insoluble oils.

Thermal treatment of the flue gas to eliminate combustible emissions can range from afterburning on a zero hearth at temperatures as low as 1,200°F for minimal odor control; to the use of a separate, properly designed, secondary combustion chamber operated at 1800°F to oxidate flue gas containing environmentally sensitive or thermally stable organics. A high-temperature secondary combustion chamber may also be required where combustible-vapor and particulate-emission concentrations are substantial.

The probable need for secondary combustion on a multiple hearth furnace can usually be determined from the chemical composition of the sludge and other bench-scale sludge characterization studies. Confirmation of a secondary combustion requirement, and the determination of data for use in design, will usually come from pilot-scale tests.

Four examples

To illustrate the importance of volatile organic content and composition of the sludge on incinerator design, four heat-and-mass balances are compared for the same sludge feed. The common basis will be 20,000 lb

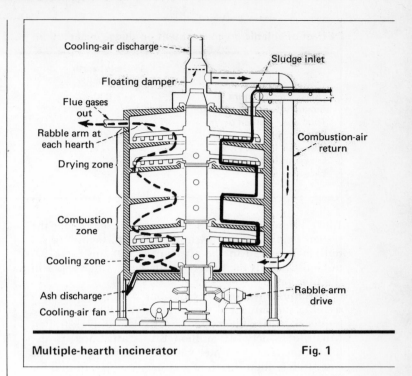

Multiple-hearth incinerator   **Fig. 1**

DSS/d at 14.5% DSS and 85.5% water. The DSS will be 77% volatile solids (VS) with a caloric content of 9,500 Btu/lb. The four incineration models are summarized in Table III.

The first heat-and-mass balance assumes that the flue gas has a temperature of 800°F as it leaves the top hearth. At this low temperature, the system heat duty is 11.1 million Btu/h. Of this total, 5 million Btu/h is contributed by auxiliary fuel. It should be noted that of the 11.1-million-Btu/h system duty, almost 8 million Btu/h leaves the furnace in the water vapor content of the flue gas. (The total flue gas is 20,500 lbs/h.) If such minor organic emissions as odor bodies are present in the flue gas from the drying hearths, a temperature of

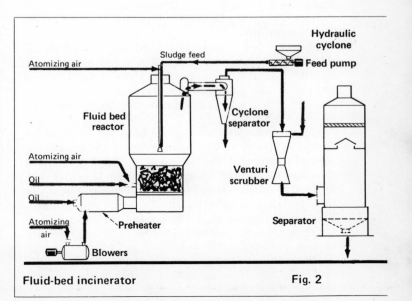

Fluid-bed incinerator   **Fig. 2**

Effect of volatile organic content on sludge incineration				Table III
	Multiple hearth (low volatile)	Multiple hearth (medium volatile)	Multiple hearth (high volatile)	Fluidized bed
Flue-gas temperature,°F	800	1,200	1,800	1,500
Auxiliary fuel, million Btu/h	5	9.8	21.6	13.3
Total duty, million Btu/h	11.1	15.4	27.3	19.4
Flue gas, lb/h	20,500	25,200	37,100	27,000
Excess air (primary), %	75	75	75	40
Excess air (secondary), %		25	25	
Excess air (total), %	75	60	45	40

1,200°F may be required for odor control, raising the system heat duty to 15.4 million Btu/h, including 4.8 million Btu fired in the afterburner. Of this total, 9.5 million Btu/h is the heat content of the water vapor in the flue gas.

If significant organic emissions are anticipated, a secondary combustion temperature of 1,800°F may be required, raising the system heat duty to 27.3 million Btu/h, of which 12.9 million Btu is attributed to the heat content of the water vapor in the flue gas. In this case, 16.6 million Btu/h of auxiliary fuel is required for secondary combustion. The last case in Table III, a fluidized-bed incinerator, will be discussed later in this section.

The melting point of the ash must be given serious consideration in incinerator-system selection and design. In a multiple hearth, if ash fusion or melting takes place on the burning hearth, large clinkers may form, which could lead to problems in conveying and ash discharge. Pilot-scale tests are recommended to evaluate the rabbling characteristics—i.e., the performance of the multiple-hearth incinerator's rabble arms.

Some variations

There are many variations of the multiple-hearth incinerator, including the rotary monohearth and the cocurrent multiple hearth. These furnaces can be used for incinerating sludges containing volatile organics, sludges with a low moisture content, or easily pyrolyzed sludges. In these units, the flue gases leave at combustion temperatures.

The fluid-bed incinerator is employed extensively for sludges, and in many cases does a satisfactory job. Excellent mixing generally takes place in the fluid-bed zone. The unit operates on low excess air, ranging from a low of 15% to a typical value of 40%, at temperatures of about 1,500°F.

Sand is generally used as the fluidized medium for sludge incineration, with the nonvolatile solids leaving the incinerator with the flue gas and being removed in a gas-cleaning train. This type of system is modeled in Table III and shown in Fig. 3. Supplemental heat can be supplied by preheating the fluidizing and combustion air and/or by direct fuel injection into the bed.

For many industrial applications, fluid-bed incinerators are operated at 1,350–1,500°F, or below this range. At these temperatures, adequate combustion usually can be achieved while operating below the melting point of the sludge ash. If partial softening, fusion, or

melting of the bed takes place, defluidization of the bed can occur. In addition, the outlet flue of the fluid bed can plug due to buildup of clinkers, and solidification of salt fumes. A thorough understanding of the variations in the physical and chemical characteristics of the ash is critical before deciding to install a fluidized-bed incinerator.

Sludge-disposal systems can be integrated with certain existing energy-producing units. For example, the exhaust gases from a steam boiler have been used to dry waste-activated sludge in a flash dryer, with the resulting dried sludge-solids fired back into the boiler to recover heat. If the boiler system is equipped with gas cleaning, this alternative may be applied.

Another alternative is sludge destruction by pyrolysis. To our knowledge, this is not widely practiced. Pyrolysis would be of interest when fusible salts are present. In theory, the sludge would be reduced to an inert residue that could contain large quantities of fixed carbon. The process would be carried out at relatively low temperatures, making it possible to stay below the fusion or softening point of the salts in the ash. Pyrolysis gas would be thermally oxidized in a separate chamber.

Waste-sludge characterization

Complete characterization of the sludge is essential in order to determine: (a) the most suitable type of incinerator, design requirements, and necessary operating conditions; (b) the need for and type of sludge pretreatment; (c) requirements for satisfactory air-pollution control of the flue gas; (d) ash handling requirements; and (e) requirements for satisfactory handling of liquid effluents from the system. Table IV outlines the most important factors in sludge characterization. A discussion of each factor follows:

Source—The source of the sludge (municipal, industrial, or a combination) will establish the importance of the other characteristics listed. Obviously, it would not be necessary to determine, for most municipal sludges, some of these characteristics. Also, knowing all the sources of an industrial sludge simplifies characterization by permitting the determination of only significant characteristics.

Sludge-solids composition—1. Volatile oils. The composition and quantity of volatile oils in the sludge solids can have a significant impact on incinerator design. For example, a multiple-hearth unit with countercurrent flow might require secondary combustion of the effluent

Waste sludge characterization factors	Table IV
Source	
Sludge-solids composition:	
Volatile oils	
Environmentally sensitive and toxic compounds	
Moisture content	
Gross heat of combustion—dry solids	
Ultimate analysis	
Ash content—dry solids	
Metal scan on ash	
Phosphorus, arsenic, selenium, mercury	
Halogens	
Sulfur	
Total dissolved solids in water phase	
Carbonates, hydroxides, water of hydration	
Thermal characteristics of ash	
Thermal decomposition characteristics—dry solids	
Solubility of ash	

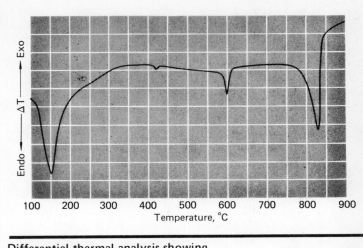

Differential thermal analysis showing endotherms for various components   Fig. 3

gas if significant quantities of volatile oils were present. Composition and quantity of volatile organics can be determined by steam distillation, or solvent extraction of the sludge, followed by a combined gas chromatograph/mass spectrographic analysis of the distillate or extract.

2. Toxic and environmentally sensitive compounds. The presence of extremely toxic or environmentally sensitive compounds—e.g., pesticides, cyanide—in the sludge will require much greater emphasis on complete combustion, and possibly additional treatment of the flue gas. Sometimes it is possible to track down and eliminate such problem compounds at the source, particularly if it is a small fraction of the total makeup of the sludge.

Moisture content—The impact of moisture content has already been discussed in some detail.

Gross heat of combustion, dry solids—The gross heat of combustion of the DSS determines the fraction of the total-system energy requirement that will be supplied by the sludge feed. This is determined by a bomb calorimeter.

Ultimate analysis—This provides necessary data for heat-and-material balances for the incinerator and auxiliary equipment.

Ash content, dry solids—Ash content of the dry solids is generally determined by standard methods at 500–550°C. Care must be taken that ashing is complete. The procedure should be carried out until constant ash weight is achieved. The ash content will determine handling requirements, and influence the selection of ash-disposal methods.

Metal scan of the ash—The presence of metals, such as Cr, Cu, Zn and Fe, will have a significant impact on the water system (quenching, scrubbing) and on the feasibility of ash disposal by landfill from the standpoint of leaching of these components. For example, Cr will be converted to the more soluble hexavalent form. These metals are generally determined by x-ray fluorescence, emission spectroscopy, or atomic absorption.

Analysis of phosphorus, arsenic, selenium and mercury—Phosphorus, arsenic, and selenium content of the solids is significant because these elements, on oxidation, may form the volatile oxides P_2O_5, As_2O_3, and SeO_3. They can be determined from a sample of the dry solids. Mercury, which is volatile at relatively low temperatures, should be determined from samples of wet sludge.

Halogens—The presence of halogens, primarily chlorine, is significant for several reasons: (1) halogens have a large impact on corrosion problems in the incinerator and auxiliary equipment; (2) they are an important factor in determining requirements for air-pollution-control equipment, (3) salts have an impact on the properties (for example, melting point) of the ash, and (4) halogens have an impact on the total dissolved solids of the water effluents from the system.

Relatively low levels of components can have a major effect on sludge characteristics. For example, sludge containing 0.9% by weight of NaCl can result in ash having a much higher (about 15% by weight) NaCl content. This is enough to cause low-melting problems.

Sulfur compounds—Sulfur in the sludge will, of course, show up as SO_2, SO_3 or SO_4 in the effluents.

Total dissolved solids (TDS) in the water phase—Since the sludge or the wet filter cake contains a high percentage of water, the effect of dissolved solids can be significant. This was illustrated in the pressure filtration example. The TDS must, of course, also be taken into account when disposing of the water phase.

Carbonates, hydroxides, water of hydration—The example below, using heats of formation, illustrates the significant effect that carbonates present in dry solids can have on the overall heat balance.

$CaCO_3$ calcination:

$CaCO_3 \rightarrow CaO + CO_2$
−288.49   −151.9 −94.05 kcal/g mol

Products − reactant $= (−94.05) + (−151.9) −$
$(−288.49) = +42.54$ kcal/g mol

Example I

Effect of chemical conditioning on characteristics of sludge and ash. Bases for calculation:

(a) 0.6 lb hydrated lime/lb DSS and 0.32 lb $FeCl_3$/lb DSS.

(b) 30% dry cake solids in filter cake (including chemicals).

(c) Assume 0.8 lb vs/lb DSS.

(d) Assume 4% DSS thickened slurry before chemical conditioning.

(e) Assume filter cake is not washed.

Lime and ferric chloride react as follows:

Lb/lb DSS

$$\underset{325}{\overset{0.32}{2\ FeCl_3}} + \underset{222}{\overset{0.22}{3\ Ca(OH)_2}} \rightarrow \underset{214}{\overset{0.21}{2\ Fe(OH)_3}} + \underset{333}{\overset{0.33}{3\ CaCl_2}}$$

After reaction	lb/lb DSS
$Ca(OH)_2$ remaining	0.38
$Fe(OH)_3$ formed	0.21
$CaCl_2$ formed	0.33

$CaCl_2$ added to wet cake:

$$\frac{4\ \text{lb DSS}}{96\ \text{lb water}} \times \frac{0.33\ \text{lb } CaCl_2}{\text{lb DSS}} =$$

$$13{,}750\ \text{ppm } CaCl_2 \text{ in wet cake}$$

$$\frac{70\ \text{lb water}}{30\ \text{lb dry cake solids}} \times \frac{13{,}750\ \text{lb } CaCl_2}{10^6\ \text{lb water}} =$$

$$\frac{0.032\ \text{lb } CaCl_2}{\text{lb dry cake solids}}$$

Composition of dry cake solids

1. Chemicals

$$\begin{array}{l} 0.38\ \text{lb } Ca(OH)_2/\text{lb DSS} \\ +\underline{0.21\ \text{lb } Fe(OH)_3/\text{lb DSS}} \\ 0.59\ \text{lb chemicals/lb DSS} \end{array}$$

2. $CaCl_2$

$$\frac{0.3\ \text{lb dry cake solids}}{\text{lb wet cake}} \times \frac{.032\ \text{lb } CaCl_2}{\text{lb dry cake solids}} =$$

$$0.96\%\ CaCl_2 \text{ in wet cake}$$

$$\begin{array}{l} 30.00\%\ \text{dry cake solids} \\ -\underline{\ 0.96\%\ CaCl_2} \\ 29.04\%\ \text{chemical and dry sludge solids (CDSS)} \end{array}$$

$$29.04\%\ \text{CDSS} \times \frac{.59\ \text{lb chemical}}{1.59\ \text{lb CDSS}} =$$

$$10.78\%\ \text{chemical solids in wet cake}$$

3. Dry sludge solids

$$\begin{array}{l} 29.04\%\ \text{CDSS} \\ -\underline{10.78\%\ \text{chemical solids}} \\ 18.26\%\ \text{DSS} \end{array}$$

$$18.26\%\ \text{DSS} \times .80\%\ \text{vs} = 14.61\%\ \text{vs}$$

Nonvolatile solids (NVS) in DSS $= 18.26 - 14.61 = 3.65\%$ NVS

$$\frac{42.54\ \text{kcal/g mol}}{100\ CaCO_3/\text{g mol}} \times 1{,}800 = 765\ \text{Btu} \frac{\text{absorbed}}{\text{lb } CaCO_3}$$

Thermal characteristics of ash—Fig. 3 illustrates the use of differential thermal analysis to detect possible problems occurring with low-melting-point components in the ash, which could cause loss of fluidization at certain temperatures in a fluid-bed incinerator, or excessive clinker formation in a rotary hearth or rotary kiln. If this problem is anticipated, the selection of a different type of incinerator, or another way to adjust the temperature profile, may be indicated. If the low melting component is water soluble, as with NaCl, washing of the filtered sludge prior to incineration may be beneficial.

Thermal-decomposition characteristics—*dry solids*—The purpose of this analysis is to determine the fractions of volatile and fixed carbon in the dry solids. This will influence such factors as temperature, excess air, and residence time necessary for complete combustion, and will aid in the selection of the most suitable type of incinerator. For example, a fluid-bed incinerator might be advantageous if the sludge solids contained a high concentration of fixed carbon.

Solubility of ash—Solubility of ash components has a major impact on the selection of a suitable method of ash disposal (landfill or subsequent handling). For example, a high $CaCO_3$ content in the sludge can result in a high CaO alkalinity in the ash.

Conclusion

This paper has discussed the relationship between sludge characteristics, feed conditioning, and dewatering requirements, and the characteristics of specific types of incinerators used for sludge disposal. The economic factors involved have not been discussed in detail. In most cases where suitable alternatives exist, the final selection of the process, equipment, and conditions will be determined largely by economic factors. However, in order to make a sound economic evaluation of alternatives, it is first necessary to conduct a complete sludge characterization. Each sludge incineration system, particularly if industrial sludges are involved, should be custom-designed to meet the specific requirements of each application.

The authors

R. G. Novak is manager of solids, thermal, air, at Hydroscience, Inc., Knoxville, Tenn. Formerly employed by Dow Chemical Co. in a variety of waste-treatment-related jobs, he holds a B.S. in chemical engineering from the Illinois Institute of Technology.

J. J. Cudahy is an environmental specialist at Hydroscience, Inc. He is responsible for solid-waste technology in the areas of incineration, air pollution control, recycle water systems, resource recovery and energy recovery. He holds an M.S. in chemical engineering from the University of Delaware.

R. L. Standifer is a process specialist in solids, thermal, air, at Hydroscience, Inc. He has a B.S. in chemical engineering from the University of Colorado, and is a registered professional engineer in the state of West Virginia.

W. E. Wass is an environmental specialist at Hydroscience, Inc. Previously employed by Dow Chemical Co. in waste-control activities, he holds a B.S. in chemical engineering from the University of Michigan.

M. B. Denove is a designer of systems for industrial-wastewater treatment and sludge handling at Hydroscience's Westwood, N.J., office. He holds a B.S. in civil engineering and an M.S. in environmental engineering, both from Manhattan College.

Incineration of industrial wastes

This article describes an incineration facility that has successfully dealt with problems such as handling of toxic materials, incinerator design, and maintenance and energy considerations. *

C. Randall Lewis, Richard E. Edwards and Michael A. Santoro, 3M Co.

☐ The 3M Co. incinerates chemical wastes from manufacturing operations in a 23,000 kcal/h (90 million Btu/h) rotary kiln incinerator that has air pollution control equipment. This article deals with the essential features that have resulted in a successful incineration operation. The features include waste materials handling considerations, design aspects, and maintenance requirements. A brief summary of a waste-heat-recovery pilot study is also included.

After studying many alternatives, the 3M Co. decided that the best long-range answer to the disposal of hazardous wastes was incineration because of the following reasons:

1) Incineration is an excellent disposal method for all types of solvent-contaminated wastes. This is a critical factor, since scrap characteristics within the 3M Co. vary considerably due to many types of manufacturing processes.

2) Incineration eliminates the groundwater pollution potential from the scrap. Complete oxidation of waste materials is the most reliable method available to produce an inert residue.

3) Anticipated pollution-control regulations can be met by incineration. As the landfilling of hazardous wastes becomes more and more restricted, incineration will continue to provide a solution to the disposal problem.

A description of the four basic components of the incinerator facility follows:

1. *Materials handling system.* The purpose of this building and equipment is the proper handling of the scrap materials so that wastes can be charged to the incinerator in a satisfactory manner. This involves pumpable scrap blending and mixing, and the movement of scrap materials to the proper feeding areas.

2. *Incineration components.* These are the primary and secondary combustion chambers used to oxidize the waste material.

3. *Air-pollution control equipment.* This equipment scrubs the exhaust gases before emission to the atmosphere.

*This article was presented at the National Waste Processing Conference & Exhibit, sponsored by the American Soc. of Mechanical Engineers in Boston, Mass., May 23–26, 1976.

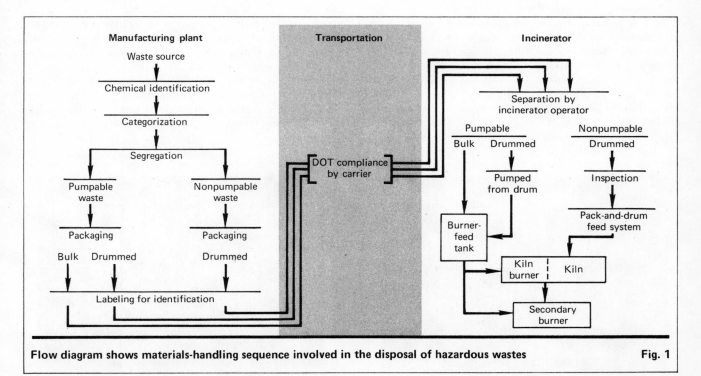

Flow diagram shows materials-handling sequence involved in the disposal of hazardous wastes   **Fig. 1**

Originally published October 18, 1976.

4. *Water-pollution control equipment.* This is for treating scrubber water before discharging it to the receiving stream.

Generally, there has been a reluctant attitude toward incineration. The main areas of concern have been on the materials handling methods, the incinerator design, the maintenance associated with operating such a facility, and energy considerations. The purpose of this paper is to describe an incineration facility that has overcome these concerns and has provided a safe, economical, and efficient means to a hazardous-waste-disposal problem.

This industrial incinerator as designed will provide the necessary temperature and residence time necessary to oxidize those hazardous organics for which Environmental Protection Agency recommendations have been published. However, organometallics that contain metals like cadmium, mercury, etc., pose a problem because these elements are intrinsically hazardous regardless of their oxidation state, or nature of bonding. They are either discharged with the ash residue, presenting a potential groundwater contamination problem; trapped in the scrubber water; or emitted to the air. For this reason it is not recommended that large quantities of organometals be incinerated in a facility of this kind.

Materials handling

There are seven basic steps involved in the proper disposal of industrial wastes, steps which require understanding and cooperation between the personnel at the waste generation source and the personnel at the incinerator facility. These seven steps are: (1) chemical identification, (2) categorization, (3) segregation, (4) packaging, (5) labelling, (6) transportation, and (7) handling and disposal. Steps 1–6 are in-plant functions, while Step 7 is carried out by the incinerator personnel. A materials-handling flow-diagram is shown in Fig. 1.

Chemical Identification—Since the personnel at the waste generation source have the greatest knowledge of the major chemical constituents in the waste, the components can best be identified prior to shipment. The identification of waste at the source facilitates compliance with the U.S. Department of Transportation

(DOT) regulations, ensures the maximum safety of all personnel involved in the waste processing, and permits the proper precautions to be taken in order to protect the physical integrity of the incineration equipment.

Categorization—Three broad categories are used to describe waste material: (1) dry scrap, (2) wet scrap, and (3) extra-hazardous scrap. Dry scrap is any dry material, such as wood, paper, and rags, which exhibits no flammable vapor hazards. Wet scrap is composed of two subcategories: pumpable wet scrap and nonpumpable wet scrap. The first one is any liquid material that can be pumped or poured into a drum or other container. Nonpumpable wet scrap is any solvent-contaminated item that cannot be pumped or poured into a drum. This includes such items as polymerized adhesives or resins, solvent-soaked rags, gloves, filter cartridges, polybags, films, and chemical powders. Extra-hazardous scrap is any material with unusual hazard such as flammability, toxicity, extreme chemical reactivity, or obnoxious odors.

Segregation—The waste material is segregated in such a way that dry scrap, nonpumpable wet scrap, and pumpable wet scrap are not mixed in any single shipping container. The dry scrap and nonpumpable wet scrap are charged to the kiln, while the pumpable wet scrap is charged to the primary and secondary burners. This usage of pumpable wet scrap is required to maintain proper combustion temperatures within the incineration system. The need for the difficult and labor-intensive process of segregating waste materials upon arrival at the incinerator is eliminated if segregation is done in-plant.

Packaging—Waste materials are packed in reconditioned, 17 H open-head drums and 17 E closed-head drums. These afford a most convenient waste-materials container, since they are common to most manufacturing operations, and comply with all DOT requirements. A 10-mil, antistatic drum liner is used with all nonpumpable scrap, so that the waste materials can be mechanically removed and the drum reclaimed. Before shipping, the liner is gathered at the top, doubled over, and securely taped. Liners are not used with pumpable materials since they inhibit the pumping operation by

Drum conveyor, with feeder in background   Fig. 2

Room where pumpable wastes are sent to storage tanks   Fig. 3

plugging the piping system. The drum lids are sealed with a fiber ring gasket prior to shipping.

Labelling—Attached to each drum of waste materials are the appropriate DOT labels and an intracompany label which categorizes the waste materials into pumpable or nonpumpable wet scrap, and indicates the health, fire, and instability hazards associated with the waste materials. The label also indicates the major chemical component and whether the material is chlorinated or not. From this information, the Btu content and compatibility characteristics of the waste can be deduced.

This labelling procedure allows the incinerator operator to easily identify the nature of the drum contents, so that proper disposal techniques can be implemented. Since the waste materials have been identified at the source prior to shipment, there is no need for an extensive, costly, and time-consuming sampling and analytical program at the incinerator facility.

Transportation—The most common method of transporting waste materials to the incinerator facility is by commercial truck lines. The drums are loaded one-high and four to a pallet. Normally, a truckload consists of 72–76 drums. The preferred method of shipping large quantities of pumpable wet scrap is by bulk tanker, since the labor required to process the waste is much less. Shipment by rail would also be very feasible if appropriate accommodations were provided.

Handling and Disposal—The handling systems for waste materials at the incinerator facility are simple and flexible. Only two are necessary—one that processes nonpumpable waste materials, and another that handles pumpable materials.

The nonpumpable system consists of a pack-and-drum feeder, a double-door airlock, and a drum conveyor (Fig. 2). The drums of nonpumpable waste materials are placed on the roller-type conveyor that moves the drums in sequence to the pack-and-drum feeder mechanism. While the drums are on the conveyor, operating personnel remove the drum lids and visually inspect the contents. The drums are then charged one at a time into the kiln, with the charging rate determined by the Btu value of the contents. As each drum enters the air-lock, a vise-type device automatically grasps it. The operator then has the option of tipping the drum to discharge the contents, or releasing the full container into the kiln.

Although many drums are routinely reclaimed, some are unavoidably charged to the kiln because improper packaging of the waste material prevents discharge by tipping. Since many 3M manufacturing processes generate adhesive-type waste materials, caution must also be taken to avoid contaminating the inside of the drum or the exterior of the liner with adhesive material that would prevent drum discharge via tipping.

The pumpable system consists of a pumping room, blend tanks, and storage tanks. Pneumatic diaphram pumps are used to transfer the pumpable wastes from the drums into storage tanks (Fig. 3). If the material is too viscous to pump, the drum is tipped and allowed to gravity-drain into the storage tank. Care must be taken to avoid mixing pumpable materials that react, solidify, or polymerize when mixed. The only answer to such an occurrence is to manually remove the material from the storage tanks. This unpleasant situation occurred several times when the incinerator facility was started-up, but experience and knowledge in liquid segregation have eliminated this problem.

The pumpable wet scrap is burned through solvent burners in both the kiln and the secondary combustion chamber. In order to achieve a uniform fuel quality, the pumpable material is mixed in blend tanks prior to incineration. All piping is recirculated to prevent settling, and mechanically comminuted to destroy any

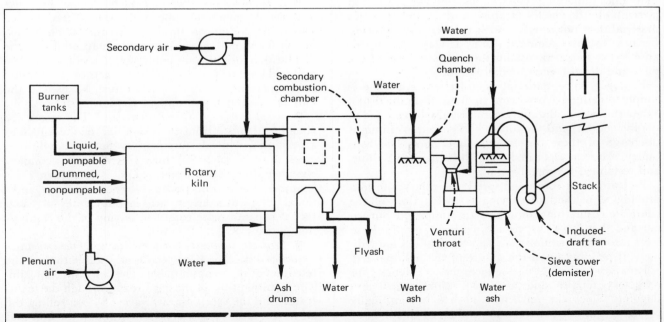

Incinerator facility features a rotary kiln, and a secondary combustion chamber for particulates   Fig. 4

agglomerations that would cause plugging problems.

This seven-step program must include a rigorous followup plan with the plants to ensure that personnel at the waste generation source follow the procedures set forth. The followup should emphasize such benefits as safety of incinerator operating personnel, physical well-being of equipment, and capability to comply with all applicable regulations. In general, the more effort put forth on steps 1–6 of the disposal process, the easier and safer the actual incineration process of Step 7 becomes.

Combustion features

The purpose of incineration is to produce stable oxides that can be returned to the environment without causing detrimental effects. In recent years, one more dimension has been added to incineration—the air-pollution aspect. It is not enough to run an incinerator that performs well only with respect to oxidation; air emission standards must also be considered.

The key to the success of this facility (Fig. 4) is the use of a rotary kiln for the primary combustion chamber. The kiln is 11 m (35 ft) long, with a 0.3 m (11 in) refractory lining of super-duty firebrick (Fig. 5). This refractory provides a desirable combination of economy, chemical resistance, and mechanical durability.

Material is fed into the kiln in 210-l (55-gal) drum quantities. The charges have a weight range of 70–230 kg (150–500 lb), and an average of 80 kg (180 lb). A most important aspect is that a rotary kiln continuously exposes new surfaces for oxidation. Its tumbling action prevents sintering of the waste materials, so complete oxidation of the charge results.

The rotary kiln provides continuous ash removal. This is important when incinerating solvent-contaminated inorganic material, especially when the material is contained in steel drums. Incineration of the organic constituents occurs in the kiln, with only the inert inorganics remaining. Continuous removal of this ash prevents shutdowns for cleaning and ensures that this material does not interfere with the oxidation process. Because 3M uses standard drums as waste containers, an effort is made to reclaim them through use of the pack-and-drum feeder previously described.

However, some material polymerizes, and some is simply too adhesive or viscous to dump from the drum. When this occurs, the container must also be charged to the incinerator, and the rotary kiln ensures continuous discharge of these burned-out containers. The containers are then separated from the other ash residue, and reclaimed for metal scrap.

By controlling rotational speed, the rotary kiln also provides a method of varying retention time of the charge to ensure that containers are completely burned out and that loose charges are oxidized completely to inert ash. The retention time can be adjusted immediately, depending upon the nature of the material fed.

Rotation of the kiln also reduces the requirement for refractory repairs due to flame impingement and slagging. Since the refractory surface is continually changing spatially, there is no prolonged direct-flame impingement on one specific portion of refractory. Naturally, prolonged flame impingement would cause the refractory to deteriorate prematurely. The formation of slag is spread over a large area, and is easily removed by raising the kiln temperature to the melting point of the slag. Caution should be exercised not to exceed the melting point of the refractory.

Erosion and thermal spalling are the only unfavorable considerations associated with rotary-kiln incineration. The erosion is a result of the abrasion caused by waste material tumbling inside the kiln. Thermal spalling occurs at the discharge end of the kiln, and is caused by the thermal shock created by the inrush of air at the end-plate seal. This spalling requires the periodic replacement of a small section of castable. Neither of these two unfavorable considerations will result in excessive maintenance.

The drum-feed concept is important in that it provides a relatively consistent feed. Materials charged to the incinerator have a large variation in heat of combustion and volatility. Once a charge is fed, the only remaining control on the temperature rise of the system is to increase the air flow. The heat and mass release are not controllable after the batch is charged. By staggering drums of material with low heat of combustion and low volatility with those of high heat of combustion and high volatility, a much more consistent feed can be achieved. Thus, the kiln temperature and the retention time of the combustion gases can be kept within acceptable limits.

A secondary combustion chamber is provided to allow for the oxidation of combustible particulate matter suspended in the gas stream. This chamber, which is also lined with refractory brick, allows a one-second retention period of the gases at 870–890 °C (1,600–1,800 °F). This is sufficient to allow complete oxidation of one-micron combustible particles.

Successful incineration has depended upon the ability to achieve the following four basic operating features:

■ *A relatively consistent temperature required for proper oxidation.* As mentioned above, the rate of feed can be varied, depending on the heat of combustion, which allows for some temperature control. In addition, the combustion of the pumpable scrap is automatically adjusted by direct control of the burners to compensate for temperature changes. Temperature is sensed at the kiln exit and at the secondary chamber exit.

■ *Complete mixing of combustion gases.* The secondary chamber is offset from the centerline of the kiln and air-pollution control equipment, so that the combustion gases must make a 90° turn to enter the secondary chamber after exiting from the kiln. Before leaving this chamber, the combustion gases encounter two additional 90° turns. This physical layout creates increased turbulence and more complete mixing of the combustion gases.

■ *Adequate retention to permit the kinetics of the combustion reaction to occur.* The kiln speed is adjustable to vary the retention of the nonpumpable material within the kiln. The retention time of the gas stream through the incinerator, and the excess air, are varied by controlling the air flow into the system. Air flow through the kiln and the secondary chamber is induced by the fan downstream of the wet-scrubber air pollution equipment.

Control is by a variable throat in the venturi scrubber and by the louvers in the air-intake duct that runs to the head end of the kiln. Both are controlled from the operator's control room. It is important to point out that the induced-draft fan has the capability to make large changes in air flow. This is provided by the specific inlet design that permits variation of flow through the fan. If this were not so, flow separation from the fan blades would cause the fan to vibrate; this is known as a "starved fan." As shown later, such a condition is of concern.

■ *Proper oxygen supply to maximize the reaction without excessive cooling of the combustion products.* This is also accomplished by the variable air-flow control.

Air-pollution control

Satisfactory combustion in the primary and the secondary chambers is the real key to air pollution control, but strict standards on particulate emissions require additional controls. The 3M incinerator is restricted to a particulate emission standard of 0.23 g per standard cubic meter of dry exhaust gas. This figure is adjusted to a 12% carbon dioxide concentration as required by the regulation. The air-pollution control system is of the water-scrubber type, and consists of five major components: a quench chamber, a venturi scrubber, a mist separator, an induced-draft fan, and a 60-m (200 ft) stack (Fig. 6).

The quench chamber uses a water spray to cool the gas stream from 870–980°C (1,600–1,800°F) to about 80°C (180°F). Because the exhaust gases are quenched, refractory-type lining is not required in the remaining chambers. The quench tank, however, is lined with an acid-resistant brick and mortar. Since some of the materials incinerated contain halogenated hydrocarbons, halogen acids such as hydrochloric acid are present in the gas stream. In addition to the quenching process, the quench chamber removes some particulates.

The venturi scrubber has been specified for the removal of particulates down to 0.1 micron. Since a high-efficiency unit was needed, a water-spray header with atomizing nozzles was added to the venturi throat. A venturi with a 0.76-m (30 in) water-gage pressure-drop is adequate. The present system complies with the air emission regulations, but venturi scrubbers must be carefully designed for each particular application.

The mist separator or demister removes the fine water droplets generated in the venturi and entrained in the gas stream. The chamber consists of a countercurrent flow of water and air, with the water cascading over plastic plates. Since the gas stream provides the necessary mixing as it passes up through the plates, a most important aspect of this chamber is the plate area. The initial demister design contained more plate area than needed, and some of the porous plates had to be replaced with solid sections to avoid short-circuiting and channelling of the gas stream.

An induced-draft fan is required for any large venturi scrubber because of the energy drop across it. In this system, the venturi throat is the principal control on the air flow through the combustion train and, therefore, the induced-draft fan must be capable of handling varying amounts of gas.

The existing fan has an inlet damper that permits compensation for variations in the gas flow. At first, the inlet damper was improperly used because of insufficient operating data. In addition, the fan collected wet particulates that passed through the venturi. The combination of the particulate buildup and the incorrect inlet-damper setting caused the fan to run out of balance most of the time. When purchased, the fan had 0.76 mm (0.003 in) clearance between the shaft and the wheel hub. The imbalance caused considerable wear and operating difficulties.

The fan has been modified by providing an interference fit between the hub and the shaft. The fan has also been provided with a water-spray system to reduce particulate buildup. The inlet damper has been adjusted to prevent fan "starving" at the most frequent air-flow rates.

Two fan wheels were purchased for this facility. One is made of a Hastelloy formulation, and the other of rubber-covered steel. The Hastelloy fan is normally used, and the rubber-covered wheel serves as a spare. This has been used on a trial basis, and one serious defect has been noted. Several rubber pads were provided for balancing, and under the stress of operation these rubber pads delaminated.

Kiln that serves as primary combustion chamber   **Fig. 5**

Air-pollution control equipment has five units   Fig. 6

The scrubber water from the air pollution equipment requires acid neutralization, chemical treatment, and sedimentation before its discharge to the receiving stream. For neutralization, ammonia was originally selected. A sparge pipe was placed in the sewer just ahead of the lift station. Since the sewer line did not flow full, much of the ammonia simply bubbled through the water, and was sluiced out of the sewer. As a result, the cast-iron sewage pumps and force main were destroyed by the acidic scrubber-water within a year of operation. To correct the situation, neutralization was improved by blocking the sewer with a weir so that the sewer pipe was completely filled, allowing more contact between the ammonia and the scrubber water. In addition, the pumps were replaced with horizontal chemical-process pumps designed for service in halogen acids at a pH of 3 or higher. The force main was replaced with fiberglass-reinforced plastic pipe. This modified system has been operating satisfactorily.

Maintenance

Primary combustion chamber—The two main maintenance concerns in this chamber are refractory wear and replacement, and slagging of inorganic salts. Because of the high abrasiveness of the steel drums rotating within the kiln, and the high and fluctuating temperatures, the super-duty fore-bricks and insulating bricks must be replaced about once every two years. This is normally a 250-manhour job.

The insulating bricks initially used were made of compressed diatomaceous earth. Subsequently, it has been found that other refractory bricks of similar heat-conductive properties function comparably. One important concern is that the hardness of the two bricks be somewhat the same, so that abrasive wear between the two is at a minimum.

Slagging of inorganic salts normally occurs where the heat in the kiln is the highest, i.e., the area to which the flame-tip reaches. The slag layer can achieve a 70 to 230 mm (3 to 9 in) thickness. As expected, this slag keeps burned ash from traveling through the kiln. The slag also acts to cut brick life, reducing its density and refractory properties. The slag ring in the kiln is normally maintained at about 50 to 70 mm (2 to 3 in.), and is controlled by slowly raising the temperature to the required melting point.

Secondary combustion chamber

The major recurring problem in this chamber is ash accumulation. Because there is not a continuous ash-removal system, the ash must be physically cleaned out. Naturally, as ash volumes build up, the efficiency of the secondary combustion chamber decreases, but noticeable affects are not evident until after about two to three months of operation. This period varies, of course, depending upon the ash content of the pumpable-wet-scrap fuel.

Generally, throughout the primary and secondary chambers and connecting sections, as particles tend to settle out on all horizontal surfaces, all areas must be cleaned periodically so that air-flows and detention times are not affected.

Air-pollution control equipment

The major recurring maintenance problem related to the air pollution control equipment is corrosion. Because the scrap materials incinerated contain some chlorinated hydrocarbons, the gas stream contains hydrochloric acid. The concentrations vary, naturally, as a function of the levels of chlorinated hydrocarbons within the pumpable and nonpumpable scrap. In addition to the acid content itself, the corrosiveness of the scrubber water becomes greater because of dealkalinization. Corrosion rates are increased even further by the erosion effects brought on by the particulate in the air stream and the water stream, and in some areas by the high-velocity water-flow itself. A major effort has been placed on developing coating systems and neutralization improvements to reduce these corrosion effects.

Quench elbows and quench chambers—These were at first lined with acid-resistant brick and mortar. Because of the high corrosion from the water and air flows, much of the mortar dissolved to the point that the brick fell out of the chamber. In addition, at times the air stream also took on alkaline properties that the mortar could not withstand.

Two things were done to resolve these problems. First, the brick was replaced using a furan-resin cement as the mortar. This mortar was particularly chosen because of its high ability to resist strong acid attack and mild alkaline conditions. Secondly, the entire interior surface of the chamber was coated with a $1/8$-in layer of the cement. This provided better protection and made repairs significantly less time-consuming and costly. The cement bonds extremely well to masonry surfaces as well as to itself.

The ceiling of the quench chamber has a rubberized coating. This coating is temperature-sensitive, and degradation starts at approximately 80°C (180°F). The location and efficiency of the water-spray nozzles is very critical in prevention of gas channelling, which results in hot spots and deterioration of the lining.

Venturi chamber—The venturi chamber was equipped with a butyl rubber lining. This lining was chosen to protect the steel structure from acid attack, and to act as a resilient counter-force to the high velocities present within the venturi. Unfortunately, neither purpose was fulfilled completely by the rubber. The acid in the air stream has a tendency to penetrate with water vapor through the 3.2-mm rubber coating and condense between the steel and rubber layer, causing steel corrosion and a blistering effect on the rubber. Corrosion occurs to a small degree at first, until the acid has been completely neutralized. As the blister enlarges, there is a higher porosity through the rubber, and more acid penetrates to the steel.

In the coated areas of the venturi that are flushed with water, and where the coated walls do not experience high gas velocities, the rubber lining seems to be intact. Without a water-protective layer, however, the lining is unsatisfactory. First, an attempt was made to repair it. Unfortunately, field application and repair is quite difficult, as new rubber does not bond well to cured rubber. A 100% solids resin product was used next. The basic resin compound is acid-resistant and bonds well to most common surfaces, such as metal, masonry and rubber. The product used is applied similar to plaster,—i.e., troweled, to a 3.2-mm thickness.

Primary exhaust stack—Basically, all the acid is removed by the air pollution control train before it reaches the fan and stack, so it would be safe to assume that the corrosion levels within the 60-m steel exhaust-stack would be quite small. This assumption, however, is not correct. The gas stream at the point of the stack inlet is for the most part saturated or supersaturated with water vapor. As the gases rise through the stack, condensation occurs with the decrease in temperature. The condensate returns to the bottom of the stack, and is revaporized by the warmer temperature of the incoming gas stream.

Essentially, the stack acts like a reflux condenser that tends to concentrate the small quantity of acid present into a very high strength. Samples of water collected from the stack water drain contain acid levels so high that the pH measurement registers zero. Because of this high acid condition, the steel stack requires protection.

The original coating was an epoxy resin applied with a 5–10-mil thickness. The particular coating picked was not adequate for protection because of the extremely high acid levels. Blistering and cracking of the coating occurred, especially at the weld seams and at the opposite end of the stack discharge where erosion contributed to the deterioration. Patching the coat was somewhat fruitless because proper surface preparation was difficult. Tapering or feathering attempts on the existing coated areas also failed. The coating did not bond well to itself in a cured state, and thus random patchwork only postponed the inevitable task of complete refinishing.

The coating system that was studied, tested and used for the complete stack recoating was a five-part epoxy-resin series. Two separate resins were used, and until now, they have extended the coating life about fourfold.

Two major facts became apparent during the coating tests: the importance of a multinumbered coating system and the importance of a more-than-adequate surface preparation.

Conclusion—Even though the corrosion potential of the airstream and the scrubber wastewater is high, the proper coatings and their application have been successful in minimizing the effect. Coatings within the air-pollution-control water-scrubbing systems are now being applied to properly prepared surfaces and in thickness levels necessary to overcome erosion, the probability of inadequate covering, and coating deterioration.

Bearings, pumps, hydraulic systems, etc., all require various amounts of repair, but these are for the most part not unique to incineration systems.

Generally, maintenance costs have averaged about 5–6% of the capital cost on an annual basis. In terms of unit cost per disposal unit this figure averages about $14 per ton. Generally, maintenance costs are significant, but comparable with those of other waste-disposal systems.

Energy considerations

The incinerator expends energy to maintain the temperatures necessary for proper combustion. Presently, about 75% of a drum of liquid scrap is needed for the incineration of one drum of nonpumpable scrap. If the pumpable scrap volumes for one week are lower than normal, auxiliary fuel is required to continue operation. Each manufacturing operation will be different regarding the amount of auxiliary fuel needed to supplement the liquid-scrap quantities. As presently operated, a minimum amount of energy is required, regardless of the form of that energy.

In addition to the auxiliary-fuel requirement, another energy consideration that has been studied is the possibility of utilizing the heat generated in the incinerator to produce steam. 3M employed a consultant to make a feasibility study of waste-heat recovery from the facility. A pilot-size boiler was installed at the secondary combustion chamber, and a small fan induced a flow of the 870–980°C (1,600–1,800°F) gas stream into a fire-tube boiler. Steam was produced from this unit, and operating data was collected for an economic assessment.

Basically, the project is not economically attractive because a sizable distribution system is required. Two other major problems are corrosion and particulate buildup in the boiler tubes. In addition, a backup fuel source or a complete boiler would be needed just to be on the safe side. For this system, therefore, waste-heat recovery is probably not feasible at the present time.

Meet the authors

C. Randall Lewis is an advanced environmental engineer whose responsibilities with the 3M Co. (3M Center, St. Paul, MN 55101) include the design, construction, startup and operation of pollution control facilities. He has an M.S. degree in chemical engineering from the University of Kentucky.

Richard E. Edwards is a senior environmental engineer in the environmental engineering and pollution control department of 3M Co. (3M Center, St. Paul, MN 55101). He has an M.S. degree in sanitary engineering from the University of Missouri.

Michael A. Santoro is supervisor of the incineration facility and of dry-scrap disposal for the 3M Co. (3M Center, St. Paul, MN 55101). He has a B.S. in chemistry from the University of Cincinnati and an M.S. in sanitary engineering from Purdue University.

How To Burn Salty Sludges

Fluid-bed incineration can handle the sludges left from most waste treatments.
The trick is to avoid eutectic melting points in the fluid beds.

CLARENCE J. WALL, JOHN T. GRAVES and ELLIOT J. ROBERTS, Dorr-Oliver Inc., Stamford, Conn.

The fluid bed is now accepted as a common unit operation—Dorr-Oliver alone has made over 550 commercial installations of fluidized beds worldwide. In the chemical and metallurgical industries, these fluid beds have been applied to such services as roasting, calcining, drying, sizing and incineration.

The last of these, incineration, is perhaps the fastest growing today, because of the importance of environmental control. Liquid-waste treatment processes and equipment are well developed and generally known to industry. But all the biological, chemical and mechanical methods of waste treatment have one bad thing in common: Sludge!

When the final finished and polished effluent passes through the decorative fountain or goldfish pond on its way to the nearest lake or river, it has left behind a load of wet bothersome solids. A 100,000-bbl/day refinery, for example, will generate from 85 to 190 tons/day of sludge, of which 3-6 tons may be oil, 4-8 solids, and the other 92% water.

The fluidized-bed reactor permits environmental control engineers to burn this sludge in a safe, simple, enclosed system, without generating any additional air, water or land pollution, oxidizing all unusable organics, and avoiding generation of CO and NO_x, to produce an ash that is either inert or a potentially valuable chemical byproduct. Furthermore, since fluid-bed incineration uses the heat value in the wastes, proper sludge preparations may permit it to run continuously without burning a drop of recoverable oil, and even generating byproduct steam.

To illustrate the growth of fluid-bed incineration: Dorr-Oliver entered this field with pilot-plant work at New Rochelle, N.Y., in 1960, and built its first fluid-bed incinerator (a small 4-ft-dia. unit for handling 210 lb/h of primary sewage sludge) at Lynwood, Wash., in 1962. Since then, we have supplied a total of 93 units, including:

municipal sewage sludge	74
carbon black waste	1
NSSC paper mill waste liquor	3
pharmaceutical waste	1
grease from domestic sewage	1
industrial waste activated sludges	5
spent coffee grounds and tea leaves	1
domestic garbage	1
oil refinery waste sludges	6

The 74 municipal-sewage-sludge systems include: primary sludge, secondary or activated sludge, digested sludge, heat-treated sludge, sludge containing precipitated chemicals resulting from phosphorus removal from the sewage, and various combinations of these. Plant capacities per incinerator unit vary from 210 to 3,500 lb/h of total dry solids. Total dry solids in the sludge feed range from 10% to 40%. The reactors' internal diameters range from 4 to 21 ft; and plants are operating in Canada, England, France, Germany, Italy, Japan, South Africa and the U.S.

Problem: Complex Mixtures Can Melt

These wastes are chemically quite complex. In addition to organic or petroleum hydrocarbons, they may contain such elements as Na, K, Mg, S and P. Also, various chemicals are added in wastewater treatment, or en-

FLUID-BED INCINERATORS like this can burn sludge in a safe enclosed system without generating pollution.

Originally published April 14, 1975.

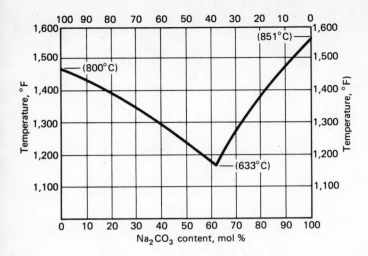

MELTING POINTS of NaCl-Na₂CO₃ system—Fig. 1

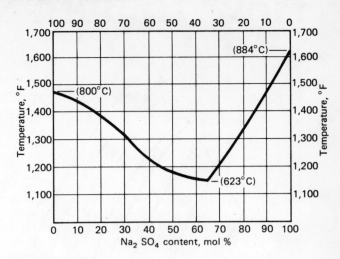

MELTING POINTS of NaCl-Na₂SO₄ system—Fig. 2

ter with rainwater runoff, etc., so that elements such as Fe, Al and SiO_2 (sand and silt) and Al_2O_3 (clay) may also be present. For seacoast refineries, seawater is a probable component; and sulfur may be present in auxiliary oil fuel supplied to the incinerator.

Such complex mixtures can bring on problems, as is perhaps best illustrated by the incineration of Neutral Sulfite Semi-Chemical (NSSC) waste liquor from paper mills. The pelletized ash product of these incinerators is primarily Na_2SO_4, but also contains some Na_2CO_3, depending on the Na/S ratio and the possible addition of sulfur in the waste liquor. Also present are minor amounts of other compounds derived from impurities in the makeup chemicals and the wood.

Now, pure Na_2SO_4 has a melting point of 1,623°F, and pure Na_2CO_3 has a melting point of 1,564°F—both high enough to keep fluid-bed particles dry and hard during incineration.

However, mixtures of these two compounds form a solid solution having a melting point lower than either; and at 47% Na_2SO_4, 53% Na_2CO_3, the melting point is 1,552°F. Moreover, for NSSC pulp mills, the makeup chemicals contain chloride impurities, which appear in the ash product as NaCl; and in combination with Na_2CO_3, NaCl forms a eutectic mixture having a melting point as low as 1,172°F at 62 mol% Na_2CO_3 (Fig. 1). Likewise, mixtures of NaCl and Na_2SO_4 form a low-melting mixture having the eutectic melting point of 1,154°F at 65 mol% Na_2SO_4 (Fig. 2). And when all three of these compounds are present, a mixture with a melting point as low as 1,134°F is possible (Fig. 3).

Since NaCl has a significant vapor pressure at the normal operating range for incineration of most waste sludges (1,400-1,600°F), one would expect it to vaporize completely (Fig. 4). With both Na_2SO_4 and NaCl present, however, the two compounds form the low-melting eutectic, and if this eutectic is allowed to accumulate in a fluid bed, defluidization will take place.

Accordingly, with any substantial amount of, say, more than 500-600 ppm of Cl in the sludge, serious prob-

lems can be encountered; and some sludges may contain upwards of 50% seawater, while the water phase of other waste streams may be almost all seawater.

Although seawater contains calcium and magnesium that should tie up sulfate radicals as $CaSO_4$ and $MgSO_4$, respectively, its incineration causes up to 70% of the sulfate to appear in the ash as Na_2SO_4, which may occur according to the reactions:

$$CaSO_4 + 2NaCl + H_2O \xrightarrow{5/8} Na_2SO_4 + 2HCl + CaO \quad (1)$$

$$MgSO_4 + 2NaCl + H_2O \xrightarrow{5/8} Na_2SO_4 + 2HCl + MgO \quad (2)$$

Also, with both sulfur and seawater in the waste sludge, another reaction would be:

$$S + 11/2O_2 + H_2O + 2NaCl \xrightarrow{5/8} Na_2SO_4 + 2HCl \quad (3)$$

If the particles of a fluid bed are silica-sand, or if there is SiO_2 in the waste sludges, the Na_2SO_4 generated from these reactions will react with the silica:

$$Na_2SO_4 + 3SiO_2 \xrightarrow{5/8} Na_2O \cdot 3SiO_2 + SO_2 + 1/2O_2 \quad (4)$$

to form a very sticky viscous sodium-silicate glass, which can cause very rapid defluidization of the bed. Also, the silica can react with salt:

$$3SiO_2 + 2NaCl + H_2O \xrightarrow{5/8} Na_2O \cdot 3SiO_2 + 2HCl \quad (5)$$

Some of the phase diagrams of $Na_2O \cdot 3SiO_2$ show eutectic melting points as low as 1,175°F (Fig. 5, 6, 7).

Resolution: Make Metal Silicates

On the other hand, the reactions of Eq. (4) and (5) do eliminate Na_2SO_4 and NaCl, as well as the eutectic mixtures formed of these compounds. And if metal oxides (CaO, Fe_2O_3 and Al_2O_3) are present, these will react with the alkali silicates to form high-melting silicate compounds. This final end-product of metal oxide - sodium oxide - silicon dioxide must have, and generally does have, a high-temperature melting point well above the temperature required for vaporizing the alkali metal chlorides and operating the fluid bed.

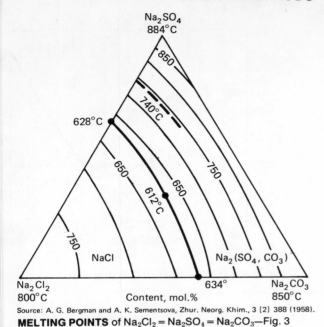

Source: A. G. Bergman and A. K. Sementsova, Zhur. Neorg. Khim., 3 [2] 388 (1958).

MELTING POINTS of $Na_2Cl_2 = Na_2SO_4 = Na_2CO_3$—Fig. 3

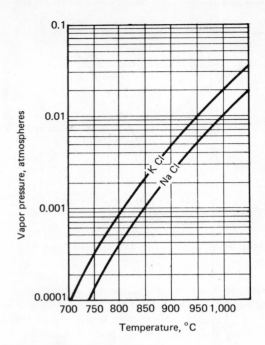

VAPOR PRESSURES of NaCl and KCl—Fig. 4

Some of the corresponding reactions are:

$$Na_2O \cdot 3SiO_2 + 3SiO_2 + 3CaO \xrightarrow{5/8} Na_2O \cdot 3CaO \cdot 6SiO_2 \quad (6)$$

This product, called devitrite, melts at 1,885°F. Also:

$$Na_2O \cdot 3SiO_2 + CaO \xrightarrow{5/8} Na_2O \cdot 2CaO \cdot 3SiO_2 \quad (7)$$

This product melts at 2,343°F. The phase diagram for Na_2O-CaO-SiO_2 is shown (Fig. 8).

When iron oxides are present, the reaction may be:

$$Na_2O \cdot 3SiO_2 + Fe_2O_3 + SiO_2 \xrightarrow{5/8} Na_2O \cdot Fe_2O_3 \cdot 4SiO_2 \quad (8)$$

Acmite, the product of this reaction, melts at 1,751°F.

For such reactions to be effective, both the SiO_2 and the metal oxides must be so finely divided as to react. In many cases, both the SiO_2 and the metal oxides are present in the waste sludge, and if of sufficiently small particle size, may combine in the manner required. In other cases, the compounds will have to be added to the sludge feed.

Clay is a natural mixture of hydrous aluminum silicates, with SiO_2/Al_2O_3 generally 2/1 to 3/1, and normally available in a very fine state. Clay, therefore, is a very convenient source of both SiO_2 and Al_2O_3, supplying both the SiO_2 for the decomposition of Na_2SO_4 to make $Na_2O \cdot 3SiO_2$ and the Al_2O_3 to react with this and form crystalline high-melting-point sodium-aluminum silicates. The reaction may be represented by:

$$Na_2O \cdot 3SiO_2 + Al_2O_3 + 3SiO_2 \xrightarrow{5/8} Na_2O \cdot Al_2O_3 \cdot 6SiO_2 \quad (9)$$

Albite, the product of this reaction, melts at 2,026°F, but if there is no SiO_2 available, the reaction may be:

$$Na_2O \cdot 3SiO_2 + Al_2O_3 \xrightarrow{5/8} Na_2O \cdot Al_2O_3 \cdot 2SiO_2 + SiO_2 \quad (10)$$

The SiO_2 formed by this reaction may go to produce albite, while nepheline, the other product, may combine with albite to form eutectic melting at 1,954°F. The phase diagram for the $Na_2O = Al_2O_3 = SiO_2$ system is shown (Fig. 9).

With kaolin clay, the clay first dehydrates:

$$Al_2O_3 \cdot 2SiO_2 \cdot 2H_2O \xrightarrow{5/8} Al_2O_3 \cdot 2SiO_2 + 2H_2O \quad (11)$$

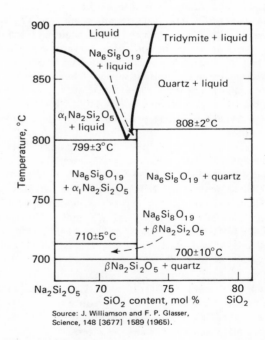

Source: J. Williamson and F. P. Glasser, Science, 148 [3677] 1589 (1965).

PHASE DIAGRAM for Na_2O-SiO_2 mixtures—Fig. 5

The dehydrated clay then reacts with NaCl and H_2O:

$$Al_2O_3 \cdot 2SiO_2 + 2NaCl + H_2O \xrightarrow{5/8} 2HCl + Na_2O \cdot Al_2O_3 \cdot 2SiO_2 \quad (12)$$

This reaction Eq. (12) occurs "in flight," and ties up the NaCl directly as a high-melting-point nepheline without going through any of the other intermediate reactions. Reactions with small amounts of potassium chlo-

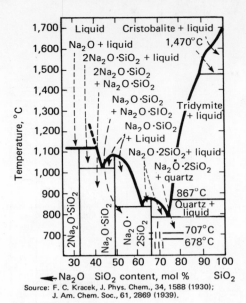

PHASE DIAGRAM for Na₂O-SiO₂ mixtures—Fig. 6

Source: F. C. Kracek, J. Phys. Chem., 34, 1588 (1930);
J. Am. Chem. Soc., 61, 2869 (1939).

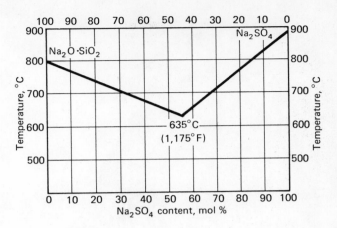

MELTING POINTS of Na₂O·SiO₂-N₂SO₄—Fig. 7

|||||||||||||||||||||||||||||||||||||||||||||||||||||||||||||||||||||||||||||

Oily Waste Mixture — Table I

Source	Water, %	Combustibles Tons/D	Total Tons/D
Flotation clarifier skimmings	90	20	200
API Separator sludge	85	6	40
Tank cleanings & misc. oils	50	35	70
Biological treatment sludge	70	2.1	7
Spent caustic	90	—	5.6*

*5.6 TPD Caustic

|||||||||||||||||||||||||||||||||||||||||||||||||||||||||||||||||||||||||||||

Analyses of Typical Bed Product — Amoco, Whiting, Indiana — Table II

Chemical Analyses

Water Soluble		Water Insoluble	
Element	Amount	Component	Amount
Na	5.34%	Al_2O_3	12.62
Cl	150 ppm	Fe_2O_3	6.69
S^{+6}	4.12%	SiO_2	23.73
Ca	810 ppm	Na	1.94
Mg	1 ppm	Ca	15.58
Al	37 ppm	Mg	2.76
K	3641 ppm	P_2O_5	2.58
Cr	250 ppm	LOI	2.44
		S^{+6}	5.79
Total	9.95%		74.13

Probable Compound Composition

Water Soluble		Water Insoluble	
Compound	Weight %	Compound	Weight %
Na_2SO_4	16.45	$CaSO_4$	24.62
K_2SO_4	0.80	$Ca_3(PO_4)_2$	4.77
NaCl	0.02	$Na_2O·Fe_2O_3·SiO_2$	16.31
$CaSO_4$	0.27	$CaSiO_3$	10.32
		$MgSiO_3$	10.00
		CaO	2.02
		Al_2O_3	10.66
Total	17.54		78.70

|||||||||||||||||||||||||||||||||||||||||||||||||||||||||||||||||||||||||||||

ride from seawater would be the same, so that this in-flight reaction between the clay and the alkali chloride avoids the Na₂SO₄, Na₂O·SiO₂, Na₂SO₄-NaCl mixture, and only the CaSO₄, MgSO₄, MgO and Na₂O·Al₂O₃·3SiO₂ compounds are formed.

Alternative: Avoid Low Temperatures and Cl

In plants where the waste sludge contains valuable chemicals, it may be desirable to avoid forming the high melting compounds so as to produce a pelletized ash that can be sold for its chemicals. In such instances, fluid-bed incineration operations can be carried out with limited amounts of chlorides and by limiting the incineration temperature, as evidenced by plants operated by American Oil Co. at Whiting, Ind., and by the Green Bay Packaging Corp., at Green Bay, Wis.

The Amoco incinerator at Whiting, on the shores of Lake Michigan, has been operating since September 1972. This system has a design capacity of more than 370 tons/d of wet sludge of 85-95% water. It is the largest fluid-bed incinerator in the world. The 40-ft-high incinerator is 20 ft I.D. at the base, tapering outward to 28 ft at the top. The waste sludge (Table I) comprises several waste streams, including: API separator sludge, water-oil emulsions from slop-oil recovery, tank cleanings, sludges developed in flotation clarifiers in processing sanitary wastewater, low-strength caustic, and other waste liquids. In addition to hydrocarbons, this sludge also contains such alkali elements as Ca, Mg, K and Na, and also some sulfur, Fe, Al and phosphates.

The original silica-sand bed has been replaced by particles of chemical solids, which are nodules composed mainly of aluminum oxide (derived from alum used as flocculant in the wastewater-treatment plant) but also containing oxides of the various elements, including Fe, which gives the product a red color (Table II).

Because of the chemical composition of this material, the fluid bed is maintained at a temperature of about 1,330°F to avoid softening or melting the bed product.

There has been a problem of a deposit of a relatively soft ash, which builds up on the hot-gas side of the heat-

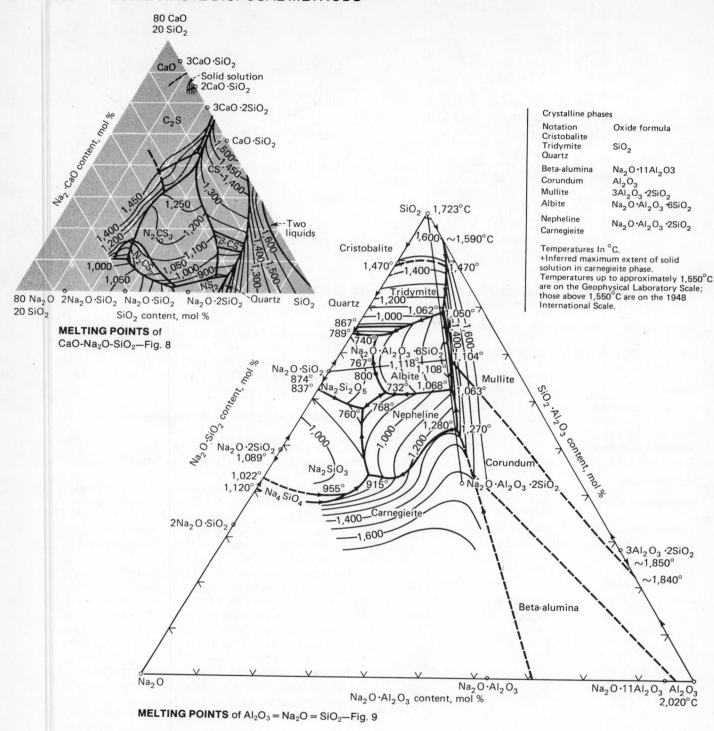

MELTING POINTS of CaO-Na₂O-SiO₂—Fig. 8

Crystalline phases	
Notation	Oxide formula
Cristobalite	
Tridymite	SiO₂
Quartz	
Beta-alumina	Na₂O·11Al₂O₃
Corundum	Al₂O₃
Mullite	3Al₂O₃·2SiO₂
Albite	Na₂O·Al₂O₃·6SiO₂
Nepheline	Na₂O·Al₂O₃·2SiO₂
Carnegieite	

Temperatures in °C.
+Inferred maximum extent of solid solution in carnegieite phase.
Temperatures up to approximately 1,550°C are on the Geophysical Laboratory Scale; those above 1,550°C are on the 1948 International Scale.

MELTING POINTS of Al₂O₃ = Na₂O = SiO₂—Fig. 9

exchanger tubes. Possibly, a small amount of liquid phase (0.1-0.2%) in the fine ash in the incinerator exit gases could be the cause of this sticking—i.e., a liquid eutectic mixture is present at the temperatures involved.

At Green Bay Packing Corp., spent NSSC liquor (a mixture of digested organics representing about one-quarter the weight of the wood, together with the salt residue of the cooking liquor) is incinerated in a fluid-bed system. The organics are completely burned, and sulfur is added, so as to produce a salt cake of 90-95% Na₂SO₄ and some Na₂CO₃ as pellets for a kraft-pulp mill.

In this operation (Fig. 10), concentrated waste liquor is fed directly into an upper fluid bed where it is oxidized to CO₂ and H₂O, and the ash is formed into pellets. A lower fluid-bed stage cools the product pellets and provides some air preheating. Hot (1,350°F) exhaust gases from the combustion compartment go directly to a venturi scrubber-evaporator, where they are cleaned and cooled in heat exhange with incoming waste liquor.

Two additional installations of this type have recently gone onstream. Plant capacities for the three installations range from 6,000 to 12,500 lb/h of total solids at a feed concentration of about 40%, which corresponds to a water load of 9,000-18,750 lb/h. The fluosolids reactors range from 14 to 20 ft I.D.

At Green Bay Packing Corp., the chloride content of

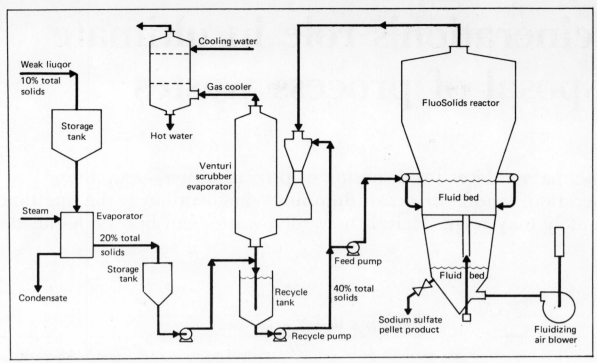

FLUOSOLIDS INCINERATOR FLOWSHEET for neutral sulfite semi-chemical waste liquor—Fig. 10

the waste liquor is adjusted so as to maintain about 3,000-3,500 ppm Cl in the Na_2SO_4-Na_2CO_3 ash product, which normally contains about 93% Na_2SO_4. Under these controlled conditions, the fluid bed is operated in the 1,350-1,380°F temperature range to maintain efficient combustion of the organics with a minimum of excess air. When the chloride content goes above this value, the pellets become soft and sticky, requiring lower temperatures, with a deterioration in combustion and more excess air required. Also, there occur scale deposits, with eventual plugging of the exit-gas duct.

Summary

Recently completed pilot-plant tests have shown that the addition of kaolin clay essentially eliminates the buildup of molten salts and stickiness in fluidized beds. The reactions between the sodium salts and clay are essentially 100% complete, while the amount of insoluble sodium salts retained by the fluid bed is very low. If the sodium salts in the feed to the incinerator can be completely reacted with the clay before they leave the reactor, the sticky compounds can be avoided in the exit gas, allowing for a heat exchanger for heat recovery.

When a sand bed is used, or when there is coarse silica in the waste-feed stream, the addition of clay leads to the formation of nepheline, which melts at about 2,336°F. If any Na_2SO_4 is present, the addition of clay leads to formation of nepheline and the liberation of SO_2.

With the metal oxides of CaO and Fe_2O_3, it is possible that some of the sticky sodium compounds such as Na_2SO_4, Na_2SO_4-NaCl mixture, and $Na_2O\cdot3SiO_2$ will form, because Eq. (6) and (7) for CaO, or Eq. (8) for Fe_2O_3 can only take place after Eq. (4) has occurred. If

Eq. (6) or (7) for CaO or Eq. (8) for Fe_2O_3 are not completed before the ash-laden incineration gases leave the reactor, some duct scale may be formed. The same situation applies as per Eq. (9) and (10), if Al_2O_3 is used with SiO_2 as a separate additive.

If for some reason clay is not used, the fluid-bed reactor would be designed with a special gas-outlet duct to avoid or minimize duct scale.

Investigation of duct-plugging problems has traced their cause to the presence of alkali elements and chlorides in the sludge feed.

Meet the Authors

J. T. Graves   E. Roberts   C. J. Wall

John T. Graves is Manager, Eastern Sales Region, Thermal Processing, at Dorr-Oliver Inc., 77 Havermeyer Lane, Stamford, CT 06904. He joined Dorr-Oliver after gaining experience as a sanitation engineer with Chicago's Water Purification Div. He has a BS in chemical engineering, and has been active in research and development since starting his career with Dorr-Oliver.

Elliott Roberts is actively involved with Dorr-Oliver Inc. as a consulting physical chemist. He joined the Dorr Company in 1928, as a chemist, and eventually became Chief Scientist before retiring. He holds a Ph.D. in physical chemistry from Yale, and is active in a number of professional societies and organizations.

Clarence J. Wall is a senior engineer in Dorr-Oliver's Thermal Products Technology Div. He joined the company in 1946 as a junior technician. He holds a BS in metallurgical engineering from South Dakota Mines U. Tech., and is active in the American Institute of Mechanical Engineers and several fraternal organizations.

Incineration's role in ultimate disposal of process wastes

Of the main choices for disposing of process wastes—recycling, incineration, landfill or ocean dumping—incineration is the one that ultimately may reign. Here is how such wastes can best be handled.

Kenneth S. Dunn, The Incinerator Co.

☐ Solid and process wastes are mostly disposed of today by incineration and sanitary landfill. As environmental protests mount over land-use allotments for landfill, pressure will build on industry to either recycle process wastes or incinerate them. Many cannot be economically recycled; here is a discussion of how they can be incinerated.

The incineration process should be a clean, high-temperature operation, particularly in the efficient destruction of industrial wastes—many of which are complex hydrocarbon compounds that will only break down under high-temperature conditions. Because stack emissions give rise to public concern, the combustion engineer must have many basic considerations in mind when designing incineration equipment.

For instance, smoke is combustible particulate matter, either solid or liquid, in the gaseous products of combustion, and since visibility depends on the absorption and scattering of light, a large number of very small particles will produce a smoke far more dense than a gas stream containing a smaller number of large particles. When volatiles are driven off from newly charged incinerable wastes they contain not only gases, but liquids in aerosol form. In a high-temperature environment with adequate air and turbulence to ensure good mixing, the oxidation process of combustion is completed very rapidly. If, however, this is not the case, then the liquid content will condense out into small droplets which, in turn, can be converted to solid particles through a combination process of pyrolysis and thermal cracking. Once formed, carbon particles in the gas stream require a temperature in excess of 1,470°F (800°C) to oxidize them. Even at these temperatures the speed of reaction with air is slow, and prevention is thus far better than cure [1].

Afterburning sections, in modern incinerators, provide a useful extension to the high-temperature gas path, and the use of such burners can be minimized if furnace design is tailored for the wastes to be properly burned. In smoke prevention, there is no substitute for the proper management of fuel and air, and it is in this area that combustion becomes an engineering art rather than a science. Odor too can, in most cases, be dealt with satisfactorily at high temperature, i.e., in the range 1,650–2,190°F (900–1,200°C).

As with any piece of engineering equipment, incinerator design includes many factors that can affect the final result. If an incinerator plant does not operate satisfactorily, the basic problem may be that of poorly designed equipment. On the other hand, well-designed equipment may be for some reason unsuited to an application, and here lies the key to the solution of many of the problems.

Incinerators have often been represented, quite wrongly, as saving little and producing nothing. However, it is not so long ago that equipment in industrial boiler houses was regarded in much the same light—that is, until the cost of producing steam became a really significant percentage of total production costs. This type of cost situation in highly industrialized countries is affecting the choice of disposal methods. It is a situation of special significance in the oil and petrochemical industries in view of their inherently difficult wastes and the increasingly stringent requirements for disposal in an environmentally acceptable way.

It is now no longer a question of "Shall we dump or shall we burn," but rather, "Can we reclaim a product or recover waste heat from the waste?" It is in this context that incineration as a final disposal process comes into its own.

Originally published October 6, 1975.

Assessing the requirement

Incinerators, particularly those for firing solid wastes, can be notoriously accommodating in attempting to dispose of whatever may be pushed into them. Unlike other engineering equipment that will "blow a fuse," fail to run if fed with the incorrect fuel or break down completely if abused, the incinerator will keep on burning—if only inefficiently. Under these conditions, emissions will become a problem. This situation can be avoided at the outset by a careful investigation of the sources of waste material likely to be involved. It is surprising just how many departments in a factory will suddenly find that they have "a little waste to dispose of" once a unit is installed on a site, after fruitless efforts of a plant manager to interest them in the project at the outset.

In every particular instance, the prerequisite for the successful application of a special-purpose incinerator is an accurate assessment of the types and quantities of waste to be fired, coupled with a careful study of the handling problems involved. This careful study requires close collaboration between user and supplier, for unless there is, at the outset, a thorough understanding of all the problems involved, then any scheme can be doomed to failure even before it leaves the drawing board.

Range of wastes

It is often found that detailed consideration has not been given to the disposal of all the plant wastes. Of course, in some cases, materials are self-excluding on an economic or technical basis, but in others the firing of increased quantities of wastes can make a considerable difference to the viability of an onsite incineration project. Quite often a wide variety of wastes—namely solids, liquids, gases and sludges—can effectively be destroyed in a single plant using "clean"

liquids for ignition and/or afterburning purposes. Additionally the cost of gas-washing equipment is less expensive for a combined unit than for a number of individual ones, and operational staffing is simplified.

In any event, the method used to determine the size and type of an incinerator for any particular purpose is much the same in each case. For every item to be burned it is necessary to know:
- The weight and volume produced daily and over a period.
- Whether it is produced on a batch or continuous basis.
- What methods of collection and handling are proposed.
- What methods of feeding are required.
- The intended daily firing period.
- Whether waste-heat recovery is required.

Basics of sound design

Having briefly described some of the pitfalls, we can establish that no furnace will satisfactorily destroy organic material unless:

(1) the furnace volume is sized for the physical bulk of material, as well as its calorific value (CV), bearing in mind that in the case of solids, the average waste contains, say, 60% volatiles and 10% ash, enabling only 30% of combustion of waste to take place on the grate;

(2) the correct combustion-air requirements are met and velocities selected to ensure thorough mixing through turbulence in the combustion zone. Entrainment of ash particles in the gas stream must be kept to a minimum;

(3) the high-temperature zones have a gas path of sufficient length and volumetric capacity to complete the combustion of volatiles. Theoretically 0.5–0.75 s is required to ensure combustion under correct conditions of temperature and mixing;

Ultimate and proximate analysis of some plastics materials	Polyethylene	Polystyrene	Polyurethane	Polyvinyl chloride
Higher heating value, Btu/lb	19.687	15.816	11.203	9.754
Ultimate analyses in percentage by weight				
Moisture	0.20	0.20	0.20	0.20
Carbon	84.38	82.13	63.14	45.04
Hydrogen	14.14	7.14	6.25	5.60
Oxygen	0.00	9.72	17.61	1.56
Nitrogen	0.06	0.32	5.98	0.08
Sulfur	0.03	0.02	0.02	0.14
Chlorine	tr	tr	2.42	45.32
Ash	1.19	0.47	4.38	2.06
	100.00	100.00	100.00	100.00
Proximate analyses in percentage by weight				
Moisture	0.20	0.20	0.20	0.20
Volatile matter	98.54	99.22	87.12	86.89
Fixed carbon	0.07	0.11	8.30	10.85
Ash	1.19	0.47	4.38	2.06
	100.00	100.00	100.00	100.00

Sequence of incinerator design procedures　　　　　　　　　　　　　　　　T/II

Category	Analysis	Handling and combustion	Construction materials	Emission control
Solids	Calorific value	Waste transfer	Refractories	Afterburners
Semisolids	Chemical composition	Blending	Insulation	Fume destruction
Sludges	Quantity	Feeding	Casing	Trapping incombustibles
Liquids	Physical characteristics	Firing		Stack design
Gases		Furnace volume		Ash removal
		Air, temperature, time		Washwater treatment
		Instrumentation		
		Overall control		

(4) full account is taken of where the process of oxidation is to be carried out, i.e., grate or gas space. It has already been pointed out that the CV of industrial wastes varies widely but so, of course, does their chemical composition. For example, with many plastics, the volatile content is very high and in these circumstances almost all combustion takes place in the gas space. Table I (T/I) sets out the characteristics of some of the well-known materials.

A consideration of the above points will illustrate how unlikely is is that any standardized design will be found suitable for a specialized application. Almost invariably, successful installations are those that are specifically tailored to suit a particular requirement. The sequence to be followed in arriving at an optimum design for an incinerator for a particular waste or mixture of wastes is indicated in T/II, and although incinerator design is obviously within the province of a particular n.anufacturer, it is given here to alert the prospective user to be aware of what the supplier needs to know in order to establish parameters for a design.

Types of incinerators

A guide to the suitability of various types of equipment is given in T/III. While not exhaustive, it includes the well-proved designs that can be fitted with gas-cleaning equipment.

Opinions, of course, vary as to the suitability of particular designs for a range of applications, but it is generally considered that two types, namely the multichamber, multicell units and the rotaries, are the most flexible. In terms of plant arrangement for mixed-waste firing, the "inline" units have a distinct advantage, for not only does the design allow the products of combustion from one cell to be used to bring succeeding cells up to temperature, but it is also capable of providing considerable flexibility for the arrangement of firing equipment and ash removal on either side of the plant. In Britain, such systems are favored and combination units have been in operation, burning a wide variety of wastes, for well over 20 years [2]. Fig. 1 (F/1) shows the flowsheet for a paint-residues plant. There are other similar types of incinerators that are installed in chemical complexes to destroy a great variety of mixed wastes. In the U.S., the first composite "inline" units are now in operation [3] and more are planned.

Rotary-kiln-type furnaces have been used with success in Europe and the U.S., and are now firing mixed residues that include up to 30% w/w of plastics in the total material fired. The plant consists of a refractory-lined cylinder slightly inclined to the horizontal at an angle usually variable between 2–5° and rotating at slow speed (4–5 rpm). Often both the speed of rotation and the inclination of the furnace are variable so that

Types of incinerators suitable for various wastes　　　　　　　　　　　　　　　T/III

Solids	Liquids	Sludges	Combined liquids and sludges	Drummed wastes
Single-cell	Multicell	Multicell (with lagoon hearths and sludge burners)	Multicell	Multicell (with added drum cell)
Multicell	Submerged combustors (for concentration of waste waters)		Rotary	
Rotary		Rotary (for dewatered sludges)		
Fluidized bed	Vortex			
Multiple hearth		Multiple hearth		

the flow of material through the cylinder and the retention time for combustion can be controlled. If liquids are being fired as well as mixed solids then these are usually counter-fired at the opposite end of the furnace from the solid waste feed. Afterburning facilities can be incorporated in a separate chamber, and the equipment generally lends itself to flexible plant layout.

By rotation, these furnaces offer the advantage of a gentle and continuous mixing of the incoming wastes, but capital and maintenance costs are high. These derive from the mechanical design requirements, e.g., rigidity of the cylinder and close tolerances for the roller path drive as well as the high-temperature seals between fixed and moving parts. Another major disadvantage is that the cascading action of the burning waste, as the kiln rotates, gives all the fine ash in the charge ample opportunity to become entrained in the exhaust gases. The linings suffer both from abrasion and from contact with distillation products that can penetrate the surface before burning, thus making it necessary to install dense refractory material of higher grade than normally required in stationary furnaces.

Rotaries can, in some circumstances, be extremely suitable for sludge burning, and some very interesting comparisons between their performance and costs against rotary hearths and fluidized beds will be found in a paper by Hescheles and Zeid [4]. There are also examples to be found of rotaries installed in combination with separate liquids furnaces, providing a higher degree of flexibility in the overall design of a multipurpose destruction unit [5].

Fluid-bed incineration

The fluidized bed incinerator is simple and compact, using sand for the bed medium. If solid waste is to be fired, it must be shredded or otherwise broken up before being introduced; because of the use of sand, the bed temperature must not exceed 2,000°F (1,090°C). Additionally, the gas flowrate must be less than the terminal velocity of the sand. Advantages can be claimed for this type of unit. The enriched oxygen of the bed coupled with mixing of the sand and waste ensures complete combustion, prevents carbon monoxide emissions and minimizes hydrocarbon emissions; the low uniform temperature within the bed limits the formation of nitrogen oxides. The facility to add limestone to the bed material means that the emission of HCl and sulfur oxides, if present, can be eliminated.

To date, the principal field of fluidized-bed incinerator application has been in the refinery and paper industries where, despite high installed costs, the economics, with waste-heat recovery, are acceptable. In other fields, such as burning of municipal sludge [4], operating costs have proved to be very high in comparison with other designs and, in consequence, their use is presently restricted. However, much development work is at present being undertaken in order to apply the theories to refuse-burning boiler plants, and there is no doubt that in the course of this work, acceptable solutions will be found to some of the existing problems. As a method, the fluidized bed process cannot be considered, at this time, to have a high degree of flexibility in handling a wide range of wastes, and power consumption will of necessity remain inherently high.

Multiple-hearth adapted

The multiple-hearth incinerator found many successful applications in the metallurgical industry before being applied to municipal and industrial wastes. It consists of vertically stacked hearths with rabble arms

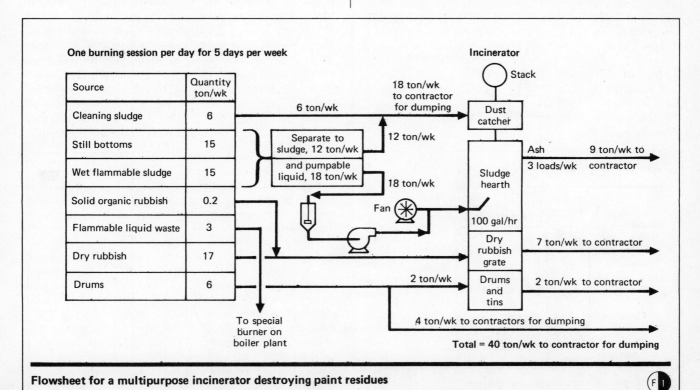

Flowsheet for a multipurpose incinerator destroying paint residues

attached to a central shaft that is rotated by a drive at the base of the unit. Wastes are fed in at the top and are moved around each hearth by means of the rabble arm, to an opening through which they are dropped to the hearth below. There is thus some retention time on each hearth, and the process is one of drying and then of burning. Flue gases flow upwards counter to the wastes, which are completely incinerated by the time they are discharged from the lowest hearth. The principal use for these incinerators has been for the destruction of municipal and industrial sludges and, like the fluidized bed, it does not have the flexibility of the rotary and multipurpose inline units.

Submerged combustors are perhaps the most specialized of the designs listed in T/III. They were originally developed to improve thermal efficiency of liquid-heating and evaporative processes, using gaseous as well as liquid fuels. Exceptionally high overall thermal efficiencies in the range 85–95% can be achieved by arranging the combustion chamber immediately above or just below the liquid level, thus forcing the combustion products to pass through the liquid by discharging them below the surface. This equipment offers very significant technological advances whenever difficult evaporative processes must be carried out, and is used as an evaporator to concentrate waste liquors that are then atomized and incinerated with auxiliary fuel where necessary. Submerged combustors have been successfully applied to HCl recovery—an area in which very few plants are currently operating in the U.S. [6]. The recent unprecedented rise in the cost of caustic for washwater dosing should make this a field of increasing interest in the disposal of large quantities of chlorinated hydrocarbons [7]. F/2 shows a submerged combustor.

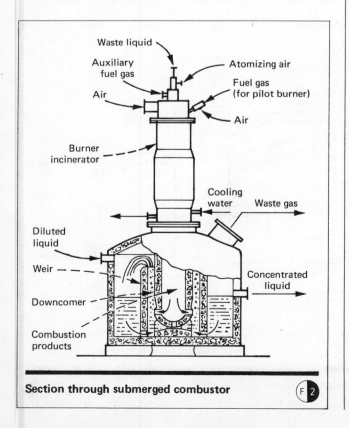

Section through submerged combustor   (F 2)

Waste liquid
Auxiliary fuel gas
Atomizing air
Air
Fuel gas (for pilot burner)
Air
Burner incinerator
Cooling water
Waste gas
Diluted liquid
Weir
Concentrated liquid
Downcomer
Combustion products

Methods of feeding and firing

Solids—The handling of waste products, from production to ultimate disposal, is often neglected in the formulation of an overall scheme for an incineration unit; this is particularly true in the case of solid wastes. Chemical engineers are often more familiar with the problem of moving liquids within a complex, but with landfill now increasingly unacceptable as a solution, their involvement with solid wastes would seem inevitable. Segregation of solid wastes by category is expensive and often impractical, hence the chosen type of furnace must not only have built-in flexibility, but its feeding equipment should be chosen with care. Continuous feeding of the furnace at a controlled rate is preferable to batch loading because it enables control of furnace combustion conditions to be more readily established. This point is often overlooked by those used to operating single or even multiple-stream liquid incinerators.

In such installations, the control of furnace excess air is straightforward because it is via the burner air registers, whereas in a solids unit, large volumes of cold air can be introduced every time a feed door is raised or lowered. One way of tackling the problem is with a hopper feed, by conveyor, coupled with ram-feeding of controlled quantities. Within the furnace this can be allied to a specialized form of reciprocating grate that uses refractory slabs instead of firebar castings, and so overcomes the difficulty of low-ash wastes and the blocking of small air spaces by melting plastic. An example of this arrangement, adapted for the firing of pathological and paper wastes, has been applied with considerable success to the firing of mixed factory waste, carbon paper and paper products.

Liquids—Many different types of burners have been produced to handle the more difficult viscous materials that may also contain particles up to $\frac{1}{8}$-in. dia. in any plane. Typical applications have been in the firing of acid tars, formic, succinic and hydrochloric acids, waste paints and solvents. As already mentioned, these burners provide a sealed system for introducing wastes to the furnace and are equipped with all standard safety devices for flame failure, flow-pressure switches and temperature controllers. Many incorporate precombustion chambers and, if not designed as high-temperature combustors, operate generally on this principle, but with reduced exit velocities from the precombustion chamber. In any application the essential point is the maintenance of a minimum CV to the burner by blending of residues, coupled with monitoring of minimum safe temperature and shutdown on flame failure.

Sludges and semisolids—Apart from their disposal on multiple-hearth units in rotary kilns, smaller quantities—particularly where other high-CV liquid residues are available to assist in the evaporation of surplus water—can be burned very satisfactorily and economically in lagoon-type hearths. Quantities of up to 2,000 lb/h can be handled by this method. Alternatively, sludges can be blended with other liquids to form a mixture that can be fired through a burner, but the economics of blending and handling, together with the relative calorific value of the materials, will govern the practical considerations. Small quantities of sludges contained in drums can be tipped automatically into a furnace through drum ports equipped with hydraulically operated tipping equipment. This method also lends itself to drummed materials that are semisolid when at ordinary temperatures. In this case the drum contents are exposed to furnace temperature and the melted material is then deposited in the lagoon.

Refractories

Most industrial incinerators can expect arduous conditions of service to be imposed upon them, for seldom will a unit be used to destroy the same range of wastes 2 or 3 years from entry into service. Under most circumstances, process wastes change as the process is modified, and the important point to be realized is that, unlike in fired heaters working at controlled temperatures and known gas conditions, the heat flux produced within a furnace will vary with the blend of materials being fired, while the gas composition in terms of contaminants will also depend upon the relative quantities in the mix. Considerable advances have been made in the production of refractory materials. Brick was the first choice for incinerator work a few years ago; now the use of monolithics, which can be defined as refrac-

tories taken into service in an unfired state, is extensive. Some of the advantages offered are the elimination of special shapes, speed of installation, freedom from joints (other than for expansion) and excellent resistance to thermal shock [8].

Responsible manufacturers respond to the need for care in selecting appropriate refractories that offer a lining contoured to give the best life at an economical cost, with minimun downtime for repairs. In terms of contaminants, alkaline noncombustible inorganics—particularly metals such as sodium, potassium, vanadium and titanium—can cause problems, as do also the trace elements such as halogens. Reducing atmospheres or conditions where, for example, a deep firebed restricts the air available for combustion to virtually stoichiometric quantities, can produce very high combustion temperatures.

An example of this can be found in the burning of certain types of plastic film that will produce, under these conditions, a temperature of 4,320°F (2,380°C), which is very much higher than the usual service temperature for a standard industrial unit. These problems have been encountered in service when burning 1.5 tons/h on a two-shift basis, and a satisfactory material

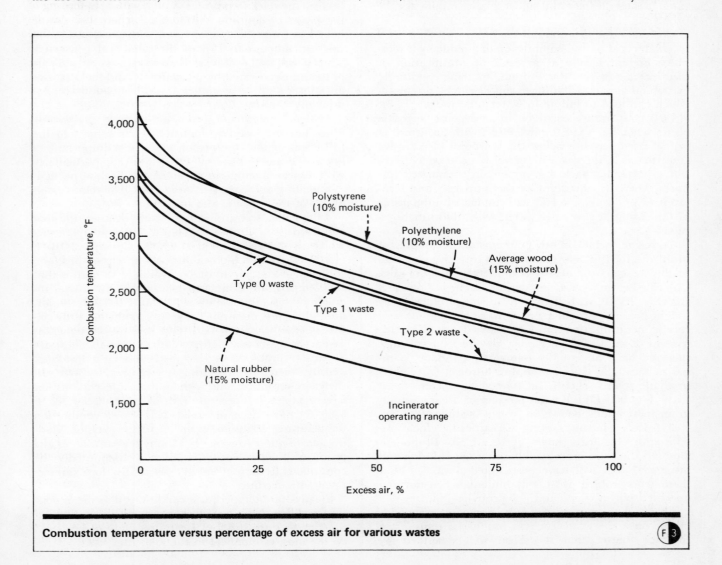

Combustion temperature versus percentage of excess air for various wastes

has now been found for the lining on the unit. F/3 shows the variation in temperature that can be anticipated for a range of materials with varying quantities of excess air. A particular problem with the selection of refractories is found in the application to rotary kilns. Considerable abrasion is present and experience suggests that monolithics are far more suitable than brick for some applications. However, for others, particularly where high-density materials are required to avoid penetration by liquids or vapors, brick is still used despite higher cost. This is an important point to be kept in mind if the furnace is to be pressurized.

Corrosion and erosion

There are at least four types of corrosion to which incinerators are subject, and the risks are greatest when waste-heat recovery boilers are incorporated. These are:

(1) high-temperature liquid,
(2) nonuniform furnace atmosphere,
(3) low-temperature or dew-point corrosion,
(4) HCl.

High-temperature corrosion usually occurs above 895°F (480°C) and is thought to be caused by molten alkali sulfates. The avoidance of high metal temperatures in superheaters and boiler plants minimizes the problem.

Corrosion due to a nonuniform atmosphere is caused by products of partial combustion in a reducing atmosphere present locally as a result of stratification or incorrect air or fuel distribution. In these conditions, carbon monoxide or hydrogen sulfide can be produced, which can lead to failure in water tube walls.

Low temperature corrosion or dew-point corrosion occurs when flue gases are in contact with surfaces that are at a temperature below the dewpoint of corrosive constituents in the gas. This condition can be avoided, but a particular hazard is present in pressurized furnaces, where if penetration of the refractory and insulating layer occurs, the HCl can condense on the inside of the casing at shutdown, often with unfortunate results.

Corrosion by HCl in other circumstances can present problems, for this very reactive and corrosive chemical readily reacts with alkalis and metals to form salts. Chlorine-bearing plastic materials, such as polyvinyl chloride (PVC), with high percentages of chlorine by weight present particular problems at the washer, and careful neutralization must be carried out. Chloride content of the wash water must also be reduced, since this can have an equally damaging effect on stainless materials. Plastics containing nitrogen (e.g., polyurethane foams) should be treated with care, as at 1,830°F (1,000°C) the main product of decomposition in an inert atmosphere is hydrogen cyanide [9]. Care is also needed in the burning of plastics to which fire-retardant chemicals have been added. Prominent amongst these are bromine, phosphorous, chlorine and nitrogen but it will have been noted from F/1 that there is no problem with contaminants when burning uncoated polyethylene and polypropylene film.

The mechanism of corrosion by flue gases, even with fairly consistent waste such as municipal refuse, is a complex matter that has been well described by Miller and Krause [10]. It is only after some considerable research that methods for dealing with it are now becoming available. This does not mean that mixed industrial wastes do not lend themselves to incinerator–waste-heat recovery systems as described by Thoemen [11]. With mixed-waste incinerators, one important point should always be borne in mind: they should be designed with a built-in degree of flexibility. However, no system can be designed for complete flexibility, and if there is a fundamental change in the mixture of wastes to be fired, or in the nature of them, the suppliers should be consulted. Where heat recovery is also incorporated, the need for caution is more than doubled.

Waste-heat recovery

Limitations—It is obvious that not all the heat generated by the burning of waste can be used. For example, neither the heat required to evaporate the moisture in the material, nor the heat required to evaporate the moisture formed during combustion, can be recovered. While it is necessary to reduce the temperature of the combustion gases to enable gas scrubbing or cleaning to be carried out, the temperature after cleaning must still be high enough to give lift to the gases leaving the stack to promote good dispersal before cooling causes precipitation of moisture. There will also be heat loss from the system; even with good insulation this can amount to 15% of the total heat generated. Thus it will be possible, at best, to recover only a part of the heat generated by incineration, and the performance for a number of materials has been admirably set down by McRee [12].

Flue-gas characteristics and heat-exchange equipment— Since heat-recovery equipment will be subject to the full effect of the combustion gases, it is important to know the gases' physical characteristics, particularly with respect to temperature, constancy of flow, particulate loading and corrosion properties. The relative volatility of the waste is also important.

Taking the first and last of these together, the first constraint is a limitation on the entry temperature, which is between 1,830°F (1,000°C) and 2,010°F (1,100°C). With a high-calorific-value waste, therefore, sufficient excess air must be available to ensure that this temperature limitation is observed. Industrial incinerators usually operate with about 100% excess air, and while this does not cause any problems with cellularic-type wastes, it can have serious consequences where the waste is a highly volatile material. These are particular problems of the wastes having a low bulk density, therefore presenting a large surface area. In these circumstances, and with air available, combustion is very rapid. Temperature rises of more than 1,380°F (750°C) have been observed. F/4 shows combustion temperatures versus excess air for various wastes. Modulation of the excess air is an obvious answer to the problem, but the elegant solution is undoubtedly the continuous feeding of volatile wastes by, for example, pneumatic feeding.

Steady-state temperature conditions also mean constancy of flow. Most heat exchangers will accept fluctuations in temperature, but steady recovery of heat will be the most economical in the long run.

Other considerations

The particulate loading of the flue gases will determine whether a fire-tube or water-tube boiler will be used, the latter being able to handle higher loadings than the former where the limit is generally accepted to be 0.15 grains/ft^3 without automatic tube cleaning.

A particulate loading less than this will normally result in the tube surface being free from fouling, but in no case should particulate loading exceed 0.5 grains/ft^3. Most industrial wastes give loadings that are less than this, and fouling is not usually a problem except in the case of the flame-retardant polymers where antimony oxides and chlorides are easily carried over with the gases. Even in such cases, water-tube boilers can be applied.

In respect to wastes containing alkali sulfates and chlorinated hydrocarbons, mention has been made of high- and low-temperature corrosion conditions. The problems can be avoided if metal temperatures are kept below 895°F (480°C) and above 445°F (230°C) at the outlet where dew-point problems for both hydrogen chloride and sulfur oxides are avoided. The mention of boiler plant presupposes that there are substantial quantities of heat available and that there is a use for the steam or hot water, but additionally, there is often the case for a simple heat exchanger for preheating combustion air (particularly useful when dealing with lower-CV wastes, and those with high water content), reducing the need for support fuel.

One further medium for heat recovery is the use of a thermal fluid instead of water. While the use of such fluids is advantageous in that high temperatures can be achieved at normal pressures (whereas with steam the higher the temperature the higher the pressure), there are disadvantages. The thermal fluids are mostly hydrocarbons and, to avoid thermal degradation and polymerization, very steady-state conditions of temperature and gas flow are necessary. Such systems are therefore only suitable where it is possible to achieve steady feeding of a consistent waste such as blended liquids and gases [13].

Emissions—types of collectors

As a rough guide, the cost of the collection system usually varies directly with the gas volume, and effective control at the smallest volume is the aim. The particle size of the dust is the most important factor to consider in choosing the type of dust collection to give the correct cleaning efficiency. Additionally, acids in gases must be accurately identified in order to assess the need for special materials in the construction of the collector.

With the increasing stringency of legislation, most applications that contain fume (i.e., submicron-sized particles) or fine dust particles below 5 microns in size require high-efficiency collectors. T/IV gives the approximate collecting efficiency of various dry and wet collectors. The EPA standards were set at 0.2 grains/std. ft^3 of dry flue gas corrected to 12% CO_2 (without the contribution of CO_2 from auxiliary fuel). In addition, the guide requires emissions to be measured by procedures described in specifications for testing at federal facilities and defines particulate matter as "any material except uncombined water which is suspended in the gas stream as a liquid or solid at standard conditions." Standard conditions were defined as "a gas temperature of 70°F and a gas pressure of 14.7 lbs absolute" [14]. Hence, wet-scrubbing systems have a distinct advantage in being able to tackle the combined problem of acid gases and particulates. High-efficiency scrubbers also find application in odor control where either high temperature or catalytic incineration does not prove wholly successful. Careful selection of the right type of equipment is necessary, since the horsepower requirements of the high-efficiency scrubbers are substantial.

In the venturi scrubber, which is perhaps the most popular type at the present time, gases are passed through a venturi tube to which water is added at the throat. Gas velocities in the throat are high and usually in the order of 20,000 ft/min. As a result, operating pressure drops are high and generally in the region of 20–50 in. water gage. In spite of short contact time, extreme turbulance provides high collection efficiency. Water consumption rates vary, depending on the operating pressure drops necessary, but are in the range of 3–20 gal/min. Water is recirculated, and provision must be made in the system for neutralization and filtration [15].

Recently a new development, which utilizes a two-phase jet system, has made its appearance. The waste gases are passed through a heat exchanger, the hot water from which is used in the nozzle. Water temperature and water/mass ratios are the two principal design parameters. Very high efficiency on submicron particles at 80% of installed cost of comparable systems is claimed [16].

The control of air pollution is, in itself, a vast subject insofar as sources are concerned, but it should be re-

Approximate efficiencies for wet and dry collectors				
Type of collector	Efficiency on standard dust * %	Approximate efficiencies, microns		
		10 %	5 %	1 %
High-efficiency cyclone	84.2	85	67	10
Small multiunit cyclone	93.8	96	89	20
Low-pressure-drop cellular collector	74.2	62	42	10
Spray tower	96.3	96	94	35
Self-induced spray collector	93.5	97	93	32
Wet-impingement scrubber	97.9	99	97	88
Venturi high-pressure-drop scrubber	99.7	99.8	99.6	94
Dry electrostatic precipitator	94.1	98	92	82

* Standard dust: 80% at 60μ ; 30% at 10μ ; 10% at 2μ .
Note: In some cases the equipment listed can be designed for higher efficiencies than those shown.

membered that in 1968 a Public Services study in the U.S. indicated that of total atmospheric pollution, 60% was attributable to transport, 19% to industry, 12% to power generation, 6% to space heating and only 3% to incineration [14]. This is not to suggest that there is no room for improvement in terms of emissions from industrial incinerators—there is. But at the same time, the percentages give a perspective to the relative sources of air pollution.

Plant costs and payoff

It is impossible to generalize on the benefits of onsite incineration either with or without waste-heat recovery, for so much depends on the type of waste (hazardous or otherwise), the quantities involved and the present costs of disposal. It will be apparent that for very small quantities of less than 500 lb/h, the economics of onsite disposal will be doubtful: the more difficult the waste, the less attractive will be the comparison with costs for contractor disposal of such quantities. For some materials, such as heavily chlorinated wastes with more than 50% chlorine, the economics of gas scrubbing cannot be justified in terms of maintenance costs, and in this area HCl recovery processes by submerged combustors are finding favor. In one process by Nittetu, the cost of recovering one metric ton of hydrochloric acid, including installation depreciation, labor maintenance and all other expenses would be $33 (1972 estimates) [6]. The throughput quantities involved would be around four metric tons/h. As size increases, so does the viability of onsite disposal, and McRee quotes annual return-on-investment for heat recovery—at 125 psi steam—on Type 1 waste of 23.6, 30.3, 34.8 and 36.7% on incinerators of 1,000, 1,500, 2,000 and 2,500 lb/h capacity. The calculations take into account all charges except shipping and installation costs, and give an equivalent to payoff times of 4.2, 3.3, 2.9 and 2.7 years respectively [12]. Savings are shown graphically in F/4. The degree of difficulty between the first and second example above bears no comparison, but between the extremes lie a whole range of possibilities for the application of flexible incineration units with or without waste-heat recovery. There seems little doubt that future economics will favor the onsite disposal of solids and sludges, particularly of the types that are more difficult to handle, because these wastes may occupy valuable space in a production area. Invariably the space wasted in the storage of solids and sludges is far greater than the requirements of the more easily stored and handled liquid residues. Such applications can be found to offer worthwhile returns on capital invested and, provided they are well designed with low chamber-gas velocities, can be installed with low-pressure-drop scrubbers to produce a standard of emissions that will satisfy legislative requirements.

Present and future developments

Is incineration the best way of dealing with waste? Undoubtedly, it has its part to play, for even with increasing interest in the recycling of that which can be recycled, we have not yet reached the stage where such processes are carried out regardless of cost. The rising prices of new materials can, however, change a

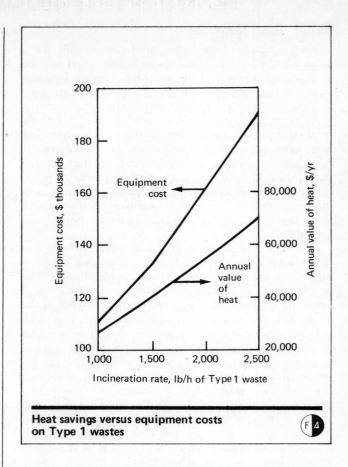

Heat savings versus equipment costs on Type 1 wastes   F 4

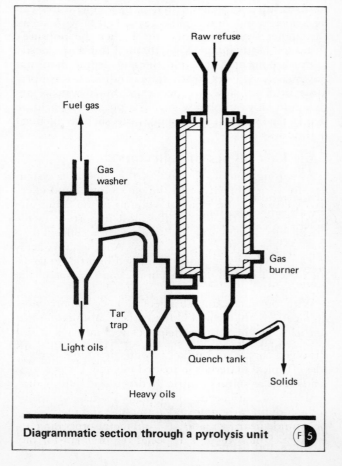

Diagrammatic section through a pyrolysis unit   F 5

recovery situation overnight to such an extent that even an old technology can be used economically.

Environmentally, there is bound to be much more emphasis on recovery, which means that such wastes as are left—and for which there is no alternative but incineration—will be the difficult ones. This will place an increasing responsibility on the incinerator designer. Incineration is not a field where major breakthroughs have been made in design. There has been novelty and innovation in recent years, but the problems of working at high temperature with the waste materials produced by an ever-expanding range of manufacturing processes, require progress to be slower than some would like to see.

Of the new systems under development, pyrolysis seems to be the most promising for applications in the future. This is a gasification process where waste is fed into a retort and heated in the range of 1,110°F (600°C) to 1,830°F (1,000°C) in the absence of air and, depending on the materials fed to it, produces a range of products and carbon char. The ability of the system to produce a storable source of energy for subsequent use makes it highly desirable. F/5 shows a diagrammatic section through a plant. Many plastic products lend themselves to the pyrolysis process, the products of which can either be repolymerized or used as fuel gases.

Further development of fluidized-bed techniques will no doubt overcome some of the present problems of stickiness in the bed; already there are a number of boiler units under construction that will use this technique. Inherently high power consumption against a background of rapidly rising costs could limit its application for industrial, as opposed to specialized, applications.

Suspension burning of finely divided materials in a vortex-type furnace chamber could also find future prominence, for this can provide high heat release in a relatively small volume without the necessity for a grate or hearth. One drawback is the need for carefully controlled size reduction of the wastes.

The waste disposal complex

The need for control in the handling and disposal of waste is well recognized and is a strong point in favor of the centralized disposal facility. A number of these are already operational in the U.S. and in Europe. The operation can be run either by a group of companies, a government agency or a single contractor, but the important point is that the system can be designed to recover usable products from both organic and inorganic wastes. Noncombustible inorganic wastes are subjected to wet chemical processing, either to convert toxic components into nontoxic derivatives (such as cyanide to nontoxic cyanate) or to convert the toxic components into water-insoluble derivatives (such as dissolved copper and nickel to water-insoluble hydroxides). Typical materials so treated include: the common acids such as sulfuric, nitric, hydrochloric, phosphoric, hydrofluoric and acetic; basic materials such as caustic soda, sodium carbonate and ammonia; and metal and/or cyanide-bearing materials from metal-processing operations. Unless there is a special requirement for bio-logical treatment, organic wastes are disposed of by high-temperature incineration.

Before any material is accepted for processing, samples must be submitted for testing. Waste liquid materials are blended in tanks before firing. Laboratory control is the keynote, with continuous monitoring of both discharge to drain and emissions to atmosphere.

Conclusions

Incineration is not a process to be applied indiscriminately, but it does have a valuable part to play in the safe disposal of a proportion of the total wastes produced by an industrial society. The state of the art is such that well-engineered solutions are commercially available to solve those problems in the areas in which burning is the right answer. In these areas, environmental standards can be met, provided always that one does not seek a solution where the incinerator must produce less total pollution than the fleet of trucks hauling the material away to a landfill.

References

1. Orning, A. A., Principles of combustion in Corey, R. C., "Principles and Practices of Incineration," Wiley Interscience, New York, 1969.
2. Sherratt, S., others, Chemical wastes and paint residues, Institute of Fuel Conference, Brighton, 1969—The Incineration of Municipal and Industrial Wastes.
3. Versatile waste system, *Mod. Plant Operation Maint.*, Winter 1974.
4. Hescheles, C. A., and Zeid, S. L., Investigation of three systems to dry and incinerate sludge, ASME 1972 National Incinerator Conference Proceedings, p. 265.
5. Hescheles, C. A., and Bonner, R. F., Jr., Ultimate disposal of wastes by pyrolysis and incineration, ASME 1974 National Incinerator Conference Proceedings, p. 321.
6. Hot option for disposal of hydrocarbon wastes, *Chem. Week*, Apr. 19, 1972.
7. Santoleri, J. J., Chlorinated hydrocarbon waste recovery and pollution abatement, ASME 1972 National Incinerator Conference Proceedings, p. 66.
8. Westland, G. J., and Clegg, J. A. M., Desirable properties of materials and application of insulating refractories, Institute of Fuel Refresher Course Notes, "Developments in refractories and insulating materials," 1974.
9. Robertson, C. A. M., Incineration of waste materials with special reference to plastics and the recovery or treatment of associated gaseous products, UMIST 1974.
10. Miller, P. D., and Krause, H. H., Corrosion of carbon and stainless steels in flue gases from municipal incinerators, ASME 1972 National Incinerator Conference Proceedings, p. 300.
11. Thoemen, K. H., Contribution to the control of corrosion problems on incinerators with water-wall steam generators, ASME 1972 National Incinerator Conference Proceedings, p. 310.
12. McRee, R. E., Waste heat recovery from packaged incinerators, ASME Incinerator Division Meeting, Arlington, Va., Jan. 1974.
13. Dunn, K. S., and Tomkins, A. G., Waste heat recovery from the incineration of process wastes, The Institution of Mechanical Engineers—Process Engineering Group Conference, Jan. 1975.
14. Squires, B. J., Detection and control of atmospheric pollution, *Environ. Pollut. Manage.*, 1973.
15. Harrington, W. M., Incineration: Environmental whipping boy?, *APWA Reporter*, Feb. 1972.
16. The Aronetics System, U.S. Pat. No. 3,613,333.

The author

Kenneth S. Dunn is Managing Director of The Incinerator Co., Howard Road, Eaton Socon, Huntingdon, England. He joined Incinco from Robey & Co. of Lincoln, where he was an Engineering Director with a special responsibility for boiler development. He is a Chartered Engineer and Fuel Technologist, a member of the Institute of Fuel and a member of the National Council of the Clean Air Soc. and Secretary of the British Environmental Protection Equipment Manufacturers Assn. (E.P.E.M.A.).

Ultimate disposal of spilled hazardous materials

Industrial and governmental agencies are now expert at cleaning up spilled chemicals. But what is to be done with the sludges and slurries that are recovered during this clean-up operation? Here's what.

Alfred W. Lindsey, Hazardous Waste Management Div., Office of Solid Waste Management Programs, U.S. Environmental Protection Agency

☐Over 13,000 spills representing nearly 20 million gallons of potentially damaging and dangerous materials occur yearly in the U. S. [1] They occur usually as a result of an accident or an equipment malfunction, but they can also be the result of a natural catastrophe such as a flood or an earthquake or be an intentional act such as indiscriminate dumping or discharging of bilge and ballast wastes (Fig. 1). Discharges vary from a few gallons to more than a million gallons and include not only the much-publicized spills into rivers, lakes and oceans but also discharges directly onto the land, where runoff and percolation can cause secondary contamination of surface streams and subsurface aquifers. Major sources include ship, truck and train accidents, ruptures of lagoons and tanks at industrial and storage sites, and pipeline breaks. [2]

More than 60% of the reported spilled materials are of an oily nature. [2] Much of the remainder consists of potentially hazardous substances such as pesticides and other toxic substances, acids and corrosive materials, solvents and other flammables, explosives, and radioactive materials. Much work has been and is being done in developing and refining means of removing spilled wastes from the water. [3] Government (the U.S. Environmental Protection Agency and the U.S. Coast Guard) and private industry "emergency response teams" are well trained and have demonstrated success in containing and cleaning up spills of hazardous materials. [1] However, once the spilled materials are removed from the water, or excavated in the case of terrestrial spills, the problem arises of what to do with the resultant matter.

Disposal of such matter can be quite complicated. The material may be in any number of forms, including debris, contaminated earth, liquids, slurries, or "gunks." It may be in pits, ponds, or temporary lagoons, tank trucks, barge holds or 55-gallon drums. Within the past few years, it has become obvious that the quick, readily available disposal options may cause secondary pollution problems rivaling the initial spill. Open burning will cause air pollution, including potential contamination with toxic gases due to emission of chemical breakdown products. The potential emission of phosgene from incomplete combustion of some heavily chlorinated hydrocarbons is an example. Plowing, dumping, lagooning, and even landfilling may present a real pollution threat to groundwaters due to leaching and percolation. Pollution of wells and surface streams can have serious health effects even though symptoms might not show up for a generation or more. As a result of burying arsenic-containing pesticides in Minnesota in the mid-1930s, eleven persons developed symptoms of arsenic poisoning in 1972, after drinking contaminated well water. Two of the victims required hospitalization and treatment. [4]

Attacking the ultimate disposal problem

The decision-maker on the spill-response team directing cleanup must find an environmentally sound yet practical method of disposing of the contaminated material. Once the material is removed from the water or contained, he should proceed to:

1. Notify the appropriate state environmental protection agency, and the Regional Administrator of the

Originally published October 27, 1975.

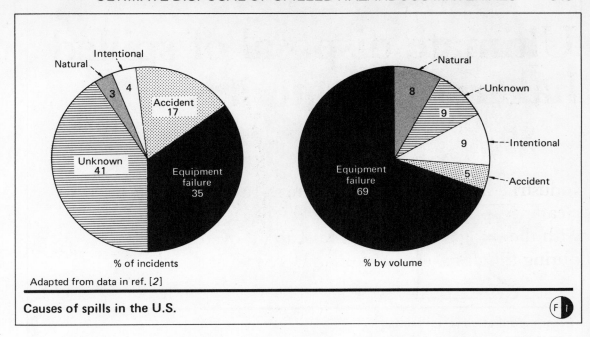

% of incidents

% by volume

Adapted from data in ref. [2]

Causes of spills in the U.S.

U.S. Environmental Protection Agency (EPA), if they are not already involved. Chances are, these agencies are already involved and are actually directing the cleanup. Responsibilities and liabilities relative to cleanup are spelled out in Section 311 of the Federal Water Pollution Control Act (P.L. 92-500) and the resulting National Contingency Plan for Removal of Oil and Hazardous Substances. [5]

2. Secure the waste in temporary storage facilities (tanks, trucks, barges, rail cars, temporary lagoons, etc.) to minimize immediate acute hazard.

3. Analyze the waste in terms of quantities and constituents in as much detail as possible and assess the relative hazard posed.

4. Analyze constraints imposed by the situation.

5. Analyze available disposal options in relation to the constraints identified above, in order to identify the practical options.

6. Choose the most environmentally sound of the practical options.

The Oil and Special Materials Control Div. of EPA's Office of Water Program Operations and the Hazardous Waste Management Div. of EPA's Office of Solid Waste Management Programs may be of major assistance in determining relative hazard of the waste materials and in advising on the environmental adequacy of the practical options.

Analyzing constraints

Although one can conceive of an almost limitless number of factors which could bear on available disposal options, the constraints likely to be encountered in cleaning up a spill can be analyzed as follows:

Transportation availability: Is the site accessible to heavy-duty trucks? Is suitable hardware (pumps, loading equipment, etc.) available to permit transferring the waste from temporary storage to the trucks? Can a sufficient number of truckers with proper equipment (glass-lined tankers perhaps) be obtained to permit

transporting the waste volume? If the answer to any of these questions is "no," and the situation can not be rectified practically, then on-site disposal may be necessary. In certain localized situations, transport by barge or rail may be practical, but suitable transfer equipment will still be necessary.

Availability of funds: Under the provisions of PL 92-500, cleanup costs for spills into waterways or adjacent shorelines are usually borne by the generator of the spill. If the discharger refuses to clean up the spill, necessary funds are obtained from a revolving fund specified by law. The fund is replenished by the funds recovered from the owner or operator of a vessel or facility held responsible for the spill. Where spill responsibility can not be determined, or the discharger is unable to pay, cost recovery is not possible and the revolving fund suffers a deficit. These provisions also apply to treatment and disposal as well as spill cleanup.

Since the revolving fund is not inexhaustible ($20 million funded at inception) and since many companies have limited resources, it behooves the on-scene coordinator to minimize treatment and ultimate disposal costs, consistent with achieving acceptable protection of public health and the environment.

Land spills which do not directly threaten surface water are not covered under PL 92-500. Generally, cleanup costs are still the responsibility of the facility or transporter creating the spill, but in this case, the source of interim emergency funds is not so clear cut.

Suitability of temporary storage facilities: Hopefully, the wastes can be made secure in temporary storage facilities such that no environmental or health hazard exists. However, in some cases this may not be possible. Consider the situation where contaminated earth or debris has been piled or unlined lagoons are filled with sludges, slurries, or contaminated water. A hazard may exist to surface or groundwaters or to human health which requires quick action. If temporary storage can not be made more secure by drumming, lining lagoons,

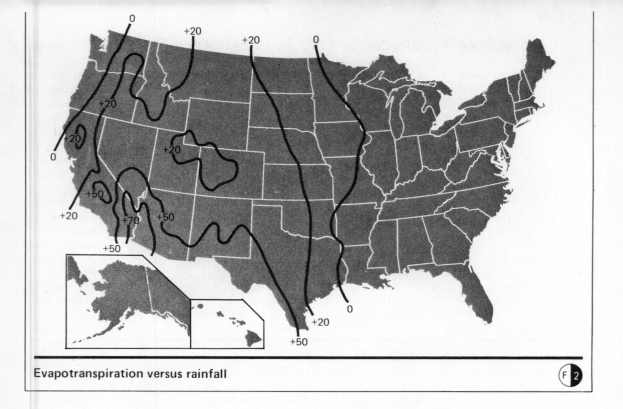

Evapotranspiration versus rainfall   (F) 2

or placing the material in tanks pending a well-thought-out disposal decision, it may be necessary to locate the most immediately available treatment or disposal option which shows promise for reducing the immediate hazard. Potentially available options are covered later.

Citizen concern: Where hazardous chemicals are involved in a spill near a populated area, the public is likely to be aware and may, understandably, voice serious concern. This concern may take the form of a public outcry in the press or elsewhere concerning the hazard or nuisance, real or imagined, posed by the temporary storage arrangements. Unless appropriate measures are taken to inform the public, citizen concern may take the form of opposition to certain disposal options such as incineration or secure burial near their neighborhood or farm. Such opposition can effectively remove from consideration an otherwise satisfactory solution.

Local, state and federal regulations covering transportation and disposal: At present there are no Federal regulations pertaining to hazardous waste disposal except where disposal of radioactive wastes and pesticides are involved. [6] Many states and some counties and cities have hazardous waste regulations and statutes that may affect disposal options, and all of these must be observed. The responsible state department or the regional office of EPA should be able to identify such regulatory constraints. EPA has generated guidelines for incineration and land disposal of solid wastes. [7] These are oriented toward municipal solid waste, however, and are thus of only peripheral interest here. The guidance and recommendations expressed here are based on currently available information and the current thinking of the author and should not be construed to be guidelines or regulations sanctioned by EPA. The Department of Transportation, on the other hand, has very specific and definite regulations regarding transportation, labeling, and packaging of hazard-

ous materials. [8] Questions requiring immediate advice on transportation matters should be telephoned to the nearest Department of Transportation Regional Hazardous Materials Specialist.

Identifying available options

Because of the almost limitless variety of physical and chemical characteristics of the wastes, and the various degrees of hazard involved, it is not possible to formulate unequivocal guidelines for all situations.

There are two basic options available: disposal on-site or nearby, or the use of existing facilities. The advantages of using existing facilities are:

a. Disposal is likely to be quicker, since facility construction is not necessary.

b. Opposition by the local populace is not likely to be high, except perhaps along the transportation route in the case of extremely hazardous wastes.

c. It is only necessary to know what type of wastes can be disposed of by which processes; it is not necessary to know how to construct the facilities.

The disadvantages include:

a. Transportation, perhaps over a long distance, is usually necessary and may increase costs considerably.

b. Adequate transportation and/or transfer facilities may not be available.

c. The available treatment and disposal facilities may not be adequate for handling the particular type of hazardous waste resulting from the spill.

On-site disposal may be the only practical alternative, and it is possible that the present site of the waste, or a site nearby, may prove suitable.

Existing disposal facilities

Existing facilities can be subdivided into three categories: (1) municipal waste facilities, (2) commercial treatment and disposal sites handling industrial wastes, and (3) industrially owned and operated facilities.

Municipal Waste Facilities: The most common existing

facilities are land disposal sites used for municipal wastes. Typically, these are owned by the local municipality or regional authority, although many are privately owned and operated, usually by refuse collection firms. Typically, many of these sites can be classified as open dumps and are often accompanied by open burning. An open dump can be characterized as land disposal with little or no regard for human safety or the environment. EPA, therefore, does not sanction disposal of any type of waste in open dumps; certainly disposal of hazardous spill cleanup materials in this manner should be avoided. It is estimated that about 90% of the routinely operating land disposal sites in the country can be classified as open dumps.

The preferred method of land disposal of municipal solid waste, the sanitary landfill, can be described as an engineered disposal method that minimizes environmental hazards by spreading the solid wastes in thin layers, compacting the layers to the smallest practical volume, and applying cover material at the end of each working day. [7] EPA's current view of the sanitary landfill requires careful site selection, design, and operation to prevent leaching of waste residuals into groundwater. While this technique is recommended for municipal solid wastes, its use in hazardous waste disposal will normally be satisfactory only if the objective is to contain insoluble solids or to prevent wind dispersal of powders. Potential problems include:

- Percolation of toxic liquids to groundwaters.
- Dissolution of solids followed by leaching to groundwater as rain or groundwater moves through the fill.
- Dissolution with subsequent leaching of solid hazardous materials, such as heavy metals, by acidic leachates generated by decomposing municipal refuse.
- Potential for undesirable chemical reactions within the fill, creating explosive situations or releasing toxic gases.
- Volatilization of wastes, releasing toxic or explosive gases or vapors to the atmosphere.
- Corrosion of drums containing hazardous materials, thus releasing contents.

Therefore, unless the wastes involved are completely insoluble and solid, or unless the location is very arid (where evaporation considerably exceeds precipitation), sanitary landfilling should be avoided. A few materials which could be considered for routine land disposal include calcium fluoride, magnesium oxide, sulfur, and vanadium pentoxide.

A number of land disposal sites in California (Class I sites) [9] and a few elsewhere are capable of receiving many hazardous wastes, including liquids and leachable wastes. These can be classified as "chemical waste" landfills, and should have no direct hydraulic connection with surface or groundwaters, thus greatly reducing the danger of contamination. [6] Wastes received are separated from each other to preclude deleterious reactions and are covered with earth or inert material as received.

In many larger cities, municipal solid waste and sewage sludges are incinerated. Incineration of hazardous waste as a pretreatment prior to land disposal is an excellent means of protecting the land environment by completely destroying the organic fraction, leaving only inorganic ashes. Volume reduction is frequently very high depending on the relative concentration of organic and inorganic materials. In many cases, no ash or particulate material remains for land disposal. Destruction of the organic molecular structure also destroys toxicity, unless the toxicity is due to the incorporation of toxic elements such as heavy metals.

Unfortunately, municipal solid waste incinerators, as they are currently installed and operated, are not designed to destroy hazardous material. Their shortcomings frequently include:

- Inability to attain high enough operating temperatures and dwell times to ensure the destruction of hazardous wastes.
- Lack of adequate emission-control equipment to remove hazardous wastes or breakdown products from flue gases.
- Improper materials of construction for withstanding corrosive off-gases, which may be generated during combustion.
- Improper materials-handling equipment, which presents a safety hazard during operation.
- Operators untrained in handling hazardous materials.
- A design or operation that results in unburned or only partially burned residues.
- Inadequate disposal techniques for potentially hazardous residues.
- Lack of sufficient instrumentation to permit careful monitoring of operating conditions and combustion efficiencies.

Multiple-hearth sewage-sludge incinerators do have application for the destruction of some hazardous wastes, by virtue of their specialized design, long retention times, capability for elevated temperatures, and adequate scrubbing facilities. EPA has conducted a full-scale evaluation of the destruction of DDT and 2,4,5-T pesticides via this technique at the municipal sewage sludge incinerators in Palo Alto, California. [10] Destruction ratios better than 99.97% were consistently achieved.

Commercial Facilities for Industrial Waste: In the past few years, a number of commercial disposal operations have been developed to treat and/or dispose of industrial wastes, usually on a profit basis. Most are private; a few are run by regional authorities interested in attracting or holding industrial establishments. These facilities are prime candidates for ultimate disposal of most hazardous wastes and should be investigated prior to deciding on the disposal of cleanup wastes. Although EPA has not evaluated the environmental adequacy of all of these facilities, most operate under state sanction and permits. EPA has published a list of known commercial facilities of this type, and this list can be obtained by writing the Technical Information Staff, Office of Solid Waste Management Programs, Environmental Protection Agency, Washington, DC 20460. [11] A more detailed directory will also be available soon. [12]

Many commercial facilities have their own trucks, which may be available to pick up hazardous spilled materials either in bulk form or in drums, thus simplifying the disposal problem. Potential drawbacks to using these facilities are:

■ Transportation will have to be obtained if it is not provided by the facility.

■ Higher costs may result from possible long-distance transportation and fees charged by the facility.

■ Large volumes of dilute materials may tax the facility's capacity and charges can be high, since these facilities receive largely concentrated wastes.

■ Normally, extensive testing for compatibility with the equipment is required, thus delaying disposal.

Commercial industrial waste facilities usually consist of a secure landfill, such as the California Class I sites, or of high-temperature incinerators of various types. Many of the larger operations include both, and some also have processes for chemical detoxification, such as hydrolysis, oxidation, etc., or for recovery of various materials. Copper recovery and the distillation regeneration of contaminated solvents are examples.

Many of these facilities rely heavily on incineration to destroy, detoxify or greatly reduce the volume of the waste for disposal. The general lack of reliable information covering adequate time and temperature criteria for destruction of each waste, coupled with the bewildering variety of incinerator models, and types, makes it very difficult for those not intimately familiar with incineration of hazardous wastes to evaluate the adequacy of any incinerator for destruction of a specific waste. Thus, in the case of spill cleanup materials, it will probably be necessary to take the advice of the facility owner concerning suitability of his facility for the job at hand. However, consideration of the following basic criteria for hazardous waste incineration should contribute to intelligent discussion:

■ Generally speaking, only organic materials should be considered candidates for incineration, although some inorganics can be thermally degraded.

■ Chlorine-containing organics form extremely cor

rosive hydrogen chloride gas upon incineration. Other halogenated hydrocarbons behave similarly. Materials of construction should be chosen accordingly, and suitable emission scrubbers provided (packed tower, venturi, or equivalent). Scrubbing solutions of caustic soda appear to be the most suitable. Systems have been developed which are capable of recovering usable hydrogen chloride from incinerator emissions.

■ Organic materials containing dangerous heavy metals (mercury, arsenic, selenium, lead, cadmium) should not be incinerated unless the fate of the metal components in the environment is known or can be satisfactorily controlled by air and water pollution control equipment.

■ Sulfur-containing organic materials will normally form sulfur oxides on incineration. Care should be taken to remove these materials from the stack gas if sulfur is present in appreciable concentrations.

■ Formation of nitrogen oxides can be minimized by keeping temperatures below 1,100°C (2,000°F).

■ The destruction ratio of a given material by incineration is dependent largely on the temperature, the dwell time at temperature, the air/feed ratio, and the turbulence. The higher the temperature used, the shorter the necessary dwell time. As a general rule, most organic hazardous materials can be almost completely destroyed in 2 seconds at 1,000°C (1,850°F); many are completely destroyed at lower temperature/dwell time conditions; but a few require more rigorous conditions. Optimal excess air conditions will depend on the facility type but may vary from 30% to 100%. Turbulence is a function of burner design but will be affected by throughput and excess air.

■ The hazard posed by incineration of hazardous materials is largely a function of the amounts, nature, and distribution of the combustion products entering the environment. These products may be undergraded hazardous material, partially degraded toxic break-

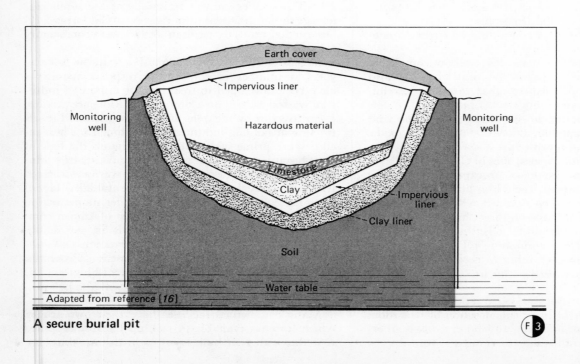

A secure burial pit

Ⓕ 3

down products, or hazardous end-products such as sulfur oxides, hydrogen chloride, or heavy metals. They may exit as a gaseous or particulate emission to the atmosphere, or in the scrubber water, or as ash residue. Care must be taken that these materials are not dispersed or disposed of in concentrations capable of affecting health or damaging property or the environment. All applicable federal, state, and local air-emission and water-effluent standards and regulations should be met.

■ Incinerators burning hazardous materials should be equipped with automatic feed-cutoff provisions in the event of either a flame-out or a reduction in reactor temperature below that known to give complete combustion. Online instrumentation capable of determining hydrocarbon or carbon monoxide levels can also be used to indicate relative combustion efficiencies. This will help to prevent emission of unburned or only partially degraded materials that may result from equipment malfunction.

Additionally, EPA plans a series of full-scale test burns to evaluate the environmental adequacy and economic viability of various techniques for hazardous waste incineration. [13] EPA has recently published an information document entitled "Incineration in Hazardous Waste Management." [14]

Industrial Facilities: A number of industrial plants have developed waste-disposal facilities, which may be adequate for hazardous-spill cleanup wastes. These may include secure landfills or high-temperature chemical incinerators. Normally, wastes from external sources are not accepted at industrial plants; but under emergency conditions, local companies may be willing to accept and dispose of spill cleanup materials. In considering the use of such facilities, or any existing facility, one should investigate sufficiently to determine that the facilities will not create health or environmental problems if used to dispose of the specific wastes resulting from the spill.

On-Site Disposal

Normally, disposal of spill cleanup materials in an existing facility is recommended, provided an environmentally sound facility can be found for the particular waste and provided transportation logistics and economics make such an undertaking practical. Where this is not practical, then on-site or nearby disposal will be necessary. Incineration will not be available due to the impracticality of constructing an incinerator on-site for a one-shot disposal operation. Open burning, ocean dumping, and open dumping cannot be sanctioned on environmental grounds. The remaining options are consequently reduced to a choice among: (1) land disposal, (2) chemical treatment, and (3) long-term storage.

Land Disposal: Burial with immediate cover is the equivalent of the sanitary landfill for on-site disposal of spill cleanup material. Simple burial requires only the digging of a pit or trench and covering the waste materials with at least two feet of earth cover. If biological decomposition can be expected, a perforated plastic pipe should be placed horizontally along the top of the buried waste (preferably in a gravel layer) with one end

bent vertically through the cover material to vent accumulated gas. In some cases, disposal of organic materials in dilute mixtures by injection into the soil plow layer may be suitable. However, experts knowledgeable on the biodegradation and leaching potential of these materials should be consulted.

Sites should be selected so that burial does not occur in the water table and so that washouts from floods and heavy rains will not release wastes to the environment. Simple burial will *not* be suitable for most hazardous wastes unless they are solid and essentially insoluble in water, and/or the area is very arid. To determine whether a region is sufficiently arid, one must determine if the potential evapotranspiration rate is significantly greater than the cumulative rainfall. If so, there will be very little migration of the wastes. Such conditions predominate only in the Southwest, the Mountain States, and the Western Plains (F/II).

For most hazardous wastes, it will be necessary to incorporate at least some aspects of secure landfill design to make on-site burial suitable (F/III). As in a secure landfill, contaminated leachate must be prevented from reaching groundwater supplies, through site selection and engineering design techniques:

■ The distance from the pit bottom to groundwater table should be maximized. The greater this distance, the more opportunity for soil attenuative processes (sorption, and ion exchange) to remove hazardous materials from the leachate. Ten feet has been suggested as a bare minimum even with very tight soils. [15].

■ An area with a high clay content is preferable. [15] clays are tight structures with low permeability which therefore reduce the rate of flow of leachates through the soil. Clay permeabilities of 10^{-8} cm/sec or less are desirable.

■ The site should be as far as possible from wells, surface water intakes, and active faults. One group of researchers suggests the following criteria: 2,000 ft. to any well, 5 miles to municipal wells or static water intakes, 1 mile upstream of a river intake, and 1 mile from an active fault. [16] Consideration should include potential damage from seismic or flood conditions.

■ The pit should be lined with a rolled clay of a tight nature such as bentonite. Not only do clays cut down the rate of leachate percolation but they also have many absorptive and ion exchange surfaces which retain heavy metal ions and organics. [15]

■ It will normally be desirable to line the pit with limestone to increase the pH of leachates. Metal hydroxides and other metal compounds are generally more insoluble at higher pH levels. Also, attentuation of metals by clays is more effective at higher pH levels and microbiological activity is normally greater under neutral or somewhat basic conditions. However, if heavy metals are not present and the material is largely organic but biodegradation is not expected to occur in the fill, then higher pH levels are not desirable since attenuation of organics by clays is more effective at a lower pH. [15]

■ The pit should be equipped with an impervious liner, such as concrete, asphalt, rubber, or plastic, or combinations of these. [15] Hypalon and polyurethane have been used widely for lining lagoons and should be effective. Another promising liner and one which will reportedly flow to plug any rupture is a mixture of 70 percent sand, 20 percent Bentonite clay, and 10 percent cement. Asphalt will also flow to fill voids. [17] A primary concern in choosing a liner is compatability of the material with the wastes to be deposited so that the liner will not be dissolved. One group of researchers found the following relative durabilities among materials in contact with various acids and organic solvents [15]:   concrete—plastic—rubber—asphalt.

■ Not only are impervious bottom and sides necessary but an impervious cover should also be provided of the same or similarly effective material to prevent infiltration by rainfall. A final covering with earth should be deep enough to support vegetation but should be mounded high enough to avoid creation of a rainfall-collecting depression upon settling of the wastes. [7]

■ One or more monitoring wells should be sunk to groundwater (preferably one above and one below the site in the direction of flow) and routine testing periodically carried out to ensure liner integrity. [15]

It may not always be possible or necessary to duplicate all of the above criteria. However, one should comply as closely as is necessary, consistent with the hazard posed by the wastes. The important process is to assess the environmental risks of the available options and choose the soundest approach consistent with other considerations (available time, money, transportation, imminent hazard, citizen pressure, state and local regulations, etc.)

Additional precautions include:

■ Extreme care must be taken to prevent cracks or holes in the liner.

■ Materials capable of emitting toxic fumes or gases should not be disposed of in this manner.

■ The site should be permanently marked and registered with state and local officials. [6]

■ Large volumes of acutely toxic materials should not be included, since the hazard posed by liner breakdown, accident, or earthquake is simply too great. Other options or permanent storage are preferred.

If construction of a secure landfill appears impractical, it may be possible to containerize small volumes of waste so as to encapsulate them prior to burial. Encapsulation involves the permanent sealing of a material and its container in another impervious container of plastic, glass, or other material which is unaffected by the waste material. This container may in turn be sealed within a durable container made of steel, plastic, concrete, or other material of sufficient thickness and strength to resist physical damage during and subsequent to burial. [6] In Germany, experiments are being conducted encapsulating wastes in nonreactive plastic liners within steel drums and surrounded by several inches of polyurethane foam. The latter precludes moisture attack on the steel and subsequent release of contents. [18] Encapsulated materials can be buried, but a permanent record of location and contents should be filed with the appropriate local office of legal jurisdiction and with the state environmental protection agency and with the appropriate EPA Regional Office. The practicality of this approach for ultimate disposal of spilled materials will depend on the volumes

to be disposed of and on the resulting costs.

Chemical Treatment: It may be desirable to chemically treat wastes prior to disposal, in order to detoxify or solidify them. Neutralization with such common neutralizers as soda ash and lime is obvious and widely practiced in spill cleanup situations. Other not-so-apparent alternatives include:

■ *Solidification*—Liquid wastes can be solidified for land disposal by a variety of techniques including adsorption on bentonite or vermiculite, incorporation in cement through use of the liquid as makeup water, or treatment with lime and flyash. Heavy metals can be insolubilized to a certain extent through formation of metal silicates. A number of firms can provide solidification services on demand but normally require volumes greater than 100,000 gal. Their services are thus applicable mainly to very large spills.

■ *Precipitation*—many dissolved hazardous materials can be precipitated to form sludges suitable for land disposal. Fluorides can be precipitated with excess lime to form a calcium fluoride sludge, which can be buried or placed in a sanitary landfill. Liquid fluorides containing ammonium, potassium, or sodium wastes can be handled in this way. Aluminum or barium fluorides will form a metal hydroxide as well as the calcium fluoride. Barium containing supernatants should be treated with sulfuric acid prior to discharge in order to precipitate barium sulfate. Mixed sludge disposal should be in a secure landfill. [19]

Heavy metal solutions can be treated to precipitate the metal hydroxide by raising the pH to 9.5 or better with caustic soda, soda ash, or other alkaline chemicals. Metallic hydroxides are gel-like and do not settle well, but aluminum sulfate can be used to increase settling. The sludges can be disposed of in a secure landfill. Since acidification will solubilize the heavy metals, care must be taken to contain all leachates or to prevent acidification. [20]

The use of these techniques will be limited to specific instances where the form of the waste is suitable (liquid, wet earth, or wet debris) and where there are means of mixing in chemicals and separating sludges.

■ *Pesticide decomposition*—A number of organic pesticides, including DDT, lindane, parathion, malathion, and demeton undergo accelerated degradation via hydrolysis when treated with highly alkaline materials. [21] Small terrestrial spills can be partially deactivated by spreading caustic soda over the surface and wetting down lightly. Solids can be partially deactivated by mixing them into caustic solutions or sandwiching them between layers of highly caustic chemicals and wetting slightly. Liquids should be causticized to a pH of 11 or higher. The method is not recommended for concentrated materials due to the liberation of volumes of noxious hydrogen chloride. Aqueous solutions of aldrin and heptachlor can be chemically oxidized by adding an excess of potassium permanganate. [22] Breakdown via the above techniques is accelerated but not necessarily rapid. Testing to determine completeness of the reaction should be carried out before complete detoxification is presumed. Some of the intermediate breakdown products may be as toxic as the original pesticide.

There are a variety of other chemical detoxification or treatment techniques usable with hazardous waste materials. However, they require sophisticated equipment or chemicals and are thus not considered adaptable to spill cleanup situations.

Storage: If none of the available disposal options appear environmentally adequate, then provisions should be made for long-term storage. Adequate storage involves placing the waste in appropriate containers in safe areas. This approach will probably be necessary only with very hazardous materials such as concentrated arsenic-containing materials. Since waste treatment and disposal technology is developing rapidly, a definite schedule should be set for review of the treatment and disposal options available. This should be done approximately on an annual basis.

The following should be given consideration when storage is planned: [6]

- Flood plains should be avoided as storage areas.
- Soil texture and structure, and geologic and hydrologic characteristics should be such as to prevent water system contamination via runoff; dikes may be required.
- Storage rooms should be dry, well ventilated, and fireproof.
- Signs should be placed on storage buildings, a fence should surround the site, and doors should be kept locked.
- Fire-fighting equipment and appropriate safety devices (respirators, etc.) should be prominently placed. The fire department should be made aware of the quantities and characteristics of the stored materials.
- Emergency instructions should be relevant to the materials stored and displayed prominently.
- Frequent checking is required.

Deteriorating containers must be replaced promptly. Development work done by EPA has resulted in a polyurethane-foam method for plugging leaking containers as a temporary measure. This should help make it possible to assure safety and still keep to a schedule for container replacement. [22]

In the process of deciding which of the means of ultimate disposal will be used, the need for environmental safeguards must contend against constraining factors such as availability of facilities, imminent hazard, lack of safe transportation, excessive costs, citizen pressure, and state and local regulations. It may not be always possible to adopt the approach that is best in terms of long-term environmental risks, but a conscientious evaluation of the alternatives and their environmental adequacy will lead to the safest and soundest overall decisions.

As stated above, the use of an existing facility is to be recommended for spill cleanup, providing the facility is suitable and available. Where disposal facilities are not available, on-site disposal will usually be necessary.

References

1. "Oil Spills and Spills of Hazardous Substances," U.S. Environmental Protection Agency, Office of Water Program Operations, Washington, D.C., Mar. 3, 1975.
2. Department of Transportation, United States Coast Guard, "Polluting Incidents In and Around U.S. Waters; Calendar Year 1973," U.S. Government Printing Office, Washington, D.C., 1974.
3. Dahm, D. B., Pilie, R. J., and Lafornara, J. P., Technology for Managing Spills on Land and Water, *Environ. Sci. and Tech.*, vol. 8, no. 13, pp. 1076–1079, Dec. 1974.
4. Feinglass, E. J., Arsenic Intoxication from Well Water in the United States, *New Eng. J. of Med.*, vol. 288, no. 16, pp. 828–830, Apr. 19, 1973.
5. Federal Water Pollution Control Act, U.S.C.A. § 1151 *et seq.*
6. U.S. Environmental Protection Agency, Pesticides and Pesticide Containers: Regulations for Acceptance and Recommended Procedures for Disposal and Storage, *Fed. Reg.*, vol. 39, no. 85, pp. 15235–15241, May 1, 1974; and Pesticides; EPA Proposal on Disposal and Storage *Fed. Reg.*, vol. 39, no. 200, pp. 36847–36950, Oct. 15, 1974.
7. U.S. Environmental Protection Agency, Thermal Processing and Land Disposal of Solid Waste: Guidelines, *Fed. Reg.*, vol. 39, no. 158, pp. 29327–29338, Aug. 14, 1974.
8. Hazardous Materials Regulations Board, *Code of Federal Regulations*, Title 49, (Transportation), Pts. 100–199, 1973.
9. Ottinger, R. S., et al., Recommended Methods of Reduction, Neutralization, Recovery or Disposal of Hazardous Waste, Vol. III, Ultimate Incineration, U.S. Environmental Protection Agency, 1973. (Distributed by National Technical Information Service, Springfield, Va., as PB-224 582.)
10. Whitmore, F. D., Durfee, R. L., A Study of Pesticide Disposal in a Sewage Sludge Incinerator, draft report, EPA Contract No. 68-01-1587, VERSAR, Inc., Springfield, Va., 1975.
11. Hayes, A. J., Hazardous Waste Management Facilities in the United States, Environmental Protection Publication SW-146, U.S. Environmental Protection Agency, Washington, D.C., Dec. 1974.
12. Farb, D., Ward, S. D., Hazardous Waste Management Facilities, Environmental Protection Publication SW-145, U.S. Environmental Protection Agency, Washington, D.C., Feb. 1975.
13. A Study to Evaluate the Environmental, Operational, and Economic Feasibility of Destructing Hazardous Wastes in Different Types of Commercial-Scale Incinerators, Request for Proposal WA 74-R423, U.S. Environmental Protection Agency, Contracts Management Division, Research and Development Procurement Section, Washington, D.C., 1974.
14. Scurlock, A. S., et al., Incineration in Hazardous Waste Management, Environmental Protection Publication SW-146, U.S. Environmental Protection Agency, Washington, D.C., 1975.
15. Saint, P. K., et al., Effect of Landfill Disposal of Chemical Wastes on Groundwater Quality, paper presented at the Annual Meeting of the Geological Soc. of Amer., Minneapolis, Nov. 14, 1972.
16. Program for the Management of Hazardous Wastes for Environmental Protection Agency, Office of Solid Waste Management Programs; final report, Battelle Memorial Institute, Richland, Wash., July 1973.
17. Private communication, Dr. J. Zeller, ETH, to J. P. Lehman, U.S. Environmental Protection Agency, Office of Solid Waste Management Programs, Oct. 18, 1973.
18. Private communication, B. Lindenmaier, to J. P. Lehman, U.S. Environmental Protection Agency, Office of Solid Waste Management Programs, Oct. 25, 1973.
19. Ottinger, R. S., et al., Recommended Methods of Reduction, Neutralization, Recovery or Disposal of Hazardous Waste, Volume XII, Inorganic Compounds, U.S. Environmental Protection Agency, 1973. (Distributed by National Technical Information Service, Springfield, Va., as PB-224 591.)
20. Ottinger, R. S., et al., Recommended Methods of Reduction, Neutralization, Recovery or Disposal of Hazardous Waste, Volume XIII, Inorganic Compounds (cont.), U.S. Environmental Protection Agency, 1973. (Distributed by National Technical Information Service, Springfield, Va., as PB-224 592.)
21. Ottinger, R. S., et al., Recommended Methods of Reduction, Neutralization, Recovery or Disposal of Hazardous Waste, Volume V, Pesticides and Cyanide Compounds, U.S. Environmental Protection Agency, 1973. (Distributed by National Technical Information Service, Springfield, Va., as PB-224 584.)

Meet the Author

Alfred W. Lindsey is Program Manager, Technology Assessment, Hazardous Waste Management Programs, at the U.S. Environmental Protection Agency, Washington, D.C. 20460, where he has been employed since 1972. Prior to joining the EPA, he worked for Paterson Parchment Paper Co., Bristol, Pa., (1967–72) where he held various positions, and for Union Camp Corp., Savannah, Ga., and Prattville, Ala. (1964–67). He holds a B.S. in Pulp and Paper Technology (1964) from North Carolina State University and has done graduate work at Drexel and George Washington University. He is a member of the Technical Assn. of the Pulp and Paper Industry.

INDEX